饲 料 学

（汉英双语）

主　编　王梦芝　丁洛阳
副主编　杨红建　王金荣　齐智利
审　订　王洪荣　Dominique Blache（澳）

东南大学出版社
SOUTHEAST UNIVERSITY PRESS
·南京·

内容简介

本教材在简要梳理国外饲料业发展状况及未来趋势的基础上,分析了我国饲料行业的发展概况,剖析了我国饲料行业存在的主要问题,并从饲料学研究的未来方向及主要内容等角度阐述了其对饲料行业发展的重要指导作用。在上述内容的基础上,本教材详细分析、阐述了各类饲料的化学成分、营养成分及提升质量的科学方法,对促进我国饲料行业科技的进步,进而促进养殖业、农业科技水平的提升及发展,并进一步促进人民生活质量和健康水平的提高及更高水平的小康社会的建成具有重要意义。

本教材适用于各级各类学校农、畜牧、动物等相关专业,也可供专业工作者及研究人员使用。

图书在版编目(CIP)数据

饲料学:汉英对照/王梦芝,丁洛阳主编. — 南京:东南大学出版社,2022.12

ISBN 978-7-5766-0185-5

Ⅰ.①饲… Ⅱ.①王… ②丁… Ⅲ.①饲料-双语教学-高等学校-教材-英、汉 Ⅳ.①S816

中国版本图书馆 CIP 数据核字(2022)第 137242 号

责任编辑:李 贤 刘 坚(liu-jian@seu.edu.cn) 责任校对:张万莹
封面设计:王 玥 责任印制:周荣虎

饲料学(汉英双语) Siliaoxue(Han-ying Shuangyu)

主　　编	王梦芝　丁洛阳
出版发行	东南大学出版社
社　　址	南京市四牌楼 2 号(邮编:210096　电话:025-83793330)
经　　销	全国各地新华书店
印　　刷	江阴金马印刷有限公司
开　　本	787 mm×1092 mm　1/16
印　　张	21.25
字　　数	550 千
版　　次	2022 年 12 月第 1 版
印　　次	2022 年 12 月第 1 次印刷
书　　号	ISBN 978-7-5766-0185-5
定　　价	69.00 元

本社图书若有印装质量问题,请直接与营销部联系,电话:025-83791830。

本书编委会

绪　　论：王梦芝（扬州大学）　　　　　高　健（中国农业大学）
第 一 章：杨红建（中国农业大学）　　　陈连民（格罗宁根大学）
第 二 章：吴文旋（贵州大学）　　　　　胡良宇（瓦赫宁根大学）
第 三 章：郑　琛（甘肃农业大学）　　　欧阳佳良（扬州大学）
第 四 章：齐智利（华中农业大学）　　　甄永康（扬州大学）
第 五 章：成艳芬（南京农业大学）　　　经语佳（南京农业大学）
第 六 章：张　晶（吉林大学）　　　　　陈逸飞（扬州大学）
第 七 章：崔小燕（扬州大学）　　　　　丁洛阳（西澳大学）
第 八 章：沈宜钊（河北农业大学）　　　王逸凡（贵州大学）
第 九 章：黎观红（江西农业大学）　　　裴江兰（扬州大学）
第 十 章：郭雪峰（塔里木大学）　　　　张振斌（扬州大学）
第十一章：王金荣（河南工业大学）　　　吴非凡（扬州大学）

前言

畜牧生产的本质就是饲养的家畜将饲料中的营养物质转化成畜产品的过程。在畜牧生产过程中,饲料成本占据全部生产成本的70%左右。饲料的合理利用,直接影响畜牧生产的效益与成败。因此,饲料学这一研究饲料的化学组成以及饲料在不同动物中消化吸收与利用的学科在畜牧生产过程中一直发挥着极其重要的作用,学习和掌握饲料学这一学科也是培养畜牧专业人才的必经之路。因此,饲料学一直是我国动物科学、动物营养及饲料加工等相关专业的高校专业课程中的必修课程。

为适应经济全球化和教育国际化,教育部高教司于2001年提出高等学校在本科教学中应创造条件对部分专业课程实行双语教学,以培养国际化的人才;《中共中央关于制定国民经济和社会发展第十四个五年规划和二〇三五年远景目标的建议》明确提出,要培养具有国际竞争力的青年科技人才后备军。为此,很多综合性大学开始了双语教学的探索。我国作为畜牧业大国,推动我国畜牧业生产与世界畜牧业的接轨必须依靠农业院校所培养的国际化的综合型人才,因此促进农业院校开展专业课的双语教学就尤为重要。然而目前开展的双语教学课程中多参考国外英文原著教材,其具有很强的系统性和理论性,这对于统计学和计算机等一类公共课来说能大大提高学生的基础知识,同时还能提高学生的外语水平。然而对于饲料学这一类与畜牧生产活动紧密结合的专业课程,其课程内容的设计应紧密围绕我国畜牧业的发展特点,国外英文原著虽然覆盖面广,但是却不一定能反映并适用于我国畜牧业发展现状。因此,亟需一本既能总结概括我国饲料行业发展现状,且在囊括饲料学基础理论知识的基础上,补充国外饲料学论著的一些比较新的理论的双语教材,用以指导我国畜牧类高校开展饲料学学科的双语教学工作。

然而,目前我国的教材市场上,类似饲料学这一类的专业课程的双语教材尚属于空缺状态。因此,在扬州大学王梦芝教授和中国农业大学杨红建教授主导下,后经由扬州大学、中国农业大学、南京农业大学、华中农业大学、东南大学、河北农业大学、河南工业大学、甘肃农业大学、贵州大学、江西农业大学、吉林大学、塔里木大学、西澳大学、格罗宁根大学、瓦赫宁根大学等单位长期从事饲料学教学和研究人员共同编撰、翻译了这本饲料学双语教材。统稿工作经由两位主编及三位副主编共同完成。在教材编写过程中,本教材有幸经由扬州大学王洪荣教授及西澳大学 Dominique Blache 教授两位专家审订,并提出了宝贵的修改意见。

本教材共包含十二章,其中绪论章节梳理总结了国内外饲料业发展现状及其未来饲料行业的发展趋势,并且概述了饲料学的基本研究方法。本书第一章详细介绍了饲料原料的基本化学组成及其特点。本书第二章介绍了国内外饲料分类的方法及我国饲料的基本种类。本书第三章至第十一章分别详细归纳总结了我国常规青干草饲料与粗饲料、青绿饲料、青贮饲料、能量饲料、蛋白质饲料、矿物质饲料、维生素饲料、饲料添加剂及配合饲料的具体种类、加工方法、质量评定标准及畜牧生产中应用时的注意事项等,为学生教学工作提供了必需的基础理论及基本方法。此外,为迎接畜牧生产国际化发展趋势,加大教育改革力度,提高本科生人才培养质量,本教材各章节内容还提供了与中文内容对应的英文译本,帮助更好地开展双语教学工作。目前,本教材尚属饲料学专业课首本中英文对照的教材,可用于更好地指导相关高校开展饲料学专业课程的双语教学。

本教材除适用于动物科学、动物营养及饲料加工等相关专业的高校专业课程的教学,也可供畜牧及饲料企业的工作及研究人员使用。

本教材获扬州大学出版基金资助。

本教材配备相关教材课件,可扫描下方二维码链接后下载。限于编者水平,教材中如有不当之处,恳请读者批评指正。

绪论 ·· 1
Introduction

第一节　国外饲料业发展概况 ·· 2
Section 1　Development of feed industry abroad

一、国外饲料业发展的萌芽阶段 ······································· 2
　　The rudimentary stage of foreign feed industry development

二、世界饲料工业的发展阶段 ·· 3
　　Development stages of world feed industry

三、牧草和粗饲料的生产及利用是世界饲料业的重要内容 ··· 8
　　Production and utilization of forage and roughage as an important part of the world feed industry

第二节　中国饲料业发展概况 ······································· 10
Section 2　Development of Chinese feed industry

一、中国饲料业发展史 ·· 10
　　Development history of feed industry in China

二、当前我国饲料业中存在的主要问题及对策 ················ 19
　　Main problems and countermeasures of feed industry in China

第三节　未来饲料行业发展趋势 ···································· 26
Section 3　Future development trend of feed industry

一、未来饲料科学的研究方向 ······································· 27
　　Research direction of feed science in the future

二、未来饲料业发展趋势 ··· 28
　　Development trend of feed industry in the future

第四节　饲料学的性质、任务、研究内容和研究方法 ………………………………………… 30
Section 4　Properties, tasks, research content and research methods of feed science

一、饲料学的性质 ………………………………………………………………………… 31
Properties of feed science

二、饲料学的任务 ………………………………………………………………………… 32
Tasks of feed science

三、饲料学的研究内容 …………………………………………………………………… 32
Research content of feed science

四、饲料学的研究方法 …………………………………………………………………… 33
Research methods of feed science

第一章　饲料化学 ……………………………………………………………………………… 35
Chapter 1　Feed chemistry

第一节　碳水化合物 ……………………………………………………………………………… 35
Section 1　Carbohydrates

一、单糖 …………………………………………………………………………………… 36
Monosaccharides

二、低聚糖 ………………………………………………………………………………… 37
Oligosaccharides

三、多糖 …………………………………………………………………………………… 40
Polysaccharides

第二节　含氮化合物 ……………………………………………………………………………… 44
Section 2　Nitrogen compounds

一、蛋白质的性质与分类 ………………………………………………………………… 45
The property and classification of protein

二、氨基酸 ………………………………………………………………………………… 49
Amino acids

三、寡肽 …………………………………………………………………………………… 50
Oligopeptides

四、其他含氮化合物 ……………………………………………………………………… 51
Other nitrogen compounds

第三节　脂类 ·· 52
Section 3　Lipids

一、脂类的分类 ··· 52
Classification of lipids

二、脂肪 ·· 53
Fat

三、类脂 ·· 56
Lipoid

第四节　矿物质 ·· 58
Section 4　Minerals

一、常量元素与微量元素 ·· 58
Major and trace elements

二、必需矿物质元素 ·· 58
Essential mineral elements

三、动物矿物质元素营养与环境 ··· 59
Animal mineral element nutrition and environment

第五节　维生素 ·· 60
Section 5　Vitamins

一、概述 ·· 60
Overview

二、维生素的分类与命名 ·· 61
Classification and naming of vitamins

第六节　水分 ··· 62
Section 6　Water

一、水分的存在形式 ·· 62
Existing form of water

二、水分的活性度 ··· 62
Water activity

第七节　其他成分 ··· 63
Section 7　Other ingredients

　一、抗营养因子 ··· 63
　　Antinutritional factors

　二、饲料的色素 ··· 64
　　Pigments of feed

　三、饲料的味嗅物质 ··· 67
　　Taste and smell substances of feed

第二章　饲料分类 ··· 69
Chapter 2　Feed classification

第一节　饲料的习惯分类法 ··· 69
Section 1　Customary feed classification

　一、根据饲料营养价值分类 ·· 70
　　Classification of feed's nutritive value

　二、根据饲料主要成分分类 ·· 70
　　Classification of feed's main ingredients

　三、根据饲料来源分类 ·· 71
　　Classification of feed's sources

　四、其他分类 ··· 71
　　Other classifications

第二节　国际饲料分类法 ·· 72
Section 2　International Feed Classification

　一、粗饲料 ·· 72
　　Roughage

　二、青绿饲料 ··· 73
　　Green forage

　三、青贮饲料 ··· 73
　　Silage

　四、能量饲料 ··· 73
　　Energy feed

五、蛋白质饲料 ·· 73
　　Protein feed

六、矿物质饲料 ·· 74
　　Mineral feed

七、维生素饲料 ·· 74
　　Vitamin feed

八、饲料添加剂 ·· 74
　　Feed additives

第三节　中国饲料分类法 ·· 74
Section 3　Chinese feed classification

一、青绿多汁类饲料 ·· 75
　　Green fresh forage

二、树叶类饲料 ·· 75
　　Tree leaves feed

三、青贮饲料 ··· 75
　　Silage

四、块茎、块根、瓜果类饲料 ··· 76
　　Tuber, root tuber, melon and fruit feed

五、干草类饲料 ·· 76
　　Hays forage

六、农副产品类饲料 ·· 76
　　Agricultural byproduct feed

七、谷实类饲料 ·· 77
　　Cereal grain

八、糠麸类饲料 ·· 77
　　Milling byproduct feed

九、豆类饲料 ··· 77
　　Beans feed

十、饼粕类饲料 ·· 78
　　Cake meal feed

十一、糟渣类饲料 ………………………………………………………………… 78
　　Distiller dried grains with soluble feed

十二、草籽、树实类饲料 …………………………………………………………… 78
　　Seed of grass and tree feed

十三、动物性饲料 ………………………………………………………………… 79
　　Animal feed

十四、矿物质饲料 ………………………………………………………………… 79
　　Mineral feed

十五、维生素饲料 ………………………………………………………………… 80
　　Vitamin feed

十六、饲料添加剂 ………………………………………………………………… 80
　　Feed additives

十七、油脂类饲料及其他 …………………………………………………………… 80
　　Oil feed and others

十八、中国饲料分类编码 …………………………………………………………… 80
　　Chinese feed classification encoding

第三章　青干草与粗饲料 …………………………………………………………… 82
Chapter 3　Green hay and roughage

第一节　青干草与草粉 ……………………………………………………………… 83
Section 1　Green hay and grass powder

一、青干草调制原理与方法 ………………………………………………………… 83
　　The principle and method of green hay processing

二、青干草的营养价值 ……………………………………………………………… 89
　　The nutritive value of green hay

三、草粉的生产与应用 ……………………………………………………………… 90
　　Production and application of grass powder

四、干草产品的质量评定 …………………………………………………………… 92
　　Quality assessment of hay products

第二节 稿秕与饲用林产品饲料 ·· 95
Section 2　Straw, hull and forest products for feeding

　一、秸秆饲料 ·· 95
　　Straw feed

　二、秕壳饲料 ·· 98
　　Hull

　三、树叶和其他饲用林产品 ·· 99
　　Leaves and other forage forest products

第三节 粗饲料的加工调制及品质评定 ··· 99
Section 3　Processing and quality evaluation of roughage

　一、物理加工 ··· 100
　　Physical processing

　二、化学处理 ··· 102
　　Chemical treatment

　三、生物学处理 ·· 105
　　Biological treatment

　四、综合利用 ··· 108
　　Comprehensive utilization

　五、粗饲料品质评定 ·· 109
　　Quality evaluation of roughage

第四章　青绿饲料 ·· 110
Chapter 4　Green forage

第一节 青绿饲料的营养特性及影响因素 ····································· 110
Section 1　Nutritional characteristics and influencing factors of green forage

　一、青绿饲料的营养特性 ·· 110
　　Nutritional characteristics of green forage

　二、影响青绿饲料营养价值的因素 ·· 112
　　The factors affecting the nutritive value of green forage

第二节 主要青绿饲料 ·· 115
Section 2　Main green forage

　一、天然牧草 ··· 115
　　Natural herbage

二、栽培牧草 …………………………………………………………………… 116
　　Cultivated herbage

三、青饲作物 …………………………………………………………………… 129
　　Green forage crops

四、叶菜类 ……………………………………………………………………… 132
　　Leaf vegetables

五、非淀粉质根茎瓜类 ………………………………………………………… 137
　　Non-starchy rhizome melons

六、水生植物 …………………………………………………………………… 141
　　Aquatic plants

七、树叶类 ……………………………………………………………………… 141
　　Leaves

第五章　青贮饲料 ……………………………………………………………… 146
Chapter 5　Silage

第一节　青贮饲料的优点及其发展 …………………………………………… 146
Section 1　Advantages and development of silage

一、青贮饲料的优点 …………………………………………………………… 147
　　The advantages of silage

二、青贮饲料的发展 …………………………………………………………… 149
　　The development of silage

第二节　青贮发酵原理及基本过程 …………………………………………… 150
Section 2　Principle and basic process of silage fermentation

一、青贮发酵原理 ……………………………………………………………… 150
　　The principle of silage fermentation

二、青贮发酵的基本过程 ……………………………………………………… 151
　　The basic process of silage fermentation

第三节　青贮饲料的原料及调制 ……………………………………………… 154
Section 3　Raw materials and preparation of silage

一、青贮原料和添加剂 ………………………………………………………… 154
　　Silage raw materials and additives

二、青贮类型 …………………………………………………………………… 160
　　Silage type

三、调制过程 ·· 162
The modulation process

第四节 青贮饲料品质鉴定及利用 ··· 165
Section 4　Quality identification and utilization of silage

一、感官评定 ·· 165
Sensory evaluation

二、实验室评定 ·· 166
Laboratory evaluation

三、饲用价值评定 ·· 167
Evaluation of feeding value

四、青贮饲料的利用 ·· 167
The utilization of silage

第六章　能量饲料 ·· 170
Chapter 6　Energy feed

第一节　谷实类饲料 ·· 171
Section 1　Cereal grain

一、营养特点 ·· 171
Nutritional characteristics

二、主要谷实类饲料的特点 ·· 173
The characteristics of main cereal grains

第二节　谷实类加工副产物饲料 ·· 180
Section 2　By-product feed of the grains

一、营养特点 ·· 180
Nutritional characteristics

二、主要加工副产物饲料 ·· 181
Main processing by-product feed

第三节　块根、块茎及瓜果类饲料 ······································ 185
Section 3　Root, tuber, melon and fruit feed

一、营养特点 ·· 185
Nutritional characteristics

二、常见的块根、块茎类饲料的特点 ·································· 186
The characteristics of main root tuber, tuber feed

9

第四节　其他能量饲料 ··· 190
Section 4　Other energy feeds

　　一、脂肪 ··· 190
　　Fat

　　二、糖蜜 ··· 194
　　Molasses

　　三、乳清粉 ··· 195
　　Whey powder

　　四、其他 ··· 196
　　Others

第七章　蛋白质饲料 ··· 198
Chapter 7　Protein feed

第一节　植物性蛋白饲料 ··· 198
Section 1　Plant protein feed

　　一、豆类籽实 ··· 199
　　Bean seeds

　　二、饼粕类 ··· 203
　　Cake and meal

　　三、其他植物性蛋白质饲料 ··· 216
　　Other plant protein feeds

第二节　动物性蛋白质饲料 ··· 220
Section 2　Animal protein feed

　　一、水产加工副产物饲料 ··· 220
　　By-product feed of aquatic processing

　　二、畜禽副产物饲料 ··· 226
　　Livestock and poultry by-product feed

第三节　非蛋白氮饲料 ··· 231
Section 3　Non-protein nitrogen feed

　　一、尿素 ··· 232
　　Urea

　　二、胺盐类 ··· 233
　　Amine salts

第八章 矿物质饲料
Chapter 8　Mineral feed

第一节　常量矿物质饲料 ································ 235
Section 1　Macroelement feed

一、钙源性饲料 ································ 236
Calcium source feed

二、磷源性饲料 ································ 240
Phosphorus source feed

三、钠源性饲料 ································ 247
Sodium source feed

四、其他常量矿物质饲料 ································ 251
Other macroelement feeds

第二节　天然矿物质饲料 ································ 253
Section 2　Natural mineral feed

一、沸石 ································ 253
Zeolite

二、麦饭石 ································ 255
Maifan stone

三、膨润土 ································ 256
Bentonite

四、凹凸棒石 ································ 257
Attapulgite

五、其他天然矿物质饲料 ································ 257
Other natural mineral feeds

第九章　维生素饲料 ································ 261
Chapter 9　Vitamin feed

第一节　脂溶性维生素饲料 ································ 263
Section 1　Fat-soluble vitamin feed

一、维生素 A 和 β-胡萝卜素 ································ 263
Vitamin A and β-carotene

二、维生素 D ··· 266
Vitamin D

三、维生素 E ··· 267
Vitamin E

四、维生素 K_3 ··· 268
Vitamin K_3

第二节　水溶性维生素饲料 ·· 269
Section 2　Water-soluble vitamin feed

一、硫胺素（维生素 B_1） ··· 270
Thiamine（Vitamin B_1）

二、核黄素（维生素 B_2） ··· 271
Riboflavin（Vitamin B_2）

三、泛酸 ··· 272
Pantothenic acid

四、维生素 B_5（烟酸和烟酰胺） ·· 273
Vitamin B_5（niacin and nicotinamide）

五、维生素 B_6 ··· 274
Vitamin B_6

六、生物素 ··· 274
Biotin

七、维生素 B_{12} ··· 275
Vitamin B_{12}

八、叶酸 ··· 276
Folic acid

九、胆碱 ··· 277
Choline

十、维生素 C ··· 278
Vitamin C

十一、肌醇 ·· 279
Inositol

第三节　维生素饲料的合理应用 ………………………………………………………… 279
Section 3　Reasonable application of vitamin feed

　　一、饲粮中维生素添加量的确定 …………………………………………………… 279
　　　　Determination of dietary vitamin supplementation

　　二、维生素饲料的选择 ……………………………………………………………… 280
　　　　Selection of vitamin feed

　　三、维生素饲料的配伍 ……………………………………………………………… 280
　　　　Compatibility of vitamin feed

　　四、维生素饲料的添加方法 ………………………………………………………… 281
　　　　Method of adding vitamin feed

　　五、维生素饲料产品的包装贮存 …………………………………………………… 281
　　　　Packaging and storage of vitamin feed products

第十章　饲料添加剂 ……………………………………………………………………… 282
Chapter 10　Feed additives

第一节　非营养性饲料添加剂 …………………………………………………………… 282
Section 1　Non-nutritive feed additives

　　一、药物添加剂 ……………………………………………………………………… 282
　　　　Medicated additives

　　二、益生素 …………………………………………………………………………… 286
　　　　Probiotics

　　三、酶制剂 …………………………………………………………………………… 290
　　　　Enzyme preparations

　　四、酸化剂 …………………………………………………………………………… 293
　　　　Acidulants

　　五、中草药添加剂 …………………………………………………………………… 296
　　　　Chinese herbal medicine additives

　　六、饲料品质调节剂 ………………………………………………………………… 300
　　　　Feed quality regulators

13

第十一章　配合饲料产品 ·········· 305
Chapter 11　Formulated feed products

第一节　饲料配合概述 ·········· 305
Section 1　Overview of feed composition

一、饲料配合的概念 ·········· 305
The concept of feed composition

二、配合饲料的特点 ·········· 306
Characteristics of formulated feed

三、饲料配合产品的种类和组成 ·········· 307
Types and composition of feed compound products

第二节　配合饲料的生产工艺 ·········· 309
Section 2　Production technology of formulated feed

一、概述 ·········· 309
Summary

二、原料的接收、清理与储存 ·········· 310
Reception, cleaning and storage of raw materials

三、原料的粉碎 ·········· 314
Crushing of raw materials

四、配料及混合 ·········· 315
Batching and mixing

五、制粒或膨化 ·········· 317
Granulation or expansion

六、包装与贮存 ·········· 320
Packaging and storage

绪论
Introduction

凡能被动物采食又能提供给动物某种或多种营养素且无毒的物质，即能够被动物摄取、消化、吸收和利用的物质称为饲料。饲料是动物赖以生存和生产的物质基础。人类在长期的畜牧业生产实践中，通过大量的动物营养和饲料科学的理论与应用研究，对动物生产需要的饲料及其营养价值（nutritive value）逐渐有了更科学、更全面、更深化的认识。饲料的种类不断增加，新型饲料资源不断产生，饲料的利用方式从传统的、经简单加工的单一使用发展到科学的、适宜加工配合的配合利用，饲料利用效率显著提高，并不断采取各种新的技术措施，生产出产品质量高、营养更全价的饲料。现在，饲料工业已成为关系国计民生的重要产业。随着科学技术的进步和生活水平的提高，人类更注重自身的健康和生活的质量，不仅要求肉、奶、蛋等畜产品数量足够，而且要求质量更高，因此，动物食用的饲料正逐渐向生产"有机食品"和"绿色食品"方向发展。

饲料学是一门研究动物饲料的化学组成、营养理化性质及其影响因素和应用技术的一门学科。饲料学通过对饲料营养价值的评定、饲料加工和日粮配合等应用技术的研究，最终达到扩大饲料资源的开发利用、保障饲料安全、获得理想动物生产性能和产品品质的目标。饲料

Feed, food given to domestic animals in the course of animal husbandry, is a substance that can be eaten by an animal and provided with one or more nutrients, i. e. a substance that can be ingested, digested, absorbed, and used by animals. Thus, feed is the material basis for the survival and production of animals. In the long-term practice of animal husbandry, the knowledge about animal nutrition and nutritive value of feedstuff becomes more scientific, comprehensive, in-depth understanding. Feed types continue to increase and new feed resources continue to produce. The pattern of feed utilization was changed from the traditional, simple processing of a single utilization to the scientific, suitable processing with coordinated utilization. Feed utilization efficiency is significantly improved and a variety of new technical measures are taken to produce high-quality and nutritive feed. Now, the feed industry has become an important industry related to the national economy and people's livelihood. With the progress of science and technology and the improvement of living standards, human beings pay more attention to their own health and quality of life, which requires not only enough meat, milk, eggs and other animal products, but also the animal products containing higher qualities. Therefore, animal feed is gradually developing towards "organic products" and "green products".

Feed science is a subject that studies the chemical composition of feed, physicochemical properties of feed and their influencing factors, and the application technology of animal feed. Feed science aims to expand the development and utilization of feed resources, to ensure the safety of feed, and to obtain ideal animal performance and high-quality product through the evaluation of feed nutritive value, research on feed processing and ration matching and other

学作为一门独立的学科，经历了相当漫长的历史发展过程，极大地推动了畜牧业和整个国民经济的发展，并且随着历史进程的推移，它在发展畜牧业进而在整个国民经济中的地位愈来愈重要。本章内容主要介绍了国内外饲料工业的发展概况，以及饲料学的主要研究内容和研究方法等内容。

application technologies. Feed science, as an independent subject, has experienced a long history, which has greatly promoted the development of animal husbandry and the whole national economy. With the development of history, feed science plays an increasingly important role in the development of animal husbandry and the whole national economy. This chapter mainly introduces the development of feed industry at home and abroad, as well as the major research content and research methods of feed science.

第一节 国外饲料业发展概况
Section 1 Development of feed industry abroad

国外饲料业尤其是西方发达国家的饲料业在近代和现代取得了巨大的成就，从此带动了畜牧业乃至国民经济和人民生活水平的不断提高。虽然发展中国家的饲料业有其独特之处，但美国和欧盟等发达国家的饲料业代表着世界饲料业的前进方向。世界饲料业发展起步较晚，但近200多年来，由于近代和现代动物营养和饲料科学在西方国家的迅速兴起，尤其是对饲料营养价值和饲养标准的研究推动了饲料工业、牧草业等的蓬勃发展，使得饲料业迅速成为一个独立的产业，饲料科学成为一门独立的学科。

Foreign feed industry, especially the feed industry in Western developed countries, has made great achievements in modern times, which has led to the continuous improvement of animal husbandry, national economy and people's living standards. Although the feed industry of developing countries has its unique features, the feed industry of developed countries such as the United States and the European Union represents the direction of the world feed industry. The development of feed industry in the world started relatively late, but in the past 200 years, due to the rapid rise of modern animal nutrition and feed science in western countries, especially the research on feed nutritive value and feeding standards, the feed industry and forage industry have developed vigorously, making the feed industry rapidly become an independent industry and feed science an independent subject.

一、国外饲料业发展的萌芽阶段
Rudimentary stage of foreign feed industry development

19世纪70年代以前，是国外饲料业发展的萌芽阶段。这一时期饲料工业还未起步，饲料业尚未成为一个独立的产业。在此期间，许多科学家对饲料及其营养价值、饲料标准的研究为以后饲料业的发展，尤其是饲料业的建立做了大量

The time before the 1870s was the embryonic stage of the development of foreign feed industry. During this period, the feed industry has not yet started, and the feed industry has not yet become an independent industry. Besides, many scientists have done a lot of researches on the animal feed, the nutritive value of feed and the standard of animal feed to create the conditions for the future development of

的准备工作,创造了条件。早在18世纪以前,人类对饲料的营养价值就有了感性的认识。如:在罗马时代,普利尼就认识到了"适时收割的干草比成熟收割的好",并指出"改进饲养才能获得良好的家畜生产效益"。进入18世纪以后,随着动物营养实验科学的发展,加上物理、化学和生物学等学科的推动才使饲料学发展有了质的飞跃。1807年,英国的科学家Pordyce证实,需要给产蛋鸡补充钙,从此开始了饲料矿物质营养的研究。1810年,德国科学家Thaer提出了"干草当量"体系,以此方法制定出了各种饲料的相对营养价值。1864年,Henneberg与Stohnann在德国Weende试验站创建了饲料概略养分分析法。在此基础上,1874年,Wolff首次提出了以总消化养分(TDN)为基础的饲养标准。早期的饲料加工设备比较简单。在1900年以前,手勺铲是饲料加工厂使用的基本混合工具,木桶是主要的产品贮存容器。1848年,Chast制袋公司首次加工出布袋,代替了木桶。有关其他设备的利用少见文章记载。在这个阶段,人们对饲料的认识,尤其对饲料营养价值的认识还很肤浅,畜牧业仅仅是一项副业。饲喂畜禽所用的基本上是谷物精料或秸秆、饲草等粗饲料,饲养水平非常低下。

feed industry, especially the establishment of feed industry. Before the 18th century, human beings had a perceptual understanding of the nutritive value of feed. For example, in Roman times, Pliny realized that the hay harvested in good time is better than the hay harvested at full maturity, and pointed out that improving feeding patterns can obtain good livestock production efficiency. After the 18th century, with the development of animal nutrition experimental science and the promotion of physics, chemistry and biology, the development of feed science has made a qualitative leap. In 1807, Pordyce, an English scientist, confirmed that it was necessary to supplement calcium for laying hens, and the research on feed mineral nutrition began afterwards. In 1810, Thaer, a German scientist, proposed the hay equivalent system, by which the relatively nutritive value of various feeds was formulated. In 1864, Henneberg and Stohnann established the methods of feed nutrient analysis in a German research station named Weende. Based on these analysis methods, Wolff firstly proposed the feeding standard based on total digestible nutrients (TDN) in 1874. Early feed processing equipment was relatively simple. Before 1900, scoop shovels were the basic mixing tools used in feed processing, and barrels were the main product storage containers. In 1848, Chast bag making company firstly produced the cloth bags to replace the wooden barrels. There were few articles about the utilization of other equipment. At this stage, people's understanding of feed, especially the nutritive value of feed, was still very superficial. Animal husbandry was only a sideline. Grain concentrate and straw, forage or other roughage were main feed for livestock and poultries and the feeding level was very low at that time.

二、世界饲料工业的发展阶段
Development stages of world feed industry

动物营养和饲料科学研究的不断深入,对饲料业的快速发展起到了巨大的推动作用。国外饲料工业尤其是发达国家的饲料工业的发展大体经历了四个阶段。

Researches on animal nutrition and feed science become constant in-depth, which has played a great role in promoting the rapid development of animal feed industry. The development of feed industry in foreign countries, especially in developed countries, has gone through four stages.

（一）世界饲料工业的开端
The beginning of world feed industry

1875年，John Barwell在美国伊利诺伊州沃基根市创建了全球第一家饲料厂Blathford's，生产犊牛饲料，它的建立标志着世界饲料工业的开始。但当时由于动物营养与饲料研究，特别是对饲料营养价值评定的局限性，其饲料产品只考虑了干物质和可消化总养分两项质量指标。

In 1875, John Barwell founded Blathford's feed factory in Waukegan (Illinois, USA) to produce calf feed and the Blathford's feed factory was the first feed factory in the world, which marked the beginning of the world feed industry. But at that time, due to the limitation of animal nutrition and feed research, especially the evaluation of feed nutritive value, only two quality indexes were considered for its feed products including dry matter and digestible total nutrients.

（二）饲料工业的起步阶段
The initial stage of feed industry

从世界上第一家饲料厂建立到20世纪初期是饲料工业的起步阶段。在这一时期，饲料加工设备不断出现。从19世纪70年代欧洲研制出对辊式磨粉机开始，到1909年S. Howes公司制造出分批卧式混合机，1911年英国的Sizer有限公司研制出第一台商品制粒机，以及之后立式混合机、糖蜜饲料制粒机的相继问世，标志着饲料加工能力不断得到增强，这一时期也是饲料原料种类逐渐得到开发和应用的阶段。19世纪80年代，农民还不知道利用皮、肉骨粉、棉籽饼粕作饲料，几乎全依赖谷物和粗饲料饲养畜禽。之后随着对制粉、肉类、油料及其他加工业副产品饲用价值的认识，原来不能被畜牧业利用的副产品相继被开发利用作为饲料。如：1888年，玉米蛋白粉在芝加哥广泛生产；1890年，肉骨粉被作为蛋白质补充料用于猪和鸡的饲粮；1890年以后，亚麻饼粉、苜蓿粉、骨粉在畜禽生产中到应用；1915年，鱼粉被广泛销售；1922年，美国首次生产大豆粕；1939年，尿素被用作反刍家畜的合成蛋白质来源。当今世

The initial stage of feed industry was the time from the establishment of the first feed factory in the world to the beginning of the 20th century. During this stage, processing equipment was constantly designed and manufactured. The counter roller mill was developed in Europe in the 1870s and the batch horizontal mixer was manufactured by S. Howes Company in 1909. The first commercial granulator was developed by British Sizer Co., Ltd. in 1911, and then the vertical mixer and molasses feed granulator came out one after another. These events indicated that the feed processing ability had been continuously enhanced, and this period was also the stage when the types of feed raw materials had been gradually developed and applied. In the 1880s, farmers did not know how to use skin, meat and bone meal, cotton seed meal as feed, and almost all farmers relied on grain and roughage to feed livestock and poultry. Later, with the understanding of the feeding value of flour, meat, oil and other by-products of processing industry, the by-products that could not be used by animal husbandry were developed and used as feed. For example, corn gluten meal was widely produced in Chicago in 1888. Meat and bone meals were used as a protein supplement for pigs and chickens in 1890. The flaxseed meal, alfalfa meal and bone meal were used in livestock and poultry production after 1890. In 1915, the fish meal was widely sold in the world. In 1922, the soybean meal was firstly produced in the United States. In 1939, urea was used as a synthetic protein source for

界所用的主要大宗饲料原料在当时均被开发出来并得到了应用。

当时,动物营养与饲料科学的研究也十分活跃。1898年,Henry将以总消化养分为基础的饲养标准修订为以可消化总养分为基础的饲养标准。1894年,Kuhn首次提出按能量直接衡量饲料的营养价值。此后,德国Kellner淀粉价以及北欧大麦饲料单位、苏联燕麦单位和美国的Aralsby的净能体系相继问世,维生素、微量元素、矿物质的研究及应用日益受到关注。1912年,波兰化学家Funk在谷壳中发现了维生素B_1;1913年,美国学者在鱼肝油和奶油中发现了维生素A。20世纪30至40年代,大部分脂溶性维生素和水溶性维生素相继被发现并被合成。1925年,美国学者Hart等人研究表明,铁和铜同时使用才能治愈大鼠的缺铁性贫血;1935年,一些学者发现钴是瘤胃活动的必需物质;1937年,发现锰可以防治家禽飞节病,从此畜禽微量元素的研究逐渐深入。此外,氨基酸的研究也取得了一些进展。20世纪40年代提出了理想蛋白质的概念,并建立了饲料中氨基酸的微生物分析法;20世纪50年代提出了化学分析法,为以后定量评定饲料的蛋白质营养价值提供了手段。1944年,美国科学院全国研究理事会(NRC)制定了一系列畜禽饲养标准,并且10年修订一次,现被公认为制定饲料配方的基础。所有这些动物营养与饲料科学的研究,都有力地支持了饲料工业的发展。在这一时期,虽然有一些专营饲料生产、销售的饲料厂出现,但多数饲料工业仍是农场的副业,以农场主经营为主,生产饲料多用于自养的畜禽,生产规模小,品种较单一,生产的饲

ruminants. The main bulk feed materials used in today's world were developed and applied at that time.

At that time, animal nutrition and feed science research were also very active. In 1898, Henry revised the feeding standard based on total digested nutrients to that based on total digestible nutrients. In 1894, Kuhn firstly proposed that the nutritive value of feed should be directly measured by energy. Since then, the German Kellner starch value, Nordic barley unit, Soviet oat unit, and net energy system of United States had come out one after another. The research and application of vitamins, trace elements and minerals had also attracted more and more attention. In 1912, Funk, a Polish chemist, discovered vitamin B_1 in the rice husk. In 1913, American scholars discovered vitamin A in cod liver oil and cream. Most of the fat-soluble vitamins and water-soluble vitamins were found and synthesized from 1930s to 1940s. In 1925, American scholar Hart and others reported that iron and copper could be used at the same time to cure the iron deficiency anemia in rats. In 1935, some scholars found that cobalt was an essential material for rumen activities. In 1937, manganese was found to prevent fowl flying joint disease, from which the research of trace elements in livestock and poultry gradually deepened. In addition, some progress had been made in the study of amino acids. In the 1940s, the concept of ideal protein was proposed, and the microbial analysis method of amino acids in feed was established. In the 1950s, the chemical analysis method was proposed, which provided the methods for quantitative evaluation of protein nutritive value of feed in the future. In 1944, the National Research Council of the American Academy of Sciences (NRC) formulated a series of animal feeding standards, which were revised once every 10 years. And the NRC feeding standards are recognized as the basis for formulating feed formula at present. All these researches on animal nutrition and feed science have strongly supported the development of feed industry. During this period, although there were some feed factories specializing in feed production and sales, most of the feed industry was still the sideline of the farm, mainly operated by farmers. The production of feed was mostly used for self-supporting livestock and poultry. The production scale was small and the variety was single. The feed produced could not meet the nutritional

料还不能满足畜禽的营养要求,生产性能不高。

requirements of livestock and poultry, and the production performance was not high.

(三) 饲料工业的成长发展阶段
The growth stage of feed industry

第二次世界大战以后,世界养殖业发生了巨大的变化,开始向集约化、专业化方向发展。与此相适应,国外饲料工业开始从农场中分离出来,饲料工业不再是一项副业,而成为一门相对独立的产业,进入成长发展时期。这一时期,冷却机、颗粒膨化机、多功能混合机问世。由于饲料机械制造业、加工业的不断发展,饲料工业的生产规模逐渐扩大。豆粕、鱼粉成为主要的蛋白质饲料;脂肪被作为能量饲料使用,并且配制出高能量的仔鸡饲粮;为了控制酸败,在脂肪中加入抗氧化剂;证明了在配制饲料时,蛋白能量比的重要性;研究出反刍动物净能体系;激素被用在牛的饲料中以促进增重;饲料营养添加剂氨基酸、维生素、微量元素,以及非营养性添加剂和合成抗生素类的研究和应用,极大地提高了饲料产品的质量和畜禽的生产性能。1950 年,美国配合饲料产量已达 2 600 多万 t;日本 1960—1970 年配合饲料产量由 243.3 万 t 增加到 1 499.7 万 t。20 世纪 60 年代,美国的饲料工业进入全国最大的 20 个工业部门之列。

After the Second World War, great changes have occurred in the world animal husbandry and the animal husbandry began to develop in the direction of intensification and specialization. To adapt to these changes, foreign feed industry began to separate from the farms and feed industry was no longer a sideline, but a relatively independent industry. This led the feed industry to enter into a period of growth and development. During this period, cooling machine, particle expander, and multi-functional mixer appeared. Due to the continuous development of feed machinery manufacturing and processing industry, the production scale of feed industry was gradually expanding. Soybean meal and fish meal became the main protein feed. Fat was used as energy feed and high-energy diet was prepared for chickens with dietary fat. In order to prevent the rancidity of fat, antioxidants were added. The importance of protein energy ratio in feed preparation was proved. Ruminant net energy system was developed. Hormones were used in the cattle feed to promote weight gain. The research and application of amino acids, vitamins, trace elements as feed additives, non-nutritive additives and synthetic antibiotics had greatly improved the quality of feed products and the performance of livestock and poultry. In 1950, the United States produced more than 26 million tons of formula feed. In Japan, the production of formula feed increased from 2.433 million tons to 14.997 million tons from 1960 to 1970. In the 1960s, the feed industry of the United States entered one of the 20 largest industrial sectors in the country.

(四) 饲料工业的现代稳定发展阶段
The modern and steady developed stage of feed industry

20 世纪 70 年代以来,由于动物科学和饲料工业技术的进步,以及电脑配方和自动化管理的引入,酶制剂、酸化剂等新型饲料添加剂不断研制和应用于生产,使技术含量高的浓缩饲料和预混料增长迅

Since the 1970s, new feed additives such as enzyme preparation and acidifiers were continuously developed and applicated due to the progress of animal science and feed industry technology and the introduction of computer formula and automatic management. This rapidly increased the production of high technical concentrated feeds and premix and

速,国外饲料工业进入现代化的稳定发展时期。在美国、德国、英国等经济发达国家,饲料工业已经发展成为一个完整的工业体系,成为与电子计算机工业等并驾齐驱的重要产业,位于工业部门经济的前10位。现代的饲料加工机械设备集团,可提供粉碎机、计算机控制配料系统、批量式混合机、颗粒机等系列配套设备。大型饲料厂自动化程度高,电脑自动抽取原料样品检验,并在极短时间(2~3 min)出查验结果,电脑自动控制配、卸料。饲料加工企业生产规模日益扩大,大型规模化企业和一体化企业逐渐成为主流。2008年,全球工业化饲料总量为7.01亿t,但占相当大比例的饲料是由数量不多的大型饲料加工企业生产的,如美国2008年工业饲料生产量为1.545亿t,其中生产能力超过1 000万t的公司就有3家,超过200万t的饲料企业就有9家,年产饲料5 020万t,占全国饲料总数的32.5%。通过集团化提高饲料生产能力和产量,不仅降低了生产成本,而且形成了品牌优势,在激烈的市场竞争中使饲料企业具有很强的抵御风险的能力;以饲料企业为龙头的一体化企业,将饲料原料生产、养殖、屠宰、肉食品加工等各个环节连接起来,减少了中间环节,大大降低了生产成本和管理成本。同时,在市场树立了统一的企业形象,为把企业做大做强打下了良好的基础。目前,在全美排名前10位的饲料企业集团中,就有3家是一体化企业。发达国家的饲料生产企业都十分注重高新技术在生产中的应用,并且视产品质量为企业的生命。如:美国在食品饲料行业中积极推行危害分析及关键控制点(HACCP),欧洲国家坚持欧盟统一

led foreign feed industry into a period of stable development and modernization. In the United States, Germany, the United Kingdom and other economically developed countries, feed industry has developed into a complete industrial system, and become an important industry in parallel with the computer industry, which is in the top 10 of the industrial sectors. Modern feed processing machinery and equipment group can provide a series of supporting equipment such as the pulverizer, computer-controlled batching system, batch mixer, granulator, etc. The large-scale feed factory has a high degree of automation. The computer automatically takes raw material samples for the detection, and the detection results can be obtained in a very short time (2-3 min). The computer automatically controls the feeding and discharging. With the increasing production scale of feed processing enterprises, large-scale enterprises and integrated enterprises have gradually become the mainstream. In 2008, the total amount of industrial feed in the world was 701 million tons, but a large proportion of feed was produced by a small number of large feed processing enterprises. For example, the industrial feed production in the United States in 2008 was 154.5 million tons, of which 3 companies had a production capacity of more than 10 million tons and 9 companies had a production capacity of more than 2 million tons, with an annual output of 50.2 million tons, accounting for 32.5% of the total feed in the country. Collectivized production improved the feed production capacity and output. It not only reduced the production costs, but also formed a brand advantage to achieve a strong ability to resist risks in the fierce market competition. Integrated enterprises with feed enterprises as the leader connected the feed raw material production, breeding, slaughtering, meat food processing and other links and reduced the intermediate links. It greatly reduced the costs of production and management. At the same time, it had established a unified corporate image in the market, which had laid a good foundation for making the enterprise bigger and stronger. At present, three feed enterprises are the integrated enterprises among the top 10 feed enterprise groups in the United States. Feed production enterprises in developed countries attach great importance to the application of high and new technology in production, and regard product quality as their life.

的饲料标准。各国饲料业十分重视开发各种饲料原料资源,并严禁使用被污染和不合格原料。为防止疯牛病的(牛海绵状脑病,BSE)发生和二恶英污染,世界上大多数国家相继采取措施,禁止所有反刍动物的肉骨粉作为反刍动物的饲料或所有动物的饲料,禁止被二恶英污染的饲料作原料。此外,由于抗生素所引起的耐药性和药物残留及激素类药物所引起的副作用,绿色饲料添加剂应运而生,越来越广泛地得到应用。

For example, the United States actively promotes hazard analysis and critical control point (HACCP) in the food and feed industry, and European countries adhere to the unified feed standard of the European Union. Feed industry all over the world attaches great importance to the development of various feed raw material resources, and strictly prohibit the use of contaminated and unqualified raw materials. In order to prevent the occurrence of bovine spongiform encephalopathy (BSE) and dioxin pollution, most countries in the world have taken measures one after another to prohibit all ruminant meat and bone meal as ruminant feed or animal feed, and prohibit dioxin contaminated feed as raw material. In addition, due to antibiotic resistance and drug residues and side effects caused byhormone drugs, green feed additives emerge as the times require and are more and more widely used.

在这一阶段,动物营养和饲料科学的研究进一步深化,许多国家相继提出了评定反刍动物饲料的理想蛋白质及蛋白质需要量新体系,反刍动物的饲养标准现已采用蛋白质新体系。对猪、禽等单胃动物的理想氨基酸模式、畜禽营养调控、饲料营养物质生物学效价的研究日益深入,以可消化氨基酸配制饲粮已应用于实践。信息技术在饲料工业中得到了广泛的应用。同时,动物营养与饲料科学同遗传学、育种学、生理学、分子生物学、化学等边缘学科的相互渗透和嫁接进一步加强。

At this stage, the research of animal nutrition and feed science has been further deepened. Many countries have put forward the new system of ideal protein and protein requirement for evaluating ruminant feed. The new system of protein has been adopted in the feeding standard of ruminant. The ideal amino acid model of monogastric animals such as pigs and poultry, the nutritional regulation of livestock and poultry, and the biological potency of feed nutrients have been increasingly studied. The preparation of diets with digestible amino acids has been applied in practice. Information technology has been widely used in feed industry. At the same time, the mutual penetration and grafting of animal nutrition and feed science with genetics, breeding, physiology, molecular biology, chemistry and other interdisciplinary subjects were further strengthened.

三、牧草和粗饲料的生产及利用是世界饲料业的重要内容
Production and utilization of forage and roughage as an important part of the world feed industry

在世界饲料业中,与饲料工业同步发展的牧草业和其粗饲料的利用也获得了长足的进步,在畜产品构成中,牛羊肉和奶类占有较大的比例,这就决定了草食家畜在世界畜牧业的发展中处于十分重要的位

In the world's feed industry, the forage industry and the utilization of roughage have also made great progress. In the composition of livestock products, beef, mutton and milk account for a large proportion, which determines that herbivores are in a very important position in the development of animal husbandry in the world. Accordingly, the

置,相应地,牧草的种植和秆等粗饲料资源的利用在饲料业中亦举足轻重。虽然各国国情不同,但无疑走精料、牧草、粗饲料三结合的道路是解决世界饲料问题的根本途径。据统计,2019年世界肉类产品的结构比例为:禽肉39.1%(1.316亿t),猪肉32.7%(1.101亿t),牛肉21.6%(7 260.4万t),羊肉4.8%(1 617.5万t),其他肉类1.8%。2019年,世界奶类总产量约为8.83亿t,人均占有量约116.5 kg。2019年,美国禽、猪、牛、羊肉所占的比例依次为:47.5%、26.1%、25.7%和0.2%,奶类总产量0.99亿t,奶类人均占有量为301.34 kg。发展草食家畜养殖和奶类生产虽然需要相当比例的精料,但这些畜产品主要由牧草和秸秆转化;在单胃畜禽的配合饲料中,紫花苜蓿草粉等草产品正成为重要的原料之一,对牧草在单胃畜禽中的营养调控正进行深入的研究。据统计,世界畜产品由青草转化的平均占55%,其中美国约占74%,德国、法国占60%,澳大利亚占90%,新西兰占100%。苜蓿为美国的第三大农作物,2019年,美国种植2 120.5万hm²苜蓿和三叶草等优质牧草,总产量1.2886亿t。其草粉、草捆、草颗粒除满足本国需要外,每年向日本、韩国等国家大量出口。2019年,我国从美国进口的苜蓿量为101.47万t,占苜蓿总进口量的74.81%。农作物秸秆是世界上最丰富的饲料来源之一。全世界秸秆年产量约29亿t,其中各种农作物秸秆所占比例为:玉米秸35%,小麦秸21%,稻草19%,大麦秸10%,黑麦秸2%,燕麦秸3%,谷草5%,高粱秸5%。秸秆的碱化、青贮、氨化、压块等技术及其作用机理自20世纪初以来逐渐普及,

cultivation of forage and the utilization of roughage resources such as straw are also important in the feed industry. Although the national conditions are different, there is no doubt that the fundamental way to solve the world feed problem is to combine concentrate, forage and roughage. According to statistic data, the proportion of world meat products in 2019 was 39.1% (131.6 million tons) of poultry, 32.7% (110.1 million tons) of pork, 21.6% (72.6 million tons) of beef, 4.8% of mutton (16.2 million tons), and 1.8% of other meats. In 2019, the total production of milk in the world was about 883 million tons, with a per capita share of about 116.5 kg. In the United States, the proportion of poultry, pork, beef, mutton in 2019 was 47.5%, 26.1%, 25.7% and 0.2% respectively. The total milk production was about 99 million tons and the per capita share of milk was 301.34 kg. Although the development of herbivorous livestock breeding and dairy production requires a considerable proportion of concentrate, these livestock products are mainly transformed by forage and straw. In the formula feed of non-ruminants, the grass products such as purple clover powder are becoming one of the important raw materials, and the nutritional regulation of forage in non-ruminants is being studied in depth. According to statistic data, 55% of the world's livestock products are converted from grass, of which 74% are from the United States, 60% from Germany and France, 90% from Australia and 100% from New Zealand. Alfalfa is the third largest crop in the United States. In 2019, the United States planted 21.205 million hm² of high-quality forage such as the alfalfa and clover, with a total yield of 128.86 million tons. Besides meeting the domestic needs, alfalfa powder, bales and granules were exported to Japan, South Korea and other countries every year. China imported 1.014 7 million tons of alfalfa from the United States, which accounted for 74.81% of the total imports of alfalfa. Crop straw is one of the most abundant feed sources in the world. The annual output of straw in the world is about 2.9 billion tons, in which the proportion of various crop straw is 35% of corn straw, 21% of wheat straw, 19% of straw, 10% of barley straw, 2% of rye straw, 3% of oat straw, 5% of cereal straw and 5% of sorghum straw. The technology and mechanism of straw alkalization, silage, ammonization and briquetting have been

某些产品进入了商业化生产阶段。在秸秆养牛方面，印度积累了丰富的经验，他们大量利用秸秆发展肉牛和水牛，解决了饲料粮不足问题。这些饲料的广泛应用极大地促进了养殖业尤其是草食家畜业的发展。

gradually popularized since the beginning of the 20th century, and some products have entered the stage of commercial production. India has accumulated rich experience in raising cattle with straw. They use straw to develop beef cattle and buffalo to solve the problem of feed shortage. The wide application of these feeds has greatly promoted the development of animal husbandry, especially the herbivorous livestock industry.

第二节　中国饲料业发展概况
Section 2　Development of Chinese feed industry

中国饲料业经历了几千年的发展，现已成为关系国计民生、与人民生活息息相关的重要产业和独立的学科。中国饲料业的发展始终与动物营养及饲料科学的发展同步。在古代发展较早而先进，在近代发展缓慢而落后，到了现代以改革开放为转折点而迅速进步并逐渐接近世界先进水平。未来的中国饲料业将可能利用数十年时间达到世界领先水平，前进道路上充满艰辛，但前景光明。

After thousands of years of development, China's feed industry has become an important industry and an independent discipline related to the national economy and people's livelihood. The development of China's feed industry has always been synchronized with the development of animal nutrition and feed science. In ancient times, the development was earlier and advanced, while in modern times, the development was slow and backward. In the future, China's feed industry will probably use decades to reach the world's leading level. The road ahead is full of hardships, but the prospect is bright.

一、中国饲料业发展史
Development history of feed industry in China

中国饲料业经历了远古、近代和现代三个不同的发展时期，它是一个相当漫长的过程。

China's feed industry has experienced three periods of development: ancient times, modern times and contemporary times.

（一）古代中国饲料业历史悠久
Long history of feed industry in ancient China

远古时期中国饲料业发展进程早于其他国家，其发展水平也高于西方国家。据史料记载，公元前6000年至公元前2000年的新石器时代，随着圈养牲畜的出现，我国开始有了饲料的萌芽。公元前2000

In ancient times, the development process of China's feed industry was earlier than that of other countries, and its development level was also higher than that of western countries. According to historical records, in the Neolithic age from 6000 BC to 2000 BC, with the emergence of captive livestock, China began to have the germination of feed. In

年至公元前771年的夏、商、周时期，在甲骨文中已出现了饲料的字样，并且在饲养管理上除重视圈养外，还采用将草切碎加上谷物喂牲畜，已经注意到粗饲料和精饲料的配合使用。公元前771年至公元前221年的春秋战国时期，我国已出现了规模化的鸡场、鸭场和养马场等，在马的饲养管理中，采用放牧和舍饲相结合的科学饲养方法。公元前221年至公元220年的秦汉时期，汉武帝派张骞出使西域时，带回了紫花苜蓿种子，首先在黄河流域试种，继而推广到全国，这可能是我国人工种植牧草的开始。这一中国饲料史上的重要事件，对推动草食家畜的发展作出了巨大的贡献。从这个时期开始，人们对饲料的加工调制和饲料资源的开发利用给予了高度重视。《国民月令》中有五、七、八月"刈当茭"的记载，即在夏秋青草丰盛季节，将青饲料刈割、晒制、贮存起来，作为冬春牲畜的饲料。在《神农本草经》中有用梓叶、桐花喂猪"肥大易养三倍"的记载。但当时人们对饲料的认识是感性的、非常肤浅的，尤其对饲料中的营养成分和营养原理不了解，因此畜牧业生产水平很低。

人们在总结以往饲养实践的基础上逐渐认识到，不同饲料原料的适当配合有利于动物的生长。秦汉时期《淮南万毕书》中记载了我国历史上第一个饲料配方："取麻子三升，捣千余杵，煮为羹，以盐一升著中，和以糠三斛（十斗或五斗为一斛）饲喂，则肥也。"这种初期的混合饲料配方尽管科学性不强，但在当时的历史条件下作为配方饲料已非常先进，它几乎早于西方国家2 000多年。唐朝（618—907）已有

the Xia, Shang and Zhou Dynasties from 2000 BC to 771 BC, the word "feed" appeared in oracle bone inscriptions. In addition to the emphasis on captive breeding, the grass was chopped up and grain was used to feed livestock. The combined use of roughage and concentrate was noticed. During the Spring and Autumn Period and Warring States Period from 771 BC to 221 BC, there were large-scale chicken farms, duck farms and horse farms in our country. In the breeding and management of horses, the scientific method of combining grazing and house feeding was adopted. From 221 BC to 220 AD during the Qin and Han Dynasties, Emperor Wu of the Han Dynasty sent Zhang Qian to the western regions to bring back alfalfa seeds. First, alfalfa seeds were planted in the Yellow River Basin, and then spread to the whole country. This may be the beginning of artificial forage planting in our country. This event has made a great contribution to the development of forage in China. Since this period, people have paid great attention to the processing and modulation of feed and the development and utilization of feed resources. In the *National Monthly Order*, there are records of "cutting when water bamboo shoots" in May, July and August, that is, cutting, drying and storing the green fodder in summer and autumn when the grass is abundant. In *Shennong Bencaojing*, it is recorded that *ziye* and *tonghua* are fed to pigs, which is "three times fatter and easier to raise". But at that time, people's understanding of feed was perceptual and very superficial, especially the nutritional components and nutritional principles of feed, so the level of animal husbandry production was very low.

On the basis of summarizing the past feeding practice, people gradually realize that the proper combination of different feed materials is conducive to the growth of animals. The first feed formula in the history of China was recorded in *Huainan Wanbishu* in the Qin and Han Dynasties, "Take three liters of hemp seeds, pound more than a thousand pestles, cook them as soup, feed them with one liter of salt, and feed them with three buckets of bran (ten or five buckets as one)." Although the early mixed feed formula was not scientific, it was very advanced as a formula feed under the historical conditions at that time. It was almost 2,000 years earlier than that in western countries. In the Tang

原始的饲养标准。公元960—1368年,我国最早的饲料添加剂开始出现。这些时期对饲料的感性认识为我国饲料业的发展提供了宝贵的经验。

Dynasty (618-907), there were original feeding standards. From 960 to 1368, the earliest feed additives appeared in China. The perceptual knowledge of feed in these periods provides valuable experience for the development of China's feed industry.

(二) 近代中国饲料业发展缓慢
Slow development of feed industry in modern China

近代饲料科学由于物理、化学、生物学、动物营养科学的发展,并被应用于饲料的研究,尤其是19世纪和20世纪国外先进的动物营养研究成果的引入,使得我国的饲料科学从理论到实践都取得了长足的进步。1926年,北京农业学校成立了"动物营养研究室",开始从动物营养理论角度研究饲料的营养价值。1939年,著名营养学家陈宰均翻译出版三册《饲料与营养》,第一次将国外动物营养与饲料科学的技术传入中国。1943年,我国留美学者王栋教授出版了《动物营养学》一书,对培养我国动物营养与饲料科学的专门人才作出了重大贡献。新中国成立初期,我国利用苏联的燕麦饲料单位及可消化粗蛋白质为主要内容的饲养标准体系。1956年,内蒙古包头市饲料公司成立,专门经营饲草饲料,以后又有天津、河北、广东等几家饲料企业问世。在此期间,我国相继开展了饲料资源调研、饲料资源开发与贮存等方面的工作。纵观我国饲料工业的近代发展历程,我们可以看到,尽管这个时期中国饲料业取得了一些进步,但千百年来把畜牧业作为一项副业,长期利用单一饲料饲养畜禽的局面基本上未有大的改变,饲料转化效率十分低下,尚未形成我国独立的饲料业体系。

Due to the development of physical, chemical, biology and animal nutrition science, modern feed science has been applied to feed research, especially the introduction of foreign advanced animal nutrition research achievements in the 19th and 20th century, which has made great progress in feed science in China from theory to practice. In 1926, Beijing Agricultural School set up "animal nutrition research laboratory" and began to study the nutritive value of feed from the perspective of animal nutrition theory. In 1939, Chen Zaijun, a famous nutritionist, translated and published three volumes of *Feed and Nutrition*, which introduced foreign animal nutrition and feed science technology to China for the first time. In 1943, Professor Wang Dong, a Chinese scholar studying in the United States, published *Animal Nutrition*, which made a great contribution to the cultivation of professionals in animal nutrition and feed science in China. In the early days of the founding of new China, China used the Soviet Union's oat feed unit and digestible crude protein as the main content of the feeding standard system. In 1956, Inner Mongolia Baotou City Feed Company was established, specializing in forage, and then several feed enterprises in Tianjin, Hebei and Guangdong came into being. During this period, our country has carried out the research of feed resources, the development and storage of feed resources. Looking at the modern development process of China's feed industry, we can see that although China's feed industry has made some progress in this period, the situation of taking animal husbandry as a sideline for thousands of years and using single feed to feed livestock and poultry for a long time has not changed greatly. The feed conversion efficiency is very low, and China's independent feed industry system has not yet been formed.

（三）现代中国饲料业迅速发展成为一门独立的学科
Feed industry in contemporary China as an independent discipline

20世纪70年代以来,是我国饲料业快速发展的时期。改革开放使饲料业充满无限生机,伴随着现代饲料工业体系的建立,动物营养与饲料科学迅速发展,牧草业、秸秆等粗饲料资源综合开发利用的开展以及饲料科学本系的建立,使饲料科学逐步成为一门独立的学科。饲料业作为一门独立的学科是我国改革开放以来畜牧业领域所取得的突出成就之一。

1. 饲料工业体系的建立

我国的饲料工业起步于20世纪70年代中后期,比发达国家晚了半个多世纪。十一届三中全会以后,党和国家不失时机地将发展饲料工业作为调整产业结构、实现国民经济生产总值"翻番"目标的重大战略措施。1978年国务院批准了国家计划委员会《关于发展我国饲料工业问题的报告》,并于1984年颁布了《1984—2000年全国饲料工业发展纲要（试行草案）》,将饲料工业建设正式纳入国民经济发展计划。1985年,国家计划委员会饲料工业办公室成立,负责全国饲料工业的统筹、规划和协调。1989年,国务院在"关于当前产业改革重点的决定"中,把饲料工业列为重点支持和优先发展的支柱产业。

经过30多年的建设,初步形成了集饲料加工业、饲料添加剂工业、饲料原料工业、饲料机械制造业,以及饲料科研、教育、推广、标准和监督检测等为一体的完整的饲料工业体系。从我国第一台饲料加工机组于1981年研制成功,1987年第一套饲料加工微机控制系统研制完

Since 1970s, China's feed industry has developed rapidly. With the establishment of modern feed industry system, animal nutrition and feed science have developed rapidly, forage industry, straw and other roughage resources have been developed comprehensively, and feed science has gradually become an independent discipline. As an independent subject, feed industry is one of the outstanding achievements in the field of animal husbandry since China's reform and opening up.

1. Establishment of feed industry system

China's feed industry started in the late 1970s, more than half a century later than the developed countries. After the Third Plenary Session of the 11th CPC Central Committee, the party and the state have taken the development of feed industry as an important strategic measure to adjust the industrial structure and achieve the goal of "doubling" the GDP. In 1978, the State Council approved the Report on the Development of China's Feed Industry issued by the State Planning Commission, and in 1984 promulgated the Outline for the Development of China's Feed Industry from 1984 to 2000 (Trial Draft), which formally incorporated the construction of feed industry into the national economic development plan. In 1985, the feed industry Office of the State Planning Commission was established to take charge of the overall planning, planning and coordination of the national feed industry. In 1989, the State Council listed the feed industry as the pillar industry with priority support and development in the "decision on the key points of current industrial reform".

After more than 30 years of development, a complete feed industry system integrating feed processing industry, feed additive industry, feed raw material industry, feed machinery manufacturing industry, as well as feed scientific research, education, promotion, standards, supervision and testing has initially come into form. Since the first feed processing unit was successfully developed in 1981 and the first set of microcomputer control system for feed processing was

成,到目前我国已完全可以自行设计、制造和安装30 t以上的大型饲料加工成套设备和微机控制系统,标志着我国经过30年的自力更生奋斗历程,已成熟地建立起自己的饲料机械制造工业,涌现出江苏正昌集团、牧羊集团等一批龙头企业。饲料机械制造企业达到200多家,粉碎机、混合机、制粒机和电控设备均达到了国际先进水平。大型企业通过扩建、收购,使行业整合速度进一步加快,此外还向养殖业等下游产业链延伸,降低企业经营风险。饲料行业培育和造就了新希望集团、通威集团等一批优秀的饲料企业,具备了成为行业领军企业的基础。2008年,35家饲料企业生产了40.6%的饲料,13家饲料添加剂企业销售额占全国的46%,优势明显。

我国饲料工业开始起步时,除部分微量元素添加剂能自行生产外,维生素、抗生素、氨基酸、酶制剂等饲料添加剂基本上依赖进口。历经30多年的发展,我国饲料添加剂工业已具相当规模。在合成氨基酸生产方面,除蛋氨酸主要依赖进口外,赖氨酸、苏氨酸和色氨酸产量分别达到56万t、4.3万t和350 t,不但满足国内市场需求,而且还成为全球重要的氨基酸供应基地。维生素方面,饲料工业所需要的14种维生素全部实现国产化,总产量37万t,占全球的50%以上。尽管某些种类饲料添加剂仍来自国外,但绝大多数的酶制剂、微量元素等营养性添加剂都能自行生产,有些产品已具备国际竞争优势。近几年来,为了生产出有利于人类健康的绿色食品,无污染、无公害、无残留、安全、绿色饲料添加剂如益生素、寡糖等抗生素替代品成为研究的热点。

completed in 1987, China has been able to design, manufacture and install large-scale feed processing complete sets of equipment and microcomputer control system with a capacity of more than 30 tons, which indicates that China has successfully established its own feed machinery system after 30 years of self-reliance. A number of leading enterprises such as Jiangsu Zhengchang Group and Shepherd Group have sprung up. There are more than 200 feed machinery manufacturing enterprises, and the pulverizers, mixers, granulators and electric control equipment have reached the international advanced level. Through expansion and acquisition, large-scale enterprises further accelerate the speed of industrial integration. In addition, they extend to the downstream industry chain such as breeding industry to reduce business risks. The feed industry has cultivated and created a number of excellent feed enterprises such as New Hope Group and Tongwei Group, and has the foundation to become a leading enterprise in the industry. In 2008, 35 feed enterprises produced 40.6% of the feed, and 13 feed additive enterprises accounted for 46% of the national sales, with obvious advantages.

At the beginning of China's feed industry, in addition to some trace element additives that can be produced by ourselves, vitamins, antibiotics, amino acids, enzymes and other feed additives basically rely on imports. With the development of more than 30 years, China's feed additive industry has a considerable scale. In the production of synthetic amino acids, in addition to methionine, the output of lysine, threonine and tryptophan reached 560,000 tons, 43,000 tons and 350 tons respectively, which not only met the domestic market demand, but also became an important amino acid supply base in the world. In terms of vitamins, all the 14 kinds of vitamins needed by the feed industry have been localized, with a total output of 370,000 tons, accounting for more than 50% of the world. Although some kinds of feed additives are still from abroad, the vast majority of enzyme preparations, trace elements and other nutritional additives can be produced by ourselves, and some products have international competitive advantages. In recent years, in order to produce green food beneficial to human health, pollution-free, residue free, safe, green feed additives such as probiotics, oligosaccharides and other antibiotic substitutes have become a research hotspot.

大宗饲料原料中,种植业由粮食作物——经济作物的二元结构逐步向粮食作物——经济作物——饲料作物和牧草的三元结构调整,极大地丰富了玉米等植物性饲料原料的供应。现有饲料原料中,除大豆、鱼粉有较大量进口外,大部分实现了自给。饲料中抗营养因子研究成效显著,如通过制油工艺技术提高饼粕质量及饲用价值,脱除有毒有害物质,提高棉、菜子饼粕蛋白质利用率,加强了芝麻饼粕、花饼粕等的开发利用,对于缓解我国饲料中蛋白质资源的紧缺状况起到了很大作用。由于谷物中非淀粉多糖(NSP)的研究和针对其设计的饲料复合酶的应用,小麦在猪中的应用日趋活跃,在玉米减产情况下确保了能量饲料的正常供应。

饲料工业的标准化也是我国饲料业健康发展的标志之一。1986年4月国家正式成立"全国饲料工业标准化技术委员会",是结束我国饲料工业无标生产的里程碑。1999年5月国务院颁布了《饲料和饲料添加剂管理条例》,标志着我国饲料工业已进入了法制管理轨道。2009年,经过20多年的建设,经过清理后的国家标准、行业标准及相关标准共449项。截至2017年11月底,我国饲料工业国家标准、行业标准共565项,为规范我国饲料产品生产及各个流通环节的产品质量管理起到了积极的作用。

由于饲料工业的健康、良性发展,我国取得了不少辉煌成就。2009年全国饲料企业已达到12 291家,饲料工业总产量达到1.4813亿t,其中配合饲料产量1.15亿t,浓缩饲料2 686万t,添加剂预混料592万t,在世界配合饲料中排行第二

Among the bulk feed materials, the planting industry has gradually adjusted from the dual structure of grain crops and cash crops to the ternary structure of grain crops, cash crops and forage, which greatly enriched the supply of corn and other plant feed materials. Most of the existing feed materials are self-sufficient except soybean and fish meal. The research on anti-nutritional factors in feed has achieved remarkable results, such as improving the quality and feeding value of meal, removing toxic and harmful substances, improving the protein utilization rate of cotton and rapeseed meal, strengthening the development and utilization of sesame meal and flower meal, etc., which has played a great role in alleviating the shortage of protein resources in feed in China. Due to the research of non-starch polysaccharides (NSP) in cereals and the application of feed complex enzymes designed for NSP, the application of wheat in pigs is becoming more and more active, which ensures the normal supply of energy feed in the case of corn yield reduction.

The standardization of feed industry is also one of the signs of the healthy development of China's feed industry. In April 1986, the National Technical Committee for Standardization of Feed Industry was formally established, which is a milestone in ending the non-standard production of China's feed industry. In May 1999, the State Council promulgated the Regulations on the Administration of Feed and Feed Additives, marking that China's feed industry has entered the track of legal management. In 2009, 449 national standards, industry standards and related standards have been sorted out after more than 20 years of construction. By the end of November 2017, a total of 565 national standards, industry standards have been sorted out in China, which has played a positive role in standardizing the production of feed products and product quality management in various circulation links in China.

Due to the healthy and benign development of feed industry, our country had made many brilliant achievements. In 2009, there were 12,291 feed enterprises in China, and the total output of feed industry reached 148.13 million tons, including 115 million tons of formula feed, 26.86 million tons of concentrated feed and 5.92 million tons of additive premix, which ranked second in the world. The

位,饲料工业产值4 266亿元,已成为中国国民经济的一个支柱产业。2019年全国工业饲料总产量达到2.29亿t,在世界饲料产量中排行第二位,其中配合饲料2.10亿t,浓缩饲料1 241.9万t,添加剂预混合饲料542.6万t,配合饲料产量上升巨大。全国10万t以上规模饲料企业621家,饲料产量1.07亿t,占全国饲料总产量的46.6%。截至2019年年底,全国共有饲料加工企业约10 000家、饲料添加剂生产企业约1 800家,总体上表现为高附加值、创新型规模化饲料添加剂企业的市场占有率不断提高,产品结构愈发多样,发展壮大势头明显,企业收购兼并步伐进一步加快。

饲料工业的不断进步,强有力地支持了畜牧业的发展,提高了我国人民的营养膳食水平。根据中国统计年鉴和中国饲料工业统计资料分析,2009年,中国肉类总产量7 642万t,连续20年居世界第一位;禽蛋总产量已达2 741万t,连续25年居世界第一;奶类总产量3 578万t,位居世界第三。2019年,中国肉类总产量7 758.8万t,居世界第一;奶类总产量3 297.6万t,居世界第三;禽蛋总产量3 309.0万t,居世界第一。

我国畜牧业产值在农业中的比重稳步上升,1978年畜牧业产值209.3亿元,占农业总产值的15%;2007年畜牧业产值增加到16 125.2亿元,占农业总产值的比重上升到33%。2019年我国畜牧业产值达到33 064.3亿元,占农业总产值的26.7%。经过40多年的改革开放,我国水产养殖业也快速增长,成为大农业中发展最快的产业之一。2007年,全国水产总量达4 747万t;

output value of feed industry is 426.6 billion *yuan*, which has become a pillar industry of China's national economy. In 2019, the total output of feed industry reached 229 million tons, including 210 million tons of formula feed, 12.419 million tons of concentrated feed and 5.426 million tons of additive premix. The yield of formula feed increased enormously. There are 621 feed enterprises which produced more than 100,000 tons of formula feed and these feed enterprises produced about 107 million tons of feed which accounted for 46.6% of the total national feed yield. By the end of 2019, there are a total of about 10,000 feed processing enterprises and about 1,800 feed additive manufacturers in China. Generally, the market share of high value-added, innovative and large-scale feed additive enterprises is constantly improving, the product structure is increasingly diverse, the momentum of development is obvious, and the pace of enterprise acquisition and merger is further accelerated.

The continuous progress of feed industry strongly supports the development of animal husbandry and improves the nutritional diet level of Chinese people. According to the analysis of China's statistical yearbook and China's feed industry statistic data, in 2009, China's total meat output was 76.42 million tons, which ranked the first in the world for 20 consecutive years, the total egg output reached 27.41 million tons which ranked the first in the world for 25 consecutive years, the total milk output was 35.78 million tons which ranked the third in the world. In 2019, China's total meat output was 77.588 million tons which ranked the first in the world, the total milk output was 32.976 million tons which ranked the third in the world, the total egg output reached 33.09 million tons which ranked the first in the world.

The proportion of production value of animal husbandry in agriculture is rising steadily. In 1978, the output value of animal husbandry was 20.93 billion *yuan*, accounting for 15% of the total agricultural output value. In 2007, the output value of animal husbandry increased to 1,612.5 billion *yuan*, accounting for 33% of the total agricultural output value. In 2019, the output value of animal husbandry reached 3,306.4 billion *yuan*, accounting for 26.7% of the total agricultural output value. After more than 40 years of reform and opening up, China's aquaculture industry has also grown rapidly, becoming one of the fastest growing

2019年，全国水产总量达6 480.4万t，相比2007年产量提高了36.5%。2019年全国渔业生产总值达12 572.4亿元，是2007年全国渔业生产总值的2.84倍。畜牧业在农村经济调整和增加农民收入的历史进程中，扮演着十分重要的角色。鉴于畜牧业成本的70%来自饲料，所以没有现代化的饲料工业就没有现代化的养殖业，就没有丰富的"菜篮子"，就不可能提高人民的生活水平。因此，饲料工业是促进农村经济协调发展和整个国民经济有序发展的前提之一。

2. 大力开发和利用牧草和秸秆等粗饲料资源

蓬勃发展的饲料工业除了主要为单胃畜禽提供配合饲料外，还在广大农区提高日益兴起的牧草种植业和秸秆等粗饲料资源的开发利用，有力地促进了草食家畜的良性发展，改善了我国人民的饮食结构。在牧草业的发展中，作为调整农业种植业结构的先锋牧草——紫花苜蓿，其发展势头强劲，种植面积已达133万hm^2。黑麦草、冬牧70、高丹草等在发展畜牧业尤其是牛羊业中亦发挥了重要的作用。近几年来，种、管、收、加工全部机械化的现代紫花苜蓿草产品发展迅速，其草粉、草颗粒作为猪、禽、兔配合饲料原料极大丰富了我国的饲料原料工业，高密度草捆正成为我国奶牛业的重要饲料。我国每年约生产6.0亿t农作物秸秆。在秸秆资源研究和利用上有重大进展，通过推广秸秆氨化、微贮等技术，改善了秸秆饲料的营养价值、适口性和消化率，因而产生了良好的饲喂效果。20世纪80年代中期，联合国粮食及农业组织

industries in large-scale agriculture. In 2007, the total yield of aquatic products in China reached 47.47 million tons. In 2019, the total yield of aquatic products reached 64.804 million tons, which was increased by 36.5% compared the total yield in 2007. In 2019, China's gross fishery product reached 1,257.24 billion *yuan*, which was 2.84 times that of 2007. Animal husbandry plays a very important role in the historical process of rural economic adjustment and increasing farmers' income. In view of the fact that 70% of the cost of animal husbandry comes from feed, there will be no modern breeding industry without modern feed industry, and there will be no rich "vegetable basket", so it is impossible to improve people's living standards. Therefore, feed industry is one of the preconditions to promote the coordinated development of rural economy and the orderly development of the whole national economy.

2. Vigorously Develop and utilize roughage such as resources of forage and straw

In addition to the booming feed industry, which mainly provides formula feed for monogastric livestock and poultry, the growing forage planting industry and the development and utilization of straw and other roughage resources in the vast agricultural areas have effectively promoted the healthy development of herbivorous livestock and improved the food structure of our people. In the development of forage industry, alfalfa, as the pioneer of adjusting the structure of agricultural planting industry, has a strong development momentum, and the planting area has reached 1.33 million hm^2. Ryegrass, Dongmu 70 and pacesetter also play an important role in developing animal husbandry, especially in cattle and sheep industry. In recent years, the modern alfalfa products with mechanization of planting, management, harvesting and processing have developed rapidly. As the raw materials of pig, poultry and rabbit compound feed, the grass powder and grass granules have greatly enriched the feed raw material industry in China. High density bales are becoming an important feed for the dairy industry in China. About 600 million tons of crop straw are produced every year in China. Great progress has been made in the research and utilization of straw resources. Through the promotion of straw ammonization, micro storage and other technologies, the nutritive value, palatability and digestibility of straw feed have

（FAO）派专家亲临我国指导秸秆氨化、青贮。1992年来，国家开始投入专项资金实施秸秆养畜项目，经过10多年的努力，我国秸秆利用规模和效率大为提升。据不完全统计，2007年，我国青贮秸秆1.8亿t（折成风干秸秆6 000万t），氨化和微贮5 000万t，养殖户通过简单切碎压块等处理后直接饲喂的约1.1亿t，每年用于反刍动物养殖的秸秆总量约为22亿t。秸秆饲料的开发利用，节约了大量粮食，推动了节粮型畜牧业的发展。由于全国牛羊业的大发展，极大地提高了牛羊肉的产量，改善了肉类的结构。2009年，肉类中猪、牛、羊肉的比重分别是64.0%、8.1%和5.2%；到2019年，肉类中猪、牛、羊肉的比重分别为54.8%、8.6%和6.3%，牛羊肉的比例上升到14.9%。

3. 动物营养与饲料科学是饲料业体系的科学支柱

动物营养学与饲料科学是现代化饲料业的主要科学支柱。这两门学科的形成和发展，有力地促进了饲料业体系的建成。20世纪70年代末期以来，我国饲料学能迅速发展成为一门独立的学科，很大程度上得益于我国动物营养与饲料科学的快速发展。为了充分开发我国的饲料资源，提高饲料利用效率，改进畜禽产品品质，我国相继开展了畜禽的营养代谢规律，饲料分子营养学与微生态营养学，畜禽的理想蛋白质模式和反刍动物的蛋白质新体系，饲料生物学效价评定，矿物质营养代谢，饲料抗营养因子，益生素和寡糖等营养与非营养性添加剂，饲

been improved, resulting in good feeding effect. In the mid-1980s, Food and Agriculture Organization of the United Nations (FAO) sent experts to China to guide straw ammonization and silage. Since 1992, our country began to invest special funds to implement the straw livestock project. After more than 10 years of efforts, the scale and efficiency of straw utilization in China have been greatly improved. Incomplete statistics data showed that the production of silage straw was about 180 million tons (equal to 60 million tons at air-dried basis), the production of ammoniated and microbe-fermented straw was 50 million tons, the amount of direct-feeding straw by farmers after simply shredding and pressing was about 110 million tons. The annual total straw for ruminant feeding is about 2.2 billion tons in 2007. The development and utilization of straw feed saved a lot of grain and promoted the development of grain saving animal husbandry. Due to the great development of cattle and sheep industry in China, the output of beef and mutton has been greatly increased, and the structure of meat has been improved. In 2009, the proportion of pork, beef, mutton was 64.0%, 8.1% and 5.2% respectively. In 2019, the proportion of pork, beef, mutton was 54.8%, 8.6% and 6.3% respectively and the proportion of beef and mutton increased by 14.9% compared with the proportion in 2009.

3. Animal nutrition and feed science is the scientific pillar of feed industry system

Animal nutrition and feed science are the main scientific pillars of modern feed industry. The formation and development of these two disciplines have greatly promoted the establishment of the feed industry system. Since the late 1970s, feed science in China has rapidly developed into an independent discipline, which is largely due to the rapid development of animal nutrition and feed science in China. In order to fully develop feed resources in China, improve feed utilization efficiency, and improve the quality of livestock and poultry products, our country has successively carried out the laws of nutrition metabolism of livestock and poultry, feed molecular nutrition and microecological nutrition, physical and ideal protein model for livestock and poultry, new protein system for ruminant, biological evaluation of feed, mineral nutrition metabolism, feed antinutritional factors, nutritional and non-nutritronal additives such as probiotics

料原料及其检测技术标准化，饲养与环境，饲料与免疫，饼粕类、糟渣类、秸秆类和树叶类的科学利用等方面的工作。现在，饲料学正向绿色饲料和绿色饲料添加剂、饲料营养与动物基因表达、牧草营养调控因子和饲料未知生长因子等微观领域方向发展。我国对有关饲料数据进行了计算机管理。1985年，完成了适用于中型计算机的饲料数据库管理系统。1986年，建立了第一个中国饲料数据库。"七五"期间又对国产饲料原料的常规成分、氨基酸、矿物质、微量元素、部分维生素及有毒有害物质进行了分析测定，取得了大量数据，连同规范化的配套饲料样本实体属性描述1 100万条一并列入中国数据库。1987年，该数据库在原来工作的基础上，又建立了集政策法规、原料产品性能等于一体的网络系统。1988年，研制成功我国第一个以线性模型为数学模型的饲料配方软件，大大提高了设计饲料配方的速度和准确性。1990年，又研发出以多目标函数建立数学模型的配方软件，降低了无解率，且可以进行多个目标决策，提高了配方质量。目前，我国已连续发布《中国饲料成分及营养价值表》31版，根据新参数配套推出的第七代优化饲料配方软件也在全国推广。

and oligosaccharides, standardization of feed raw materials and their detection technology, feeding and environment, feed and immunity, scientific utilization of cake meal, dregs, straws and leaves, etc. At present, feed science is developing towards the micro fields of green feed and green feed additives, feed nutrition and animal gene expression, forage nutrition regulatory factors and feed unknown growth factors. The data of feed were managed by computer in our country. In 1985, a feed database management system suitable for medium-sized computers in China was completed. In 1986, the first Chinese feed database was established. During the Seventh Five-Year Plan period, China also analyzed and determined the conventional components, amino acids, minerals, trace elements, some vitamins and toxic and harmful substances of domestic feed raw materials, and obtained a large amount of data. Together with the 11 million entity attribute descriptions of standardized matching feed samples, they were listed into China database. In 1987, on the basis of the original work, a network system integrating policies and regulations, raw material and product performance was established. In 1988, China successfully developed the first feed formula software with linear model as mathematical model, which greatly improved the speed and accuracy of feed formula design. In 1990, the formulation software based on multi-objective function was developed, which can reduce the rate of no solution, make multi-objective decision and improve the quality of formulation. At present, China has continuously released the 31st edition of Table of Feed Composition and Nutritive Values in China, and the seventh generation of optimized feed formula software based on the new parameters has also been promoted nationwide.

二、当前我国饲料业中存在的主要问题及对策
Main problems and countermeasures of feed industry in China

（一）饲料及牧草资源短缺
Shortage of feed and forage resources

蛋白质饲料的缺口较大。2019年，我国进口鱼粉142万t，其中进口秘鲁鱼粉77.1万t，占总进口量

There is a big gap in protein feed. In 2019, China imported 1.42 million tons of fish meal and 0.771 million tons of fish meal was imported from Peru, which accounted for

的54.3%,且价格居高不下。豆粕是我国畜禽配合饲料中的主要蛋白质补充料,2019年,我国大豆种植面积达到933.0万hm^2,大豆总产量达1 810万t,但是2019年我国大豆进口量依然高达8 851万t,继续成为全球最大的大豆进口国。2019年,我国进口的大豆大约有5 767万t来自巴西,相当于同期中国进口大豆总量的65.2%,其次是美国,进口量约为1 694万t,占同期中国进口大豆总量的19.1%。蛋白质原料的短缺和依赖进口,一方面致使其价格频繁波动,增加了饲料和养殖业经营风险,给畜产品价格波动埋下隐患;另一方面导致低质、劣质蛋白质饲料原料充斥市场,对饲料和养殖产品质量安全造成威胁。同时,国内一些饲料添加剂如蛋氨酸和一些药物依赖进口的局面还未从根本上得到改变。国内草产品尤其是苜蓿草捆、草粉和草颗粒生产严重不足,大部分奶牛饲养不得不依赖秸秆加高精料的模式,产量和品质的改善无从谈起,少量苜蓿草捆的进口不过是杯水车薪,解决不了根本问题。从农业部发布的《全国苜蓿产业发展规划(2016—2020)》可知,2015年我国优质苜蓿总供给量为300万t,其中中国产180万t,进口120万t。2020年,我国新增优质苜蓿种植面积300万亩,优质苜蓿产量达360万t,按照苜蓿进口量150万t计算,2020年我国优质苜蓿总供给量为510万t。但是根据"十三五"草食畜牧业发展规划,2020年全国优质苜蓿总需求量为690万t,仍缺口180万t,苜蓿进口量呈急剧加大的态势。随着我国饲料工业的进一步发展,饲料原料匮乏问题会更进一步突出。饲料原粮需求源于养殖产

54.3% of the total imports. The price of fish meal remained high. Soybean meal is the main protein supplement in China's animal formula feed. In 2019, China's soybean planting area reached 9.33 million hm^2, and the total soybean output reached 18.1 million tons. However, China's soybean imports in 2019 were still as high as 88.51 million tons and China continued to be the largest soybean importer in the world. In 2019, about 57.67 million tons of soybeans were imported from Brazil, equivalent to 65.2% of total soybean imports. China also imported about 16.94 million tons from the United States, accounting for 19.1% of China's total soybean imports. The shortage of protein raw materials and dependence on imports, on the one hand, leads to frequent price fluctuations, increases the operational risks of feed and breeding industry, and causes hidden dangers to the price fluctuations of livestock products; on the other hand, it leads to low-quality and inferior protein feed raw materials flooding the market, and poses a threat to the quality and safety of feed and breeding products. At the same time, the situation that some domestic feed additives such as methionine and some drugs depend on import has not been fundamentally changed. The production of domestic grass products, especially alfalfa bales, grass powder and grass particles, is seriously insufficient. Most dairy cows have to rely on the mode of straw plus high concentrate feed. The improvement of yield and quality is impossible. The import of a small amount of alfalfa bales is just a drop in the bucket, which cannot solve the fundamental problem. According to the National Alfalfa Industry Development Plan (2016-2020) released by the Ministry of Agriculture, the total supply of high-quality alfalfa in China in 2015 was 3 million tons, of which 1.8 million tons were produced in China and 1.2 million tons were imported. In 2020, China increased the planting area of high-quality alfalfa by 300,000 hm^2, and the yield of high-quality alfalfa reached 3.6 million tons. According to the alfalfa import volume of 1.5 million tons at that time, the total supply of high-quality alfalfa in China was 5.1 million tons in 2020. However, according to the development plan of herbivory animal husbandry in the 13th Five-Year Plan, the total demand of high-quality alfalfa in China was 6.9 million tons in 2020, and there was still a gap of 1.8 million tons. The import

品需求,动物性产品需求数量上的增长和养殖方式的转变,一方面推动饲料消费在总量上持续增长,另一方面也使得饲料消费结构发生变化。传统的农家饲料用量减少,以饲料粮为主的工业饲料消费增加,进一步提升了饲料粮消费需求。2019年,我国粮食总产量为6.6亿t,达到历史最高水平,其中稻谷和小麦产量合计为3.43亿t。据测算,"十一五"期间,我国人口以每年800万—1 000万的速度增长,但由于人均口粮消费量的减少,口粮消费总量基本保持在2.5亿t,今后这个数字基本上不会改变;目前饲料粮消费约占粮食消费的40%,约为2.7亿t,其中,除产生豆粕的大豆大多需要进口外,其他饲料粮均以国内自给为主。随着养殖业的发展,饲料粮的持续增加是必然的,预计到2030年,我国饲料粮的比重将达到50%,届时不仅蛋白质饲料严重不足,饲用玉米也会由平衡趋向紧平衡,进而出现短缺。例如:2019年我国玉米产量2.57亿t,消费量为2.75亿t,其中饲用玉米占总量的53.78%,为了保证国家的小麦、稻谷的粮食安全,玉米播种面积需进一步扩大,但空间有限。

解决我国发展动物生产的饲料原料不足问题要从以下几方面着手:一是要大力开发能量和蛋白质饲料资源。通过制油工艺技术改进提高饼粕质量及饲用效价,脱除有毒、有害物质,提高棉、菜饼粕氨基酸利用率;在我国传统工艺不能很

volume of alfalfa increased sharply. With the further development of China's feed industry, the shortage of feed raw materials will become more prominent. The demand for feed raw materials comes from the demand for breeding products. The growth of the demand for animal products and the transformation of breeding methods, on the one hand, promote the continuous growth of feed consumption in the total amount and change the structure of feed consumption on the other hand. The consumption of traditional farm feed is reduced, and the consumption of industrial feed mainly based on feed grain is increased, which further improves the consumption demand of feed grain. In 2019, China's total grain output was 660 million tons, including 343 million tons of rice and wheat. It was estimated that during the Eleventh Five-Year Plan period, China's population grew at the number of 8−10 million per year. However, due to the reduction of per capita grain consumption, the total grain consumption would remain 250 million tons, which will not change in the future. At present, feed grain consumption accounts for about 40% of grain consumption, about 270 million tons, of which most of the others do not need to be imported and are mainly self-sufficient in China, except for soybeans used to produce soybean meal. With the development of the breeding industry, it is inevitable that feed grain will continue to increase. It is estimated that by 2030, the proportion of feed grain in China will reach 50%. At that time, not only the protein feed will be seriously insufficient, but also the feed corn will turn from balance to tight balance, and then there will be a shortage. For example, the yield of corn in China was 257 million tons and the corn consumption was 2.75 million tons in 2019, of which 53.78% was used for animal feed. In order to ensure the national food security of wheat and rice, the space of corn planting area should be expanded but the potential is limited.

To solve the problem of insufficient feed raw materials for animal production in China, we should start from the following aspects: Firstly, we should vigorously develop energy and protein resources. The quality and feeding potency of cake meal can be improved by improving the oil production technology, and toxic and harmful substances can be removed to improve the utilization rate of amino

快提高油脂饼粕质量的前提下,要通过采取经济而有效的物理、化学、生物脱毒技术及营养调控的手段,来优化各种优质植物蛋白质资源,大力研究开发适用的无鱼粉、无豆粕饲粮,充分利用非常规的蛋白质饲料资源;重视开发"非粮"饲料,优化利用我国5亿t的针、阔树叶和5 000万t以上的糠麸,1 900万t的白酒精和4 100万t的啤酒精及近3 000万t的薯类资源,以尽可能多地节约饲料粮。二是挖掘耕地潜力,扩大饲料作物种植面积,大幅度提高饲料粮产量;加快种植业结构由二元结构向三元结构调整,增加优质牧草的种植面积,力争达到农业部制定的《全国种植业结构调整规划》的发展目标,包括保口粮、保谷物、饲草生产与畜牧养殖协调发展等。力争加大玉米、大豆的生产量和高赖氨酸、高糖、高油玉米种植,加大小麦替代玉米的力度和小麦复合酶的应用,以补充玉米的不足;扩大无毒棉和双低油菜品种的种植面积;充分利用中低产田和农区经济林间隙地大力种植优质高产紫花苜蓿等牧草,除用来饲喂草食家畜外,还可开发其草粉和草颗粒作为单胃动物的蛋白质补充饲料。三是继续做好推广应用各种青贮和氨化秸秆饲料的工作,尤其是优质农作物秸秆的高效利用,发展牛羊养殖业。四是推广集成技术,缓解蛋白质需求压力。在猪禽饲料中使用低蛋白高氨基酸配方技术,在保证生产性能的前提下,饲料配方中蛋白质含量下调2-4个百分点。目前,我国是世界上氨基酸类产品的主要生产国和出口大国,已经具备了全面推广该技术的物质基础和必要条件。

acids in the cotton and rapeseed meal. The economic and effective physical, chemical and biological detoxification technology and nutritional control measures should be adopted to optimize the quality of oil cake meal on the premise that the traditional technology cannot improve the quality of oil cake quickly. Moreover, quality plant protein resources, suitable fish free and soybean free diets and unconventional protein feed resources should also be researched, developed and fully made use of. We should pay attention to the development of "non-grain" feed, optimize the utilization of China's 500 million tons of needle and broad leaf, more than 50 million tons of bran, 19 million tons of white alcohol, 41 million tons of beer extract and nearly 30 million tons of potato resources, so as to save feed as much as possible. The second point is to tap the potential of cultivated land, expand the planting area of feed crops, and greatly improve the yield of feed grain; to speed up the adjustment of planting structure from binary structure to ternary structure, increasing the planting area of high-quality forage. Efforts should be made to achieve the developed goals set in the National Planting Industry Structure Adjustment Plan by the Ministry of Agriculture, through ensuring diet and grain production and coordinating the development of forage production and animal husbandry. China should strive to increase the production of corn and soybean and the planting of high lysine, high sugar and high oil corn, and to increase the substitution of wheat for corn and the application of wheat complex enzyme to supplement the shortage of corn, to expand the planting area of non-toxic cotton and double low rapeseed varieties, and make full use of medium and low yield fields and the gap between economic forests in agricultural areas to vigorously plant high-quality and high-yield alfalfa and other forages except for feeding grass. Besides feeding livestock, grass powder and grass granules as protein supplement feed for monogastric animals can also be developed. Thirdly, the promotion and application of all kinds of silage and ammoniated straw feed, especially the efficient use of high-quality crop straw should be focused to develop cattle and sheep breeding industry. The fourth point is to promote integrated technology to relieve the pressure of protein demand. Using low protein and high amino acid formula technology in pig and poultry feed can reduce protein content by 2-4 percentage points on the premise of ensuring production performance. At present, China is the main producer and exporter of amino acid products in the world, and has the material basis and necessary conditions to promote the technology in an all-round way.

(二) 配合饲料的使用比例低
Low proportion of formula feed

中国饲料工业虽然形成了较大的生产能力,但与养殖业巨大的饲料消耗量相比较,工业饲料占所有饲料的比例还不高,约有50%农户仍没有使用配合饲料。目前,中国的工业饲料用户主要为具有一定规模的养殖企业和养殖户,而占养殖业绝大多数份额的小型养殖户较多地使用青绿饲料、农副产品、单一的谷物及营养不全价的自配饲料,养殖业生产水平较低。因此,要提高我国养殖业的整体水平,必须努力拓展配合饲料的使用空间。

1990年以来,我国配合饲料年均增长数量和增长率逐年上升,而浓缩饲料和添加剂预混合饲料的增长率逐年下降。1990—2000年,配合饲料的增长率为6.6%,浓缩饲料和添加剂预混合饲料增长率分别高达37.7%和28.3%;2000—2009年,配合饲料的增长率提高到7.7%,浓缩饲料和添加剂预混合饲料增长率分下降为8.9%和9.9%,仅为20世纪90年代年增长率的约1/4和1/3。2000年以后,随着养殖结构的变化以及浓缩饲料和添加剂预混合饲料市场的逐渐规范,浓缩饲料和添加剂预混合饲料高速增长的现象不再出现,饲料行业逐步转为以配合饲料产量增长为主体,特别是2010年,这种结构上的变化更为明显。1999年,我国人均占有配合饲料为45 kg(美国510 kg,欧盟310 kg,世界平均水平98 kg);2009

Although China's feed industry has formed a large production capacity, compared with the huge feed consumption of the breeding industry, the proportion of industrial feed in all feeds is not high, and about 50% of farmers still do not use formulated feed. At present, the industrial feed users in China are mainly breeding enterprises and farmers with a certain scale, while the small-scale farmers, which account for the vast majority of the livestock inductry, mostly use green feed, agricultural and sideline products, single grain and self-feed with incomplete nutrition, and the production level of the livestock industry is relatively low. Therefore, in order to improve the overall level of China's breeding industry, we must strive to expand the use of formula feed.

Since 1990, the average annual growth quantity and growth rate of formula feed in China have increased year by year, while the growth rate of concentrated feed and additive premix feed has decreased year by year. From 1990 to 2000, the growth rate of formula feed was 6.6%, and that of concentrate feed and additive premix feed was 37.7% and 28.3% respectively; from 2000 to 2009, the growth rate of formula feed increased to 7.7%, and that of concentrate feed and additive premix feed decreased to 8.9% and 9.9%, only about 1/4 and 1/3 of the annual growth rate of 1990s. After 2000, with the change of breeding structure and the gradual standardization of concentrated feed and additive premixed feed market, the phenomenon of high-speed growth of concentrated feed and additive premixed feed no longer appeared, and the feed industry gradually turned to the growth of formula feed output as the main body, especially in 2010, this structural change was more obvious. In 1999, China's per capita share of formula feed was 45 kg (510 kg in the United States, 310 kg in the European Union, 98 kg in the world), and it reached about 110 kg per capita in 2009. In 2019, China's per capita share of for-

年,达到了人均 110 kg 左右;2019年,我国人均占有配合饲料为 150 kg 左右。今后配合饲料的发展空间将逐步扩大,需要加大饲料粮的生产。

mula feed was about 150 kg. In the future, the development space of formula feed will be gradually expanded, and the production of feed grain needs to be increased.

(三)单个工业饲料企业的生产规模小
The production scale of a single industrial feed enterprise being small

企业规模的大小往往是其抗风险能力强弱的重要标志,特别是加入 WTO 后竞争加剧,高科技大型饲料企业在世界饲料业竞争中具有明显优势。中国饲料工业虽然在世界饲料工业总量排名中处于第二位,但单个企业的平均规模较小,难以抵抗饲料市场尤其是发达国家大型饲料集团的竞争。要做大做强中国的饲料业,需要加快培育一批具有国际竞争力的行业领军企业。2009 年,中国有饲料企业 12 291家,生产 1.48 亿 t 饲料,每个企业仅平均生产饲料 1.2 万 t;同年美国生产了 1.55 亿 t 饲料,只有 500 家饲料加工企业,每个企业的年均生产规模为 30.9 万 t。截至 2019 年年底,全国共有饲料加工企业约10 000 家,全国工业饲料总产量达到 2.29 亿 t,在世界饲料产量中排行第二位,全国 10 万 t 以上规模饲料企业 621 家,饲料产量 1.07 亿 t,占全国饲料总产量的 46.6%,总体上表现为高附加值、创新型规模化饲料添加剂企业市场占有率不断提高,产品结构愈发多样,发展壮大势头明显,企业收购兼并步伐进一步加快。中国饲料企业要在国内外饲料业的激烈竞争中占据一定市场,必须通过改组、联合等形式,形成企业的现代化、国际化和集团化。

The scale of an enterprise is often an important sign of its ability to resist risks. Especially after China's accession to the WTO, the competition has intensified. Large high-tech feed enterprises have obvious advantages in the competition of the world feed industry. Although China's feed industry ranks second in the world's total feed industry, the average scale of a single enterprise is small, which makes it difficult to resist the competition of the feed market, especially the large feed groups in developed countries. To make China's feed industry bigger and stronger, we need to speed up the cultivation of a number of leading enterprises with international competitiveness. In 2009, there were 12,291 feed enterprises in China, which produced 148 million tons of feed, and each enterprise only produced 12,000 tons of feed on average; in the same year, the United States produced 155 million tons of feed, and had only 500 feed processing enterprises, and the average annual production scale of each enterprise was 309,000 tons. By the end of 2019, there were about 10,000 feed processing enterprises in China. The industrial feed output in China was up to 229 million tons in China, which ranked the second in the world of feed production. There were 621 feed enterprises to produce more than 100,000 tons and these feed enterprises produced 107 million tons of feed, which accounted for 46.6% of the total feed production in China. On the whole, the market share of high value-added, innovative scale feed additive enterprises is constantly improving, the product structure is increasingly diverse, the momentum of development is obvious, and the pace of enterprise acquistion and merger is further accelerated. In order to occupy a certain market in the fierce competition of feed industry at home and abroad, Chinese feed enterprises must transform towards modernization, internationalization and collectivization by means of reorganization and combination.

（四）基础研究薄弱
Weak basic research

我国对动物营养与饲料科学的基础性、前沿性研究投入较少，致使饲料营养物质的代谢规律、营养物质需要量等方面的研究水平落后于发达国家，制定的饲养标准和原料标准借鉴和参考国外的多，自己研究的少，严重影响饲料工业发展的科技水平和持续发展的后劲。某些氨基酸如蛋氨酸的生产技术和生产量、药物添加剂新品种开发等技术方面的研究不能完全适应饲料工业的发展需要，牧草加工设备严重落后，依赖进口的局面未从根本上改变。针对这些问题，应特别加强国家自然科学基金等基础性、前沿性研究的投入力度，以完成我国饲料业全面接近国际先进水平，并在我国国民经济中跃居前十大产业之一的宏伟目标。

China's investment in basic and cutting-edge research of animal nutrition and feed science is insufficient, resulting in the research level of feed nutrient metabolism law and nutrient requirement lagging behind developed countries. The feeding standards and raw material standards formulated by China draw on and refer to foreign countries more, and the research of our own is less, which seriously affects the scientific and technological level and sustainable development of feed industry. Some amino acids such as methionine production technology and production, new varieties of drug additives and other technical research cannot fully meet the needs of the development of feed industry, and forage processing equipment is seriously backward, and the situation of relying on imports has not fundamentally changed. In view of these problems, we should especially strengthen the investment in basic and cutting-edge research, such as National Natural Science Foundation of China, so as to achieve the grand goal that China's feed industry will fully approach the international advanced level and become one of the top ten industries in China's national economy.

（五）解决饲料安全问题刻不容缓
Urgence of solving the problem of feed safety

改革开放后，我国饲料工业和养殖业取得的巨大发展和人民饮食结构的极大改善是显而易见的。但是蛋白质饲料原料和饲料产品中的三聚氰胺、抗生素等生长促进剂在畜产品中的药物残留及兴奋剂等"非法添加剂"的滥用对人类健康所造成的危害已成为广大消费者疑虑的焦点。近期，转基因大豆、玉米、豆粕频频进入中国饲料市场，尽管转基因饲料原料的安全问题目前尚无法定论，但大多数人对此表现出了极大的不安和担忧。我国在加入了WTO后，获得了世界各国贸易的市场准入证，但畜产品在欧盟、日本等国设置的"绿色贸易壁垒"

After the reform and opening up, the great development of China's feed industry and breeding industry and the great improvement of people's food structure are obvious. However, the harm to human health caused by the drug residues of melamine, antibiotics and other growth promoters in protein feed raw materials and feed products and the abuse of doping and other "illegal additives" has become the focus of concern of consumers. Recently, genetically modified soybean, corn and soybean meal have frequently entered China's feed market. Although the safety of genetically modified feed raw materials is still uncertain, most people show great uneasiness and worry about it. After China's accession to the WTO, it has obtained the market access certificate for the trade of all countries in the world. However, livestock products have been frustrated in the face of the "green trade barriers" set up by the European Union, Japan

面前连连受挫，不能不使我们极大地关注食品安全问题。为了人民的健康和冲破"绿色壁垒"，提高我国畜产品在国际上的竞争力，加速我国饲料工业标准化进程迫在眉睫。为了解决食品的无污染、无公害、安全、绿色，我国首批绿色畜产品认证准则《绿色食品　动物卫生准则》、《绿色食品　兽药使用准则》和《绿色食品　饲料及饲料添加剂使用准则》已由农业部审定，正式颁布执行。在《绿色食品　饲料及饲料添加剂使用准则》中，明确规定90%的动物饲料必须来自绿色食品生产基地，禁止使用任何激素类、安眠镇静类药品，禁止使用任何已禁用的药物性饲料添加剂（包括维吉尼霉素、泰乐菌素、杆菌肽锌、螺旋霉素、卡巴氧等）；禁止用以哺乳动物为原料加工的饲料饲喂反刍动物；禁止使用工业合成的油脂和转基因方法生产的饲料原料等。同时，对兽药使用也做出了严格规定。A级绿色食品三个部颁行业标准的出台，为我国养殖业和食品业指明了方向，使企业有章可循。这将极大地缩短我国肉食品与国际水平的差距，大大提高中国肉食品在全球的竞争力。

and other countries, which makes us pay great attention to food safety issues. For the sake of people's health and breaking through the "green barriers", it is urgent to speed up the standardization process of China's feed industry in order to improve the international competitiveness of China's livestock products. In order to solve the problem of food pollution-free, pollution-free, safe and green, the first batch of green animal products certification standards in China, such as Green Food Animal Health Dtandards, Green Food Veterinary Drug Use Standards and Green Food Feed and Feed Additive Use Standards, have been approved by the Ministry of agriculture and officially promulgated and implemented. In the Green Food Feed and Feed Additive Use Standards, it is clearly stipulated that 90% of animal feed must come from the green food production base. It is forbidden to use any hormone, sleeping sedative drugs, any banned drug-based feed additives (including virginiamycin, tylosin, bacitracin zinc, spiramycin, carbamox, etc.) and to use them for lactation. It is forbidden to feed ruminants with feedstuffs made from synthetic oil or feedstuffs produced by transgenic methods. At the same time, the use of veterinary drugs has also made strict provisions. The introduction of three ministerial industry standards for A-class green food has pointed out the direction for China's breeding industry and food industry, so that enterprises have rules to follow. This will greatly shorten the gap between China's meat food and the international level, and greatly improve the competitiveness of China's meat food in the world.

第三节　未来饲料行业发展趋势

Section 3　Future development trend of feed industry

21世纪以来，人类生活条件的改善在需要更大规模发展畜牧业，同时，对提高自身健康和生活质量的要求更为强烈，这迫使我们更加关注畜产品乃至饲料产品的质量。饲料科学要适应未来饲料业的发展，并更多地利用诸多学科的相关

Since the beginning of the 21st century, the improvement of human living conditions requires the development of animal husbandry on a larger scale. At the same time, the demand for improving their own health and quality of life is stronger, which forces us to pay more attention to the quality of animal products and even feed products. Feed science should be adapted to the future development of feed industry,

研究成果为之服务。

and related research results of many disciplines should also be made fully use of.

一、未来饲料科学的研究方向
Research direction of feed science in the future

未来饲料科学的研究领域将由宏观向微观、由静态向动态方向发展。与这种发展相适应，将更注重从分子营养学水平研究饲料的营养价值。同时，动物营养和饲料科学与边缘学科的互相渗透与嫁接将进一步加强，如开展营养遗传学的研究，探讨畜禽的一些主要生产性能的遗传基础如基因表达与饲料营养的关系以及如何通过选种提高反刍动物采食量、降低能量和蛋白质维持需要量；饲料营养与免疫的相互关系受到关注，研究的内容包括营养与免疫的协同作用，营养不良或过剩与疾病的关系，营养与免疫系统细胞分化、细胞代谢物的关系，应激状态下保持动物最适免疫功能的营养需要量等；营养生理的研究是饲料营养研究的另一热点课题，通过研究将明了动物在各种不同生理状态下对各种营养需要量参数的影响及其畜禽饲料营养调控机理的研究。借助于其他的研究手段，使动物营养与饲料科学研究更深入，更能揭示与畜禽之间各种错综复杂关系的本质。对饲料营养价值评定和畜禽营养需要量的研究的重视程度将得到加强。为了满足畜牧业的发展对日益增长的饲料量的需要，饲料资源的进一步开发利用将是饲料工作者必须认真研究的问题。发达国家为实现贸易保护设置了种种"绿色贸易壁垒"，世界各国尤其是发展中国家必须采取积极的应对措施，把饲料科学的研究重点放在饲料安全上。为了生产出国际上认可

In the future, the research field of feed science will develop from macro to micro, from static to dynamic. In line with this development, more attention will be paid to the nutritive value of feed from the level of molecular nutrition. At the same time, the interpenetration and grafting between animal nutrition and feed science and marginal disciplines will be further strengthened, for example, research on nutrition genetics will be carried out to explore the genetic basis of some main production performance of livestock and poultry, such as the relationship between gene expression and feed nutrition, and how to improve ruminant feed intake, reduce energy and protein requirements through seed selection. The interaction between feed nutrition and immunity has been paid much attention. The research contents include the synergistic effect between nutrition and immunity, the relationship between malnutrition or excess and disease, the relationship between nutrition and cell differentiation and cell metabolites of immune system, the nutritional requirements for maintaining optimal immune function of animals under stress state, etc. The research on nutritional physiology is another hot topic in feed nutrition research. Through the study, the effects of different physiological states of animals on various nutrient requirements and the regulation mechanism of feed nutrition for livestock and poultry will be clarified. With the help of other research methods, animal nutrition and feed science can be further studied, and the essence of various complex relationships with livestock and poultry can be revealed. More attention will be paid to the evaluation of feed nutritive value and the research of livestock and poultry nutritional requirements. In order to meet the needs of the development of animal husbandry for the increasing amount of feed, the further development and utilization of feed resources will be a problem that feed workers must seriously study. Developed countries have set up all kinds of "green trade barriers" to realize trade protection. All countries in the world, especially developing countries,

准入的绿色畜产品,微生态制剂、寡聚糖等绿色饲料添加剂的研究和应用日趋活跃,以替代耐药性和有残留的抗生素甚至完全禁止抗生素作为预防和生长促进剂使用。

must take positive measures to put the research of feed science on feed safety. In order to produce internationally recognized green animal products, the research and application of probiotics, oligosaccharides and other green feed additives are becoming more and more active to replace antibiotic resistance and residual antibiotics, and even completely prohibit the use of antibiotics as prevention and growth promoters.

二、未来饲料业发展趋势
Development trend of feed industry in the future

伴随着动物营养与饲料科学研究的进一步深入,饲料业将呈现如下发展趋势。

With the further development of animal nutrition and feed science research, the world feed industry will show the following development trend.

(一) 配方设计更科学
More scientific formula design

随着动物营养需要量的研究和预测进一步朝着动态、准确化方向发展,各种动物营养需要、饲养新标准将陆续问世,饲料营养成分和其他营养参数将不断得到更新和完善以及数学模型在动物营养中的应用,使今后在饲料配方设计中所选用的营养参数更全面、更科学、更先进。如:动物营养研究动态数学模型可以帮助饲料生产者和猪生产者决定生长阶段与饲料的数量、季节性饲粮的使用、屠体目标体重、猪基因型选择、分性别饲养和降低污染的饲粮设计。随着合成氨基酸生产量的增加和技术的提升和完善,在猪禽饲养中通过添加合成氨基酸更为平衡其生产性能不降低的低蛋白饲粮逐渐受到重视。在奶牛饲养中,使用美国加利福尼亚大学提出的碳水化合物和蛋白质新体系,可使原料的选择及营养成分的搭配更具针对性。

With the further development of the research and prediction of animal nutrition requirements towards dynamic and accurate direction, various animal nutrition requirements and new feeding standards will come out one after another, feed nutrients and other nutritional parameters will be constantly updated and improved, and the application of mathematical models in animal nutrition will make the nutritional parameters selected in feed formula design more comprehensive, more scientific and more advanced in the future. For example, animal nutrition research dynamic mathematical model can help feed producers and pig producers to determine the growth stages and feed quantities, seasonal diet use, target carcass weight, pig genotype selection, sex-specific feeding and diet design to reduce contamination. With the increase of production of synthetic amino acids and the improvement and perfection of technology, the low-protein diet which can balance the production performance of pigs and poultry by adding synthetic amino acids has been paid more and more attention. In dairy cattle breeding, using the new carbohydrate and protein system proposed by the University of California can make the selection of raw materials and the collocation of nutrients more targeted.

(二）饲料产品的科技含量更高
The scientific and technological content of feed products being higher

近30年来在饲料机械制造、加料加工工艺、饲料添加剂以及营养调控等动物营养科学领域所取得的众多技术成果的有机结合并被应用于生产和今后上述研究的进一步深化，将显著提高动物生产各类饲料产品的技术含量，进而改善畜禽的生产性能，提高畜禽健康和畜产品质量。

In the past 30 years, many technological achievements in animal nutrition science, such as feed machinery manufacturing, feed processing technology, feed additives and nutrition regulation, have been organically combined and applied to production. The further deepening of the above research in the future will be significant. It aims to improve the technical content of animal production of various feed products, improve the production performance of livestock and poultry, and improve the health of livestock and poultry and the quality of livestock products.

（三）饲料企业规模会更大
Feed enterprise scale becoming bigger

饲料企业的规模进一步加大，大型饲料集团将逐步挤垮或吞并中小型企业，主宰世界饲料市场。饲料一体化逐渐成为当代发展主流。

The scale of feed enterprises will be further enlarged, and large feed groups will gradually crush or annex small and medium-sized enterprises to dominate the world feed market. Feed integration has gradually become the mainstream of contemporary development.

（四）饲料原料的来源更广
More sources of feed raw materials

未来动物生产的发展将面临严峻的饲料资源供求矛盾的挑战，现有可利用饲料短缺将极大地制约动物饲养业的可持续发展。利用物理、化学和生物技术对饲料原料尤其是新型饲料原料进行开发，并提高其营养物质利用率、饲料转化效率是解决饲料原料资源不足与增加配合饲料产量供求矛盾的重要途径。如：通过现代遗传技术、育种技术将使高油玉米、转基因品种等更多种类的饲料作物新品种应用于饲料生产。现代动物营养科学和饲料分析检测技术的发展，将为明了饲料中各种潜在营养物质和抗营养因子，以更好地开发和利用各种饲料资源提供可靠的技术保证。在这一方

In the future, the development of animal production will face the challenge of severe contradiction between supply and demand of feed resources. The shortage of available feed will greatly restrict the sustainable development of animal husbandry. Using physical, chemical and biological technology to develop feed raw materials, especially new feed raw materials, and improve the utilization rate of nutrients and feed conversion efficiency are important ways to solve the contradiction between the shortage of feed raw materials and the increase of the output of formula feed. For example, through modern genetic technology and breeding technology, more varieties of feed crops such as high oil corn and transgenic varieties will be applied to feed production. The development of modern animal nutrition science and feed analysis and detection technology will provide reliable technical guarantee for understanding various potential nutrients and antinutritional factors in feed, so as to better

面，欧洲国家利用来源广泛丰富的饲料原料(如粮食、油料、食品加工的副产物)，采用预加工、添加剂和膨化等技术降低原料中抗营养因子并提高其营养价值的经验值得借鉴。青绿饲料和牧草的利用将进一步加强。由于人类膳食结构的改善更多地倾向于草食畜产品的利用，所以，牛、羊肉及牛奶在食品构成中的比例会显著增加，需要生产更多的符合绿色饲料要求的青绿牧草和饲料作物。同时，蛋白质含量高、品质优良的紫花苜蓿草捆、草粉、草颗粒等饲草生产、草品产业化及其在国际上的流通将成为开发和利用的关键。

develop and utilize various feed resources. In this regard, the experience of European countries using feed materials (such as grain, oil and by-products of food processing) with extensive sources to reduce antinutritional factors and improve their nutritive value by using pre-processing, additives and puffing technologies is worth learning form. The utilization of green forage and forage will be further strengthened. As the improvement of human dietary structure is more inclined to the utilization of herbivorous animal products, the proportion of beef, mutton and milk in food composition will increase significantly, and more green forage and feed crops that meet the requirements of green feed need to be produced. At the same time, forage production, industrialization and international circulation of alfalfa with high protein content and good quality will become the key to development and utilization.

（五）安全绿色食品将是关注的焦点
The focus of safe green food

随着人类对环境保护、食品安全与健康的进一步关注，饲料业要不断发展高新技术迎接这一挑战。如：在饲料中使用酶制剂提高利用率，降低排污量；在饲料加工中避免交叉污染，推行避免残留或零残留技术；认真研究和学习国际上的全面质量管理(TQC)、良好生产规范(GMP)、ISO9000族标准和国际动物卫生法典(IAHC)等安全管理体系；研究和建立与国际接轨的饲料产品质量标准，以保证饲料的安全、高效和绿色。

With people's further attention to environmental protection, food safety and health, feed industry should continue to develop high and new technologies to meet this challenge. For example, enzyme preparation should be adopted in feed to improve utilization rate and reduce the amount of pollution. Cross pollution should be avoided in feed processing. Avoidance or zero residue technology should be implemented. The total quality control (TQC), good production practise (GMP), ISO9000 standards and international animal health code (IAHC) and other safety management systems should be studied to research and establish the food product quality standards in line with the international standards, in order to ensure the feed's safety, efficiency and green.

第四节 饲料学的性质、任务、研究内容和研究方法
Section 4 Properties, tasks, research content and research methods of feed science

饲料学作为一门学科，既有其独立性，又与其他学科密不可分，可利用相关学科的大量先进成果来发

As a discipline, feed science has its independence and is inseparable from other disciplines. It can develop itself by making use of a large number of advanced achievements of

展自己。因此,现代饲料学的性质、任务和内容有其不同于其他学科的特点。

related subjects. Therefore, the nature, task and content of modern feed science are different from those of other disciplines.

一、饲料学的性质
Properties of feed science

饲料学是一门研究饲料的营养、饲料生产、饲料加工、饲料配合、人畜卫生、畜产品品质以及环境保护等的一门学科,同时也是一门涉及农业、工业、食品、医药、机械、内外贸等十多个行业的综合性学科,是畜牧养殖业的主要科学支柱之一。为了人类生存的健康永续发展,这门学科正朝着与物理、化学、生理、生化、遗传、育种、免疫等边缘学科相互渗透、嫁接,揭示营养物质在动物体内的代谢规律,调控饲料营养物质在动物体内的合理吸收、代谢与分配,探讨动物机体的外界饲养条件与内部微生态环境的关系以及饲料如何向安全绿色方向发展,它的发展对提高人类的生活和健康水平、促进国民经济的发展乃至社会的稳定,都有至关重要的作用。

养殖业成本中的70%左右来自饲料,饲料是发展畜牧业的主要物质基础。饲料学对动物生产的发展至关重要,它不仅是培养动物生产、动物营养与饲料科学专业人才的一门重要学科,也是推动动物生产不断发展的理论与技术基础,因此在理论和实践上都具有重要的地位。

就饲料学这门学科的性质而言,它直接服务于畜牧业,是高等农业院校动物科学专业的一门专业基础课和动物营养与饲料科学专业的一门专业课。

Feed science is a subject that studies feed nutrition, feed production, feed processing, feed formula, human and animal health, animal product quality and environmental protection. It is also a comprehensive discipline involving more than ten industries such as agriculture, industry, food, medicine, machinery, domestic and foreign trade. It is one of the main scientific pillars of animal husbandry. For the healthy and sustainable development of human survival, this discipline is penetrating and grafting with physics, chemistry, physiology, biochemistry, genetics, breeding, immunity and other marginal subjects, revealing the metabolic law of nutrients in animals, regulating the reasonable absorption, metabolism and distribution of feed nutrients in animals, and exploring the relationship between the external feeding conditions of animal body and the internal micro-ecology environment and the safe green development of feed, which plays a vital role in improving human life and health level, and so promoting the development of national economy and even social stability.

About 70% of the cost of animal husbandry comes from feed, which is the main material basis for the development of animal husbandry. Feed science is very important to the development of animal production. It is not only an important subject to cultivate professionals in animal production, animal nutrition and feed science, but also the theoretical and technical basis to promote the continuous development of animal production. Therefore, it has an important position in theory and practice.

As far as the nature of feed science is concerned, it directly serves animal husbandry and is a professional basic course of animal science majors in agricultural colleges and a professional course for animal nutrition and feed science majors.

二、饲料学的任务
Tasks of feed science

饲料为动物的一切生命活动提供营养物质,动物的整个生命过程,离不开饲料及其包含的营养物质。饲料学的基本任务是:运用现代生物科学和农业科学先进技术和成果,深刻揭示饲料的营养代谢规律及各种营养物质间的相互关系、饲用价值及生理功能,最终达到为畜牧生产提供优质、高效、安全、符合现代环保要求的配合饲料,提高畜禽的生产性能和保证畜产品质量的目标。

Feed provides nutrients for all life activities of animals, and the whole life process of animals can't be separated from feed and the nutrients contained in it. The basic task of feed science is to use the advanced technology and achievements of modern biological science and agricultural science to deeply reveal the nutrient metabolism law of feed, the relationship between various nutrients, feeding value and physiological function, so as to provide high-quality, efficient, safe formula feed that meets the requirements of modern environmental protection for animal production, improve the production performance of livestock and poultry and ensure the livestock yield objective of product quality.

三、饲料学的研究内容
Research content of feed science

根据饲料学的性质和任务,包括如下研究内容:

1. 饲料化学 研究与动物生产有关的饲料中各种营养物质的种类、生理及生物学功能,是饲料学研究的基础内容。

2. 饲料营养价值评定 研究饲料营养价值评定的原理与方法,评定各种饲料对不同动物的营养价值。它是科学合理利用饲料的依据。

3. 饲料分类 研究建立饲料分类的方法,对饲料资源进行科学的分类和编号,便于各种饲料的合理利用和管理。

4. 饲料原料 研究各类饲料的分类、营养特性、加工方法、质量标准及饲用价值,并提出科学开发饲料资源的方法和途径。

5. 饲料与畜产品质量安全 研究饲料中各种营养物质与畜产品

According to its nature and task, feed science includes the following research content:

1. Feed chemistry It studies the kinds, physiological and biological functions of various nutrients in feed related to animal production, which is the basic content of feed science research.

2. Feed nutritive value evaluation It studies the principles and methods of feed nutritive value evaluation, and evaluates the nutritive values of various feeds to different animals. It is the basis of scientific and rational use of feed.

3. Feed classification It studies and establishes the methods of feed classification, and scientifically classifies and numbers feed resources, which is convenient for rational utilization and management of various feeds.

4. Feed raw materials The classification, nutritional characteristics, processing methods, quality standards and feeding values of feed raw materials are studied, and the methods and ways of scientific development of feed resources are put forward.

5. Quality and safety of feed and animal products It studies the relationship between various nutrients in feed and

品质及风味的关系,揭示饲料中各种有毒有害物质、抗营养因子对畜禽生产性能及环境的影响,寻求为了人畜安全钝化和减少饲料中的有害因子的方法,为生产无公害优质安全的畜产品提供理论依据及技能,使饲料生产和养殖业可持续发展。

6. 饲料配合 研究如何运用动物对各种营养物质的需要量和饲养标准配制饲料,阐明科学配制饲粮的原则及科学设计不同种类畜禽不同生长阶段饲料配方的方法和手段。

the quality and flavor of animal products, revealing the effects of various toxic and harmful substances and antinutritional factors in feed on the production performance of livestock and poultry and environment, and seeking the methods of passivation and reduction of harmful factors in feed for human and animal safety. The theoretical basis and technical support are provided for the production of pollution-free, high-quality and safe animal products through studies, which can make feed production and breeding industry sustainblly develop.

6. Formula feed It studies how to use animal requirements for various nutrients and feeding standards to prepare feed, and expounds the principle of scientific preparation of feed and the methods and means of scientific design of feed formula for different kinds of livestock and poultry at different growth stages.

四、饲料学的研究方法
Research methods of feed science

饲料学的研究方法可分为化学分析法、体外消化模拟法以及动物试验。

化学分析法主要通过检测饲料中各种营养物质的含量,以判断饲料的营养价值。其中最经典和最常用的是德国 Weende 试验站的 Hunneberg 和 Stohman 两位科学家在 1860 年创立的饲料概略养分分析法,也称 Weende 饲料分析体系。该方法可以测定饲料中六种概略养分的含量,包括水分、粗蛋白、粗脂肪、粗纤维、粗灰分和无氮浸出物,从而评价饲料的营养价值。Van Soest 等在 1996 年将粗纤维进一步分解为中性洗涤纤维、酸性洗涤纤维和酸性洗涤木质素,进一步分析了粗饲料的营养价值。

体外消化模拟法通过在体外条件下模拟饲料在动物消化道内的消化过程,从而估测饲料在动物体内的消化率,包括一步法(单一酶

The research methods of feed science can be divided into chemical analysis method, in vitro digestion simulation method, and animal experiment.

The chemical analysis method mainly detects the content of various nutrients in the feed to determine the nutritive value of the feed. One of the most classic and most commonly used methods is the feed summary nutrient analysis method established in 1860 by two scientists Hunneberg and Stohman at the Weende Experimental Station in Germany, also known as the Weende feed analysis system. This method can determine the content of six summary nutrients in feed, including moisture, crude protein, crude fat, crude fiber, crude ash, and nitrogen-free extract so as to evaluate the nutritive value of feed. Van Soest et al. further decomposed crude fiber into neutral detergent fiber, acid detergent fiber and acid detergent lignin, and further analyzed the nutritive value of roughage in 1996.

The in vitro digestion simulation method simulates the digestion process of feed in the digestive tract of the animal under in vitro conditions to estimate the digestibility of feed in the animal body. It includes one-step method (single

法）、胃-小肠两步法和胃-小肠-大肠三步法，以及反刍动物的瘤胃模拟技术。

动物试验研究饲料营养价值主要通过动物饲养试验进行。将饲料以不同配比、加工等处理后，饲喂给试验动物，通过消化试验、代谢试验、平衡试验等试验设计，研究饲料对动物的生长性能、生产性能、健康状况等方面的影响。

enzyme method), stomach-small intestine two-step method, and stomach-small intestine-large intestine three-step method, and rumen simulation techniques for ruminants.

Animal experiments to study the nutritive value of feed are mainly carried out through animal feeding experiments. The feed is fed to experimental animals after being processed in different proportions and processing, and the effects of feed on growth performance, production performance and health status of animals are studied through the design of digestion experiments, metabolism experiments, and balance experiments.

第一章 饲料化学
Chapter 1　Feed chemistry

养殖生产中的动物必须不断地从外界摄取各种营养物质。这些营养物质主要来自于各种植物性和动物性的饲料,因此,了解构成饲料的各种营养物质的化学基础,是我们进一步学习饲料营养价值评定、饲料原料、饲料资源开发与饲粮配置技术的基础。

Animals in aquaculture production must constantly take in various nutrients from the outside world. These nutrients mainly come from various plant and animal feed. Therefore, understanding the chemical basis of various nutrients that constitute feed is the basis for us to further study the evaluation of feed nutritive value, feed raw materials, feed resource development and feed configuration techniques.

第一节　碳水化合物
Section 1　Carbohydrates

碳水化合物是自然界分布最广的一类有机物质,更是植物性饲料的一项重要组成成分,其含量一般约占植物体干物质总重的50%-80%。碳水化合物的名称来源于法语 hydrate(氢氧化物,水化物) de carbone,最初用于含有C、H、O元素的化合物。碳水化合物主要由碳、氢、氧三大元素遵循C:H:O为1:2:1的结构规律构成基本糖单位,可用通式[$(CH_2O)_n$]描述不同碳水化合物分子的组成结构。但少数碳水化合物并不遵循这一结构规律。饲料中所含碳水化合物种类较多,但根据单糖的聚合度,主要分为3大类,即单糖(不能被水解的简单化合物)、低聚糖(单糖聚合度≤10的碳水化合物,又称寡糖)和高聚糖(单糖聚合度>10的复杂碳水化

Carbohydrates are the most widely distributed organic substances in nature, and also an important component of plant feed. Its content generally accounts for about 50%-80% of the total dry matter of the plant. The name of carbohydrate comes from the French hydrate (hydroxide, hydrate) de carbone, originally used for compounds containing C, H, O elements. Carbohydrates are mainly composed of three major elements: carbon, hydrogen, and oxygen, and follow the structural rule of C:H:O as 1:2:1 to form basic sugar units. The general formula [$(CH_2O)_n$] can be used to describe the composition of different carbohydrate molecules. But a few carbohydrates do not follow this structural law. There are many types of carbohydrates in feed, but according to the degree of polymerization of monosaccharides, they are mainly divided into three categories: monosaccharides (simple compounds that cannot be hydrolyzed), oligosaccharides (carbohydrates with more than 10 degrees of monosaccharide polymerization, also called oligosaccharides) and high saccharides (complex carbohy-

合物，又称多糖）。此外，尚含一些糖类衍生物（如几丁质、甘油等）。

drates with less than 10 degrees of monosaccharide polymerization, also called polysaccharides). In addition, it still contains some sugar derivatives (such as chitin, glycerin, etc.).

一、单糖
Monosaccharides

单糖是最简单的一类碳水化合物，包括丙糖、丁糖、戊糖、己糖、庚糖及衍生糖，其分子结构特点是：① 1个碳原子的2个共价键分别与1个氢原子和1个羟基相连，下余2个价键再分别与其他碳原子相连；② 每个糖分子中均含有1个碳基（也称碳氧基），也就是说，从化学结构特点看，单糖属多羟基醛、酮或它们的缩合物。故有人又将单糖分成醛糖（如葡萄糖）和酮糖（如果糖）。

单糖不仅有链状结构，同样还有环状结构。一般在单糖分子中原子数达5个时可借助"氧桥"形成稳定的环（含8个碳原子的单糖少见）。单糖的链状结构和环状结构实际上是同分异构体，环状结构最重要，例如：葡萄糖在晶体状态或水溶液中，绝大部分是环状结构，在水溶液中链状结构和环状结构可以互变。

所有糖类都有不对称碳原子，故都具有旋光性。旋光性也是鉴定糖的重要指标。丙糖是最简单的单糖，比较重要的丙糖有甘油醛（丙醛糖）和二羟丙酮（丙酮糖），它们的磷酸酯是糖代谢的重要中间产物。现以具有1个不对称碳原子的最简单的单糖——甘油醛为例说明糖的旋光性。凡羟基在甘油醛的不对称碳原子右边者被称为D-型，而在左边的称为L-型。

Monosaccharides are the simplest type of carbohydrates, including trioses, tetoses, pentoses, hexoses, heptoses, and derived sugars. Its molecular structure features are: ① Two covalent bonds of a carbon atom are connected to a hydrogen atom and a hydroxyl group respectively, and the remaining 2 valence bonds are connected to other carbon atoms respectively; ② Each sugar molecule contains a carbon group (also called a carbonoxy group), that is to say, from the chemical structure characteristics, monosaccharides are polyhydroxy aldehydes, ketones or their condensates. Therefore, some people divide monosaccharides into aldose (such as glucose) and ketose (like sugar).

Monosaccharides not only have a chain structure, but also a cyclic structure. Generally, when the number of atoms in a monosaccharide molecule reaches 5, a stable ring can be formed with the help of an "oxygen bridge" (monosaccharides with 8 carbon atoms are rare). The chain structure and and ring structure of monosaccharides are actually isomers. The most important structure is the ring structure. For example, glucose in the crytalline state or in aqueous solution is mostly circular structure. In aqueous solution, the chain structure and the ring structure can be interchangeable.

All sugars have asymmetric carbon atoms, so they are all optically active. Optical rotation is also an important indicator of sugar identification. Triose is the simplest monosaccharide. The more important trioses are glyceraldehyde (propionose) and dihydroxyacetone (acetone sugar). Their phosphate is an important intermediate product of sugar metabolism. Glyceraldehyde, the simplest monosaccharide with one asymmetric carbon atom, is used as an example to illustrate the optical rotation of sugar. Where the hydroxyl group is on the right of the asymmetric carbon atom of glycerol is called D-type, and the one on the left is called L-type.

单糖中以戊糖（五碳糖）和己糖（六碳糖）最为常见,对动植物来说也是最重要的,其因戊糖中的核糖是核酸的组成成分；木糖是半纤维素和果胶的组成成分,反刍动物可借助瘤胃微生物发酵作用,使90%以上的木糖被消化利用。另外,也有报告指出,木糖及木磺酸等具有促进畜禽代谢和营养吸收的功能。己糖中的葡萄糖是动物体极易吸收的一种糖（胃壁直接吸收）,可为动物组织提供能量,在植物体内也可以以纤维素、淀粉等化合态形式贮存起来。果糖（由葡萄糖经异构化反应转化而来）是糖中最甜者,供能上虽与葡萄糖相同,但在吸收、运载等营养生理过程中却存在一定的差异,故应注意合理利用。

Among the monosaccharides, pentose (five-carbon sugar) and hexose (six-carbon sugar) are the most common monosaccharides, and they are also the most important for animals and plants, because the ribose in pentose is a component of nucleic acid; xylose is hemicellulose with the composition of pectin, ruminants can use rumen microbial fermentation to digest more than 90% of xylose. In addition, the report also points out that xylose and wood sulfonic acid have the function of promoting metabolism and nutrient absorption in livestock and poultry. Glucose in hexose is a sugar that is easily absorbed by the animal body (directly absorbed by the stomach wall) and can provide energy for animal tissues. It can also be stored in the form of cellulose, starch and other chemical forms in the plant. Fructose (converted from glucose through isomerization reaction) is the sweetest sugar. Although it has the same energy supply as glucose, there are certain differences in the nutrient and physiological processes such as absorption and transportation. Therefore, it should be rationally utilized.

二、低聚糖
Oligosaccharides

低聚糖一般是指由 2-6 个单糖通过糖苷键组成的一类糖。其中以双糖分布较广,营养意义较大。

Oligosaccharides refer to a type of sugar composed of 2 to 6 monosaccharides via glycosidic bonds. Among them, disaccharides are widely distributed and have greater nutritional significance.

（一）双糖
Disaccharide

双糖又称二糖,是由 2 分子单糖脱水缩合而成的一类糖。植物组织中存在的双糖主要有蔗糖、麦芽糖、纤维二糖、乳糖、海藻糖、蜜二糖等,而动物乳中则含有较多乳糖。双糖在动物体消化道内需经相应酶作用分解成单糖,才能被动物体吸收利用。

Disaccharides are a group of sugars formed by the dehydration and condensation of two molecules of monosaccharides. The disaccharides present in plant tissues mainly include sucrose, maltose, cellobiose, lactose, trehalose, melibiose, etc., while animal milk contains more lactose. Disaccharides need to be decomposed into monosaccharides in the digestive tract of animals by corresponding enzymes before they can be absorbed and utilized by passive objects.

1. 蔗糖

蔗糖是由葡萄糖和果糖组成的一种非还原性二糖,分布较广、含量

1. Sucrose

Sucrose is a non-reducing disaccharide composed of glucose and fructose. The plants that are widely distributed

较高的有甜菜(15%-20%)、甜高粱(10%-18%)、甘蓝(10%-15%)、枫树(3%-6%)等。各种果实、根茎类、蔬菜与树木汁液中也有不等含量。初生乳猪小肠和胰脏分泌的蔗糖酶极少,在出生后1周内只能喂乳糖或葡萄糖,若喂蔗、果糖,则会引起乳猪严重下痢。一般乳猪出生后2周才可喂少量蔗糖或淀粉,犊牛消化道内的双糖酶或胰碳水化合物酶发展更慢,2月龄时才可利用。

2. 麦芽糖

麦芽糖为淀粉与糖原的组成成分,由2分子葡萄糖缩合生成,也属一种还原性双糖,大量存在于发芽的谷物中,尤其是麦芽中。动物体内含相应的酶,故可被直接吸收利用。

3. 乳糖

乳糖是半乳糖以β-1形式、4糖苷键与葡萄糖结合而成的一种还原性双糖,主要存在哺乳动物乳中。动物种类不同,乳中含量不同,如:牛乳中含4.5%-5.5%,猪乳中含4.9%,马乳中含6.1%,山羊乳中含4.6%,人乳中含4.0%-5.0%。由于乳糖酶仅存在于哺乳动物幼畜体内,成年家畜如摄取乳糖过多时,除被肠道微生物发酵后吸收外,剩余者排出体外。

4. 纤维二糖

纤维二糖是纤维素的基本构成单位,也是其他不少多糖和糖苷的组成成分,它是2个葡萄糖分子以β-1,4糖苷键连接的。自然界中无游离态,只有当纤维素经微生物发酵、酶解或酸水解时,才会产生游离态纤维二糖,动物体内无相应水解它的酶(β-葡萄糖苷酶),故无法直接利用。

其他双糖还有蜜二糖、龙胆二糖、松二糖等。

with high content of sucrose are sugar beet (15%-20%), sweet sorghum (10%-18%), cabbage (10%-15%), maple (3%-6%), etc. The sap of various fruits, rhizomes, vegetables and trees also has varying content. Sucrose secreted by the small intestine and pancreas of newborn suckling pigs is extremely low, and they can only be fed with lactose or glucose within 1 week after their birth. If fed with sucrose and fructose, severe diarrhea in suckling pigs will occur. Generally, sucrose or starch can be fed to suckling pigs 2 weeks after their birth. Disaccharidase or pancreatic carbohydrase in the calf's digestive tract develops more slowly, and can be fed with only at 2 months of age.

2. Maltose

Maltose is a component of starch and glycogen. It is formed by the condensation of two molecules of glucose. It is also a reducing disaccharide, which is abundantly present in sprouted grains, especially malt. Animals contain corresponding enzymes, so maltose can be directly absorbed and utilized.

3. Lactose

Lactose is a reducing disaccharide formed by the combination of galactose with B-1,4 glycosidic bonds and glucose. It is mainly found in mammalian milk. The content of lactose is different in different animal species, such as 4.5%-5.5% in cow's milk, 4.9% in sow milk, 6.1% in horse milk, 4.6% in goat milk, 4.0%-5.0% in human milk. As lactose is only present in the mammalian body, and when too much lactose is ingested by adult livestock, it will be excreted from the body except for being fermented by intestinal microorganisms after fermentation.

4. Cellobiose

Cellobiose is the basic building block of cellulose, and it is also a component of many other polysaccharides and glucoside. It is connected by two glucose molecules with β-1, 4 glycosidic bonds. There is no free state in nature. Free cellobiose will only be produced when cellulose is fermented, enzymatically or acid hydrolyzed by microorganisms. Animals have no corresponding enzyme (β-glucosidase) to hydrolyze it, so they cannot be used directly.

Other disaccharides include melibiose, gentiobiose, turanbiose, etc.

(二) 其他常见的低聚糖
Other common oligosaccharides

其他常见的低聚糖主要有棉籽糖、水苏糖等。

Other common oligosaccharides mainly include raffinose, stachyose, etc.

1. 三糖

主要的三糖有棉籽糖、甘露三糖等。棉籽糖是由半乳糖、葡萄糖和果糖组成的一种无还原性三糖，棉籽中含量较高（约8.0%），大豆、成熟的甜菜及蔗糖废糖蜜中有一定含量（约0.5%）。

1. Trisaccharide

The main trisaccharides are raffinose, mannose and so on. Raffinose is a non-reducing trisaccharide composed of galactose, glucose and fructose. The content of trisaccharides in cottonseed is high (about 8.0%), and there is a certain content of trisaccharides in soybeans, mature beets, and sucrose (about 0.5%).

2. 四糖

主要四糖为水苏四糖，是一种由1分子果糖、1分子葡萄糖和2分子的半乳糖组成的四糖，常见于多种植物的根茎和籽实中，而以唇形科水苏属（Stachys）植物含量较高。

2. Tetrasaccharide

The main tetrasaccharide is stachyose, which is a tetrasaccharide composed of 1 molecule of fructose, 1 molecule of glucose and 2 molecules of galactose. It is commonly found in the roots and seeds of many plants, and high in Stachys of the Labiatae family.

3. 甘露寡糖

甘露寡糖又称甘露低聚糖或葡甘露寡聚糖，是由几个甘露糖分子或甘露糖与葡萄糖通过 α-1,6、α-1,2 或 α-1,3 糖苷键连接组成的低聚糖。甘露寡糖广泛存在于魔芋粉、瓜儿豆胶、田菁胶及多种微生物细胞壁内，目前饲料添加剂用甘露寡糖主要是通过对酵母细胞壁的提取。作为饲料添加剂用的甘露寡糖多为二糖、三糖、四糖等结构的混合物，甘露寡糖不为单胃动物消化道酶分解，可作为动物肠道内有益微生物如双歧杆菌、乳酸杆菌等的营养素，有促进消化道有益菌株的增殖和抑制有害微生物的作用，同时还有促主机体免疫力提高和促进动物生产的作用。

3. Mannooligosaccharide

Mannooligosaccharides, also known as glucomannanoligosaccharides, are oligosaccharides composed of several mannose molecules or mannose and glucose connected by α-1,6, α-1,2 or α-1,3 glycosidic bonds. Mannanoligosaccharides are widely present in the cell walls of konjac flour, guar gum, sesbania gum and many kinds of microorganisms (glucomannan). Currently, mannanoligosaccharides used in feed additives are mainly extracted from yeast cell walls. Mannanoligosaccharides used as feed additives are mostly mixtures of disaccharides, trisaccharides, tetrasaccharides and other structures. Mannanoligosaccharides are not decomposed by monogastric animal digestive tract enzymes, but can be used as nutrients for beneficial microorganisms in animal intestines, such as bifidobacteria and lactobacilli, and can promote the proliferation of beneficial bacteria in the digestive tract and inhibit harmful microorganisms. It also can promote the immunity of the host and promote the role of animal production.

其他低聚糖还有毛蕊花糖（3半乳糖＋1葡萄糖＋1果糖）及乳糖等。

Other oligosaccharides include verbasciose (3 galactose with 1 glucose and 1 fructose), lactose, etc.

（三）利用低聚糖时应注意的问题
Problems that should be paid attention to when using oligosaccharides.

1. 动物本身无消化棉籽糖和水苏糖等的酶，故无法对其直接利用。大肠微生物虽可对其进行发酵，但常产生引起胃肠酸气的气体如 CO_2 和 H_2，因此，豆类或豆类产品等食入过多时，易发生肠胃酸胀。

2. 低聚糖中，有的有还原性（如麦芽糖），有的无还原性（如蔗糖）。了解还原性的意义在于，原料或成品中单糖或还原性糖的瑞基，在加热或长期贮藏过程中，易与氨基酸或胺发生缩合反应（羧氨反应），并产生黑褐色素。此反应被称为"褐变反应"或"美拉德反应"，反应结果可降低饲料中有效氨基酸含量，使整个饲料的营养价值下降。

1. Animals themselves have no enzymes to digest raffinose and stachyose, so oligosaccharides cannot be used directly. Although large intestine microorganisms can ferment it, they often produce gaseous gas such as CO_2 and H_2 that cause gastrointestinal acid gas. Therefore, when beans or legume products are eaten too much, gastrointestinal swelling is prone to occur.

2. Among the oligosaccharides, some are reductive (such as maltose) and some are non-reductive (such as sucrose). The significance of understanding reducibility is that the aldehyde group of monosaccharides or reducing sugars in raw materials or finished products can easily undergo condensation reactions (carboxamide reactions) with amino acids or amines during heating or long-term storage, and produce dark brown pigments. This reaction is called "browning reaction" or "Maillard reaction", and the result of the reaction can reduce the content of effective amino acids in the feed and reduce the nutritive value of the entire feed.

三、多糖
Polysaccharides

多糖是由10个糖单位以上单糖分子经脱水、缩合而成，是结构复杂的高分子化合物，一般单糖数多在百个以上。多糖广泛分布于植物和微生物体内，动物体内也有少量分布（主要为糖原）。从其功能角度考虑多糖可以分成营养性多糖（贮存性多糖）和结构多糖，如：淀粉、菊糖、糖原等属营养性多糖，其余多糖属结构多糖。多糖一般不溶于水，只有水解或发酵后才能被动物吸收利用。

Polysaccharide is formed by dehydration and condensation of monosaccharide molecules with more than 10 sugar units. It is a type of polymer compound with complex structure. Generally, the number of monosaccharides is more than one hundred. Polysaccharides are widely distributed in plants and microorganisms. There are also a small amount of distribution in animals (mainly glycogen). From the perspective of their functions, polysaccharides can be divided into nutritive polysaccharides (storage polysaccharides) and structural polysaccharides. For example, starch, inulin, glycogen, etc. belong to nutritive polysaccharides, and the rest belong to structural polysaccharides. Polysaccharides are generally insoluble in water, and can be absorbed and utilized by animals only after hydrolysis or fermentation.

(一) 淀粉
Starch

淀粉是由 D-葡萄糖组成的一种多糖,以微粒形式大量存在于植物种子、块茎及干果实中,属植物体中一种贮藏物质。玉米、高粱、小麦等谷实中含量高,一般可达 60%–70%;甘薯、本薯、马铃薯中含量约 25%–30%。淀粉分直链淀粉(溶于热水)和支链淀粉(不溶于热水)两种,前者是葡萄糖以 α-1,4 糖苷键连接的链状分子;后者是除葡萄糖以 α-1,4 糖苷键结合的主链外,尚含有 α-1,6 糖苷键与主链相连的支链。来源不同,直、支链粉比例不同。一般淀粉中,直链淀粉占 15%–25%,其余为支链淀粉。而糯米、粘高粱中 99% 的淀粉属支链淀粉。豆类中淀粉虽含量不高(如大豆仅含 0.4%–0.9%),但全属直链淀粉。

淀粉的特性可概括为"三化",即糊化、老化和胶化。动物采食饲料后,饲料中的淀粉便会在淀粉酶的作用下,降解为多个长短不一的多苷链片段(统称糊精),然后再变为麦芽糖,最终以葡萄糖的形式被吸收利用。

Starch is a polysaccharide composed of D-glucose, which exists in charge quantities in the form of particles in seeds, tubers and dried fruits of plants. It is a kind of storage substance in plant body. The content of corn, sorghum, wheat and other grains is high, generally up to 60%–70%; the content of starch in sweet potatoes, native potatoes, and potatoes is about 25%–30%. The starch is divided into amylose (dissolved in hot water) and amylopectin (insoluble in hot water). The former is a chain molecule connected by α-1,4 glycosidic bonds to glucose. The later one contains a chain molecule connected by α-1,4 glycosidic bonds to glucose and α-1,6 glycosidic bond. The content ratio of straight and branched chain powder is different with different sources. In general starch, amylose accounts for about 15%–25%, and the rest is amylopectin. 99% of starch in glutinous rice and sticky sorghum is amylopectin. Although the starch content in beans is not high (for example, soybeans only contain 0.4%–0.9%), they are all amylose.

The characteristics of starch can be summarized as "three transformations", namely gelatinization, aging and retrogradation. After animals eat the feed, the starch in the feed will be degraded into multiple polyglycoside chain fragments (collectively called dextrin) of different lengths under the action of amylase, and then into maltose, and finally be absorbed and utilized in the form of glucose.

(二) 糊精
Dextrin

糊精是淀粉消化或加温水解而产生的一系列有支链的低分子化合物。据研究,支链淀粉在动物消化过程中可先分解成 α-极限糊精,α-极限糊精再在糊精酸作用下进一步水解成麦芽糖与葡萄糖,供动物利用。由于糊精是嗜酸菌的良好培养基,因而在动物消化道内尚有促进 B 族维生素合成的功效。

Dextrin is a series of branched low-molecular compounds produced by starch digestion or heated hydrolysis. According to research, amylopectin can be first decomposed into α-limit dextrin during animal digestion. Under the action of acid, α-limit dextrin is further hydrolyzed into maltose and glucose for animal use. Because dextrin is a good medium for acidophilic bacteria, it still has the effect of promoting the synthesis of B vitamins in the digestive tract of animals.

(三) 糖原
Glycogen

糖原结构与支链淀粉相似,是糖在动物体内存在的另一种形式,除酵母中含量较高外(占干物质的3%~20%),一般饲料中含量极微。糖原由葡萄糖合成。化学上把由葡萄糖合成糖原的过程称为糖原生成作用。糖原的分解是糖原首先在磷酸化酶催化下生成葡萄糖-1-磷酸,又在磷酸葡萄糖变位酶催化下转变成葡萄糖-6-磷酸,最后再在葡萄糖-6-磷酸酶催化下在肝、胃中水解为葡萄糖和磷酸,以补充血糖。

The structure of glycogen is similar to amylopectin. It is another form of sugar that exists in animals. Except for the higher content in yeast (3%-20% of the dry matter), the content of glaycogen in general feed is very small. Glycogen is synthesized from glucose. Chemically, the process of synthesizing glycogen from glucose is called glycogen production. The decomposition of glycogen is that glycogen is firstly catalyzed by phosphorylase to produce glucose-1-phosphate, and then converted into glucose-6-phosphate under the catalysis of phosphoglucose mutase, and finally glycogen is hydrolyzed into glucose and phosphoric acid in liver and stomach under the catalysis of glucose-6-phosphatase to supplement blood glucose.

(四) 非淀粉多糖(NSP)
Non-starch polysaccharides (NSP)

NSP是植物的结构多糖的总称,是细胞壁的重要成分。它主要由纤维素、半纤维素、果胶、阿拉伯木聚糖、β-葡聚糖、甘露聚糖、葡甘聚糖等组成。纤维素属于不溶性NSP,其余的属于可溶性NSP,其抗营养作用日益受到关注。

1. 纤维素

纤维素广泛存在于植物界,是植物细胞壁的主要结构成分。秸秆饲料中含量高,其含量超过其他碳水化合物的总和。纤维素的化学结构是由多个β-1,4糖苷键连接的葡萄糖聚合体,呈扁带状微纤维。由于微纤维间与氢键牢固连接,因而使纤维素具有基本不溶性和极大的抗酶性。哺乳动物体内不含β-1,4糖苷键水解酶(纤维素酶),故无法直接利用,但消化道内共生的细菌、真菌分泌的纤维素酶可分解纤维素以供动物利用。

2. 半纤维素

半纤维素是由多个高聚糖组成

NSP is the general term for the structural polysaccharides of plants and an important component of plant cell walls. It is mainly composed of cellulose, hemicellulose, pectin, arabinoxylan, β-glucan, mannan, glucomannan, etc. Cellulose belongs to insoluble NSP, and the rest belong to soluble NSP. The anti-nutritional effect of soluble NSP has attracted increasing attention.

1. Cellulose

Cellulose is widely found in the plant kingdom and is the main structural component of plant cell walls. Straw feed is high in content, and its amount exceeds the total of other carbohydrates. The chemical structure of cellulose is a glucose polymer connected by multiple β-1, 4 glycosidic bonds, which is a ribbon-shaped microfiber. Because the microfibers are firmly connected with hydrogen bonds, the cellulose is basically insoluble and extremely resistant to enzymes. The mammalian body does not contain β-1, 4 glycosidic bond hydrolase (cellulose), so it cannot be used directly. However, the cellulose enzyme secreted by bacteria and fungi symbiotic in the digestive tract can break down cellulose for animal use.

2. Hemicellulose

Hemicellulose is a heterogeneous mixture composed of

的一种异源性混合物，包括戊聚糖、己聚糖等各自的聚合体，其中戊聚糖中的木聚糖是构成植物茎叶、秸秆的骨架；而果聚糖则为植物体的贮存物质（牧草中含量较多）。虽然半纤维素总是与纤维素共同存在于植物细胞壁中，但却属于两种完全不同的高聚糖。纤维素和半纤维素虽为草食动物重要的营养物质，但对单胃动物却被列为限制因子。为了提高纤维素、半纤维素的消化性，多年来国内外学者先后研究与推广粗饲料的碱化、氨化、微贮处理以及通过添加 β-葡聚糖酶、纤维素酶等来提高饲料的利用率。

3. 果胶

果胶属胶状多糖类，是细胞壁成分之一，广泛存在于各种高等植物细胞壁和相邻细胞之间的中胶层中，具有黏着细胞和运送水分的功能。果胶溶于水，而在酒精及某些盐溶液中凝结沉淀，通常利用这一性质提取果胶。果胶为白色或淡黄褐色的粉末，微有特异臭，味微甜带酸，无固定熔点和溶解度，相对密度约为 0.7，溶于 20 倍的水可形成乳白色黏稠状液体，呈弱酸性，不溶于乙醇及其他有机溶剂，能为乙醇、甘油和蔗糖糖浆润湿，与 3 倍以上的砂糖混合后，更易溶于水。植物的果胶物质一般有 3 种形态，即原果胶、果胶和果胶酸。原果胶多与纤维素半纤维素结合，存在于细胞壁中，不溶于水，但可水解成果胶。

multiple high saccharides, including pentosans, hexosans and other respective polymers. Among them, the xylan of pentosans is the skeleton of plant stems, leaves and straws. Fructan is the storage material of the plant body (the content is more in the forage). Although hemicellulose always co-exists in the plant cell wall with cellulose, they are two completely different kinds of high saccharides. Although cellulose and hemicellulose are important nutrients for herbivores, they are listed as limiting factors for monogastric animals. In order to improve the digestibility of cellulose and hemicellulose, scholars at home and abroad have successively studied and promoted the alkalization, nitridation, and micro-storage treatment of roughage for many years, as well as the addition of β-glucanase, cellulase, etc. to improve feed utilization rate.

3. Pectin

Pectin is a colloidal polysaccharide, which is one of the composition contents of cytoderm. It is widely present in the middle glue layer between the cell walls of adjacent cells in higher plants. It has the function of adhering cells and transporting water. Pectin is soluble in water, but pectin condenses and precipitates in ethanol and some salt solutions. Pectin is usually extracted by this property. Pectin is white or light yellow-brown powder. It has a peculiar odor, slightly sweet and acidic taste, no fixed melting point and solubility, and a relative density of about 0.7. It can form a milky white viscous liquid when dissolved in 20 times of water. It is weakly acidic and insoluble in ethanol and other organic solvents. Ethanol, glycerin and this sugar syrup are moistened, and when mixed with more than 3 times the amount of sugar, it is more soluble in water. Pectin substances in plants generally have three forms, namely, protopectin, pectin and pectic acid. Propectin is mostly combined with cellulose and hemicellulose, existing in the cell wall, insoluble in water, but can be hydrolyzed into pectin.

（五）木质素
Lignin

木质素是一种高分子苯基丙烷的非晶体聚合物，与碳水化合物相比较，碳多（氢氧比并非 2∶1），且常含氮，因此严格来说它不属碳水化

Lignin is an amorphous polymer of high-molecular-weight phenylpropane. Compared with carbohydrates, it has more carbon (hydrogen-oxygen ratio is not 2∶1) and often contains nitrogen. Therefore, it is not strictly a carbohy-

合物。木质素的准确化学结构至今尚不太清楚，但经用同位素标记物研究，木质素合成是以芳香族氨基酸（苯丙氨酸、酪氨酸）开始的，即先由芳香族氨基酸形成基质物质（如松柏醇），再经细胞壁过氧化物酶氧化，最终聚合而成，并认为木质素与半纤维素之间以共价键连接。但也有报道认为，植物不同，木质素与纤维素或半纤维素连接的键合不同。

drate. The exact chemical structure of lignin is still unclear. However, after using isotope markers, the synthesis of lignin starts with aromatic amino acids (phenylalanine, tyrosine). That is, the matrix material (such as coniferyl alcohol) is first formed by aromatic amino acids, and then oxidized by cell wall peroxidase, and finally polymerized. It is believed that the lignin and hemicellulose are connected by a covalent bond. But there are also reports that different plants have different bonds between lignin and cellulose or hemicellulose.

（六）结合糖
Bound saccharides

结合糖是指糖与非糖物质的结合物，常见的有与蛋白质结合的糖蛋白。它分布广泛，具有多种生物学功能。氨基多糖又称黏多糖或糖胺聚糖，是一类含氨基糖或氨基糖衍生物的杂多糖，是由多个二糖单位形成的长链多聚体，其主链由己糖胺和糖醛酸组成，有的含有硫酸根。常见的氨基多糖有硫酸软骨素、透明质酸、硫酸皮肤素、硫酸角质素等。蛋白多糖又称黏蛋白或蛋白聚糖，这类物质由蛋白质和氨基多糖通过共价键相连接而构成。蛋白多糖是动物组织细胞间隙中的重要成分。

Bound saccharides refer to the combination of sugar and non-sugar substances. The common ones are combined with proteins, collectively called glycoproteins. It is widely distributed and has multiple biological functions. Amino polysaccharides are also called mucopolyphages or glycosaminoglycans. It is a class of heteropolysaccharides containing amino sugars or amino sugar derivatives. It is a long-chain polymer formed by multiple disaccharide units. Its main chain is composed of hexosamine and uronic acid, and some contain sulfate. Common amino polysaccharides include chondroitin sulfate, hyaluronic acid, dermatan sulfate, keratan sulfate and so on. Proteoglycan, also known as mucin or proteoglycan, is composed of protein and amino polysaccharide connected by covalent bonds. Proteoglycan is an important component in the intercellular space of animal tissues.

第二节　含氮化合物
Section 2　Nitrogen compounds

饲料中所有含氮物质统称为粗蛋白（CP），它又包括真（纯）蛋白质与非蛋白含氮物（NPN）。氨基酸是组成真蛋白质的基本单位，主要由C、H、O、N四种元素组成（约占98%），同时还有少量的S、P、Fe

All nitrogen-containing substances in feed are collectively referred to as crude protein (CP), which also includes true (pure) protein and non-protein nitrogen (NPN). Hydroxy acid is the basic unit of true protein, mainly composed of C, H, O, N elements (about 98%), and a small amount of S, P, Fe and other elements.

等元素。非蛋白含氮物又包括游离氨基酸、铵盐、肽类、酰胺、硝酸盐等。

多数饲料中蛋白质含氮量接近于16%（变幅在14.90%-18.87%），因此饲料中的粗蛋白质含量被定义为：

饲料 CP 含量 = 6.25 × 饲料含氮量

但实际上不同种类饲料间蛋白质的含氮量存在差异。另外，不同饲料的 NPN 种类与比例也存在差异，例如几种饲料 NPN 占总氮的百分比：青饲料40%、甜菜50%、青贮饲料30%-60%、马铃薯30%-40%、麦芽30%、成熟籽实3%-10%。

Non-protein nitrogenous substances include free amino acids, ammonium salts, peptides, amides, nitrates, etc.

The protein nitrogen content in most feeds is close to 16% (with a range of 14.90%-18.87%), so the crude protein content in feeds is defined as:

Feed CP content = 6.25 × feed nitrogen content

However, in fact, there are differences in the nitrogen content of protein among different types of feeds. In addition, the types and proportions of NPN of different feeds are also different. For example, the percentages of NPN in total nitrogen of several feeds are: 40% for green fodder, 50% for sugar beet, 30%-60% for silage, 30%-40% for potato, 30% for malt, 3%-10% for mature seeds.

一、蛋白质的性质与分类
The property and classification of protein

（一）蛋白质的性质
The property of protein

蛋白质同糖类一样，具有水解特性，它可在酶、酸、碱等条件下发生水解胨、多肽、氨基酸等。有些蛋白质对动物消化酶有很强的抗性，如硬蛋白，但在高温高压或酸性溶液的条件下可发生水解，如利用这一原理生产水解羽毛粉。

所有蛋白质均具有胶体性质，但它们在水中的溶解度不同，其溶解度在由角蛋白的不溶解到白蛋白的高度溶解范围内变动，溶解的蛋白质可以通过添加某些盐类如 NaCl 或 $(NH_4)_2SO_4$，使它们从溶液中沉淀出来。这是一种物理过程，蛋白质的性质并未改变，将已沉淀的溶液稀释，能使蛋白质再次溶解。

尽管蛋白质肽键内存在的是无

Proteins, like sugars, have hydrolytic properties, which can hydrolyze into peptones, peptides, amino acids, etc. under the conditions of enzymes, acids, alkalis, etc. Some proteins have strong resistance to animal digestive enzymes, such as hard proteins, but they can be hydrolyzed under conditions of high temperature, high pressure or acidic solutions. For example, hydrolyzed feather powder can be produced using this principle.

All proteins have colloidal properties, but their solubility in water is different. Their solubility varies from the insolubility of keratin to the high solubility of albumin. The dissolved proteins can be precipitated from the solution by adding certain salts such as NaCl or $(NH_4)_2SO_4$. This is a physical process. The properties of the protein have not changed. Diluting the precipitated solution can make the protein re-dissolve.

Although there are amino and carboxyl groups that do

酸碱反应的氨基和羧基,但在它们肽键的两个末端或侧链氨基酸残基上,含有许多自由氮基或自由糖基,是两性电解质。各种蛋白质显示其各种特定的等电点,蛋白质在等电点易生成沉淀,这个特性常被用作蛋白质的分离提纯,并且由于其大分子和离解度低,在维持体内蛋白质溶液形成的渗透压中起重要作用,这种缓冲和渗透作用对于血液蛋白质维持体内环境的稳定和平衡具有非常重要的意义。

not react with acid-base in protein peptide bonds, there are many free nitrogen groups or free carboxyl groups on the two ends or side chain amino acid residues of their peptide bonds. Therefore, like amino acids, proteins are also ampholytes. Various proteins show various specific isoelectric points. Proteins are easy to form precipitates at the isoelectric point. This feature is often used for the separation and purification of proteins. Because of their large molecules and low dissociation degree, they are important in maintaining the penetration pressure caused by protein solutions in the body. This buffering and osmotic effect is of great significance for blood proteins to maintain the stability and balance of the body environment.

在某些因素的作用下,蛋白质会变性。所谓蛋白质变性是指任何非水解蛋白质作用造成天然蛋白质独特结构的改变所引起的蛋白质化学性质、物理性质和生物学活性上的固定变化。如:加热作用使许多蛋白质发生凝固。除了加热以外,还有许多能引起蛋白质变性的试剂,包括强酸、碱、乙醇、丙酮、脲以及重金属盐类。

Under the action of certain factors, proteins will be denatured. The so-called protein denaturation refers to the fixed changes in the chemical properties, physical properties and biological activities of proteins caused by the changes in the unique structure of natural proteins caused by the action of non-hydrolyzed proteins. For example, heating causes many proteins to coagulate. In addition to heating, there are many agents that can cause protein denaturation, including strong acids, alkalis, ethanol, acetone, urea, and heavy metal salts.

(二) 蛋白质的分类
The classification of protein

天然饲料中蛋白质种类多,结构复杂,分类方法较多。常见的分类方法有如下几种:

按生理功能可分为:结构蛋白(如胶原纤维、肌原纤维等)、贮藏蛋白(如清蛋白、谷蛋白、酪蛋白)和生物活性蛋白(如酶、激素等)。

按蛋白分子形状可分为:球蛋白和纤维蛋白。

按加工性状表现可分为:面筋性蛋白(醇溶蛋白、谷蛋白)和非面筋性蛋白(清蛋白、球蛋白)。

按化学组成,一般多将蛋白质分为:单纯蛋白质、复合蛋白质和衍生蛋白质。

In natural feed, there are many types of protein with complex structure and many classification methods. The common classification methods are as follows:

According to physiological functions, they can be divided into structural proteins (such as collagen fibers, myofibrils, etc.), storage proteins (such as albumin, gluten, casein) and bioactive proteins (such as enzymes, hormones, etc.).

According to the shape of protein molecules, they can be divided into globulin and fibrin.

According to processing properties, they can be divided into gluten proteins (gliadin, gluten) and non-gluten proteins (albumin, globulin).

According to their chemical composition, proteins are generally divided into simple proteins, compound proteins and derived proteins.

1. 单纯蛋白质

单纯蛋白质指经彻底水解只产生氨基酸的蛋白质。若进一步按其溶解性可分为7种。

（1）清蛋白 又称白蛋白。溶于水、盐酸、碱及稀盐溶液中。加硫酸铵至饱和时，会从溶液中沉淀析出，遇热凝固。清蛋白广泛存在于动植物组织中，如血清蛋白、乳酪蛋白、豆清蛋白（豌豆中）和麦清蛋白（小麦中）等。

（2）球蛋白 不溶或微溶于水，但加少量盐、酸或碱时可缓慢溶解。加硫酸铵至饱和时，可从溶液中沉淀析出。动物球蛋白遇热凝固，植物球蛋白遇热不凝固。球蛋白广泛存在于动植物组织中，如血清球蛋白、肌球蛋白、棉籽球蛋白、大豆及豌豆球蛋白等。化学组成中以缬氨酸、亮氨酸、精氨酸含量较高，丙氨酸、色氨酸含量一般较低。

（3）谷蛋白 不溶于水、盐溶液和乙醇，但溶于稀酸、稀碱溶液，遇热凝固。仅存在于植物组织中，如小麦中的麦谷蛋白、大米中的米谷蛋白等，含谷氨酸较多。

（4）醇溶蛋白 不溶于水及盐溶液，可溶于10%酒精中，加热不凝固。仅存在于植物体中，如玉米醇溶蛋白、小麦醇溶蛋白等，化学组成上含脯氨酸较多，赖氨酸、色氨酸一般较少。

（5）精蛋白 可溶于水和稀酸，能被稀氨水沉淀，遇热不凝固，属动物性蛋白。鱼精子、卵子和胸腺等组织中含量较高（如鲑精蛋白等），化学组成中含赖氨酸、精氨酸较高，不含胱氨酸、谷氨酸。

（6）组蛋白 可溶于水和稀酸，能被氨水沉淀。组织中常与酸性物质结合成盐而存在，加热不凝

1. Simple protein refers to a protein that produces only amino acids after thorough hydrolysis. According to its solubility, it can be divided into 7 kinds.

(1) Albumin is also called ricim. It is dissolved in water, hydrochloric acid, alkali and dilute salt solution. When ammonium sulfate is added to saturation, it will precipitate out of the solution and solidify when heated. Albumin is widely found in animal and plant tissues, such as serum protein, milk protein, soy albumin (in peas) and wheat albumin (in wheat).

(2) Globulin is insoluble or slightly soluble in water, but can be slowly dissolved when a small amount of salt, acid or alkali is added. When ammonium sulfate is added to saturation, it can precipitate out from the solution. Animal globulin solidifies when heated, while plant globulin does not solidify when heated. Globulin is widely found in animal and plant tissues, such as serum globulin, myosin, cottonseed globulin, soybean and pea globulin. In chemical composition, the content of valine, leucine and arginine is relatively high, while the content of alanine and tryptophan is generally low.

(3) Glutelin is insoluble in water, salt solution and ethanol, but soluble in dilute acid, dilute alkali solution, and it coagulates when heated. It only exists in plant tissues, such as glutenin in wheat and oryzenin in rice, which contain more glutamic acid.

(4) Gliadin is insoluble in water and salt solutions, but soluble in alcohol of 10% concentration. It does not solidify when heated. It only exists in plants, such as zein and wheat gliadin, etc. The composition contains more proline, and generally less lysine and tryptophan.

(5) Spermatin is soluble in water and dilute acid, and can be precipitated by dilute ammonia, and does not solidify when exposed to heat. It is an animal protein. Fish sperm, eggs, and thymus have higher content (such as salmin, etc.). The chemical composition of it contains higher lysine and arginine, but does not contain cystine and glutamate.

(6) Histone is soluble in water and dilute acid, and can be precipitated by ammonia. The tissues are often combined with acidic substances to form a salt, which does not

固,属动物性蛋白。化学组成上组氨酸、赖氨酸、精氨酸含量较高,酪氨酸含量低。

(7) 角硬蛋白 水及盐溶液中不溶,遇热不凝固,主要存在于动物的表皮、毛、腱、角、软骨等组织中,化学组成上含胱氨酸、甘氨酸较高。该蛋白难消化,但经高温高压水解或酸碱处理(破坏二硫基)可供作饲料。

2. 复合蛋白质

复合蛋白质是由单纯蛋白和非蛋白辅基结合而成。水解时不仅产生氨基酸,而且还会产生其他物质。按辅基不同,复合蛋白一般又可分为:

(1) 脂蛋白 由蛋白质与脂肪或类脂质(磷脂、类固醇等)结合而成,血、乳、蛋黄、神经及细胞膜中含量较高。

(2) 核蛋白 由单纯蛋白与核酸结合而成,为细胞核的组分,对动物的生长、繁殖有特殊功能。

(3) 糖蛋白 由蛋白质和碳水化合物组成,如动物体中的黏蛋白等。

(4) 色蛋白 由蛋白质和色素物质组成(以色素作为辅基)。如:叶绿蛋白以镁卟啉为辅基,血红蛋白和细胞色素C以铁卟啉为辅基。

(5) 磷蛋白 由蛋白质和磷酸组成,如酪蛋白、卵黄磷蛋白。

(6) 金属蛋白 由蛋白质和金属元素组成。如除蛋白质外,铁蛋白中含$Fe(OH)_3$、细胞色素氧化酶含Fe和Cu、乙醇脱氢酶含Zn、黄嘌呤氧化酶含Mo和Fe等。

solidify when heated. It is an animal protein. In terms of chemical composition, the content of histidine, lysine and arginine is relatively high, and the content of tyrosine is low.

(7) Keratin scleroprotein is insoluble in water and salt solutions, and does not solidify when exposed to heat. It is mainly found in animal skin, hair, tendon, horns, cartilage and other tissues. The chemical composition contains higher cystine and glycine. It is difficult to digest, but it can be used as feed after high temperature and high pressure hydrolysis or acid-base treatment (destruction disulfide group).

2. Compound protein

Compound protein (onjugated protein) is formed by combining simple protein and non-protein prosthetic group. Not only amino acids are produced during hydrolysis, but other substances are also produced. According to different prosthetic groups, complex proteins can generally be divided into:

(1) Lipoprotein is a combination of protein and fat or lipids (phospholipids, steroids, etc.). The content is higher in blood, milk, egg yolk, nerve and cell membrane.

(2) Nucleoprotein is a combination of simple protein and nucleic acid. It is a component of cell nucleus and has special functions for animal growth and reproduction.

(3) Glycoprotein is composed of protein and carbohydrate, such as mucin in animal body, etc.

(4) Chromoprotein is composed of protein and pigment material (with pigment as a prosthetic group). For example, chlorophyll protein uses magnesium porphyrin as a prosthetic group, and hemoglobin and cytochrome C use iron porphyrin as a prosthetic group.

(5) Phosphoprotein is composed of protein and phosphoric acid, such as casein and yolk phosphoprotein.

(6) Metalloprotein is composed of protein and metal elements. For example, in addition to protein, ferritin contains $Fe(OH)_3$, cytochrome oxidase contains Fe and Cu, alcohol dehydrogenase contains Zn, and xanthine oxidase contains Mo and Fe, etc.

3. 衍生蛋白质

衍生蛋白质既包括蛋白质分子内部结构变化的变性蛋白质，又包括天然蛋白质经酸、碱、酶等处理后所生成的蛋白胨、蛋白脒、肽、明胶等。

3. Derived protein

Derived protein includes not only the denatured protein with changes in the internalstructure of the protein molecule, but also the peptone, peptone, peptide, gelatin, etc., that are generated by the treatment of natural protein with acid, alkali, enzyme, etc.

二、氨基酸
Amino acids

（一）组成蛋白质的氨基酸及其结构
The amino acids that make up proteins and their structures

天然存在的氨基酸有 200 余种，但常见的组成蛋白质的氨基酸仅 20 种（编码氨基酸）。由于氨基酸在种类、数量和排列顺序方面的不同，又组成了自然界多种多样的蛋白质。天然存在的氨基酸多为易于被动物吸收利用的 L 型氨基酸。

20 种氨基酸的不同点在于其侧链的 R 基团。根据氨基酸的 R 辅基化学特性，饲料中的氨基酸通常可分为四类。

There are more than 200 kinds of natural amino acids, but only 20 kinds of amino acids (coding amino acids) are commonly found in protein. Due to the different types, numbers and order of amino acids, a variety of proteins are formed in nature. Most naturally occurring amino acids are L-type amino acids that are easily absorbed and utilized by animals.

The difference between the 20 amino acids is the R group in the side chain. According to the chemical characteristics of the R prosthetic group of the amino group, the amino acids in the feed can usually be divided into 4 categories.

（二）氨基酸的理化性质
Physicochemical properties of amino acids

不同氨基酸在水中的溶解度不同，赖氨酸、精氨酸易溶于水，胱氨酸、酪氨酸等则难溶于水，但所有的氨基酸都不同程度地可溶于盐酸溶液。

氨基酸是两性化合物，既具有碱性基因，又有酸性基因；既可以不带电荷的分子状态存在，又可以带相反电荷的偶极离子（或称两性离子）状态存在，还可以前两者的混合物状态存在，氨基酸在水溶液中以偶极离子状态存在。当所存在于一定 pH 值的溶液中静电荷为 0 时

Amino acids have different degrees of solubility in water. Lysine and arginine are easily soluble in water, while cystine and tyrosine are difficult to dissolve in water. All amino acids are soluble in hydrochloric acid solution to varying degrees.

Amino acids are amphoteric compounds with both alkaline and acidic genes. They can exist as uncharged molecules, or as oppositely charged dipolar (or zwitterion) ions, or as a mixture of the previous two. Amino acids exist in the state of dipolar ions in aqueous solutions. Any nitrogenous acid that is present in a solution at a pH such that its static charge is zero is called the isoelectric point of the amino acid.

的 pH 为该氨基酸的等电点。

所有的天然氨基酸中（甘氨酸除外）都存在有不对称的α-碳原子，具有旋光性，即有 L-型和 D-型两种立体构型。目前发现存在于天然蛋白质中的氨基酸（除甘氨酸外）都是 L-型的，但在某些生物体内，特别是细菌中，D-型氨基也是广泛存在的。旋光性物质在化学反应中，发生消旋现象并将其转变为 D-型和 L-型的等当量混合物，称消旋物，用一般的有机合成方法人工合成氨基酸时，得到的都是无旋光性的 DL-消旋氨基酸。

组成蛋白质的 20 种氨基酸，一般对可见光都没有光吸收，但有几种氨基酸对紫外光有明显的吸收能力，如酪氨酸、色氨酸和苯丙氨酸，利用这一特性可测定这些氨基酸的含量。此外，氨基酸还有氧化脱氨基反应、还原脱氨基反应、脱羧基反应、非酶棕色化反应等特性。

All natural amino acids (except glycine) have asymmetric α-carbon atoms, so they all have optical rotation. There are two three-dimensional configurations, L-type and D-type. At present, the amino acids found in natural proteins (except glycine) are all L-types, but in some organisms, especially bacteria, D-type amino acids are also widely present. In the chemical reaction, the optically active substance will disappear and transform into an equivalent mixture of D-type and L-type, which is called racemate. When the synthetic amino acid is artificially synthesized by general organic synthesis methods, DL-recemic amino acids with no dazzling properties are obtained.

The 20 kinds of amino acids that make up proteins generally have no light absorption for visible light, but there are several amino acids that have obvious absorption capacity for ultraviolet light, such as tyrosine, tryptophan and phenylalanine. This feature can be used to determine the content of amino acids. In addition, amino acids have oxidative deamination reaction, reductive deamination reaction, decarboxylation reaction, amino-carboxyl reaction (Maillard reaction), etc.

三、寡肽
Oligopeptides

寡肽也称小肽，主要指由 2 个或 3 个氨基酸残基构成的二肽或三肽，由于寡肽在动物营养与生理机能上的特殊作用，愈来愈受到国内外学者们的关注。

肽有开链和环链之分，开链肽有两个末端，环链肽无末端。一般蛋白质的肽链为开链，某些抗生素中的肽（如短杆菌肽 S、酪酸杆菌肽、某些细菌的多黏菌肽和缬氨霉素等）则为环链。

除了蛋白质局部水解可以产生各种简单的肽外，自然界和生物体内还存在许多活性的寡肽。例如：动植物细胞内广泛存在的谷胱甘肽

Oligopeptides are also called small peptides, which mainly refer to dipeptides or tripeptides composed of 2 or 3 amino acid residues. Oligopeptides have attracted more and more attention from scholars at home and abroad due to their special roles in animal nutrition and physiological functions.

Peptides include open chain and cyclic chain. Open chain peptides have two ends. The general protein peptides are open-chain peptides. The peptides in some antibiotics (such as gramicidin S, butyricin, polymyxin and valinomycin of some bacteria, etc.) are cyclic chains.

In addition to partial protein hydrolysis that can produce various simple peptides, there are many active oligopeptides in nature and organisms. For example, glutathione, which is widely present in animal and plant cells, is a tripeptide;

是一种三肽；动物肌肉中存在的肌肽是一种二肽等。

寡肽中的二肽或三肽在动物肠道被直接吸收,显示出它对饲料蛋白质利用的优势,使得小肽饲料添加剂生产工业及其消化代谢机理的研究成为当今动物营养研究中的又一热点。

寡肽的理化特性与氨基酸类似,故不赘述。

the carnosine found in animal muscles is a kind of dipeptide.

The dipeptide or tripeptide in the oligopeptide is directly absorbed in the animal intestine, showing its advantage in the utilization of feed protein, making the research of the small peptide feed additive production industry and its digestion and metabolism mechanism become another aspect of animal nutrition research hot spot.

The physicochemical properties of oligopeptides are similar to those of amino acids, so they are not described.

四、其他含氮化合物
Other nitrogen compounds

(一) 酰胺类
Amides

天冬酰胺和谷酰胺分别是天冬氨酸和谷氨酸的重要酰胺衍生物,这两种酰胺已被列入氨基酸类。尿素是一种最简单的酰胺,它是哺乳动物体内氮代谢的主要尾产物。另外,禽类代谢的主要尾产物是尿酸。

Asparagine and glutamine are important amide derivatives of aspartic acid and glutamic acid respectively. These two amides have been listed in the amino acid category. Urea, one of the simplest amides, is the main tail product of nitrogen metabolism in mammals. In addition, the main tail product of poultry metabolism is uric acid.

(二) 硝酸盐
Nitrates

硝酸盐存在于植物饲料中,尤其是幼嫩的青饲料,它本身并无毒性,但在适宜的条件下,如青饲料堆放或焖煮不当,在微生物的作用下可使其转化为亚硝酸盐,动物食用后会引起中毒。对于这一点,在生产中须特别注意。

Nitrate exists in plant feed, especially young green feed. It is not toxic. However, under suitable conditions, such as green fodder stacking or improper simmering, it can be converted into nitrite under the action of microorganisms which can cause poisoning after being eaten by animals. For this, special attention must be paid in production.

(三) 核酸
Nucleic acids

核酸是一种高分子化合物,水解后生成一种有碱性含氮化合物(嘌呤、嘧啶)、戊糖(核糖或脱氧核糖)和磷酸组成的混合物。核酸在动物体内的作用是贮存遗传信息,

Nucleic acid is a polymer compound that is hydrolyzed to produce a mixture of basic nitrogen compounds (purine, pyrimidine), pentose (ribose or deoxyribose) and phosphoric acid. The function of nucleic acid in the animal body is to store genetic information and use this information in the

并通过它们将这些信息用于蛋白质的合成过程中,对动物的生命和生产起着十分重要的作用。

process of protein synthesis through them, which plays a very important role in animal life and production.

第三节 脂类
Section 3 Lipids

脂类也称脂质,指饲料干物质中的乙醚浸出物,包括脂肪(真脂肪)和类脂质。脂肪是甘油和脂肪酸组成的三酰甘油,亦称甘油三酯或中性脂肪;类脂质包括有游离脂肪酸、磷脂、糖脂、脂蛋白、固醇类、类胡萝卜素和脂溶性维生素等。

Lipids, also known as lipids, are ether extracts of dry matter in feed, including fats (true fats) and lipids. Fats, also known as triacylglycerides or neutral fats, are triacylglycerols composed of glycerol and fatty acids. Lipids include free fatty acids, phospholipids, glycolipids, lipoproteins, sterols, carotenoids, and fat-soluble vitamins.

一、脂类的分类
Classification of lipids

脂类的分类方法较多,如按可否皂化分类,按是否含有甘油分类等。下面按其化学结构特点将脂类分为以下四类:

1. 单纯脂类

包括甘油三酯和蜡质,蜡由脂肪酸和一个长链一元醇组成。

2. 复合脂

复合脂分子中除了脂肪酸和甘油外,尚含有其他化学基团。如磷脂(为含磷、氮的脂类,如磷脂酰胆碱、磷脂酸乙酰胺、磷脂酰丝氨酸等),糖脂(如半乳糖甘油酯)和脂蛋白。

3. 萜类、类固醇及其衍生物

此类一般不含脂肪酸。

4. 衍生脂

衍生脂系上述脂类的水解产物,如甘油、脂肪酸及其氧化产物、前列腺素等。

Lipids are classified in many ways, e. g., according to the degree of saponification, or the content of glycerin, etc. The lipids are divided into the following 4 categories according to their chemical structure characteristics.

1. Simple lipids

Simple lipids include triglycerides and waxes. Waxes are composed of fatty acids and a long-chain monohydric alcohol.

2. Compound lipids

In addition to fatty acids and glycerol in its molecule, there are other chemical groups, such as phospholipids (lipids containing phosphorus and nitrogen, such as phosphatidylcholine, phosphatidyl acetamide, phosphatidyl serine, etc.), glycolipids (such as galactose glycerides) and lipoproteins.

3. Terpenoids, steroids and their derivatives

Terpenoids, steroids and their derivatives generally do not contain fatty acids.

4. Derived lipids

Derived lipids are the hydrolysis products of the above lipids, such as glycerol, fatty acids and their oxidation products, prostaglandins, etc.

二、脂肪
Fat

脂肪的种类不同,油脂性状不同。如:玉米油含90%不饱和脂肪酸,室温下呈液态;牛油含饱和脂肪酸高,室温下呈固态;奶油含较多的低级挥发性脂肪酸,故熔点低于牛油。大量的研究显示饲料油脂性质和营养价值主要取决于构成它的脂肪酸,脂肪类型不同,脂肪酸组成也不同。

The type of fat is different, and the properties of oil are different. For example, corn oil contains 90% of unsaturated fatty acids and is liquid at room temperature. Cream contains high saturated fatty acids and is solid at room temperature. Cream contains more low-grade volatile fatty acids, so the melting point is lower than that of butter. A large number of studies have shown that the properties and nutritive value of feed oil are mainly determined by the fatty acids that constitute it. Different types of fats have different compositions of fatty acids.

(一) 脂肪酸的分类与组成
Classification and composition of fatty acids

当前从油脂分离出的近百种脂肪酸,除微生物界可见有碳原子为奇数呈分枝结构的脂肪酸外,其他多呈偶数碳原子结构的直链脂肪酸。直链脂肪酸又可根据其是否溶解、挥发和饱和程度分成如下3种类型。

1. 水溶性挥发性脂肪酸

即分子中碳原子数<10的脂肪酸,常温下呈液态,如丁酸、己酸、辛酸和癸酸等,常见于奶油、椰子油中。

2. 非水溶性挥发性脂肪酸

如十二碳的月桂酸。

3. 非水溶性不挥发性脂肪酸

该类型脂肪酸又可根据其是否饱和分为饱和脂肪酸和不饱和脂肪酸。

碳原子数量少的脂肪酸具挥发性,故称挥发性脂肪酸(VFA),碳原子数愈多,熔点愈高;饱和度愈低,熔点也愈低。一般来说:陆生动植物脂肪多为 C_{16} 和 C_{18} 的脂肪酸(以 C_{18} 脂肪酸居多),主要有软脂

According to research, there are currently nearly one hundred fatty acids isolated from oils and fats. Except for fatty acids with an odd number of carbon atoms in a branched structure, other straight-chain fatty acids with an even number of carbon atoms can be seen in the microbial world. Straight-chain fatty acids can be divided into the following 3 types according to whether they are dissolved, volatilized and saturated.

1. Water-soluble volatile fatty acids are fatty acids with less than 10 carbon atoms in the molecule. They are liquid at room temperature, such as butyric acids, caproic acids, caprylic acids and capric acids, which are commonly found in butter and coconut oil.

2. Non-water-soluble volatile fatty acids such as twelve-carbon lauric acid.

3. Non-water-soluble non-volatile fatty acids. This type of fatty acid can be divided into saturated fatty acids and unsaturated fatty acids according to whether they are saturated or not.

Fatty acids with fewer carbon atoms are volatile, so they are called volatile fatty acids (VFA). The more carbon atoms they have, the higher the melting point is; the lower saturation they have, the lower the melting point is. Generally speaking, the fats of terrestrial animals and plants are mostly C_{16} and C_{18} fatty acids (mostly C_{18} fatty acids),

酸、油酸和硬脂酸；海洋水产动物多为C_{20}和C_{22}的不饱和脂肪酸；淡水鱼以C_{18}不饱和脂肪酸比例较高；果仁脂肪中主要是软脂酸、油酸、亚油酸；种子中主要是C_{16}的软脂酸和C_{18}的油酸、亚油酸和亚麻酸，以C_{18}不饱和脂肪酸较多；哺乳动物乳中，除软脂酸、油酸外，尚含5%－30%的C_4—C_{10}低级脂肪酸。

在不饱和脂肪酸中，有几种脂肪酸在动物体无法合成或合成量较小，满足不了动物需要，必须由饲料供给，这些不饱和脂肪酸称为必需脂肪酸（EFA）：亚油酸（LA）、α-亚麻酸（ALA）和花生四烯酸（ARA）。

mainly palmitic acids, oleic acids and stearic acids; C_{20} and C_{22} are mainly unsaturated fatty acids in marine aquatic animals; freshwater fish has a higher proportion of C_{18} unsaturated fatty acids; the kernel fats are mainly palmitic acids, oleic acids, and linoleic acids; the seeds mainly have C_{16} palmitic acids and C_{18} oleic acids, linoleic acids and linoleic acids, with more C_{18} unsaturated fatty acids; in addition to palmitic acid and oleic acid, mammalian milk contains 5%－30% $C_4 - C_{10}$ lower fatty acids.

Among the unsaturated fatty acids, there are still several kinds of fatty acids that cannot be synthesized in the animal body or are synthesized in small amounts, which cannot meet the needs of the animal. They must be supplied from the feed. These unsaturated fatty acids are called essential fatty acids (EFA): linoleic acid (LA), α-linolenic acid (ALA) and arachidonic acid (ARA).

（二）脂肪的主要性质
The main properties of fat

1. 脂肪的水解特性

脂类可在稀酸或强碱溶液中水解成脂肪的基本结构单位——甘油和脂肪酸，脂类分解成基本结构单位的过程，也可以在微生物产生的脂肪酶催化下水解，这类水解对脂类营养价值没有影响，但水解产生的某些脂肪酸有特殊的异味或酸败味，可能影响动物适口性，脂肪酸碳链越短，这种异味越浓。脂肪在强碱溶液中水解生成的高级脂肪酸盐习惯上称为"肥皂"，因此，把脂肪在碱性溶液发生的水解反应称为"皂化反应"。皂化1 g脂肪所需KOH的毫克数称为该脂肪的皂化价，某油脂如皂化价高，说明组成该油脂的脂肪酸碳链较短；反之油脂如皂化价低，表明该油脂的脂肪酸碳链较长。

2. 不饱和脂肪酸的加成反应

不饱和键中的π键断裂，与试剂的两个原子或基团结合，这样的

1. The hydrolysis characteristics of fat

Lipids can be hydrolyzed into the basic structural units of fat—glycerol and fatty acids in dilute acid or strong alkali solution. The process of decomposing lipids into basic structural units can also be catalyzed by lipase produced by microorganisms hydrolysis. This type of hydrolysis has no effect on the nutritional value of lipids, but some fatty acids produced by hydrolysis have a special peculiar or rancid taste, which may affect the palatability of animals. The shorter the fatty acid carbon's chain is, the stronger the peculiar smell will be. The higher fatty acid salt produced by the hydrolysis of fat in a strong alkaline solution is customarily called "soap". Therefore, the hydrolysis reaction of fat in an alkaline solution is called "saponification". The mg of KOH required to saponify 1 g of fat is called the saponification value of the fat. If the saponification value of a certain oil is high, it means that the fatty acid carbon chain that composes the oil is short; on the contrary, if the saponification value is low, it indicates the fatty acid carbon chain of the oil is longer.

2. The addition reaction of unsaturated fatty acids

The π bond in the unsaturated bond breaks and combines with two atoms or groups of the reagent. This reaction

反应叫作加成反应。在镍、铂等催化剂或酶的作用下,可以将不饱和脂肪酸加入到氢中,即脂肪酸分子中碳原子双键的过程可以得到氢,变成饱和脂肪酸,称为氢化作用。氢化能增加脂肪的熔点和硬度,因此氢化又称"硬化"。食品工业上也广泛利用植物油经氢化成固态生产"人造黄油"。不饱和脂肪酸在一定条件下可与碘发生加成反应,能化合碘的数量可反映脂肪中不饱和键的多少。因此,通常用100 g脂肪或脂肪酸所能化合碘的克数——即碘价来表示脂肪或脂肪酸的不饱和程度。

青草中不饱和脂肪酸含量占脂肪酸总量的80%以上,饱和脂肪酸仅占20%,而反刍动物体脂肪,含30%-40%不饱和脂肪酸和60%-70%饱和脂肪酸并且十分稳定,其原因是氢化作用。牧草中的真脂和类脂在瘤胃中受微生物的作用发生水解,产生甘油和各种脂肪酸,其中包括饱和与不饱和脂肪酸,不饱和脂肪酸在瘤胃中经氢化作用变成饱和脂肪酸,故参与牛体脂肪代谢的脂肪酸多为饱和脂肪酸。所以饱和脂肪酸较多,且较稳定。

3. 脂肪氧化酸败

脂肪的酸败作用有两种类型,即水解型和氧化型。

(1)水解型酸败通常是微生物(如霉菌繁殖产生解脂酶)作用于脂肪,引起简单的水解反应,使之水解为脂肪酸、甘油二酯、甘油一酯和甘油的结果,这类水解对脂类营养价值并无妨碍,但水解产生的某些脂肪酸有特殊的异味或酸败味,可能影响动物适口性。脂肪酸碳链越短,这种异味越浓。

is called an addition reaction. Under the action of catalysts or enzymes such as nickel and platinum, unsaturated fatty acids can be added to hydrogen, that is, the process of double bonds of carbon atoms in fatty acid molecules can obtain hydrogen and become saturated fatty acids, which is called hydrogenation. Hydrogenation can increase the melting point and hardness of fat, thus it is also called "hardening". In the food industry, vegetable oils are also widely used to produce "margarine" through hydrogenation into solid state. Unsaturated fatty acids can undergo an addition reaction with iodine under certain conditions. The amount of iodine that can be combined can reflect the number of unsaturated bonds in the fat. Therefore, the number of grams of iodine that 100 g of fat or fatty acid can combine is called iodine value. which is usually used to express the degree of unsaturation of fats or fatty acids.

The content of unsaturated fatty acids in grass accounts for more than 80% of the total fatty acids, while the content of saturated fatty acids accounts for 20%. However, the ruminant body fat contains 30%-40% of unsaturated fatty acids and 60%-70% of saturated fatty acids and are very stable. The reason is hydrogenation. Because the true fats and lipids in the pasture are hydrolyzed by microorganisms in the rumen, glycerol and various fatty acids are produced, including saturated and non-saturated fatty acids. Saturated fatty acids and unsaturated fatty acids are hydrogenated in the rumen to become saturated fatty acids. Therefore, most of the fatty acids involved in fat metabolism in cattle are saturated fatty acids. So bovine body has more saturated fatty acids and is more stable.

3. Fat oxidation rancidity

There are two types of rancidity in fats: hydrolyzed and oxidized.

(1) Hydrolyzed rancidity is usually the result of microorganisms (such as lipase produced by mold reproduction) acting on fat, causing a simple hydrolysis reaction to hydrolyze it into fatty acids, diglycerides, monoglycerides and glycerol. The nutritive value of lipids is not hindered, but some fatty acids produced by hydrolysis have a special peculiar smell or rancid taste, which may affect the palatability of animals. The shorter the fatty acid carbon's chain is, the stronger the odor will be.

（2）氧化酸败是脂肪在贮藏过程中，受到氧气的作用自发地发生氧化，或在微生物、酶等作用下氧化，生成过氧化物，并进一步氧化成低级的醛、酮、酸等化合物，同时出现异味。按照引起脂肪氧化酸败的原因和机制，通常分为酮型酸败和氧化型酸败两种类型。

氧化酸败的结果既降低了脂肪的营养价值，也产生不适宜的气味，导致动物采食量下降，同时增加抗氧化物质的需要量。并且肠道受到刺激，引起胃肠道微生物区系发生变化，使动物胃肠道发炎或引起消化紊乱。油脂氧化酸败的程度可用酸价来表示。酸价就是指用以中和1 g 油脂中游离脂肪酸所需的 KOH 毫克数，一般酸价大于 6 的油脂不能饲喂动物。高脂肪饲料在贮藏时，贮期不宜过长，粉碎后宜加入抗氧化剂或控制每次粉碎量等。

(2) Oxidative rancour refers to the spontaneous oxidation of fat during storage under the action of oxygen, or under the action of microorganisms, enzymes, etc., to form peroxides, and further oxidized to lower aldehydes, ketones, acids and other compounds with peculiar smell. According to the cause and mechanism of fat oxidation rancidity, it is usually divided into two types: ketone rancidity and oxidation rancidity.

The results of oxidation rancidity not only reduce the nutritive value of fat, but also produce unsuitable odors. The stench causes a decrease in animal feed intake and increases the amount of antioxidants in the feed. In addition, the intestinal tract is stimulated, causing changes in the microflora of the gastrointestinal tract, causing inflammation of the animal's gastrointestinal tract or causing digestive disorders. The degree of oil oxidation acid number can be expressed by acid value. The so-called acid value refers to the number of milligrams of KOH required to neutralize free fatty acids in one fat. Generally, fats with an acid value greater than 6 cannot be fed to animals. When high-fat feed is stored, the storage period should not be too long, and antioxidants should be added after crushing or the amount of crushing should be controlled each time.

三、类脂
Lipoid

（一）磷脂与糖脂
Phospholipids and glycolipids

磷脂是动植物细胞的重要组成成分，在动植物体中广泛存在。以动物体而言，磷脂在脂肪转运中起重要作用，因此若肝脏中磷脂不足，就会使肝中脂肪运转发生障碍，使动物产生脂肪肝症。正常动物体组织可自行合成磷脂，不必由饲料供给，但若所供饲料缺乏合成磷脂的原料如胆碱、甲硫氨酸，除易导致脂肪肝症发生外，还可引发其他缺乏磷脂的代谢病变。

Phospholipids are an important component of animal and plant cells and are widely present in animals and plants. As far as animals are concerned, phospholipids play an important role in fat transport. Therefore, if the phospholipids in the liver are insufficient, the fat transfer in the liver will be impeded and the animals will develop fatty liver disease. Normal animal body tissues can synthesize phospholipids on their own and do not need to be supplied by feed. However, if the supplied feed lacks raw materials for phospholipid synthesis such as choline and methionine, other metabolism lesions resulting from lacking phospholipids may also occur to animals in addition to fatty liver disease.

糖脂类化合物分子与卵磷脂相似,其特点是不与含氮碱基的磷酸化合物结合,而是代之以与1-2个半乳糖分子结合,且结合位是在第一个碳原子上:所含脂肪酸多为不饱和脂肪酸,糖脂是禾本科、豆科青草中粗脂肪的主要成分,动物外周的中枢神经等组织中也有分布,糖脂可通过消化酶和肠道微生物分解,被动物吸收利用。

Glycolipid molecules are similar to lecithin, which is characterized by not binding to nitrogen-containing phosphate compounds, but instead binding to 1-2 galactose molecules, and the binding site is on the first carbon atom: Most of the fatty acids contained are unsaturated fatty acids. Glycolipids are the main components of crude fat in grasses and leguminous grasses, and are also distributed in the central nervous system and other tissues around animals. Glycolipids can be decomposed by digestive enzymes and intestinal microorganisms, thus be utilized by animals.

(二)萜类
Terpenoids

萜类属异戊二烯的衍生物。通常多根据其分子中异戊二烯的数目将其分为单萜、二萜等,如叶绿素中的叶绿萜为二萜,胡萝卜素为四萜,脂溶维生素A、E、K亦属萜类。因此含叶绿素多的青绿饲料其粗脂肪的营养价值就相对较低。

Terpenoids are derivatives of isoprene. They are usually divided into monoterpenes, diterpenes, etc. according to the number of isoprene in their molecules. For example, the chlorophyll terpenes in chlorophyll are diterpenes, carotene is tetraterpenes, and fat-soluble vitamins A, E, and K are also terpenoids. Thus, the nutritive value of crude fat in the green feed containing more chlorophyll is relatively low.

(三)固醇
Sterol

固醇是一类以环戊烷多氢菲为骨架的物质,广泛存在于生物体组织内。可游离存在,也可与脂肪酸结合以酯的形式存在,虽含量少,但有重要生理功能。固醇按来源可分为三种。

Sterol is a kind of substance with cyclopentane polyhydrophenanthrene as its skeleton, which is widely present in biological tissues. It can exist freely or in the form of an ester combined with fatty acids. Although the content is small, it has important physiological functions. Sterols can be divided into 3 types according to their sources.

1. 动物固醇

动物固醇在动物体内多以酯形式存在,胆固醇为其代表,是固醇类激素的合成原料。如皮肤中的7-去氢胆固醇在紫外光照射下,可转变成维生素D_3,供动物利用。

1. Zoosterol

Zoosterol mostly exists in the form of esters in animals. Cholesterol is the representative, which is the synthetic raw material for sterol hormones. For example, 7-dehydrocholesterol in the skin can be converted into vitamin D_3 under ultraviolet light for use by animals.

2. 植物固醇

植物固醇为植物细胞的主要组分,无法被动物有效利用,其中以存在豆类中的豆固醇,存在于谷物胚、油中的谷固醇为代表。

2. Phytosterol

Phytosterol is the main component of plant cells and cannot be effectively utilized by animals. Among them, stigmasterol in beans and sitosterol in grain germs and oil are typical ones.

3. 酵母固醇

酵母固醇以麦角固醇为代表，存在酵母、霉菌及某些植物中，经紫外光照射可转化成维生素 D_2 供动物利用。

3. Zymosterol

Zymosterol is represented by ergosterol. It exists in yeast, mold and certain plants, and can be converted into vitamin D_2 by ultraviolet light for animal use.

第四节　矿物质
Section 4　Minerals

矿物质元素是动物生命活动和生产过程中起重要作用的一大类无机营养素，现已发现在 107 种元素中有 60 种以上的元素能在动物组织器官中找到，其中已确定有 27 种矿物质元素为动物组织所必需的元素。按照它们在动物体内含量的不同，分为常量元素和微量元素。

Mineral elements are a major category of inorganic nutrients that play an important role in animal life activities and production processes. It has now been discovered that more than 60 of the 107 elements can be found in animal tissues and organs, among which 27 kinds of mineral elements have been identified as essential elements for animal tissues. According to their different content in animals, they are divided into macro elements and trace elements.

一、常量元素与微量元素
Major and trace elements

常量元素是指动物体内含量在 0.01% 以上的元素，包括有 Ca、P、S、Cl、K、Na、Mg 7 种；微量元素是指动物体内含量在 0.01% 以下的元素，动物体内必需的微量元素有 Fe、Cu、I、Zn、Mn、Co、Mo、Se、Cr 等。

Major elements refer to the elements with a content of 0.01% or more in animals, including 7 kinds of Ca, P, S, Cl, K, Na, and Mg. Trace elements refer to the elements with a content of less than 0.01% in animals. The necessary trace elements in the animal body are Fe, Cu, I, Zn, Mn, Co, Mo, Se, Cr, etc.

二、必需矿物质元素
Essential mineral elements

1950 年以前，知其生理作用的矿物质元素有 Ca、P、S、Cl、Na、K、Mg 7 种常量元素和 Fe、Cu、I、Zn、Mn、Co 6 种微量元素，1953 年发现了 Mo，1957 和 1959 年分别发现了 Se 和 Cr，以后又陆续发现了 F、Ni、Si、V、As、B 等 11 种元素。某些以前认为是有毒的元素，现已被证明

Before 1950, the mineral elements known for their physiological functions were seven major elements called Ca, P, S, Cl, Na, K and Mg and six trace elements called Fe, Cu, I, Zn, Mn and Co. Mo was discovered in 1953, and Se and Cr were discovered in 1957 and 1959 respectively. Later, 11 elements, such as F, Ni, Si, V, As, and B were successively discovered. Certain elements that were previously considered to be toxic have now been proved to

为动物必需的元素；也有原认为没有营养的元素，已被认定其营养价值的存在。如20世纪70年代初证明动物体内Si参与软骨与结缔组织中黏多糖的合成。近几十年来，随着营养学科的进展，对饲料中各种矿物质元素在动物生命活动中的营养作用日趋明确。作为动物的必需矿物质元素应符合以下四个条件。

1. 普遍存在于各种动物正常组织中，且在群体内分布均匀、含量稳定。

2. 该元素对各种动物的基本生理功能与代谢规律是共同的。

3. 该元素缺乏或供给过多，在各物种动物间表现出相似的生理生化失常，即相同的缺乏症和过多症。

4. 给动物补给该元素，能治疗或减轻其缺乏症。

be essential elements for animals; there are also elements that were thought to have no nutritional significance and have been recognized as having nutritional value. For example, in the early 1970s, it was proved that Si in animals was involved in the synthesis of mucopolysaccharides in cartilage and connective tissue. In recent decades, with the development of nutrition science, the nutritional role of various mineral elements in feed in animal life activities has become increasingly clear. As essential mineral elements for animals, they should meet the following four conditions.

1. It is widely found in normal tissues of various animals, with uniform distribution and stable content in the population.

2. This element is common to the basic physiological functions and metabolism of various animals.

3. The lack or oversupply of this element shows similar physiological and biochemical abnormalities among animals of various species, that is, the same deficiency and overabundance.

4. Supplying this element to animals can cure or relieve their deficiency.

三、动物矿物质元素营养与环境
Animal mineral element nutrition and environment

动物必需的矿物质元素中，有些元素由于动物对它的需要量很少，且一般饲料中所含都能满足动物需要，如F、Pb、Cd、As等，因此，一般不会缺乏，相反，在生产实际中主要涉及防止中毒的问题。

但也有多种必需元素在一般饲养条件下，易于造成动物不足或缺乏。如对于猪，Ca、P、Cl、Co、Fe、Cu、Zn、Mn、I和Se等11种元素易出现缺乏。如果我们在饲养实践中给予合理补充的话，既可以防止缺乏症的发生，又可提高动物生产力。当然，如果日粮中B_{12}充足的话，动物就不会出现Co缺乏的问题。但近些年来，有迹象表明，某些必需元素，尽管生产上常用饲料中供给已

Among the essential mineral elements for animals, some elements, such as F, Pb, Cd, As, etc., are generally sufficient. Since a small amount of them required by animals and their content of general feed can meet the needs of animals, prevention of poisoning should be considered in animals' diets.

There are also a variety of essential elements that are easy to be in shortage in the animals that are fed under general conditions. For example, for pigs, 11 elements such as Ca, P, Cl, Co, Fe, Cu, Zn, Mn, I and Se are prone to be deficient. If we give reasonable supplements in the breeding practice, we can not only prevent the occurrence of deficiency, but also increase animal productivity. Of course, if B_{12} in the diet is sufficient, animals will not suffer from Co deficiency. However, in recent years, there have been signs that some essential elements, such as Mg and Mn, may be beneficial in proper addition, although they are already sup-

经足够，但适当添加也是有益的，如Mg、Mn等。

当然，动物矿物质营养与环境之间存在密切关系，如岩石、土壤、大气、水、植物等可直接影响动物矿物质营养状况，而气候、季节、施肥与作物的田间管理、环境污染等，又能间接影响动物矿物质营养，所以与动物矿物质有关的疾病及矿物质营养本身都带有明显的地区性和季节性，尤其是在微量元素方面。

plied in the common feed used in production.

Of course, there is a close relationship between animal mineral nutrition and the environment. For example, rocks, soil, atmosphere, water, plants, etc. can directly affect animal mineral nutrition status, while climate, season, fertilization and field management of crops, environmental pollution, etc. can indirectly affect animal mineral nutrition. Therefore, diseases related to animal minerals and mineral nutrition themselves are obviously regional and seasonal, especially in terms of trace elements.

第五节　维生素
Section 5　Vitamins

一、概述
Overview

19世纪末到21世纪初研究发现，用纯的蛋白质、碳水化合物、脂肪、盐和水组成的纯合饲粮（不能满足动物需要）不能使大小鼠存活太长的时间，而在这些饲粮中加入少量乳汁，则可大大延长实验鼠类的生存时间，当时认为与乳中少量未知物有关。不久，人们已认识到至少有两种因子，即脂溶性的A因子和水溶性的B因子。到1912年，卡西米尔·冯克（Casimir Funk）提出了名词"Vitamins"，他分离出了抗脚气病的因子，推测这一因子可能含氮，属于胺类，故称为"Vitamins"（Vital amines），后来这一名词被改写为"Vitamin"。

不同学者提出的维生素的概念不完全相同，一般认为"维生素属于维持人和动物正常生理机能所必需，且需要量又极微小的一类低分子有机物质"。它对人和动物的重

From the end of the 19th century to the beginning of 21st century, studies have found that a homogenous diet consisting of pure protein, carbohydrates, fat, salt and water (which cannot meet the needs of animals) cannot keep rats and mice alive for too long, and add a small amount of milk in the fodder can greatly prolong the survival time of experimental mice. At that time, the result was thought to be related to a small amount of unknown substances in the milk. Soon, people have realized that there are at least two factors, namely fat-soluble factor A and water-soluble factor B. By 1912, Casimir Funk proposed the term "Vitamins". He isolated anti-beriberi factors. It was speculated that this factor may contain nitrogen and belong to amines. He called these factors "vitamins" (Vital amines), later this term was rewritten as "Vitamin".

The concept of vitamins proposed by different scholars is not exactly the same. It is generally believed that "vitamins are a type of low-molecular organic substances that are necessary but extremely low-required to maintain the normal physiological functions of humans and animals." Its impor-

要作用,并非表现在它的能量价值,也非在于它们作为动物的结构物质,而主要是以活化剂的形式,参与体内物质和能量代谢的各生化反应。它们在动物体内数量极少,却作用很大,而且每一种维生素都具有其特殊的作用,相互间不可替代。

tant effect on humans and animals is not manifested in its energy value, nor is it that they act as structural substances of animals, but mainly in the form of activators, which participate in the various biochemical reactions of substance and energy metabolism in the body. Their amount is small in the animal's body, but they have a great effect. Moreover, each kind of vitamin has its special function and cannot be replaced with each other.

二、维生素的分类与命名
Classification and naming of vitamins

维生素按其溶解性可分为脂溶性维生素和水溶性维生素。维生素最初被发现时,人们对其化学结构、理化特性和生理功能均无确切了解,随后的研究发现,起初认识的某种维生素并非某一特定的单一化合物,而是多种化合物的组成,于是出现了"维生素族(组)"的命名,如B组维生素。

动物体内的维生素据其来源情况可分为:外源性维生素和内源性维生素。外源性维生素指由饲料提供的维生素,内源性维生素不是由外界饲料摄入,而是在动物体内合成的。

内源性维生素有两种情况:① 消化道微生物合成的。如反刍动物瘤胃微生物和各种动物的大肠微生物可以合成B组维生素和维生素K。② 动物本身的器官或组织合成的维生素。如动物皮肤中存在的二脱氢胆固醇,经紫外线照射后可转化为维生素D_2;动物的肾上腺(还有肠及肝脏)可合成维生素C。

Vitamins can be divided into fat-soluble and water-soluble vitamins according to their solubility. When vitamins were first discovered, people had no exact understanding of their chemical structure, physical and chemical properties and physiological functions. Later studies found that a certain vitamin that was initially recognized was not a specific single compound, but a combination of multiple compounds, so the name "vitamin family (group)" appeared. For example, the B vitamins include many kinds.

Vitamins in animals can be divided into exogenous vitamins and endogenous vitamins according to their sources. Exogenous vitamins refer to the vitamins provided by the feed. The endogenous vitamins are not taken in by the external feed, but are synthesized in the animal body.

There are 2 types of endogenous vitamins: ① Vitamins synthesized by microorganisms in the digestive tract. For example, ruminant rumen microbes and various animal large intestine microbes can synthesize group B vitamins and vitamin K. ② Vitamins synthesized by the animal's own organs or tissues. For example, didehydrocholesterol present in animal skin can be converted into vitamin D_2 after ultraviolet radiation; the animal's adrenal glands (as well as the intestines and liver) can synthesize vitamin C.

第六节 水分
Section 6　Water

水是维持动植物和人类生存不可缺少的物质之一。常用来作为配合饲料原料的谷物、豆类等水分含量一般在12%-14%,但有些饲料,如青饲料水分含量可达60%-90%,有的甚至更高(如水生饲料)。

Water is one of the indispensable substances for the survival of animals, plants and humans. The moisture content in grains and beans commonly used as compound feed ingredients is generally 12%-14%, but in some feeds, such as green feed, water content can reach 60%-90%, and some are even higher (such as aquatic feed).

一、水分的存在形式
Existing form of water

饲料中的水分按其形式可分为两种,即自由水和结合水。自由水是一种具有与普通水一样的热力学运动能力的水,也称为游离水。而结合水是与饲料中的蛋白质、碳水化合物的活性基团结合而不能自由运动的水。结合水与一般液体水的性质不同,由于其牢固的结合,也有将其定义为"冷至0 ℃以下也不冻的水";同时也没有溶解的作用。把一些物理和化学性质不同的水加在一起就构成了"饲料的水分"。这些水在饲料中的比例和分布往往是不均匀的,它与饲料的加工和贮藏有着密切的关系。

The water in feed can be divided into two types according to its form: free water and bound water. Free water is a kind of water with the same thermodynamic movement ability as ordinary water, also called free water. The bound water is the water that is bound with the active group of protein and carbohydrate in the feed and cannot move freely. The nature of bound water is different from general liquid water. Because of its firm bond, it is also defined as "water that does not freeze when cooled to below 0 ℃": it also has no dissolving effect. Adding these waters with different physical and chemical properties together constitutes "feed moisture". The proportion and distribution of the water in the feed are often uneven, and it is closely related to the processing and storage of the feed.

二、水分的活性度
Water activity

饲料的水分,无论自由水或结合水,在100-105 ℃下干燥时都可以蒸发掉,因此饲料水分一般都用该温度下的失重来测定。事实上,饲料水分随环境条件的改变而变动。如果周围环境空气干燥,则水分从饲料中蒸发而逐渐干燥。反

The moisture of the feed, whether free water or bound water, can evaporate when dried at 100-105 ℃, so feed moisture is generally measured by weight loss at this temperature. In fact, feed moisture changes with environmental conditions. If the surrounding air is dry, the moisture will evaporate from the feed, and the feed water will decrease and become dry. Conversely, if the environmental humidity

之,如果环境湿度高,则干燥的饲料就会吸收空气中的水分。总之,不管是吸湿还是干燥,最终将达到平衡。通常此时的水分称为平衡水分。可以看出,饲料中的水分并不是静止的,而是活动的,因此,饲料中的水分也可以用活性度来表示,所谓饲料水分的活性度(A_w)就是指饲料所显示的水蒸气压(P)对同一温度下的最大水蒸气压(P_0)之比。即

$$A_w = \frac{P}{P_0}$$

对于纯水来说,其蒸汽 P 和 P_0 是相等的,即它的 A_w 应为 1。对饲料来说,水和蛋白质、碳水化合物等固形物在一起,而且水分相对也较少,其水蒸气压也就较小,所以其水分活性度 A_w 小于 1。水果等含水量高的物品其 A_w 值为 0.98 - 0.99,而谷物等配合饲料原料的 A_w 值为 0.60 - 0.64。微生物可以繁殖的饲料 A_w 值为:细菌不低于 0.90(多数为 0.94 - 0.99)、酵母为 0.88、霉菌为 0.80。如果饲料的 A_w 值高出这些值,饲料便容易被微生物破坏。

is high, the dry feed will absorb the moisture in the air. In short, whether it is moisture absorption or drying, it will eventually reach equilibrium. Usually, the moisture at this time is called equilibrium moisture. It can be seen that the water in the feed is not static, but active. Therefore, the water in the feed can also be expressed in terms of activity. The so-called water activity (A_w) is the ratio of water vapor pressure (P) shown in the feed to the maximum water vapor pressure (P_0) at the same temperature.

$$A_w = \frac{P}{P_0}$$

For pure water, its steam P and P_0 are equal, that is, its A_w should be 1, but for feed, water is integrated with solids such as protein and carbohydrates, and the water content is relatively small. The vapor pressure is also small, so the water activity A_w of the feed must be less than 1. For example, the A_w value of high water content items such as fruits is 0.98 - 0.99, while the A_w value of the compound feed ingredients such as grains is 0.60 - 0.64. The A_w value of the reproduced feed is not less than 0.90 for bacteria (mostly 0.94 - 0.99), 0.88 for yeast, and 0.80 for mold. If the A_w value of the feed department is higher than these values, the feed is easily destroyed by microorganisms.

第七节　其他成分
Section 7　Other ingredients

一、抗营养因子
Antinutritional factors

饲料可提供动物赖以生存及生产所必需的各种营养成分,但有些饲料存在某些能破坏营养成分或以不同机制阻碍动物对营养成分的消化、吸收和利用并对动物的健康状况产生负作用的物质,这些物质被

Feed can provide various nutrients necessary for the survival and production of animals. However, some feeds have certain substances that can destroy nutrients or hinder the digestion, absorption and utilization of nutrients by animals, and have negative effects on animal health. These substances are called feed antinutritional factors (ANFs).

称为饲料抗营养因子（ANFs）。有些饲料还可能存在对动物主要产生毒性作用的物质——毒物（或毒素），但在实践中 ANFs 和毒物间并无特别明显的界限，毒物也通常表现出一定的抗营养作用，有些抗营养因子也表现出一定的毒性作用。

Some feeds may also contain substances that have a major toxic effect on animals—poisons (or toxins), but in practice, there is no particularly obvious boundary between ANFs and poisons, and poisons usually show certain antinutritional effects, and some antinutritional factors also show certain toxic effects and cause certain damage to animals.

二、饲料的色素
Pigments of feed

饲料色素不仅是判定饲料品质的感观指标之一，而且在一定程度上还影响着动物产品的质量和价值。比如：使牛奶中的黄油增色，禽蛋蛋黄和皮肤增色，以及影响某些水产类动物皮肤颜色（尤其是观赏鱼）和肉的色泽。因此，色素研究也是饲料科学中不可忽视的。

Feed pigment is not only one of the sensory indicators for judging feed quality, but also affects the quality and value of animal products to a certain extent. For example, it can enhance the color of butter in milk, the color of poultry egg yolks and skin, and affect certain aquatic animal skin color (especially ornamental fish) and meat color. Therefore, pigment research is also an issue that cannot be ignored in feed science.

（一）天然色素与结构
Natural pigments and structure

饲料中的天然色素都是由发色基团和助色基团组成的。发色基团是指凡有机分子在紫外光及可见光区域内（200 nm – 700 nm）有吸收峰的基团，如：═C═C═、═C═O、—CHO、—COOH、—N═N—、—N═O、—NO$_2$、═C═S 等。但当分子中只含有 1 个发色基团时，物质并不呈色（因吸收光波长仅在 200 nm – 400 nm），只有当 2 个或 2 个以上生色基团共轭，其吸收光波段移至可见光区域内时，物质才会呈色。助色基团是指本身吸收波段在紫外区，若将其接到共轭体系或发色基团上，则可使共轭键或发色基团的光吸收波段移向长波方向的基团。这种基团包括：—OH、—OR、—NH$_2$、SH、—Br 和 —Cl 等。随着共轭双键数目的增加，吸收光带向波长增加方向移动便由无色变

According to research, natural pigments in feed are composed of chromogenic groups and auxochrome groups. The chromophore group refers to a group that has an absorption peak for organic molecules in the ultraviolet light and visible light region (200 nm to 700 nm). The chromatic groups are ═C═C═, ═C═O, —CHO, —COOH, —N═N—, —N═O, —NO$_2$, ═C═S, etc. It is worth mentioning that when the molecule contains only one chromophore, the substance does not show color (because the wavelength of light absorbed is only between 200 – 400 nm), only when two or more chromophores share the yoke, the material will only show color when the light absorption band moves to the visible light region. The auxochromic group means that its own absorption band is in the ultraviolet region. If it is connected to a conjugated system or chromophore group, the light absorption band of the conjugated bond or chromophore group can be shifted to the long wave direction. Such groups include —OH, —OR, —NH$_2$, SH, —Br, and —Cl, etc. As the number of co-moment double bonds increases, the absorption band moves in the direction

有色,且逐渐加深。

of increasing wavelength and changes from colorless to colored, and the color gradually deepens.

(二) 饲料中的天然色素
Natural pigments in feed

饲料中天然色素种类很多,与营养有关的主要有如下三类:

1. 吡咯衍生物

属于4个吡咯环的α-碳原子通过次甲基相连而成的一类复杂共轭体系,通常称卟吩,中间有金属原子以共价键或配位键与之相结合。

(1) 叶绿素 为一切绿色植物的绿色来源,是绿色植物细胞叶绿体中的一种色素。由于叶绿素不稳定,易提色,故常将绿色残存度作为评定牧草品质的感观指标之一。加之其镁原子在酸性条件下可被氢原子取代,变成脱镁叶绿素(呈褐绿色),故又将脱镁叶绿素与叶绿素统称为色原,色原在消化道内不被吸收。

(2) 血红素 由一个铁原子与卟啉环构成的铁卟啉化合物,在生物体内与蛋白质结合共存。血粉中由于该铁已变成三价,消化道难以吸收,故生产中不宜用血粉供铁源。

2. 异戊二烯衍生物-类胡萝卜素

该类衍生物是以异戊二烯残基为单元,以共轭键为基础组成的一类色素,广布于自然界,但仅部分具着色功能。

(1) 胡萝卜素类 结构为共轭多烯烃。可分为α-胡萝卜素、β-胡萝卜素、γ-胡萝卜素及番茄红素四种,其中以β-胡萝卜素最为重要,其结构、生理功能与维生素A相似,故称维生素A原,但着色效果较差。

(2) 叶黄素类 为共轭多烯烃的含氧衍生物,以醇、醛、酮、酸的形式

There are many types of natural pigments in feed, and there are three main types related to nutrition.

1. Pyrrole derivatives

This derivative belongs to a complex conjugated system formed by the α-carbon atoms of 4 pyrrole rings connected by a methine group, usually called porpan, with metal atoms in the middle by covalent bond or coordination bond combining with it.

(1) Chlorophyll is the green source of all green plants and is a kind of pigment in the chloroplast of green plant cells. Since chlorophyll is unstable and easy to improve the color, the green residual degree is often used as one of the sensory indicators to evaluate the quality of forage. In addition, its magnesium atoms can be replaced by hydrogen atoms under acidic conditions to become pheophytin (brown-green), so pheophytin and chlorophyll are collectively referred to as chromogens, and chromogens are not absorbed in the digestive tract.

(2) Heme is an iron porphyrin compound that is composed of one iron atom and a porphyrin ring. Since the iron in blood meal has become trivalent, it is difficult to be absorbed by the digestive tract. Thus, blood meal should not be used as a source of iron in production.

2. Isoprene derivatives-carotenoids

These derivatives are a type of pigment composed of isoprene residues as a unit and based on conjugated bonds. It is widespread in nature, but only part of it has a coloring function.

(1) Carotene structure is conjugated polyolefin. It can be divided into 4 kinds: α-carotene, β-carotene, γ-carotene and lycopene. Among them, β-carotene is the most important. Its structure and physiological function are similar to vitamin A, so it is called provitamin A, but the coloring effect is poor.

(2) Xanthophylls are oxygen-containing derivatives of conjugated polyolefins, which exist in the form of alcohols,

存在,可分为:① 叶黄素($C_{40}H_{56}O_2$),化学名称为3,3′-二羟基-α-胡萝卜素,广泛分布在绿色植物和玉米中,着色效果较佳;② 玉米黄素($C_{40}H_{56}O_2$),化学名称同叶黄素,黄色玉米中含量较高;③ 隐黄素($C_{40}H_{56}O_2$),化学名称同叶黄素,主要存在于黄玉米与南瓜中;④ 番茄黄素($C_{40}H_{56}O$),化学名称为3-羟基番茄黄素,番茄中含量高;⑤ 辣椒红素($C_{40}H_{56}O_3$),主要存在于红辣椒中,着色效果较好;⑥ 柑橘黄素($C_{40}H_{56}O$),化学名称为5、8环氧β-胡萝卜素,主要存在于柑橘皮中,着色效果较好;⑦ 虾黄素($C_{40}H_{56}O_4$),化学名称为3,3′-二羟基-4,4′-二酮基-β,β′-胡萝卜素,主要存在于虾、蟹、牡蛎、昆虫等动物体内,与蛋白质结合时呈蓝色(虾青素),煮熟后因蛋白变性,被氧化成虾红素(即3,3′-二羟基-4,4′-二酮基-β,β′-胡萝卜素),对虾、蟹及鱼类着色效果较好;⑧ 茜草色素,存在于蕈类和鳟鱼中,具有较佳着色效果。

3. 多酚类色素

结构中最基本的母核是苯环和γ-吡喃环,自然界常见的有花青素、黄酮素和儿茶素。由于三者具有相同碳架,苯环上有2个或2个以上羟基,故同为植物性饲料中的水溶性色素,但对动物产品无着色效应。其中有的属营养物质,如芦丁;有的属抗营养因子,如单宁等。此外,属该类色素的有酮类衍生物(红曲色素、姜黄色素)、醌类衍生物(如虫胶色素)和甜菜红素等。

饲料色素除具有判定饲料品质、影响苗禽产品质量和价值等作用外,尚可对一些动物的采食量产生一定影响。据报道,红色对虹鳟鱼有良好诱食力,黄土色和红色对虾有良好的诱食效应。

aldehydes, ketones, and acids. They can be divided into: ① Lutein ($C_{40}H_{56}O_2$), whose chemical name is 3, 3′-dihydroxy-α-carotene, is widely distributed in green plants and corn, and its coloring effect is better; ② Zeaxanthin ($C_{40}H_{56}O_2$), whose chemical name is the same as lutein, with highly exists content in yellow corn; ③ Cryptoxanthin ($C_{40}H_{56}O_2$), with chemical name the same as lutein, is mainly found in yellow corn and pumpkin; ④ Lycopene ($C_{40}H_{56}O$), with chemical name of 3-hydroxy lycopene, highly exists in tomato; ⑤ Capsanthin ($C_{40}H_{56}O_3$), mainly exists in red peppers, and its coloring effect is better; ⑥ Citrusflavin ($C_{40}H_{56}O$), whose chemical name is 5 or 8 epoxy β-carotene, mainly exists in citrus peel, and its coloring effect is better; ⑦ Astaxanthin ($C_{40}H_{56}O_4$), whose chemical name is 3,3′-dihydroxy-4,4′-diketone -β, β′-carotene, is mainly found in animals such as shrimps, crabs, oysters, insects, etc. It appears blue when combined with protein (astaxanthin). When cooked, it is oxidized into astaxanthin (i. e. 3,3′-dihydroxy-4,4′-diketone-β, β′-carotene) due to protein denaturation. Its coloring effect of shrimp, crab and fish is better; ⑧ Madder pigment, present in mushrooms and trout Among fish, has better coloring effect.

3. Polyphenol pigments

The most basic parent nuclei in the structure is the benzene ring and the γ-pyran ring. Anthocyanins, flavonoids and catechins are commonly found in nature. Since the three have the same carbon frame and 2 or more hydroxyl groups on the benzene ring, they are all water-soluble pigments in plant feeds, but they have no coloring effect on animal products. Some of them are nutrients such as rutin and some are antinutritional factors such as tannins. In addition, there are derivatives of this type of pigments (monascus pigment, curcumin), quinone derivatives (such as shellac pigment) and betaine.

In addition to determining feed quality and affecting the quality and value of seedling and poultry products, feed pigments can also have a certain impact on the feed intake of some animals. It is reported that red has a good attractive effect on rainbow trout, ocher and red have a good attractive effect on shrimp.

三、饲料的味嗅物质
Taste and smell substances of feed

饲料味嗅是指动物以味觉、嗅觉为基础，对饲料气味和滋味产生的一种综合感应力。

1. 味觉概述

饲料滋味融入唾液，刺激口腔及舌表面味蕾，通过神经纤维传导至大脑味觉中枢，大脑分析后产生味觉。动物不同，口腔和舌上味蕾数不同，对各种滋味的嗜好不同。研究已表明在味觉上家禽有区分甜、苦、咸味的能力。小鸡不仅能区分戊糖、己糖、双糖和三糖甜度，而且可区分糖水与糖精水，拒饮含2%的盐水。在味感嗜好上，牛喜欢甜味及挥发性脂肪酸味，对苦味耐受力低，拒食大茴香味。山羊喜挥发性脂肪酸味，对苦味耐受力强，拒食过酸物质。马的味、嗅觉均灵敏，喜食苹果甜味，拒食霉变味和鱼肝油味。断奶仔猪喜食新鲜乳汁甜味、玉米香味、味精味、柠檬酸味和鱼溶浆味。成猪对有机酸、柑橘、鲜肉与糖蜜味偏好，厌食肉骨粉、化学药品及重金属味。鱼类对巧克力、乳酪、牛乳、鱼肉味偏好。

2. 味觉与呈味物质

（1）酸味与酸味物质　酸味是由H^+刺激舌黏膜引起，因此凡在溶液中能解离出H^+的化合物均称酸味物质。酸味物质不仅可提高饲料适口性，降低胃内pH，激活消化酶，提高消化吸收能力，而且可减少肠内细菌对营养物质的竞争力。酸味物质酸味阈值常因无机、有机酸不同而异。一般无机酸味阈值在

Feed taste and smell refers to the comprehensive induction of animal feed smell and taste based on taste and smell.

1. Overview of taste

The taste of feed is incorporated into saliva, stimulates the taste buds on the surface of the mouth and tongue, and is transmitted to the taste center of the brain through nerve fibers. The brain produces taste after analysis. Different animals have different taste buds in the mouth and tongue, and have different tastes for various tastes. Studies have shown that poultry has the ability to distinguish between sweet, bitter, and salty in taste. The chicken can not only distinguish the sweetness of pentose, hexose, disaccharide and trisaccharide, but also distinguish sugar water and saccharin water, and refuse to drink 2% of salt water. In terms of taste preference, cattle like sweet taste and volatile fatty acid taste, have low tolerance to bitter taste, and refuse to eat anise scent. Goats like the taste of volatile fatty acids, have strong tolerance to bitterness, and refuse to eat too acidic substances. Horse's taste and sense of smell are all sensitive. It likes the sweet taste of apple, but refuses the taste of moldy and cod liver oil. Weaned piglets like to eat fresh milk sweetness, corn flavor, MSG flavor, lemon sour flavor and fish pulp flavor. Adult pigs have a preference for organic acids, citrus, fresh meat and molasses, and anorexia of meat and bone meal, chemicals and heavy metals. Fish have a preference for chocolate, cheese, milk, and fish flavors.

2. Taste and flavoring materials

（1）Sourness and sour substances　Sourness is caused by H^+ which stimulates the tongue mucosa. Therefore, all compounds that can dissociate H^+ in the solution are called sour substances. Sour substances can not only improve the palatability of feed, lower the pH in the stomach, activate digestive enzymes, improve digestion and absorption capacity, but also reduce the competitiveness of intestinal bacteria for nutrients. The sourness threshold of sour substances often varies with inorganic and organic acids. Generally, the

pH 3.5-4.0之间,有机酸味阈值在pH 3.7-4.9之间。酸味物质的稀溶液与口腔黏膜接触才会感到酸味,但由于舌黏膜可中和H^+使酸味逐渐消失,所以酸味的持久性除与酸味物质多少有关外,尚与H^+的解离度和舌黏膜的中和缓冲效应有关。常见的酸味物质主要有乳酸、醋酸、柠檬酸、酒石酸、苹果酸、琥珀酸和磷酸等。

(2)甜味与甜味物质　甜味物质分天然与合成两大类。一般认为,凡具有甜味感的物质都有1个负电性原子A(氧或氮),与该原子上的1个质子(H^+)以共价键相连接,即A—H,如(—OH),(=NH),(—NH$_2$)等基团,并从A—H基团的原子起2.5-4.0 A的距离内。

threshold of inorganic acidity is between pH 3.5-4.0, and the value of organic acidity is between pH 3.7-4.9. The dilute solution of sour substances will only feel sour when it comes into contact with the oral mucosa. However, because the tongue mucosa can neutralize H^+ and make the sourness gradually disappear, the persistence of sourness is not only related to the amount of sour substances, but also the dissociation degree of H^+ and tongue mucosa. The neutralization effect is related to the buffer effect. Common sour substances mainly include lactic acid, acetic acid, citric acid, tartaric acid, malic acid, succinic acid and phosphoric acid.

(2) Sweetness and sweet substances　Sweet substances are divided into two categories: natural and synthetic. It is generally believed that all substances with a sweet taste have a negatively charged atom A (oxygen or nitrogen), which is connected to a proton (H^+) on the atom by a covalent bond, namely A-H, such as (—OH), (=NH), (—NH$_2$) and other groups, and within the distance of 2.5-4.0 A from the atom of the A-H group.

第二章 饲料分类
Chapter 2 Feed classification

随着现代饲料学的发展,新型饲料不断被开发,饲料的种类繁多,养分组成和饲料价值各异。为合理利用饲料,在了解各种饲料特点的基础上对饲料进行分类很有必要。饲料分类即对每种饲料确定一个能够反映其特性和营养价值的标准名称。属于同一标准名称的饲料,其特性、成分与营养价值基本相似。

目前,世界各国饲料分类方法尚未完全统一。各国按照饲料来源、饲喂动物、营养价值等的习惯分类法在实际生产中应用较广。美国学者哈里斯(Harris)等人(1956)提出的饲料分类原则和编码体系,应用计算机技术建立国际饲料数据库管理系统,在国际上有近30个国家采用或赞同,并逐渐发展为当今饲料分类编码体系的基本模式,被称为国际饲料分类法。而多数国家仍采取国际饲料分类与本国生产实际相结合的分类方法。我国在20世纪80年代,依据国际饲料分类原则与我国传统分类体系,提出了我国的饲料分类法和编码系统。

With the development of modern feed science, new feeds are constantly being developed. There are many types of feeds, with different nutrient compositions and nutritive values. For the sake of the proper use of feeds, it is necessary to classify feeds according to their characteristics. Feed classification is to determine a standard name for each feed to reflect its characteristics and nutritive value. For feeds belonging to the same standard name, their characteristics, composition and nutritive value are basically similar.

At present, the classification of feed in the world has not been completely unified. Customary feed classification in accordance with the feed sources, feeding animals, nutritive value, etc., is widely used in practice in many countries. The feed classification principles and coding system, proposed by American scholar Harris et al. (1956), applied computer technology to establish an international feed database management system that has been adopted or approved by nearly 30 countries in the world, and gradually developed into the basic model of today's feed classification and coding system, known as the International Feed Classification. However, most countries still adopt the classification method that combines the international feed classification with the actual production. In the 1980s, according to the principles of international feed classification and our traditional classification system, our feed classification and coding system were put forward.

第一节 饲料的习惯分类法
Section 1 Customary feed classification

饲料分类的基本原则是实用、

The basic principles of feed classification are practical,

简便,具有科学性。实际生产中常根据营养价值、主要成分、饲料来源等对饲料进行分类。

simple and scientific. In practical production, feeds are often classified according to their nutritive value, main ingredients, and sources, etc.

一、根据饲料营养价值分类
Classification of feed's nutritive value

1. 粗饲料

粗饲料是体积大、纤维丰富、可消化养分少的饲料。如:秸秆、荚壳、干草,也包括多汁饲料。

2. 精饲料

精饲料是体积小、纤维少、可消化养分丰富的饲料。如:谷实类、饼粕、加工副产品等。

3. 特殊饲料

特殊饲料是其他一切不属于精、粗饲料的饲料种类。如:营养及非营养型添加剂、矿物质补充料、激素等。

1. Roughage

Roughage refers to feeds that are large in size, rich in fiber and less digestible nutrients, such as straw, pod shells, hay, and succulent forage is also included.

2. Concentrate

Concentrate refers to feeds that are small in size, low in fiber, digestible and nutritional, such as cereals, cake flour, processing by-products, etc.

3. Special feed

Special feed refers to all other feed types that are not classified as roughage and concentrate, such as nutritional and non-nutritional additives, mineral supplements, hormones, etc.

二、根据饲料主要成分分类
Classification of feed's main ingredients

1. 蛋白类饲料

包括植物性蛋白饲料、动物性蛋白饲料、微生物类蛋白饲料。

2. 脂肪饲料

包括豆类、油籽、米糠类。

3. 淀粉饲料

包括谷实类、薯类等。

4. 纤维饲料

包括草、干草、秸秆等。

5. 矿物质饲料

包括食盐、钙粉、石粉等。

6. 维生素饲料

包括单体维生素和复合维生素。

7. 多汁饲料

包括鲜草、青绿及青贮饲料等。

8. 氨基酸及非蛋白氮类饲料

1. Protein feed includes vegetable protein feed, animal protein feed, and microbial protein feed.

2. Fat feed includes beans, oilseeds, and rice bran.

3. Starch feed includes cereals and potatoes, etc.

4. Fiber feed includes grass, hay, and straw, etc.

5. Mineral feed includes salt, calcium powder, and stone powder, etc.

6. Vitamin feed includes monomer vitamin feed and multivitamin feed.

7. Succulent plant feed includes fresh grass, green forage and silage, etc.

8. Amino acid and non-protein nitrogen feed.

三、根据饲料来源分类
Classification of feed's sources

1. 植物性饲料
如青绿饲料、青贮饲料、干草、块根、籽实加工副产品等。

2. 动物性饲料
如鱼粉、乳清粉、血粉、水解羽毛粉、昆虫蛋白、蛆虫等。

3. 微生物饲料
如益生菌、酵母等。

4. 天然矿物质饲料
如石粉、膨润土等。

5. 人工合成、半合成饲料
包括合成氨基酸、维生素、尿素及其衍生物、矿物质补充料以及各种营养、非营养性添加剂等。

1. Plant feed includes green forage, silage, hay, root tuber, and seed processing by-products, etc.

2. Animal feed includes fish meal, whey powder, blood meal, hydrolyzed feather meal, insect protein, maggots, etc.

3. Microbial feed includes probiotics, and yeast, etc.

4. Natural mineral feed includes rock flour, bentonite, etc.

5. Artificial synthetic and semi-synthetic feed includes synthetic amino acids, vitamins, urea and its derivatives, mineral supplements, and various nutritional and non-nutritional additives.

四、其他分类
Other classifications

1. 根据动物特性分类
如猪饲料、鸡饲料等；生长料、育肥料等。

2. 根据饲料的形态分类
包括固态、液态两种。

3. 根据获得饲料的手段分类
根据饲料的经济特性，分为商品饲料、自配饲料。其中商品饲料又可分为：全价配合饲料、浓缩料、预混料。

1. Classification according to the animal characteristics
Feeds can be classified by feeding animal species, such as pig feed, chicken feed, etc. Feeds can be also classified by stages of feeding animals, such as feeds special for animal growth, feeds special for breed, etc.

2. Classification according to the feed form
There are two forms of feed, including solid feed and liquid feed.

3. Classification according to the acquisition means
Feeds can be divided into commodity feeds and self-prepared feeds according to the economic characteristics of feeds, and commodity feeds can be further divided into compound feeds, concentrate, and premix.

第二节　国际饲料分类法

Section 2　International Feed Classification

国际饲料分类法是美国学者哈里斯（Harris）等人于1963年根据营养成分特性提出来的。它将饲料分为粗饲料、青绿饲料、青贮饲料、能量饲料、蛋白质饲料、矿物质饲料、维生素饲料、饲料添加剂共8大类，并提出"3节，6位数"对每类饲料冠以相应的国际饲料编码（IFN）。首位数表示饲料归属的类别，后5位数按照饲料的重要属性给定编码，表示为△-△△-△△△。

国际饲料分类法具有以下特点：① 主要以饲料营养价值分类，具有量的规定，因而能更好地反映各类饲料的营养特性及在畜禽饲粮中的地位；② 规定了每种饲料均需要描述来源、种及变种、饲用部分、调制处理方法等8个商品特点，因此能更好地反映影响饲料营养价值的因素；③ 每种饲料都具备一个标准编号，每一类饲料可提供99 999种饲料编号用，便于计算机管理和配方设计。

The International Feed Classification was proposed by American scholar Harris et al. (1963) on the basis of the characteristics of nutritional composition. It divides the feed into eight categories: roughage, green forage, silage, energy feed, protein feed, mineral feed, vitamin feed, and feed additives. And a six-digit with three-part International Feed Number (IFN) is assigned to each feed description. The first digit of this IFN denotes the class of feed, and the last 5 digits are coded according to the importance of the feed, expressed as △-△△-△△△.

There are several characteristics for International Feed Classification. Firstly, it is capable of better reflecting the nutritional characteristics of each feed and their significance ordering because of nutritional value-based classification and quantification. Secondly, the International Feed Classification consist of all 8 descriptions applicable to the feeds, such as feed origin, part fed to animals, process and treatment to which the part has been subjected, etc. So it can better reflect the factors affecting feed nutrition. Lastly, each feed is assigned to a standard IFN number for classification, providing a maximum 99 999 feed number available for computer management and feed formulation.

一、粗饲料
Roughage

粗饲料是指天然水分含量在60%以下，干物质中粗纤维含量大于或等于18%的饲料原料。此种饲料以风干物形式饲喂，如：干草类、农作物秸秆、稻草等。IFN形式为1-00-000。

Roughage refers to the feed raw materials with the natural moisture content below 60% and the crude fiber content in dry matter no less than 18%. This kind of feed is fed in the form of air-dried, such as hay, stover, straw, etc. INF is expressed as 1-00-000.

二、青绿饲料
Green forage

青绿饲料是指天然水分含量在60%以上的青绿牧草、饲用作物、树叶类及非淀粉质的根茎、瓜果类。IFN 形式为 2-00-000。

Green forage includes green pastures, cultivated fodder crops, tree leaves, roots and crops with moisture content more than 60%. The IFN is expressed as 2-00-000.

三、青贮饲料
Silage

青贮饲料是指以天然新鲜青绿植物性饲料为原料,在厌氧条件下,经过以乳酸菌为主的微生物发酵后调制成的饲料。具有柔软多汁、气味酸香、适口性好、营养丰富的特点,有利于长期保存。如:玉米青贮、豆类秸秆青贮。IFN 形式为 3-00-000。

Silage is a type of fodder made from green foliage crops which have been preserved by acidification, achieved through fermentation under anaerobic condition with lactic acid bacteria as the main microorganisms. Silage is soft, juicy and sour with good palatability and enriched nutrition, which is good for long-term storage. It can be made from many field crops, such as corn silage, straw silage. The IFN is expressed as 3-00-000.

四、能量饲料
Energy feed

能量饲料是指饲料绝干物质中粗蛋白质含量低于20%,粗纤维含量低于18%的饲料,如:谷实类、糠麸类、淀粉质块根块茎类、糟渣类等。一般每千克饲料物质含消化能在1.05 MJ 以上的饲料原料均属能量饲料。IFN 形式为 4-00-000。

Energy feed refers to the feed with crude protein content less than 20% and crude fiber content less than 18% in absolute dry matter, such as cereals, bran, starchy roots and tubers, etc. Generally, feeds with digestible energy above 1.05 MJ per kilogram of feed dry matter are energy feeds. The IFN is expressed as 4-00-000.

五、蛋白质饲料
Protein feed

蛋白质饲料是指自然含水率低于45%,干物质中粗纤维低于18%,而干物质中粗蛋白质含量不低于20%的饲料,如:豆类、饼粕类、鱼粉等。包括植物性、动物性蛋白饲料、单细胞蛋白饲料和非蛋白氮饲料四

Protein feed refers to the feed with less than 45% of natural moisture content, less than 18% of crude fiber (dry basis), and no less than 20% of crude protein (dry basis), such as beans, cakes, fish meal, etc. It can be divided into four categories: vegetable protein feed, animal protein feed, single-cell protein feed and non-protein nitrogen feed. Its

大类。IFN 形式为 5-00-000。

IFN is expressed as 5-00-000.

六、矿物质饲料
Mineral feed

矿物质饲料是指可供饲用的天然的、化学合成的或经特殊加工的无机饲料原料或矿物质元素的有机络合物原料，如：石灰石粉、沸石粉、膨润土、动物骨粉、贝壳粉、磷酸氢钙等。IFN 形式为 6-00-000。

Mineral feed refers to natural, chemically synthesized or specially processed inorganic feed raw meterials or organic complex with mineral elements for feeding, such as limestone powder, filler powder, bentonite, animal bone meal, shell powder, calcium hydrogen phosphate, etc. Its IFN is expressed as 6-00-000.

七、维生素饲料
Vitamin feed

维生素饲料是指由工业合成或提取的单一或复合维生素制品，但不包括富含维生素等天然青绿饲料。分为脂溶性维生素饲料和水溶性维生素饲料。IFN 形式为 7-00-000。

Vitamin feed refers to the industrial synthesis or extration of single or complex vitamin products, but does not include feeds rich in vitamins and other natural green forage. It consists of fat-soluble vitamin and water-soluble vitamin. Its IFN is expressed as 7-00-000.

八、饲料添加剂
Feed additives

饲料添加剂是指在饲料生产加工、使用过程中添加的少量或微量非营养性物质，对强化基础饲料营养价值，促进动物生长繁殖，保证动物健康，节省饲料成本，改善畜产品品质等方面有明显的效果，包括抗生素、色素、香料、激素和药物等。IFN 形式为 8-00-000。

Feed additives refer to trace of non-nutritive substances added in feed during feed production, processing and use. It has obvious effects on strengthening the nutritive value of basic feed, promoting animal growth and reproduction, ensuring animal health, improving the quality of livestock products, and saving feed costs, including antibiotics, colouring materials, flavours, hormones and medicants, etc. Its IFN is expressed as 8-00-000.

第三节　中国饲料分类法
Section 3　Chinese feed classification

中国饲料数据库情报网中心（1987）根据国际饲料分类法与我国传统分类原则，建立了中国饲料

China Feed Database Information Network Center (1987) established the Chinese feed database management system and feed classification based on the combination of

数据库管理系统及饲料分类方法。我国的饲料分类法和编码系统在国际饲料八大类分类法的基础上,结合中国传统饲料分类习惯共划分为17亚类,使用中国饲料编码(CFN)进行分类编号。

the International Feed Classification and the customary feed classification of China. The feed classification and coding system in China is divided into 17 subclasses based on the eight international feed classification methods, combined with the traditional Chinese feed classification habits, and Chinese feed code (CFN) is used for classification and numbering.

一、青绿多汁类饲料
Green fresh forage

凡天然水分含量大于或等于45%的栽培牧草、草地牧草、野菜,及部分未完全成熟的谷物植物等皆属此类。CNF形式为2-01-0000。

This category includes cultivated pastures, grassland pastures, wild vegetables with 45% or more of natural moisture content, and partially immature cereal plants. Its CNF form is 2-01-0000.

二、树叶类饲料
Tree leaves feed

树叶类饲料有两种类型:① 采摘的新鲜树叶,饲用时的天然水分含量在45%以上,属于青绿饲料,CFN形式为2-02-0000。② 采摘的新鲜树叶风干后饲喂,干物质中粗纤维含量大于或等于18%,如风干的乔木、灌木、亚灌木的树叶等,属于粗饲料,CFN形式为1-02-0000。

There are 2 types of tree leaves feed: ① Fresh leaves with more than 45% of moisture content. It belongs to green forage with CFN being 2-02-0000. ② Fresh leaves picked after air drying are fed with no less than 18% of crude fiber (dry basis), such as the leaves of air-dried trees, shrubs, sub-shrubs, etc., which belong to roughage, and the CFN is expressed as 1-02-0000.

三、青贮饲料
Silage

青贮饲料有三种类型:① 常规青贮,由新鲜的植物性饲料调制成的青贮饲料,一般含水量在65%-75%,CFN形式为3-03-0000。② 低水分青贮饲料,亦称半干青贮饲料,是用天然水分含量为45%-55%的半干青绿植株调制成的青贮饲料,CFN形式为3-03-0000。③ 谷物湿贮,以新鲜玉米、麦类籽实为主要原料,不经干燥即贮于密闭的青贮设

There are 3 types of silage: ① Conventional silage, which is made from fresh plant with 65%-75% of moisture content. The CFN form is 3-03-0000. ② Low moisture silage, also known as semi-dry silage, is a silage made from semi-dry green plants with 45%-55% of natural moisture content. Its CFN form is 3-03-0000. ③ Grain silage is mainly made from fresh corn and wheat seeds. It is fermented with lactic acid and stored in a closed silage facility without drying with 28%-35% of moisture content. It belongs to energy feed in view of nutrient content, but in view

备内,经乳酸发酵制成,其水分含量为28%-35%。根据营养成分含量,属于能量饲料,但从调制方法分析又属于青贮饲料,CFN形式为4-03-0000。

of the preparation method, it belongs to silage feed. Its CFN is expressed as 4-03-0000.

四、块茎、块根、瓜果类饲料
Tuber, root tuber, melon and fruit feed

天然水分含量不低于45%的块茎、块根、瓜果类饲料,如胡萝卜、芜菁、饲用甜菜等,鲜喂时CFN形式为2-04-0000。当此类饲料脱水后,干物质中粗纤维和粗蛋白质含量都较低,干燥后属于能量饲料,如甘薯干、木薯干等,干喂时CFN形式为4-04-0000。

For tuber, root tuber, melon and fruit feed with no less than 45% of natural moisture content, such as carrots, turnips, beets, etc., the CFN form is 2-04-0000 when fresh. When fed after drying with a low content of crude fiber and crude protein (dry basis), such as dried sweet potato, dried cassava, etc., it belongs to energy feed with CFN being 4-04-0000.

五、干草类饲料
Hays forage

干草类包括人工栽培或野生牧草的脱水或风干物,其水分含量在15%以下。水分含量在15%-25%的干草压块亦属此类,有两种类型:① 干物质中的粗纤维含量大于或等于18%者,属于粗饲料,CFN形式为1-05-0000;② 干物质中的粗蛋白质含量大于或等于20%而粗纤维含量又低于18%者,如一些优质豆科干草、苜蓿叶粉、早期刈割的苜蓿和紫云英调制的草粉等,属于蛋白质饲料,CFN形式为5-05-0000。

Hay includes the dehydrated or air-dried forms of cultivated or wild forages with less than 15% of moisture content. Hay briquettes with 15%-25% of moisture content also fall into this category with two types: ① Hay with no less than 18% of crude fiber (dry basis) belongs to roughage with CFN being 1-05-0000; ② Those whose crude protein (dry basis) is no less than 20% and crude fiber (dry basis) is less than 18%, such as some high-quality legume hay, alfalfa leaf meal, early cut alfalfa and grass meal mixed with Chinese milkweed, belong to protein feed with CFN being 5-05-0000.

六、农副产品类饲料
Agricultural byproduct feed

农副产品类有三种类型:① 干物质中粗纤维含量大于或等于18%者,如秸、荚、壳等,属于粗饲料,CFN形式为1-06-0000;② 干物

There are 3 types of agricultural byproducts: ① Feeds with no less than 18% of crude fiber (dry basis), such as straws, pods, shells, etc., belong to roughage with CFN being 1-06-0000; ② Feeds with less than 18% of crude fi-

质中粗纤维含量小于18%,而粗蛋白质含量也小于20%者,属于能量饲料,CFN形式为4-06-0000;③ 干物质中粗纤维含量小于18%,而粗蛋白质含量大于或等于20%者,属于蛋白质饲料,CFN形式为5-06-0000。后两者较为罕见。

ber (dry basis) and less than 20% of crude protein (dry basis) belong to energy feed with CFN being 4-06-0000; ③ Feeds with less than 18% of crude fiber (dry basis) and no less than 20% of crude protein (dry basis) belong to protein feed with CFN being 5-06-0000.

七、谷实类饲料
Cereal grain

谷实类饲料的干物质中,一般粗纤维含量小于18%而粗蛋白质含量也小于20%者,如玉米、稻谷等,属于能量饲料,CFN形式为4-07-0000。

In the dry matter of cereal grain, those with less than 18% of crude fiber and less than 20% of crude protein, such as corn and rice, are energy feed with CFN being 4-07-0000.

八、糠麸类饲料
Milling byproduct feed

糠麸类饲料有两种类型:① 饲料干物质中粗纤维含量小于18%,粗蛋白质含量小于20%的各种粮食的碾米、制粉副产品,如小麦麸、米糠等,属于能量饲料,CFN形式为4-08-0000;② 粮食加工后的低档副产品,如统糠、生谷机糠等,其干物质中的粗纤维含量多大于18%,属于粗饲料,CFN形式为1-08-0000。

There are 2 types of milling byproduct feed: ① Rice whitening and flour milling byproducts of various grains with less than 18% of crude fiber (dry basis) and less than 20% of crude protein (dry basis), such as wheat bran, rice bran, etc., belong to energy feed with CFN being 4-08-0000; ② Low-quality byproducts after grain processing with more than 18% of crude fiber (dry basis), such as rice mill byproduct, belong to roughage with CFN being 1-08-0000.

九、豆类饲料
Beans feed

豆类饲料有两种类型:① 豆类籽实干物质中粗蛋白质含量大于或等于20%而粗纤维含量又低于18%者,属于蛋白质饲料,如大豆等,CFN形式为5-09-0000;② 个别豆类籽实的干物质中粗蛋白质含量在20%以下,如江苏的爬豆,属于

There are 2 types of beans feed: ① Beans with no less than 20% of crude protein (dry basis) and less than 18% of crude fiber (dry basis) belong to protein feed, such as soybeans. The CFN is expressed as 5-09-0000; ② Beans with less than 20% of crude protein (dry basis) belong to energy feed, such as Cucumber bean from Jiangsu. Its CFN is expressed as 4-09-0000. It is rare that the crude fiber (dry

能量饲料,CFN 形式为 4-09-0000。豆类饲料干物质中粗纤维含量大于或等于 18% 者罕见。

basis) of beans is 18% or more.

十、饼粕类饲料
Cake meal feed

饼粕类饲料有三种类型:① 干物质中粗蛋白质大于或等于 20%,粗纤维含量小于 18% 者,属于蛋白质饲料,大部分饼粕属于此类,CFN 形式为 5-10-0000;② 干物质中粗蛋白质含量大于或等于 20%,而粗纤维含量大于或等于 18% 者,属于粗饲料,如含壳量多的葵花籽饼及棉籽饼,CFN 形式为 1-10-0000;③ 干物质中粗蛋白质含量小于 20%,粗纤维含量小于 18%,如米糠饼、玉米胚芽鞘等,属于能量饲料,CFN 形式为 4-10-0000。

There are 3 types of cake meal feed: ① Most of cakes and meals have less than 18% of crude fiber (dry basis) and no less than 20% of crude protein (dry basis), which belong to protein feed with CFN being 5-10-0000; ② Cakes and meals with no less than 20% of crude protein (dry basis) and no less than 18% of crude fiber (dry basis), such as sunflower cake and cotton meal, belong to roughage with CFN being 1-10-0000; ③ Cakes and meals with less than 20% of crude protein (dry basis) and less than 18% of crude fiber (dry basis), such as rice bran cake, corn germ meal, etc., belong to energy feed with CFN being 4-10-0000.

十一、糟渣类饲料
Distiller dried grains with soluble feed

糟渣类饲料有三种类型:① 干物质中粗纤维含量大于或等于 18% 者,属于粗饲料,CFN 形式为 1-11-0000;② 干物质中粗蛋白质含量低于 20%,粗纤维含量也低于 18% 者,属于能量饲料,如优质粉渣、醋糟、甜菜渣等,CFN 形式为 4-11-0000;③ 干物质中粗蛋白质含量大于或等于 20%,而粗纤维含量小于 18% 者,属于蛋白质饲料,如含蛋白质较多的啤酒糟、豆腐渣等,CFN 形式为 5-11-0000。

There are 3 types of distiller dried grains with soluble feed: ① Feeds with no less than 18% of crude fiber (dry basis) belong to roughage with CFN being 1-11-0000; ② Feeds with less than 20% of crude protein (dry basis) and less than 18% of crude fiber (dry basis) belong to energy feed, such as high-quality powder residue, vinegar lees, beet residue, etc. Its CFN is expressed as 4-11-0000; ③ Feeds with no less than 20% of crude protein (dry basis) and less than 18% of crude fiber (dry basis), such as brewer's dried grain and bean curd residue, belong to protein feed with CFN being 5-11-0000.

十二、草籽、树实类饲料
Seed of grass and tree feed

草籽、树实类饲料有三种类型:

There are 3 types of seed of grass and tree feed:

①干物质中粗纤维含量大于或等于18%者,属于粗饲料,如灰菜子等,CFN形式为1-12-0000;②干物质中粗纤维含量在18%以下,粗蛋白质含量小于20%者,属于能量饲料,如干沙枣等,CFN形式为4-12-0000;③干物质中粗纤维含量在18%以下,而粗蛋白质含量大于或等于20%者,属于蛋白质饲料,但较为罕见,CFN形式为5-12-0000。

① Feeds with less than 18% of crude fiber (dry basis), such as gray rapeseed, etc., belong to roughage with CFN being 1-12-0000; ② Feeds with less than 18% of crude fiber (dry basis) and less than 20% of crude protein (dry basis), such as dried jujube, etc., belong to energy feed with CFN being 4-12-0000; ③ Feeds with less than 18% of crude fiber (dry basis) and no less than 20% of crude protein (dry basis) belong to protein feed, but relatively rare. Its CFN is expressed as 5-12-0000.

十三、动物性饲料
Animal feed

动物性饲料均来源于渔业、畜牧业的动物性产品及其加工副产品,有三种类型:①干物质中粗蛋白质含量大于或等于20%者,属于蛋白质饲料,如鱼粉、动物血、蚕蛹等,CFN形式为5-13-0000;②干物质中粗蛋白质含量小于20%,粗灰分含量也较低的动物油脂,属于能量饲料,如牛脂等,CFN形式为4-13-0000;③干物质中粗蛋白质含量小于20%,粗脂肪含量也较低者,主要以补充钙磷为目的,属于矿物质饲料,如骨粉、贝壳粉等,CFN形式为6-13-0000。

Animal feed is derived from animal products of fishery, animal husbandry and their processing by-products. There are 3 types of animal feed: ① Feeds with less than 20% of crude protein (dry basis) belong to protein feed, such as fish meal, animal blood, silkworm pupae, etc. Its CFN is expressed as 5-13-0000; ② Animal fats with less than 20% of crude protein (dry basis) and low crude ash content, such as tallow, etc., belong to energy feed with CFN being 4-13-0000; ③ Feeds with less than 20% of crude protein (dry basis) and low crude fat content belong to mineral feed, mainly in purpose of supplying calcium and phosphorus, such as bone meal, shell meal, etc. Its CFN is expressed as 6-13-0000.

十四、矿物质饲料
Mineral feed

矿物质饲料有两种类型:①可供饲用的天然矿物质,如石灰石粉等;或化工合成无机盐类和有机配位体与金属离子的螯合物、络合物,如磷酸氢钙、硫酸铜等,CFN形式为6-14-0000。②来源于动物性饲料的矿物质也属此类,如骨粉、贝壳粉等,CFN形式为6-13-0000。

There are 2 types of mineral feed: ① Natural minerals for feeding, such as limestone powder, etc., or chemically synthesized inorganic salts and chelate or complexes of organic ligands and metal ions, such as calcium hydrogen phosphate, copper sulfate, etc with CFN being 6-14-0000. ② Minerals derived from animal sources also belong to this category, such as bone meal, shell meal, etc. with CFN being 6-13-0000.

十五、维生素饲料
Vitamin feed

维生素饲料是指由工业合成或提取的单一或复合维生素制剂,如琉胺素、核黄素、胆碱、维生素A、维生素D、维生素E等,但不包括富含维生素的天然青绿多汁饲料。CFN形式为7-15-0000。

Vitamin feed refers to single or multi-vitamin products synthesized or extracted by industry, such as thiamine, riboflavin, choline, vitamin A, vitamin D, vitamin E, etc., but no natural green fresh foliage rich in vitamins are included. Its CFN is expressed as 7-15-0000.

十六、饲料添加剂
Feed additives

饲料添加剂是指为了补充营养物质,保证或改善饲料品质,提高饲料利用率,促进动物生长和繁殖,保障动物健康而掺入饲料中的少量或微量营养性及非营养性物质。我国将饲料添加剂分为两种类型:① 营养性饲料添加剂,如用于补充氨基酸的工业合成赖氨酸、蛋氨酸等,CFN形式为5-16-0000;② 非营养性添加剂,如生长促进剂、饲料防腐剂、饲料黏合剂、驱虫保健剂等非营养性物质,CFN形式为8-16-0000。

Feed additives refer to a small amount or trace nutritive or non-nutritive substances incorporated into feed to supplement nutrients, ensure or improve feed quality, improve efficiency of feed utilization, promote animal growth and reproduction, and ensure animal health. There are 2 types of feed additives in China:① Nutritive feed additives, such as synthetic lysine and methionine used to supplement amino acids. The CFN is expressed as 5-16-0000;② Non-nutritive additives, such as growth accelerators, feed preservatives, feed binders, insect repellents and health care agents. The CFN is expressed as 8-16-0000.

十七、油脂类饲料及其他
Oil feed and others

油脂类饲料是以补充能量为目的,以动物、植物或其他有机物为原料,经压榨、浸提等工艺制成的饲料,属于能量饲料,CFN形式为4-17-0000。

Oil feed is made of animals, plants or other organic materials through pressing, extracting and other processes for the purpose of supplying energy. It belongs to energy feed with CFN being 4-17-0000.

十八、中国饲料分类编码
Chinese feed classification encoding

中国饲料编码共"3节,7位数",首位为IFN分类编号,第2、3

Feeds number of China consists of "3 sectrons with 7 digits", the first digit denotes the International Feed Num-

位为 CFN 亚类编号,第 4-7 位为具体饲料顺序号,表示为 △-△△-△△△△。如:青绿多汁类饲料的 CFN 饲料编码为 2-01-0000。此类分类方法的特点有:① 增加了 2、3 位码的层次,既可以根据国际饲料分类原则判定饲料性质,又可以根据传统习惯从亚类中检索出饲料资源出处,对 IFN 系统进行了补充及修正;② 较 IFN 系统容纳量提高,此系统最多能容纳 $8 \times 16 \times 9\,999 = 1\,279\,872$ 种饲料。

ber, and the second and third digits are the CFN subclass numbers, and the fourth to seventh digits are the specific feed numbers, expressed as △-△△-△△△△. For example, the CFN of green fresh forage is 2-01-0000. This specific feed number system has several advantages from different aspects. ① It not only shows the nature of feed according to the principles of IFN in the second part of CFN, but also retrieves the feed subclasses based on customary feed classification. The IFN system is supplemented and modified; ② Compared with the IFN system, the capacity of CFN system is increased with a maximum of $8 \times 16 \times 9\,999 = 1\,279\,872$ feed number.

第三章 青干草与粗饲料
Chapter 3　Green hay and roughage

粗饲料是指自然状态下水分含量在60%以下、干物质中粗纤维含量不低于18%，能量价值低的一类饲料，主要包括干草类、农副产品类、树叶、糟渣类等。粗饲料中的粗纤维含量很高，一般可达25%-45%，且主要由纤维素、半纤维素、木质素、果胶、多糖醛和硅酸盐等组成，但粗纤维的组成因粗饲料的类型不同或者植物的生长阶段不同而具有一定的差异。粗纤维大多不易消化，所以粗饲料中可消化营养成分含量较低，有机物的消化率在70%以下，而且粗饲料质地较粗硬，适口性较差，但是它是家畜不可缺少的饲料，尤其是草食家畜不可缺少的饲料种类，同时对单胃动物也有促进肠胃蠕动和增强消化力的作用。

粗饲料的来源很广、数量庞大，主要来源是农作物的秸秆秕壳，而且总量是粮食产量的1-4倍之多。据不完全统计，目前全世界每年农作物秸秆产量约为38亿t，我国每年产量超过8亿t，位居世界第一。野生的禾本科草本植物量更大。在这些无法为人食用的生物总量中，却蕴藏着巨大的潜在营养和能量。因此，若对其进行适当的加工处理并应用于畜牧生产，必将获得巨大的生产、生态、经济和社会效益。

Roughage refers to a kind of feed with less than 60% of moisture content in natural state and 18% or more of crude fiber content based on dry matter of feed and low energy value, mainly including hay, by-products from agricultural processing, leaves and dregs, etc. Usually, the crude fiber content is high up to 25%-45% in roughage, including cellulose, hemicellulose, lignin, pectin, polysaccharides and silicate, etc. However, the composition of crude fiber varies with different types of roughage or different growth stages of plants. Most of crude fiber is hardly digested, hard to eat, resulting in lower digestible nutrients and 70% digestibility of organic matter in roughage. In addition, roughage has rough texture and poor palatability, but it is an indispensable feed for domestic animals, especially for herbivorous animals. At the same time, roughage promotes gastrointestinal motility and enhances nutrients digestibility for monogastric animals.

The source of roughage is very wide and the quantity is huge. The main source is the crop straw or shell, and their total yield is approximately 1 to 4 times by the grain output. According to incomplete statistics, the annual crop straw output around the world is about 3.8 billion tons, and Chinese annual output is more than 800 million tons, ranking first in the world. Moreover, the quantities of wild grasses are more abundant. These inedible resources, however, maintain considerable potential in providing the nutrition and energy to domestic animals. Therefore, if these precious ingredients would be properly processed and applied to livestock breeding, substantial production, ecological, economic and social benefits will begained.

第一节 青干草与草粉
Section 1 Green hay and grass powder

在植物的生长季节,青草是放牧和舍饲最好的饲草。但在植物的非生长季节,由于天气寒冷,牧草的地上部分便枯萎死亡,遗留在地面上的枯草,其营养价值较夏秋青绿牧草下降60%-70%,特别是优良的豆科和禾本科牧草,其营养价值几乎损失殆尽。对于冬春季节舍饲期的草食家畜,单依靠营养价值较低的秸秆,远满足不了其营养需要。因此,在夏秋牧草旺盛生长期,在调制青贮料的同时,应调制贮备好优质青干草供冬春季节利用。

优质草粉是近几年兴起的配合饲料的重要原料,在畜禽尤其是单胃畜禽养殖业中发挥着愈来愈重要的作用,需要我们给予足够的重视。

During the growing season, grass is the best forage for grazing and feedstock. However, in the non-growing season, due to the cold weather, the above ground parts of the herbage wither and die, and the nutritive value drops by 60%–70% compared with that of them in summer and autumn, especially the fine legumes and grass, their nutritive is almost lost. In winter and spring, herbivores are fed by low nutritional straw solely is far from enough to meet their nutritional needs. Therefore, in the vigorous growth period of herbage, both high quality silage and green hay should be stored properly and prepared well for use in winter and spring.

As an important raw material of formulated feed, high quality grass powder plays an increasingly important role in livestock production, especially monogastric animals industry, which needs our enough attention.

一、青干草调制原理与方法
The principle and method of green hay processing

青干草是将牧草及禾谷类作物在质量和产量最好的时期刈割,经自然或人工干燥调制成长期保存的饲草。青干草可常年供家畜饲用。优质的干草,颜色青绿,气味芳香,质地柔松,叶片不脱落或脱落很少,绝大部分的蛋白质、脂肪、矿物质和维生素被保存下来,是家畜冬季和早春不可缺少的饲草。调制青干草方法简便,成本低,便于长期大量贮藏,在畜禽饲养上有重要作用。随着农业现代化的发展,牧草的刈割、搂草、打捆实现了机械化,青干草的质量在不断提高。

干草是草食家畜必备的饲草,

Green hay is a kind of forage grass and cereal crop which is harvested in proper period with good quality and yield and then is prepared by natural or artificial drying, and then can be used all year round. High quality hay is green in color, fragrant in smell, soft in texture, with no or little shedding of leaves. Most of the protein, fat, minerals and vitamins are preserved, and it is an indispensable forage for livestock in winter and early spring. Making hay plays an important role in raising livestock and poultry due to some advantages, such as simple processing, lower cost, convenient for long-term storage. With the development of agricultural modernization, the cutting, raking and baling of herbage have been mechanized, and the quality of green hay has been constantly improved.

Hay is necessary for herbivores and can't be taken

是秸秆等不可替代的饲料种类。不同类型的畜牧生产实践表明,只有优质的青干草才能保证家畜的正常生长发育,才能获得优质高产的畜产品。

place by straw. The real breeding efficiency in different types of livestock production shows that only the high-quality green hay ensure the normal growth and development of livestock, and high quality and yield of livestock products.

(一) 青干草调制过程中营养物质的变化规律
The changes of nutrients in green hay processing

在青草干燥调制过程中,草中的营养物质发生了复杂的物理和化学变化,一些有益的变化有利于干草的保存,一些新的营养物质产生,与此同时,一些营养物质被损失掉。结合调制过程中营养物质变化的特点,干草的调制应尽可能地向有益方面发展。为了减少青干草的营养物质损失,在牧草被刈割后,应该使其迅速脱水,促进植物细胞死亡,减少营养物质不必要的分解浪费。

1. 牧草干燥水分散失的规律

正常生长的牧草水分含量为80%左右,青干草达到能贮藏时的水分则为15%-18%,最多不得超过20%,而干草粉水分含量为13%-15%。为获得这种含水量的青干草或干草粉,必须将植物体内的水分快速散失。刈割后的牧草散发水分的过程大致分为两个阶段:

第一阶段:也称凋萎期。此时植物体内水分向外迅速散发,良好天气经5-8 h,禾本科牧草含水量减少到40%-50%,豆科牧草减少到50%-55%。这一阶段从牧草植株体内散发的是游离于细胞间隙的自由水,散失水的速度主要取决于大气含水量和空气流速,所以干燥、晴朗有微风的条件能促使水分快速散失。

第二阶段:是植物细胞酶解作用为主的过程。这个阶段植物体内的水分散失较慢,这是由于水分的

During the grass drying process, the nutrients undergo complex physical and chemical changes, among which some beneficial changes are conducive to the preservation of hay. Some new nutrients are produced, and meanwhile, some nutrients are lost in the process. Combined with the characteristics of nutrient changes in the process of hay preparation, hay preparation should be developed to the beneficial aspects as far as possible. In order to reduce the nutrients loss, the herbage should be rapidly dehydrated after harvest to promote the death of plant cells and reduce the unnecessary waste of nutrients.

1. The regularity of moisture loss in herbage drying procedure

The moisture content of normal growth herbage is about 80%, the moisture of green hay which can be stored is 15%-18%, and is not more than 20%, while the moisture of hay powder is 13%-15%. In order to obtain this water content of green hay and hay powder, the water in the plant must be rapidly dissipated. The process of moisture distribution from the herbage after cutting includes two stages.

The first stage is also known as wilting period. During the wilting period, moisture rapidly distributes from plant. In sunny day, moisture drops to 40%-50% in Gramineae or 50%-55% in Leguminous grass after 5-8 h. In this stage, moisture released from herbage is named as free water which is floating in the intercellular space, and the rate of water loss depends mainly on the atmospheric water content and air velocity. Therefore, dry, sunny and breezy conditions promote the rapid loss of water.

The second stage is the process of enzymatic hydrolysis of plant cells. In this stage, the water loss speed is slow because the evaporation of the cuticle instead of transpiration

散失由第一阶段的蒸腾作用为主，转为以角质层蒸发为主，而角质层有蜡质，阻挡了水分的散失。使牧草含水量由 40%-55% 降至 18%-20%，需 1-2 天。

为了使第二阶段水分快速散失，可采取勤翻晒的方法。不同植物其保水能力也不相同，豆科牧草比禾本科保水能力强，所以其干燥速度比禾本科慢，这是由于豆科牧草含碳水化合物少、蛋白质多，影响了蓄水能力的缘故。另外，幼嫩的植物纤维素含量低，而蛋白质含量高，保水能力强，不易干燥；相对枯黄的植物则相反，易干燥。同一植物的不同器官，水分散失也不相同。叶片的表面积大，气孔多，水分散失快，而茎秆水分散失慢。因此，在干燥过程中要采取合理的干燥方法，尽量使植物各个部位均匀干燥。

2. 晒制过程中其他养分的变化

在晒制青干草时，牧草经阳光中紫外线的照射作用，植物体内的麦角固醇转化为维生素 D。这种有益的转化，可为家畜冬春季节提供维生素 D，而且是维生素 D 的主要来源。另外，在牧草干燥后，贮藏时其植株体内的蜡质、挥发油、萜烯等物质氧化产生醛类和醇类，使青干草有一种特殊的芳香气味，增加了牧草的适口性。

3. 青干草干燥过程中营养物质的损失及其影响因素

（1）植物体生物化学变化引起的损失：牧草刈割以后，晒制初期植物细胞并未死亡，其呼吸与同化作用继续进行。呼吸作用的结果使水分通过蒸腾作用减少；植物体内贮藏的部分无氮浸出物水解成单糖，作为能源被消耗；少量蛋白质也被

in water loss, and wax existed in cuticle prevents the loss of water. Approximately 1-2 days will be spent in ensuring moisture drops from 40%-55% to 18%-20%.

The herbage should be tedded frequently to accelerate the moisture loss during the second stage. Different plants also have different water-holding capacities. Leguminous herbages have better water-holding capacity than Gramineae, so their drying speed is slower than Gramineae. This is because Leguminous herbages contain less carbohydrates and more proteins, which affects their water-holding capacity. In addition, the young plants have low cellulose content, high protein content, strong water retention ability, and are not easy to dry. In contrast, relatively withered and yellow plants are easy to dry. Water loss varies from organ to organ in the same plant, such as water dissipates quickly in leaf due to large surface and more stomates, while water dissipates slowly in stalk. Therefore, in the drying process, each part of plant should dry evenly as far as possible by reasonable drying method.

2. Changes of other nutrients in herbage drying procedure

During the grass drying process, ergosterol in the plant is converted into vitamin D by the solar ultraviolet radiation. This beneficial conversion provides vitamin D to livestock during winter and spring, and that is the main source of vitamin D. In addition, during storage, the waxes, volatile oils, terpene and other substances in the plants are oxidized to produce aldehydes and alcohols, which gives green hay a special aromatic smell and increases the palatability of the herbage.

3. Nutrients loss and influencing factors in herbage drying procedure

（1）Loss caused by biochemical changes in plant body. After the cutting of herbage, plant cells do not die in the early stage of drying while respiration and assimilation always continue. As a result of respiration, moisture reduces by transpiration, part of nitrogen-free extracts stored in plants are hydrolyzed into simple sugars and consumed as energy, small amounts of protein are also broken down into

分解成肽、氨基酸等。当水分降低到40%-50%时，细胞才逐渐死亡，呼吸作用才会停止。据此在田间无论采用哪一种方法晒制青干草，都应迅速使水分下降到40%-50%，以减少呼吸等作用引起的损失。

细胞死亡以后，植物体内仍继续进行着氧化破坏过程，参与这一过程的既包括植物本身的酶类，又包括微生物活动产生的分解酶。破坏的结果使糖类分解成氧化碳和水，氨基酸被分解成氨而损失，胡萝卜素在体内氧化酶和阳光的漂白作用下遭到破坏，该过程直到水分减少到17%以下时才会停止。因此，调制过程中，应注意暴晒方法和时间，既使水分迅速降到17%以下，又要尽量减少其氧化破坏。

（2）机械作用引起的损失：干草在晒制和保藏过程中，由于受接草、翻草、搬草、堆垛等一系列机械操作的影响，不可避免地会造成部分细枝嫩叶破碎脱落。据报道，一般叶片损失20%-30%，嫩枝损失6%-10%。豆科牧草的茎叶损失比禾本科更为严重。鉴于植物叶片和嫩枝所含的可消化养分多，因此机械损失不仅使干草产量下降，而且使干草品质降低。为了减少机械损失，按调制需要，当牧草水分降至40%-50%时，应马上将草堆成小堆进行堆内干燥，并注意减少翻草、搬运时叶子的破碎脱落。

（3）阳光的照射与漂白作用的损失：晒制干草时主要是利用阳光和风力使青草水分降至足以安全贮藏的程度。阳光直接照射会使植物体内所含的胡萝卜素、叶绿素遭到破坏，维生素C几乎全部损失。叶绿素、胡萝卜素遭到破坏，导致叶色变浅，且光照愈强，曝晒时间愈长，

peptides, amino acids, etc. When water content drops to 40%-50%, cells gradually die and respiration stops. Accordingly, no matter what kind of method is adopted to make hay in the field, the moisture should be quickly reduced to 40%-50%, in order to reduce the loss caused by respiration and other effects.

After cell's death, the process of oxidative destruction continues in plants, which involves both plant enzymes and enzymes produced by microbial activities. In turn, sugars are degraded into carbon monoxide and water, amino acids are lost by breaking into ammonia, and carotene is destroyed by oxidase and sunlight bleaching until the water drops below 17%. Therefore, attention should be paid to the method and time of exposure, which is to not only quickly reduce moisture below 17%, but also to minimize oxidation damage during herbage drying procedure.

(2) Loss caused by mechanical action. In the process of drying and preservation of hay, due to the influence of a series of mechanical operations, such as grass grafting, turning over, moving grass, stacking, etc., it is inevitable that part of the twig young leaves will break and fall off. It has been reported that 20%-30% of leaf loss and 6%-10% of twig loss are common. The loss of stem and leaf in Leguminous herbage is more serious than that of Gramineae. Because plant leaves and twigs contain more digestible nutrients, mechanical loss reduces not only hay yield but also grass quality. In order to reduce the mechanical loss, when the grass moisture drops to 40%-50% according to the requirements of regulation, the grass should be immediately piled into small piles for in-pile drying, and attention should be paid to reduce the breakage and falling off of leaves during turning over and handling.

(3) Loss caused by sunlight exposure and bleaching effect. The main sources used to reduce the moisture content of grass under the lowest content for safe storage are sunlight and wind in hay making process. Direct exposure under sunlight leads carotene, chlorophyll to be destroyed, and vitamin C is almost totally lost in plants, which leads to lighter leaf color, and the stronger the light is, the longer the exposure time is, the greater the blenching loss will be.

漂白作用造成的损失愈大。据测定,干草曝露田间一昼夜,胡萝卜素损失75%,若放置一周,96%的胡萝卜素即遭破坏。所以,为了减少阳光对胡萝卜素及维生素C等营养物质的破坏,应尽量减少曝晒时间。后期采取小堆干燥,不仅可减少机械损失,也减少了阳光漂白作用。

(4)雨水淋洗作用造成的损失:晒制干草过程中如遇阴雨,则可造成可溶性营养物质的大量损失。试验表明,雨水淋洗可使40%可消化蛋白质受损,50%热能受损。若遇阴雨连绵加上干草霉烂,则营养物质损失甚至可达一半以上。

总之,晒制干草过程中营养物质的损失较大,总的营养物质要损失20%-30%,可消化蛋白质损失在30%左右,维生素损失50%以上。

(二)干草调制的方法
The method of hay processing

1. 田间干燥法

田间晒制干草可根据当地气候、牧草生长、人力及设备等条件,采取平铺晒草法、小堆晒草法或平铺小堆结合晒草法,以达到更多地保有养分的目的。

作物、牧草种类不同,刈割期不同。一般栽培的豆科牧草在初花期、禾本科牧草在抽穗开花期刈割。天然牧草可在夏秋季刈割,但以夏季刈割调制的青干草品质较优。人工栽培牧草应尽量采取非雨季节调制干草的方法。如:河南、河北、山东、陕西等省栽培苜蓿,可用第一茬(5月份)晒制,第二、三茬(正处于7-9月份雨季)作青饲料或青贮料用。

平铺晒草法虽然干燥速度快,

It is determined that when hay is exposed to the field for one day, 75% of carotene will be lost. If it is left for one week, 96% of carotene will be destroyed. Therefore shortening exposure time under sunlight is a good way to reduce the damage of carotene and vitamin C, and other nutrients as well. In the later stage, the small pile drying way should be adopted to reduce both the mechanical loss and sunlight bleaching.

(4) Loss caused by rainwater leaching. In case of overcast rain during hay making, a large amount of soluble nutrients can be lost. Tests show that 40% of the digestible protein and 50% of the heat energy are damaged by rain. In case of continuous rain and hay mildew, the loss of nutrients can even reach more than half.

In a word, the loss of nutrients in hay making process is greater, during which the total nutrients loss is up to 20%-30%, digestible protein loss is about 30%, and vitamin loss is more than 50%.

1. The field drying

The proper hay making method in field, such as tile curing, small pile curing, or tile plus small pile curing, should be adopted depending on the local climate, herbage growth, manpower and equipment conditions, to prevent nutrients loss.

Different kinds of crops and herbage should be mown at different growth period. Generally, cultivated Leguminous herbage should be mown at the early flowering stage and Gramineae should be mown at the heading and flowering stage. Wild herbage can be mown during summer and autumn, but the quality of green hay prepared by mowing in summer is better, and the hay should be made in non-rainy seasons. For example, alfalfa is cultivated in Henan, Hebei, Shandong, Shaanxi and other provinces. The first crop can be used to make hay in May, and the second and third crops can be used as fodder or silage during July to September.

Although the tile method is fast in drying, but the loss

但养分损失大,故目前多采用平铺与小堆结合晒草法。具体方法是:青草刈割后,即可在原地或另选一地势较高处将其摊开曝晒,每隔数小时翻草一次,以加速水分蒸发。一般早上刈割,傍晚叶片已凋萎,其水分估计已降至50%左右,此时就可把青草集成约1 m的小堆,每天翻动1次,使其逐渐风干。如遇天气恶化,草堆外层宜盖草苫或塑料布,以防雨水冲淋。待天气晴朗时,再倒堆翻晒,直至干燥。

田间平铺与小堆结合干燥法的优点是:① 初期干燥速度快,可减少植物细胞呼吸作用造成的养分损失;② 后期接触阳光曝晒面积小,能更好地保存青草中的胡萝卜素,同时因堆内干燥,可适当发酵,产生一些酯类物质,使干草具有特殊香味;③ 茎叶干燥速度趋于一致,可减少叶片嫩枝的破损脱落;④ 遇雨时便于覆盖,不致受到雨水淋洗,造成养分的大量损失。

2. 草架干燥法

在潮湿地区或多雨季节晒草,放在地面干燥容易导致牧草腐烂和养分损失,故宜采用草架干燥。用草架干燥,可先在地面燥4-10 h,待含水量降到40%-50%时,然后自下而上逐渐堆放。草架干燥方法虽然要花费一定经费建造草架,并多耗费一定劳力,但能减少雨淋的损失,通风好、干燥快,能获得品质优良的青干草,营养损失也少。

3. 化学制剂干燥法

应用较多的有碳酸钾、碳酸钾+长链脂肪酸混合液、碳酸氢钠等。其原理是这些化学物质能破坏植物体表面的蜡质层结构,促进植物体内的水分蒸发,加快干燥速度,减少豆科牧草叶片脱落,从而减少蛋白

of nutrients is large. As a result, tile plus small pile method is mostly used. Briefly, after mowing, the grass is exposed in the air in situ or another higher terrain, and turned over per few hours to accelerate the evaporation of water. Generally, the leaves wither in the evening if they are cut in the morning, and the water content is estimated to be about 50% at this time. Then, the grass can be integrated into a small pile of 1 m, which can be turned once a day to gradually air dry. In case of weather deterioration, the outer layer of the haystack should be covered with grass cover or plastic sheeting to prevent rain. When the weather is clear, pile them up and let them dry.

The merits of combination of tile technology and small pile come as follows: ① the drying speed is fast during in early period, reducing nutrients loss caused by plant cell respiration; ② in the late stage, carotene in the grass is better preserved due to the small area of exposure to sunlight, and the drying condition in pile facilitates hay to ferment properly during which a variety of esters can be produced, leading to the particular fragrance of hay; ③ the drying speed of stem and leaf tends to be the same, which can reduce the damage of leaf twig falling off; ④ it is easier to cover hay in case of rain to prevent a large loss of nutrients.

2. The frame drying

In wet areas or rainy seasons, grass drying on the ground can easily lead to the decay of herbage and nutrients loss, so it is advisable to use hay frame drying method. Firstly, grass should be dried on the ground for 4-10 h, when the water content drops to 40%-50%, then the grass piles up gradually from bottom to top using the hay frame. Although it costs a certain amount of money to build the frame and consumes a certain amount of labor, the drying method can reduce the nutrients loss by rain and lead to good ventilating and quicker drying speed, consequently, the higher quality green hay is made.

3. The drying using chemical agent

The chemical agents used are potassium carbonate, potassium carbonate + long chain fatty acid mixture, sodium bicarbonate, etc. The principle is these chemicals can destroy the structure of the waxy layer on the surface of the plant body, promote the evaporation of water in the plant body, accelerate the drying speed, reduce the leaf twig fall-

质、胡萝卜素和其他维生素的损失。但成本较田间干燥和草架干燥方法高,适宜在大型草场进行。

4. 人工干燥法

通过人工热源加温使饲料脱水,温度越高,干燥时间越短,效果越好。150 ℃干燥20-40 min即可;温度高于500 ℃,6-10 s即可。高温干燥的最大优点是时间短,不受雨水影响,营养物质损失小,能很好地保留原料本色。但机器设备耗资巨大,一台大型烘干设备安装至利用需数百万元,且干燥过程耗能多,故应慎用。

ing off, and thus reduce the loss of protein, carotene and other vitamins. However, the cost is higher than that of field drying and frame drying, so it is suitable to be carried out in large grasslands.

4. The artificial drying

The artificial hay drying is a grass dehydration method by heating with artificial heat source. The higher the temperature is, the shorter the drying time will be, and the better the effect will be. The drying time is 20-40 min when it is heated 150 ℃, and 6-10 s, above 500 ℃. The advantage of high heat drying is shorter during process and thus not affected by rain, lower nutrients loss, and preservation of grass characteristics. However, drying machinery and equipment cost a lot. A large drying equipment costs millions of RMB from installation to usage, and needs much energy to maintain normal function, so it should be used with caution.

二、青干草的营养价值
The nutritive value of green hay

青干草的营养价值与原料种类、生长阶段、调制方法有关。多数青干草的消化能值为8-10 MJ/kg,少数优质干草的消化能值可达到12.5 MJ/kg,还有部分干草的消化能值低于8 MJ/kg。干草粗蛋白质含量变化平均在7%-17%,个别豆科牧草高达20%以上。粗纤维含量高,为20%-35%,但其中纤维的消化率较高。此外,干草中矿物质元素含量丰富,一些豆科牧草中的钙含量超过1%,足以满足一般家畜需要,禾本科牧草中的钙也比谷类籽实高。维生素D含量可达到16-150 mg/kg,胡萝卜素含量为5-40 mg/kg。

干草营养价值的高低还与其利用有关。干草利用好坏,涉及干草营养物质利用的效率和经济效益。利用不好,可使损失超过15%。猪禽等单胃动物只宜利用高质量或粗

The nutritive value of green hay depends on plant species, growth stage and processing method. The digestible energy of most green hay is 8-10 MJ/kg, while some high-quality hay is as high as 12.5 MJ/kg and some low-quality hay, less than 8 MJ/kg. The crude protein content in hay varies from 7% to 17% on average, and that of some Leguminous herbage, up to 20%. The crude fiber content in hay is 20%-35%, but the fiber digestibility is high. In addition, hay is rich in mineral elements. The calcium content in some Legumes exceeds 1%, which is enough to meet the needs of common livestock. The calcium content in Gramineae is also higher than that in grain seeds. Vitamin D content in hay is 16-50 mg/kg, and carotene content is 5-40 mg/kg.

The nutritive value of hay also depends on its utilization, which relates to the nutrient utilization efficiency and economic benefit. Poor utilization can result in a nutrients loss of more than 15%. Monogastric animal, for example, pigs and poultries only are fed with high quality hay with

纤维含量较低的一些干草，如紫花苜蓿、紫云英等，且需限量饲喂，粉碎拌以精饲料饲喂为宜。牛羊利用干草可不受限制，但要注意采食过程中的浪费，最好适当切短，高低质量干草搭配饲喂，用饲槽让其随意采食较好。有条件的情况下，可将干草制成颗粒饲用，可提高干草利用率。粗蛋白质含量低的干草可配合尿素使用，有利于补充牛羊粗蛋白质摄入的不足。

lower crude fiber content, such as alfalfa and Astragalus sinicus L., etc., which should be limited in feeding amount and combined with concentrations by grind. For ruminants, hay feeding amount is not restricted, but it is better to pay attention to the waste in the process of feeding. It is better to cut the hay properly, feed the hay with high and low quality, and use the feeding trough to allow it to eat at will. If conditions permit, hay can be made into granules which can obviously improve nutrients' utilization. Hay with low crude protein content can be used in combination with urea, which is beneficial to supplement the deficiency of crude protein intake of cattle and sheep.

三、草粉的生产与应用
Production and application of grass powder

草粉及青干草粉与草捆、草颗粒一样，属于草产品，而且是一种主要的牧草产品形式，在发展畜牧业尤其是猪、禽、鱼养殖业中有重要的作用。

Grass powder and green hay powder, like bales and granules, belong to grass products and are main forms of forage products. They play an important role in the development of animal husbandry, especially in the breeding of pigs, poultry and fish.

（一）草粉生产
Grass powder production

加工草粉的原料主要是紫花苜蓿、三叶草等优质豆科牧草以及豆科与禾本科混播的牧草，优良的黑麦草、黑麦、羊草等禾本科牧草也可作为原料。生产草粉时对牧草的质量要求较高，故对刈割期的选择尤为重要，一般在牧草蛋白质和维生素含量以及产量较高的时期刈割，具体刈割期与青干草基本类同。采用先平铺后小堆的田间干燥或人工烘干法有利于保持草粉的绿色和良好的品质。牧草干燥至水分含量为13%-15%时，用锤片式粉碎机粉碎。粉碎的粒度依据饲养畜禽的种类而定，一般在鱼类饲料中应用粉碎细度为过0.30 mm筛，至少过0.45 mm筛；禽类和仔猪饲料中比鱼类稍粗些，草屑长度1 mm左右

The raw materials for processing grass powder are mainly high-quality Leguminous grass such as alfalfa and clover, as well as grass mixed with Legumes and Grass. Excellent Gramineae such as rye grass and wild rye can also be used as raw materials. In the process of producing grass powder, the quality requirement of herbage is high, so it is particularly important to select the mowing period. Generally, the grass should be mown at the period with high protein and vitamin content and high yield, which is basically similar to green hay processing. It is helpful to keep the green and good quality of the powder by using the method of field drying by tile plus small pile or artificial drying. The grass will be crushed by hammer crusher when the moisture of grass drops to 13%-15%. The granularity of crushing depends on the species of livestock and poultry. Generally, the fineness of crushing is over 0.30mm, or at least over 0.45 mm in fish feed; the fineness of crushing is 1 mm for poultry and piglets; and the fineness of crushing for fatte-

为宜;育肥猪和母猪饲用的草屑可长至 2 mm 左右。为了减少草粉在贮存过程中的营养损失和便于运输,生产中常把草粉压制成草颗粒。一般草粒的密度为草粉的 2－2.5 倍,可减少草的运输体积,同时减少与空气的接触面积,从而减少养分的损失,并且在压制过程中,还可加入抗氧化剂,以减少胡萝卜素及其他维生素的破坏。

ning pigs and sows is about 2 mm. In addition, grass powder is often pressed into grass particles to reduce the nutrients loss in storage and facilitate transportation. Generally, the density of grass particles is 2 − 2.5 times of grass powder, which can reduce the transport volume of grass and the contact area with air, thus reducing the loss of nutrients. And in the process of suppression, the antioxidants can be added to reduce the damage of carotene and other vitamins.

(二) 干草粉的饲用价值
Feeding value of hay powder

优质的豆科、禾本科或豆科和禾本科混播的牧草草粉,具有蛋白质、维生素、β-胡萝卜素含量高的特点,可在反刍动物和单胃动物饲粮中应用。如:在现蕾至初花期刈割并且调制良好的优质紫花苜蓿草粉,在雏鸡和产蛋鸡饲粮中可用至 5%,青年鸡饲料中可用至 15%;育肥猪和母猪饲粮中可分别用至 10%－15% 和 15%－30%;兔饲粮中可用至 20%－50%。国外也有维生素草粉等干草产品。

维生素草粉有很高的营养价值。当维生素干草粉含粗蛋白质达到 19% 时,每千克草粉中含有的氨基酸和矿物质为:赖氨酸 11.6 g、蛋氨酸 2.1 g、色氨酸 2.9 g、胱氨酸 3.8 g;钙 21.6 g、磷 3.5 g、钾 14.9 g、钠 0.79 g。虽然草粉的能值较低而纤维素含量偏高,但是蛋白质、胡萝卜素和矿物质的含量则大大优于谷物,故优质草粉是配合饲料良好的补充剂,对畜禽饲粮的营养平衡作用很大。在国际市场上,优质牧草的价格相当于黄玉米的价格。欧美国家生产大量的维生素草粉,美国每年产百余万吨。

High quality Leguminosae, Gramineae or Leguminosae and Gramineae mixed grass powder, with high content of protein, vitamin and beta carotene, can be used in the diets of ruminants and monogastric animals. For instance, good quality alfalfa powder is milled from budding to early flowering and is well prepared. It can be used up to 5% in the diets of chicks and laying hens and up to 15% in the feed of young chickens; it can be used up to 10%−15% in the diets of fattening pigs and up to 15%−30% in the diets of sows; it can be used up to 20%−50% in the rabbit's diets. There are also vitamin grass powder and other hay products abroad.

Vitamin grass powder has high nutritive value. When the crude protein content of vitamin grass powder reaches 19%, the amino acids and minerals contained in each kilogram of grass powder are as follows: 11.6 g of lysine, 2.1 g of methionine, 2.9 g of tryptophan, 3.8 g of cystine; 21.6 g of calcium, 3.5 g of phosphorus, 14.9 g of potassium, 0.79 g of sodium. Although the energy value of grass powder is low and the content of cellulose is high, the content of protein, carotene and mineral is much better than that of grain. Therefore, high quality grass powder is a good supplement of mixed feed and plays a great role in the nutrient balance of livestock and poultry diets. In the international market, the price of high-quality pasture is equivalent to the price of yellow corn. Europe countries and the United States produce a large number of vitamin grass powder, while the United States produces more than a million tons every year.

影响草粉质量的因素很多。首先与原料的种类有关；其次是刈割期,过早刈割时牧草质量好但产量低,过迟刈割虽产量高,但木质化严重影响草及草粉的品质。牧草的年收获次数不同,草粉的营养成分亦不同。

牧草干燥方法对干草营养物质的损失率有决定性影响。当采用快速人工干燥牧草时,损失率最小,胡萝卜素损失不高于3%-10%,其他营养物质损失不高于3%-8%。

因此,美国、俄罗斯、新西兰、德国、丹麦、法国等国家多用滚筒式高温快速烘干机来干制牧草。初始被干燥牧草与400-1 150 ℃的热气流接触,终端热气流温度为90-120 ℃。对草茎段干燥时间为5-25 min,草叶段为0.2-2 min。滚筒式烘干机可以通过调整工作参数,将牧草烘干至某个最终含水量。不过,随最终含水量的不同,干草的质量指标与烘干机的热效率也将变化。

Many factors affect the quality of grass powder: the first is the type of raw materials, and the second is the mowing period. The quality of herbage is good but the yield is low in the early mowing period, while the yield is high in the late mowing period, but the quality of grass and powder is seriously affected by lignification. The annual harvest frequency of herbage is different, and the nutrient composition of grass powder is also different.

The drying method of herbage has a decisive influence on the nutrient loss of hay. The loss rate of carotene is not higher than 3%-10% and other nutrients are not higher than 3%-8% when rapid artificial drying is used.

Therefore, in the United States, Russia, New Zealand, Germany, Denmark, France and other countries roller type high temperature fast dryer is used to dry grass. At the beginning, the dried herbage is in contact with the hot gas flow of 400-1,150 ℃, and the temperature of the hot gas flow at the end is 90-120 ℃. The drying time of the stem segment is 5-25 min, and the leaf segment is 0.2-2 min. By adjusting the working parameters, the drum dryer can dry the pasture section to a certain final water content. However, the mass index of hay and the thermal efficiency of the dryer vary with the final moisture content of the grass segment.

四、干草产品的质量评定
Quality assessment of hay products

要合理利用干草产品,必须首先了解其品质。在生产上,品质良好的干草,可以广泛地应用,达到节省精料、提高生产力的目的。而品质低劣的干草产品不能用来饲喂动物,应及时处理,否则会影响家畜健康,造成经济损失。

To make rational use of hay products, one must first understand their quality. In production, good quality hay can be widely used to save concentrate and improve productivity. Poor quality hay products, however, cannot be fed to animals, and should be handled in a timely manner, otherwise it will affect livestock in health, causing economic loss.

（一）干草和干草粉的质量标准
Quality standards for hay and hay powder

优质干草和干草粉在外观上要求均匀一致,不霉烂或结块,无异味,色泽浅绿或暗绿,洁净而爽香,

High quality hay and hay powder are required to be uniform in appearance, with no mildew or cake or no peculiar smell, and should be light green or dark green in color,

不混入砂石、铁钉、塑料废品、破布等有害物质。评定干草的质量,合理的标准非常重要。在我国,大多采用优质的苜蓿为原料生产苜蓿草粉,我国饲料用苜蓿草粉质量标准为 NY/T 140—2002。

(二) 评定方法
Assessment method

1. 植物学组成

植物种类不同,营养价值差异较大。按植物学组成,牧草一般可分为豆科草、禾本科草、其他可食草、不可食草和有毒有害草共五类。

天然草地刈割晒制的干草,豆科比例大者为优等草;禾本科和其他可食草比例大者,为中等草;不可食草比例大者为劣等草;有毒有害植株超过 10% 者,则不可供作饲料。人工栽培的单播草地,只要混入杂草不多,就不必进行植物学组成分析。

2. 干草的颜色和气味

颜色和气味是干草品质好坏的重要标志。凡绿色程度越深的干草,表明胡萝卜素和其他营养成分含量越高,品质越优。此外,芳香气味也可作为评定干草品质优劣的标志之一。按绿色程度可把干草品质分为四类:

(1) 鲜绿色:表示青草刈割适时,调制过程未遭雨淋和阳光强烈曝晒,贮藏过程未遇高温发酵,较好地保存了青草中的成分,属优质干草。

(2) 淡绿色:表示干草的晒制和贮藏基本合理,未遭雨淋发霉,营养物质无重大损失,属中等品质干草。

(3) 黄褐色:表示青草刈割过晚,或晒制过程遭雨淋或贮藏期内

clean and fragrant with no harmful substances such as sand, iron nails, plastic waste, rags, etc. Reasonable criteria are very important in evaluating the quality of hay. In China, alfalfa powder is produced using high quality alfalfa as raw material. The quality standard of alfalfa in China is NY/T 140—2002.

1. Botanical composition

Different plant species have great differences in nutritive value. According to the botanical composition, herbage can be divided into five types: Leguminous grass, Gramineae grass, other herbivorous grass, non-herbivorous grass and poisonous and harmful grass.

The hay made by cutting and drying from natural grassland is superior grass with large proportion of Legumes. Grass with large proportion of Gramineae and other herbivores is medium grass. The large proportion of non-herbivore is inferior grass. More than 10% of poisonous and harmful plants should not be used as feed. In cultivated monoculture grassland, as long as there are few weeds mixed, botanical composition analysis is not necessary.

2. The color and smell of hay

The color and smell of hay are important indicators of hay quality. The darker the green degree is, the higher the content of carotene and other nutrients and the better the quality of hay will be. In addition, aroma can also be used as a marker to evaluate the quality of hay. According to the degree of green, the quality of hay can be divided into four categories:

(1) Bright green: Grass is timely mowed, and not exposed to rain and strong sunlight in the preparation process, and not fermented at high temperature in the storage process. It is considered as high-quality hay in that the ingredients are preserved well.

(2) Light green: Grass is regarded as medium quality hay because of reasonable drying and storage process, without being rot by rain or loss of nutrients.

(3) Yellowish brown: Grass is mowed late, or exposed to rain during the drying process, or fermented at

经过高温发酵,营养成分虽受到重大损失,但尚未失去饲用价值,属次等干草。

(4) 暗褐色:表示干草的调制与贮藏不合理,不仅受到雨淋,且发霉变质,不宜再作饲用。

3. 含叶量

一般来说,叶子所含的蛋白质和矿物质比茎多1-1.5倍,胡萝卜素多10-15倍,而粗纤维比茎少50%-100%。因此,干草含叶量也是评定其营养价值高低的重要标志。

4. 刈割期

刈割期对干草的品质影响很大,一般栽培豆科牧草在现蕾开花期、禾本科牧草在抽穗开花期刈割比较适宜。就天然草地野生牧草而言,可按优势的禾本科、豆科牧草确定刈割期。在禾本科牧草的穗中只有花而没有种子时,这时属于花期刈割;然而,当大部分穗中含有种子或者留下护颖时,则属于刈割过晚;豆科草如在茎下部的2-3个花序中仅见到花,则属花期刈割,如果草屑中有大量种子则属刈割过晚。

5. 含水量

含水量高低是决定干草在贮藏过程中是否变质的主要标志。干草按含水量一般分为四类:干燥(≤15%)、中等干燥(15%-17%)、潮的(17%-20%)、湿的(≥20%)。

生产中测定干草含水量的简易方法是:手握干草一束轻轻扭转,草茎破裂不断者为水分合适(17%左右);轻微扭转即断者,为过干象征;扭转成绳茎仍不断裂开者为水分过多。

6. 总评

凡含水量在17%以下,毒草及有害草不超过1%,混杂物及不可食草在一定范围之内,不经任何处

high temperature during the storage period. Although it suffers great loss of nutrients, it has not lost its feeding value yet. It belongs to the second-class hay.

(4) Dark brown: Hay is not properly prepared and stored. It is not only exposed to rain, but also deteriorated, so it is not suitable for feeding.

3. The leaf content of hay

Generally, the leaf contains 1-1.5 times of more protein and mineral, 10-15 times of more carotene, and 50%-100% less of crude fiber than the stem. Therefore, the leaf content of hay is also an important indicator to evaluate its nutritive value.

4. The mowing period

The mowing period has a great influence on the quality of hay. For example, it is more suitable to cut Leguminous herbage at the budding stage and, to cut Gramineae at the heading stage. As far as the wild herbage of natural grassland is concerned, the mowing period could be determined according to the dominant Gramineae and Leguminous herbage. It is the flowering mowing period if the grass is mown when it has only flowers and no seeds in the ear. Mowing while most of the ear has seeds or left to protect the glume belongs to the late mowing. For legumes, clipping while 2 to 3 inflorescence can only be seen in the lower stem, it is flowering clipping; it is a late clipping if lots of seeds can be seen in the grass cuttings.

5. The moisture content of hay

The moisture content is the main marker to determine whether hay deteriorates during storage. Hay is generally divided into four categories according to water content, including dry (≤15%), moderately dry (15%-17%), humid (17%-20%), wet (≥20%).

The simple method to determine the moisture content of hay in production is to hold a bundle of hay and twist it gently, and if the straw is fractured but not broken, then the moisture content is appropriate (about 17%); if the hay is broken with slight twist, it is too dry; if it is not broken after being twisted into a rope, it has too much water.

6. General evaluation

Qualified hay (or grade hay) shall be deemed as qualified hay (or grade hay) if the water content is less than 17%, the poisonous grass and harmful grass is not more

理即可贮藏或者直接喂养家畜者，可定为合格干草（或等级干草）。含水量高于17%，有相当数量的不可食草和混合物，需经适当处理或加工调制后，才能用于喂养家畜或贮藏者，属可疑干草（或等外干草）。严重变质、发霉，有毒有害植物超过1%以上，或泥沙杂质过多，不适于用作饲料或贮藏者，属不合格干草。对合格干草，可按前述指标进一步评定其品质优劣。

than 1%, the mixture and non-herbivorous grass are within a certain range, and can be stored without any treatment or directly fed to livestock. The hay is substandard variety if its water content is above 17%, or a significant amount of non-herbivorous and non-herbivorous mixture needs to be properly treated or processed before it can be used for livestock or hoarders. Hay with more than 1% of serious metamorphism, mildew, toxic and harmful plant, or too much sediment impurities, or not suitable for use as feed or storage, is unqualified. The quality of qualified hay can be further evaluated according to the above-mentioned indexes.

第二节　稿秕与饲用林产品饲料
Section 2　Straw, hull and forest products for feeding

稿秕饲料即农作物秸秆秕壳，其来源广、数量多，总量是粮食产量的1-4倍之多。这类饲料最大的营养特点是：粗纤维含量高，一般都在30%以上；质地坚硬，粗蛋白质含量很低，一般不超过10%；粗灰分含量高，有机物的消化率一般不超过60%。稿秕饲料对于草食家畜尤为重要，在某种情况下（如冬季耕牛），它们还是唯一的家畜饲料。另外，草食家畜消化道容积大，可采用秸秆等粗饲料来填充，以保证消化器官的正常蠕动，使家畜有饱感。对于奶牛，饲粮中使用一定比例的秸秆饲料，可提高奶的乳脂率。

The chaff hull and crop straw are rich in sources and large in quantity, and the total quantity is 1-4 times as much as the grain output. The biggest nutritional characteristic of this kind of feed is: high in crude fiber content, generally above 30%; texture is hard and crude protein content is low, generally no more than 10%; the crude ash content is high; the digestibility of organic matter generally does not exceed 60%. Straw and hull are particularly important for herbivores, and in some cases, such as farm cattle during winter, are the only feedstuffs. In addition, herbivorous livestock digestive tract volume is large, and straw and other roughage can be used to fill it to ensure the normal peristalsis of digestive organs, so that livestock have a sense of satiety. For dairy cows, using a certain proportion of straw in the diet can improve the milk fat ratio.

一、秸秆饲料
Straw feed

我国秸秆饲料主要有稻草、玉米秸、麦秸、豆秸和谷草等。

In China, straw feed is mainly composed of rice straw, corn straw, wheat straw, soybean straw and millet straw.

（一）稻草
Rice straw

稻草是水稻收获后剩下的茎

Rice straw is the stem and leaf lefted over after rice

叶,其营养价值很低,但数量非常大。我国稻草产量为1.88亿t,应重视利用。研究表明,牛、羊对其消化率为50%左右,猪一般在20%以下。

稻草的粗蛋白质含量为3%-5%,粗脂肪为1%左右,粗纤维为35%;粗灰分含量较高,约为17%,但硅酸盐所占比例大;钙、磷含量低,分别为0.29%和0.07%,远低于家畜的生长和繁殖需要。据测定,稻草的产奶净能为3.39-4.43 MJ/kg,增重净能为0.21-7.32 MJ/kg,消化能(羊)为7.32 MJ/kg。为了提高稻草的饲用价值,除了添加矿物质和能量饲料外,还应对稻草进行氨化、碱化处理。经氨化处理后,稻草的含氮量可增加一倍,且其中氮的消化率可提高20%-40%。

harvest. Its nutritive value is very low, but the quantity is very large. The rice straw production in China is 188 million tons, which should be utilized properly. Research has shown that the digestibility of straw is about 50% in cattle and sheep, while under 20% in pigs.

The crude protein content of straw is 3%-5%, the ether extract is about 1%, and the crude fiber is 35%. The content of crude ash is high, about 17%, but the proportion of silicate is large. Low calcium and phosphorus content (0.29% and 0.07%, respectively) is far below the growth and breeding needs of livestock. It is determined that the net energy for lactation of straw is 3.39-4.43 MJ/kg, the net energy for gain is 0.21-7.32 MJ/kg, and the digestible energy is 7.32 MJ/kg in sheep. In order to improve the feeding value of straw, besides adding mineral and energy feedstuffs, ammonization and alkalization treatment should also be carried out. After ammonization, the nitrogen content of straw can be doubled, and the digestibility of nitrogen can be increased by 20%-40%.

(二) 玉米秸
Corn straw

玉米秸具有光滑外皮,质地坚硬,一般作为反刍家畜的饲料,若用来喂猪,则难于消化。反刍家畜对玉米秸粗纤维的消化率在65%左右,对无氮浸出物的消化率在60%左右。玉米秸青绿时,胡萝卜素含量为3-7 mg/kg。

生长期短的夏播玉米秸,比生长期长的春播玉米秸粗纤维少,易消化。同一株玉米,上部比下部的营养价值高,叶片又比茎秆的营养价值高。玉米秸的营养价值优于玉米芯,和玉米苞叶的营养价值相似。

为了提高玉米秸的饲用价值,一方面,在果穗收获前,在植株的果穗上方留下一片叶后削取上梢青饲用,或者制成干草、青贮料,这种做法对玉米产量的影响较小,因为它改善了通风和照明条件。另一方

Corn straw, with a smooth and hard skin, is commonly used as a feedstuff for ruminants, but is difficult to digest when fed to pigs. The crude fiber digestibility is about 65% in ruminants, and the nitrogen free extract digestibility is about 60%. When corn straw is green, the content of carotene is 3-7 mg/kg.

The corn straw with short growing period cultivated in summer has less crude fiber and is easier to digest than the long growing period corn straw cultivated in spring. The upper part of the same corn plant has more nutritive value than the lower part, and the leaf has more nutritive value than the stem. The nutritive value of corn straw is superior to that of corn cob, and similar to that of corn bract.

In order to improve the feeding value of corn straw, on the one hand, before harvest, the upper shoot of corn stalk above the leaf can be cut for direct feeding, or making hay and silage. The practice shows less influence on the corn production because it improves the ventilation and lighting conditions. After harvest, on the other hand, the upper half

面,收获后立即将全株分成上半株或上 2/3 株切碎直接饲喂或调制成青贮饲料。

or upper 2/3 of the whole stalk can be immediately chopped and then fed directly to animals or mixed into silage.

(三) 麦秸
Wheat straw

麦秸的营养价值因品种、生长期的不同而有所不同,常用作饲料的有小麦秸、大麦秸和燕麦秸。小麦秸粗纤维含量高,并有硅酸盐和蜡质,适口性差,营养价值低,小麦秸主要用于饲喂牛、羊,经氨化或碱化处理后效果较好。大麦秸的产量比小麦秸要低得多,但适口性和粗蛋白质含量均高于小麦秸,可作为反刍动物的饲料。在麦类秸秆中,燕麦秸是饲用价值最好的一种,对牛、羊、马的的消化能分别高达9.17 MJ/kg、8.87 MJ/kg 和 11.38 MJ/kg。

The nutritive value of wheat straw varies with different varieties and growing periods. The varieties commonly used as feedstuffs are wheat straw, barley straw and oat straw. Wheat straw has high content of crude fiber, silicate and wax, and so shows poor palatability and low nutritive value. Wheat straw is mainly used for feeding cattle and sheep, and the quality is improved after ammonization or alkalization treatment. The yield of barley straw is much lower than that of wheat straw, but its palatability and crude protein content are higher than that of wheat straw, so it can be used as a feedstuff for ruminants. Among wheat straw, oat straw has the best feeding value, and its digestible energy reaches 9.17 MJ/kg, 8.87 MJ/kg, and 11.38 MJ/kg in cattle, sheep, and horses, respectively.

(四) 豆秸
Soybean straw

豆秸有大豆秸、豌豆秸和蚕豆秸等。由于豆科作物成熟后叶子大部分凋落,豆秸主要以茎秆为主,茎已木质化,质地坚硬,维生素与蛋白质也减少,但与禾本科秸秆相比较,其粗蛋白质含量和消化能都较高。风干大豆含有的消化能为:猪 0.71 MJ/kg、牛 6.82 MJ/kg、绵羊 5.99 MJ/kg。大豆秸适于饲喂反刍家畜,尤其适于喂羊。在各类豆秸中豌豆秸营养价值最高,但是新豌豆秸水分较多,容易腐败变黑,要及时晒干后贮存。在利用豆秸类饲料时,要很好地加工调制,搭配其他精粗饲料混合饲喂。

The soybean straw includes soybean stem, pea stem, and broad bean stem. Since most of the leguminous leaves fall off when leguminous crops mature, the main part of soybean stem is the stem, which is hard in texture and has less vitamins and protein. However, compared with grass straw, the crude protein content and digestibility are higher. The digestible energy of air-dried soybean stem is 0.71 MJ/kg for pigs, 6.82 MJ/kg for cattle, and 5.99 MJ/kg for sheep. Soybean stem is suitable for feeding ruminants, especially sheep. Among all kinds of soybean stem, pea stem has the highest nutritive value, but pea stem properly mowed has more moisture and is easy to rot and turn black, so it should be dried and stored in time. When soybean stem is used as a feedstuff, it should be well processed and mixed with other refined roughage.

（五）谷草
Millet straw

谷草即粟的秸秆，其质地柔软厚实，适口性好，营养价值高。在各类禾本科秸秆中，以谷草的品质最好，是马、骡的优良粗饲料，还可饲喂牛、羊，与野干草混喂，效果更好。

Millet straw is soft, palatable and nutritious. Among all kinds of grass straw, the millet straw has the best quality, which is an excellent feedstuff for horses and mules. It can also be fed to cattle and sheep and mixed with wild hay to get better results.

二、秕壳饲料
Hull

农作物收获脱粒时，除分离出秸秆外，还分离出许多包被子实的颖壳、荚皮与外皮等，这些物质统称为秕壳。由于脱粒时常沾染很多尘土异物，也混入一部分瘪的子实和碎茎叶，使它们的成分与营养价值往往有很大的变异。总的看来，除稻壳、花生壳外，一般秕壳的营养价值略高于同一作物的秸秆。

When crops are harvested and threshed, in addition to the separation of straw, a lot of glumes, pods and husks containing the bedding are also separated. These substances are collectively called hulls. As threshing is often contaminated not only with a lot of dust and foreign matter, but also mixed with a part of the flat seeds and crushed stems and leaves, so that their composition and nutritive value often have a great variation. In general, the nutritive value of the general chaff hull is slightly higher than that of the same crop straw excluding rice husk and peanut shell.

1. 豆荚类

如大豆荚、豌豆荚、蚕豆荚等。无氮浸出物含量为42%-50%，粗纤维为33%-40%，粗蛋白质为5%-10%，牛和羊消化能分别为7.0-11.0 MJ/kg、7.0-7.7 MJ/kg，饲用价值较好，尤其适于反刍家畜利用。

1. Pods

This kind of feedstuff includes bean pods, pea pods, broad bean pods, etc., whose nutrient content is as follows: 42%-50% of nitrogen free extract, 33%-40% of crude fiber, 5%-10% of crude protein, 7.0-11.0 MJ/kg of digestible energy for cattle, 7.0-7.7 MJ/kg of digestible energy for sheep. Pods have a good feeding value, especially suitable for ruminants.

2. 谷类皮壳

有稻壳、小麦壳、大麦壳、荞麦壳和高粱壳等。这类饲料的营养价值仅次于豆荚，但数量大、来源广，值得重视。其中稻壳的营养价值很差，对牛的消化能低，适口性也差，仅能勉强用作反刍家畜的饲料。但其经过适当的处理，如氨化、碱化、高压蒸煮或膨化可提高营养价值。另外，大麦秕壳带有芒刺，易损伤口腔黏膜引起口腔炎，应当注意。

2. Cereal hulls

This kind of feedstuff includes rice husks, wheat husks, barley husks, buckwheat husks and sorghum husks. The nutritive value of these ranks only second to pods, but this kind of feedstuff is worthy of attention due to its large quantity and wide sources. Among them, rice husks have poor nutritive value, low digestibility to cattle and poor palatability, which can only be barely used as ruminant's feedstuff. However, after proper treatment, such as ammonization, alkalization, high pressure cooking or puffing, the nutritive value can be improved. In addition, barley hulls have thorns, which can easily damage the oral mucosa causing

3. 其他秕壳

一些经济作物副产品如花生壳、油菜壳、棉子壳、玉米芯和玉米苞叶等也常用作饲料。这类饲料的营养价值很低，须经粉碎与精料、青绿多汁饲料搭配使用，主要用于饲喂牛、羊等反刍家畜。棉子壳含少量棉酚（约0.068%），饲喂时要防止中毒。

3. Other hulls

Some economical byproducts such as peanut shells, rape shells, cotton shells, corn cobs and corn bracts are also commonly used as feedstuffs. This kind of feedstuff has very low nutritive value, and must be used together with concentration and green forage, and mainly fed to ruminants such as cattle and sheep. Cottonseed husk contains a small amount of gossypol (about 0.068%), so the usage should be careful to prevent poisoning.

三、树叶和其他饲用林产品
Leaves and other forage forest products

树叶作为饲料，在国外已有30多年的历史。在俄罗斯、罗马尼亚、加拿大等国早已工厂化生产，且用叶粉代替草粉在全价配合饲料中应用，质优价廉，很受市场青睐。日本曾利用刺槐叶粉代替苜蓿草粉养鸡，效果很好。我国现有森林面积1.3亿多公顷，树叶产量占全树生物量的5%。每年各类乔木的嫩枝叶约有5亿多吨，薪炭林及灌木林的嫩枝叶数量也相当巨大，树木的子实也是良好的饲料。如果能合理利用这一宝贵资源，对我国饲养业的发展将会起到重要作用。

青贮技术可以降低桑叶、构树的单宁成分。

As feed, leaves have a history of more than 30 years abroad. Leaves have already been factory-produced in Russia, Romania, Canada and other countries. The usage of leaf powder instead of grass powder in the complete formula feed has been popular because of high quality and low price. The locust tree leaf powder was used instead of alfalfa powder to raise chickens in Japan and led to positive effect. In China, forest area is more than 130 million hectares, and leaves yield accounts for 5% of the total tree biomass. There are more than 500 million tons of young branches and leaves of all kinds of trees are produced each year. The number of twigs and leaves of firewood forests and shrubs is also quite large. The seed of trees is also a good feedstuff. If we can reasonably use this precious resource, it would play an important role in the development of animal husbandry.

Silage technology can reduce tannin content in mulberry leaf and paper mulberry.

第三节 粗饲料的加工调制及品质评定
Section 3 Processing and quality evaluation of roughage

粗饲料经过适宜加工处理，可明显提高其营养价值。大量科学研究和生产实践证明，粗饲料经一般粉碎处理可提高采食量7%，加工

The nutritive value of roughage is improved obviously after proper processing. Many scientific studies and production practices have proved that the feed intake of roughage is increased by 7% by general crushing treatment and 37% by

制粒可提高采食量37%;而经化学处理可提高采食量18%-45%,提高有机物的消化率30%-50%。因此,粗饲料的合理加工处理对开发粗饲料资源具有重要的意义。目前,粗饲料加工调制的主要途径有物理学、化学和生物学处理三种,分述如下。

processing and granulating treatment. After chemical treatment, the feed intake is increased by 18%-45%, and the digestibility of organic matter is increased by 30%-50%. Therefore, reasonable processing of roughage is of great significance for developing roughage resources. At present, there are three main ways of roughage processing and modulation: physical, chemical, and biological processing, which are described as follows.

一、物理加工
Physical processing

(一) 机械加工
Mechanical processing

机械加工是指利用机械将粗饲料铡碎、粉碎或揉碎,这是粗饲料利用最简便而又常用的方法,尤其是秸秆饲料比较粗硬,加工后便于咀嚼,可减少能耗,提高采食量,并减少饲喂过程中的饲料浪费。

Mechanical processing refers to the use of machinery to cut, crush or knead break roughage, and this is the simplest and common way of using roughage. In particular, straw feed is relatively coarse and hard, and it is easy to chew after procssing, which can reduce energy consumption, increase feed intake, and reduce feed waste in the feeding process.

1. 铡碎 利用铡草机将粗饲料切短至1-2 cm。稻草较柔软,可稍长些;而玉米秸秆较粗硬且有结节,以1 cm左右为宜。玉米秸青贮时,应使用铡草机切短至2 cm左右,以便于踩实。

1. Chop. This physical processing means using the hay cutter to chop the roughage to 1-2 cm long. Straw is softer and can be slightly longer; however, the corn stalk is coarse, hard, and nodular, which should be chopped to about 1 cm. In making silage, corn stalk should be cut to about 2 cm for easy compaction.

2. 粉碎 粗饲料粉碎可提高饲料利用率和便于混拌精饲料。冬春季节饲喂绵羊、山羊的粗饲料应加以粉碎,但不应粉碎太细,否则会影响反刍。粉碎机筛孔径以8-10 mm为宜。优质花生秧等制成干草粉作为猪、禽配合饲料的原料时,要粉碎成较细的粉状,以便充分搅拌。

2. Smash. The grinding of roughage can improve the utilization and facilitate the mixing with concentration. In winter and spring, the roughage should be crushed to feed sheep and goats, but should not be crushed too fine, otherwise it will affect the rumination. The sieve aperture of the crusher should be 8-10 mm. When the high-quality peanut seedling is crushed into hay powder as the raw feedstuff material for pigs and poultry, it should be crushed into a fine powder to be fully stirred.

3. 揉碎 揉碎机械是近几年推出的新产品。为适应反刍家畜利用粗饲料的特点,将秸秆饲料揉搓成丝条,尤其适于玉米秸的揉碎,可

3. Knead break. The kneading machine is a new product introduced in recent years. In order to adapt to the characteristics of roughage utilization in ruminants, the straw is rubbed into strips, especially suitable for the crushing of

饲喂牛、羊、骆驼等反刍家畜。秸秆揉碎不仅可提高适口性,也可提高饲料利用率,是当前秸秆饲料利用比较理想的加工方法,但秸秆揉碎的能耗较高。

corn stalks, and can be fed to cattle, sheep, camels and other ruminants. Straw crushing improves both palatability and utilization, which is the ideal processing method for straw processing. However, the energy consumption of straw crushing is higher.

(二) 热加工
Hot processing

1. 蒸煮

将切碎的粗饲料放在容器内加水蒸煮,以提高秸秆饲料的适口性和消化率。吉林省延边朝鲜族自治州的农民多年来都有蒸煮稻草的习惯,有时还添加尿素,以增加饲料中粗蛋白质的含量。据报道,在压力 2.07×10^6 Pa 下处理稻草 1.5 min,可获得较好的效果。如压力为 $7.8 \times 10^5 - 8.8 \times 10^5$ Pa,则需处理 30 - 60 min。

1. Steaming and boiling

To improve the palatability and digestibility of straw, the chopped roughage is put into a container and steamed with water. Farmers in Yanbian Korean Autonomous Prefecture, Jilin Province, have been steaming straw for years, sometimes adding urea to increase the crude protein content. It is reported that good results are obtained by treating straw under pressure of 2.07×10^6 Pa for 1.5 min. If the pressure is $7.8 \times 10^5 - 8.8 \times 10^5$ Pa, it needs to be treated for 30 - 60 min.

2. 膨化

膨化是利用高压水蒸气处理后突然降压以破坏纤维结构,对秸秆甚至木材都有效果。膨化可使木质素低分子化和分解结构性碳水化合物,从而增加可溶性成分。内蒙古提出了热喷处理工艺,麦秸在气压 7.8×10^5 Pa 下处理 10 min,喷放压力为 $1.37 \times 10^6 - 1.47 \times 10^6$ Pa。干物质消化率和动物增重速度均有显著提高。但因膨化设备投资较大,目前在生产上尚难以广泛应用。

2. Puffing

Puffing is the sudden reduction of pressure after treatment with high pressure water vapor to destroy the fiber structure, which is effective for straw and even wood. Puffing can make lignin to be lowly molecular and decompose structural carbohydrate, thus increase the soluble component. In Inner Mongolia, the wheat straw is treated at air pressure of 7.8×10^5 Pa for 10 min and the spraying pressure is $1.37 \times 10^6 - 1.47 \times 10^6$ Pa, then the digestibility of dry matter and average daily gain of animal significantly increase. However, due to the large investment for equipment, it is difficult to be widely used in production.

3. 高压蒸汽裂解

高压蒸汽裂解是将各种农林副产物,如稻草、蔗渣、刨花、树枝等置入热压器内,通入高压蒸汽,使物料连续发生蒸汽裂解,以破坏纤维素-木质素的紧密结构,并将纤维素和半纤维素分解出来,以利于牛羊消化。此法与膨化法一样实用性较差。

3. High pressure steam cracking

High pressure steam cracking is to put all kinds of agricultural and forestry by-products, such as straw, bagasse wood, branches into the hot press through high pressure steam. Then the continuous steam cracks materials, such as the cellulose-lignin tight structure, then cellulose and hemicellulose are decomposed out to facilitate digestion for cattle and sheep. In fact, this method is as poor in practicability as puffing.

（三）盐化
Salinization

盐化是指铡碎或粉碎的秸秆饲料，用1%的食盐水，与等重量的秸秆充分搅拌后，放入容器内或在水泥地面堆放，用塑料薄膜覆盖，放置12-24 h，使其自然软化，可明显提高适口性和采食量。此法在东北地区广泛利用，效果良好。

Salinization means the straw feed is cut or crushed, stirred with 1% salt water and equal weight straw, and then put into a container or piled on the cement floor. Then the treatment straw is covered with plastic film and left for 12 to 24 hours to soften naturally, which can significantly improve palatability and intake. It is widely used in northeast China with good results.

（四）其他
Others

除上述3种途径外，还可利用射线照射以增加饲料的水溶性部分，提高其饲用价值。有人曾用γ射线对低质饲料进行照射，有一定的效果，但设备造价高，难以在生产上应用。

In addition to the above three ways, the water-soluble part of the feed can be increased by exposure to radiation to improve its feeding value. Some people have used γ ray irradiation to low quality feed, which has certain effect, but because of the high cost of equipment, it is difficult to be used in production.

二、化学处理
Chemical treatment

（一）碱化处理
Alkalization treatment

碱化作用是通过碱类物质的氢氧根离子打断木质素与半纤维素之间的酯键，使大部分木质素（60%-80%）溶于碱中，把镶嵌在木质素-半纤维素复合物中的纤维素释放出来。同时，碱类物质还能溶解半纤维素，也有利于反刍动物对饲料的消化，可提高粗饲料的消化率。碱化处理所用原料主要是氢氧化钠和石灰水。

Alkalization breaks the ester bond between lignin and hemicellulose by the hydroxyl ion, so that most lignin (60%-80%) is dissolved in the alkali and the cellulose embedded in the lignin-hemicellulose complex is released. At the same time, alkali can also dissolve hemicellulose, which is also beneficial to the digestion of ruminants and can improve the digestibility of roughage. Alkalization uses sodium hydroxide and lime water as raw materials.

1. 氢氧化钠处理

1921年德国化学家贝克曼首次提出"湿法处理"，即将秸秆放在盛有1.5%氢氧化钠的溶液池内浸泡24 h，然后用水反复冲洗至中性，

1. Treatment of sodium hydroxide

In 1921, German chemist Beckmann first proposed "wet treatment", that is, straw was put into a solution pool containing 1.5% sodium hydroxide for 24 h, and then was repeatedly washed to neutral with water. Whatever wet or

湿喂或晾干后喂反刍家畜,有机物消化率可提高25%。但此法用水量大,许多有机物被冲掉,且污染环境。1964年威尔逊等提出了改进方法,用占秸秆质量4%-5%的氢氧化钠,配制成30%-40%的溶液,喷洒在粉碎的秸秆上,堆放数日,不经冲洗直接喂用,可提高有机物消化率12%-20%,称为"干法处理"。这种方法虽较"湿法"有较多改进,但畜类采食后粪便中含有相当数量的钠离子,对土壤和环境也有一定程度的污染。

2. 石灰水处理

生石灰加水后生成的氢氧化钙是弱碱溶液,经充分熟化和沉积后,可用上层的澄清液(即石灰乳)处理秸秆。具体方法是:每100 kg秸秆,需3 kg生石灰,加水200-300 kg,将石灰乳均匀喷洒在粉碎的秸秆上,堆放在水泥地面上,经1-2天后可直接饲喂牛羊。这种方法成本低,生石灰来源广,方法简便,效果明显。苏联在20世纪30-40年代就广泛应用此法,我国在许多地方也有采用此方法的。

(二)氨化处理
Ammonification treatment

氨化处理秸秆饲料始于20世纪70年代。秸秆饲料蛋白质含量低,但当与氨相遇时,其有机物与氨发生氨解反应,破坏木质素与多糖(纤维素、半纤维素)链间的酯键结合,并形成铵盐,成为牛、羊瘤胃内微生物的氮源。同时,氨溶于水形成的氢氧化铵对粗饲料有碱化作用。因此,氨化处理是通过氨化与碱化双重作用以提高秸秆的营养价值。秸秆经氨化处理后,粗蛋白质含量可提高100%-150%,纤维素

dry feeding, the organic matter digestibility increased by 25% in ruminants. This method, however, consumed a lot of water, and then wasted many organic matter and polluted the environment. In 1964, Wilson et al. proposed an improved method, which used sodium hydroxide accounting for 4%-5% of the straw mass to make a solution of 30%-40%, and then sprayed it on the crushed straw, piled it up for several days, and fed it directly without washing, which could improve the organic matter digestibility by 12%-20%, known as "dry treatment". Although this method is more improved than "wet method", considerable amount of sodium ions appear in the feces of livestock after feeding, which also pollutes the soil and environment to a certain extent.

2. Lime water treatment

Calcium hydroxide generated by adding water to quicklime, is a weak alkali solution. After full maturation and deposition, the liquid supernatant (lime milk) can be used to treat straw. The specific method is as follows: for every 100 kg straw, 3 kg of quicklime is required, and 200-300 kg of water is added. The lime milk is uniformly sprayed on the crushed straw which is piled on the cement floor. After one to two days, it can be directly fed to cattle and sheep. This method has the advantages of low cost, wide sources of quicklime, simple processing, and obvious effect. This method was extensively used in Soviet Union from 1930s to 1940s, and is popular in many places of China.

Ammonification treatment for straw feed began in the 1970s. Straw feed has low protein content, but the organic matter reacts with ammonia in ammonification. Then the ester bond between lignin and polysaccharide (cellulose and hemicellulose) chain is destroyed, and then formed ammonium salt, which becomes the nitrogen source of microorganisms in the rumen of cattle and sheep. At the same time, alkalization occurs because ammonium hydroxide dissolves in water. Therefore, ammonization treatment improves the nutritive value of straw by both ammonization and alkalization. After ammonization, the crude protein content of straw increases by 100%-150%, the cellulose content de-

含量降低10%,有机物消化率提高20%以上。氨化后的秸秆质地柔软,气味糊香,颜色棕黄,提高了饲料的适口性,增加了采食量,是牛、羊良好的粗饲料。

氨化饲料的质量受秸秆饲料本身的质地优劣、氨源的种类及氨化方法等诸多因素的影响。氨源的种类很多,国外多利用液氨,需有专用设备,进行工厂化加工或流动服务。我国广大农村多利用尿素、碳酸氢铵作氨源。靠近化工厂的地方,氨水价格便宜,也可作为氨源使用。由于氨化饲料制作方法简便,饲料营养价值提高显著,所以,近年来世界各国普遍采用。我国自20世纪80年代后期开始推广应用,尤其是小麦秸、稻草氨化较多。

creases by 10%, and the digestibility of organic matter increases by more than 20%. The ammoniated straw is soft in texture, fragrant in smell and brown in color, which improves the palatability and increases the feed intake, making it a good roughage for cattle and sheep.

The quality of ammoniated feed is affected by many factors, such as the quality of straw itself, the type of ammonia source and the ammonization method. There are many kinds of ammonia sources. Liquid ammonia is mostly used abroad. Special equipment is needed for factory processing or mobile service. The urea and ammonium bicarbonate are used as ammonia sources in Chinese rural areas. Close to chemical plants, ammonia water is cheap and can also be used as an ammonia source. The ammoniated feed is widely used in the world in recent years because of its simple production method and significant increasing in its nutritive value. Since the late 1980s, ammoniated wheat straw and rice straw have been widely used in China.

(三)酸处理
Acid treatment

使用硫酸、盐酸、磷酸和甲酸处理秸秆饲料称为酸处理,其原理和碱化处理相同,用酸破坏木质素与多糖(纤维素、半纤维素)链间的酯键结构,以提高饲料的消化率。但酸处理成本太高,在生产上很少应用。

The use of sulfuric acid, hydrochloric acid, phosphoric acid, and formic acid to treat straw feed is called acid treatment. The principle is the same as alkalinization. The acid destroys the ester bond structure between lignin and polysaccharides (cellulose and hemicellulose) chain, thus improves the digestibility of the feed. But the cost of acid treatment is too high and seldom used in production.

(四)氨-碱复合处理
Ammonia-alkali compound treatment

为了使秸秆饲料既能提高营养成分含量,又能提高饲料的消化率,理想的做法是把氨化与碱化二者的优点结合利用,即秸秆饲料氨化后再进行碱化。如:稻草氨化处理的消化率仅55%,而复合处理后则达到71.2%。复合处理虽投入成本较高,但能够充分发挥秸秆饲料的经济效益和生产潜力。

In order to improve both nutrient content and digestibility in straw feed, the ideal way is to combine the advantages of ammonification and alkalization, that is, the straw feed receives ammonification and then alkalization. For example, the digestibility of ammoniated straw is only 55%, but reaches 71.2% after compound treatment. Although the cost is high, the compound treatment gives full play to the economic benefit and production potential of straw feed.

三、生物学处理
Biological treatment

粗饲料的生物学处理主要是指利用微生物处理。其主要原理是利用某些有益微生物,在适宜培养的条件下,分解秸秆中难以被家畜利用的纤维素或木质素,并增加菌体蛋白、维生素等有益物质,软化秸秆,改善适口性,从而提高粗饲料的营养价值。微生物种类很多,但用于饲料生产真正有价值的是乳酸菌、纤维分解菌和某些真菌。应用这些微生物加工调制粗饲料与青贮饲料、发酵饲料一样,也是在厌氧条件下加入适当的水分、糖分,在密闭的环境下进行乳酸发酵。在粗饲料微生物的处理方面,国外筛选出一批优良菌种用于发酵秸秆,如层孔菌、裂褶菌、多孔菌、担子菌、酵母菌、木霉等。我国已培育出一些可供生产应用的优良菌株,并有了成型的固体培养技术,已有一定的优势。以下介绍目前常用的两种方法。

The biological treatment of roughage mainly refers to the use of microorganisms in treating roughage. The main principle is to use some beneficial microorganisms to decompose the cellulose or lignin in the straw that is difficult to be used by livestock under suitable culture conditions, and increase the beneficial substances such as microbial crude protein (MCP) and vitamins to soften the straw and improve the palatability, so as to improve the nutritive value of roughage. There are many types of microorganisms, but the truly valuable ones for feed production are lactobacillus, cellulosic and certain fungi. The application of these microorganisms to the processing of roughage is the same as silage and fermentation feed—lactic acid fermentation under anaerobic conditions with appropriate water and sugar in a closed environment. In terms of the processing of selection, several excellent strains have been screened out from abroad for fermentation, such as *Fome Lividus*, *Schizophyllum commune*, *Polyporus anceps*, *Basidi omycete*, yeast and *Trichoderma viride*, etc. In China, some excellent strains for production and application have been cultivated, and the forming technology of solid culture has some advantages. The following two methods are commonly used in current treatment of microorganisms.

(一)粗饲料发酵法
Fermentation of roughage

粗饲料发酵法分为四步进行:首先,将准备发酵的粗饲料如秸秆、树叶等切成 20－40 mm 的小段或粉碎;其次,按每 100 kg 粗饲料 1－2 g 加入用温水化开的菌种,搅拌均匀,使菌种均匀分布于料中,边翻搅、边加水,水温以 50 ℃ 为宜,水分以手握紧饲料、指缝有水珠但不流出为宜;再次,将搅拌好的饲料堆积或装缸,插入温度计,上面盖好一层干草粉,当温度上升到 35－45 ℃ 时,翻动一次;最后,堆积或装缸,压

The roughage fermentation method is divided into 4 steps. Firstly, the roughage prepared for fermentation, such as straw and leaves, is cut into small segments of 20－40 mm long or crushed. Secondly, 1－2 g of each 100 kg roughage is added with warm water to make the strains evenly distributed in the roughage. The fungus strains are evenly mixed, and the water is added while stirring. The water should be warm water at 50 ℃. Thirdly, the mixed feed is piled up or into the cylinder, inserted into the thermometer, covered with a layer of hay powder. When the temperature rises to 35－45 ℃, turn roughage once. Finally, the roughage is piled up or filled in the cylinder, compacted

实封闭1-3天即可饲喂。

在生产上,也有在粉碎的粗饲料中加入麦麸,再接种链孢霉菌制成菌丝。因为链孢霉菌体含有丰富的蛋白质、碳水化合物,还有蛋白酶、淀粉酶、脂肪酶,能促进消化,对育肥猪有良好作用。

and sealed for 1 to 3 days. Then the roughage can be fed to livestock.

In production, wheat bran is also added to the crushed roughage and then inoculated with streptomyces to make mycelia. Because streptomyces body contains rich protein, carbohydrate, protease, amylase, lipase, this type of fermented mixed feed can promote digestion and benefit fattening pigs.

(二) 粗饲料人工瘤胃发酵
Artificial rumen fermentation of roughage

人工瘤胃发酵是根据牛、羊瘤胃特点,模拟牛、羊瘤胃内的主要生理条件,即温度恒定在38-40 ℃之间,pH控制在6-8,保持厌氧环境,保证必要的氮、碳和矿物质营养,采用人工仿生制作,使粗饲料质地明显呈"软""黏""烂",汁液增多。

Artificial rumen fermentation is based on the ruminal characteristics of cattle and sheep, simulated ruminal physiological conditions of cattle, sheep rumen, namely 38-40 ℃ constant temperature, pH control in 6-8, anaerobic environment, necessary nitrogen, carbon, and minerals. Using artificial bionic production the texture of the roughage is obviously soft, sticky and rotten, more juicy and special smell as well.

1. 制作方法

可按下述程序进行。首先,采用导管法或永久瘤胃瘘管法,从屠宰牛、羊瘤胃中直接获得瘤胃液。瘤胃液要保存在40 ℃的真空干燥箱内,将瘤胃内容物粉碎,一般600 g瘤胃内容物可制得100 g菌种。其次,准备各种作物秸秆、秕壳,粉碎待用。然后,进行保温。实际应用中有三种常用保温方法。

1. Procedures

The following procedures can be followed. Firstly, the rumen fluid is obtained directly from the cattle and sheep with permanent rumen fistula, or rumen of slaughtered cattle and sheep. The ruminal fluid should be stored in a vacuum drying oven at 40 ℃, and the ruminal contents should be shredded. Generally, 100 g strains can be produced from 600 g ruminal contents. Then roughage like crop straw and chaff shell is crushed and kept warm. In practical application, there are three common methods of heat preservation.

(1) 暖缸自然保温法:在装发酵料的大缸周围和底部,填装150 mm厚的秕谷、糠麸、木屑等踏实,四周用土坯或砖砌起围墙,缸口处用土坯或砖铺平抹好,上面盖上草帘等物保温。

(1) Natural insulation method of warm cylinder. The 150 mm thick of grain, bran, and sawdust are used to be around and at the bottom of the large cylinder filled with fermentation materials, then the walls are lined with adobe or brick around them, and the head of the cylinder is paved with adobe or brick. Finally, straw curtain is covered for insulation.

(2) 加热保温法:北方可在缸下部修建火道或烟道,利用烧火的余热进行保温,为使受热均匀,可加火门调节。

(2) Heating and holding method. In north part of China, a fire way or a flue can be built at the bottom of the cylinder, and the waste heat of the fire can be used for heat preservation. In order to make the heating be even, the fire door can be added to adjust the temperature.

（3）室内保温法：利用固定的房屋，建造火墙、火炉、土暖气等方法，使室温保持在 35－40 ℃。最后，堆积或装缸，压实封闭 36 h，即可饲用。

制作瘤胃发酵饲料时，也可添加其他营养物。瘤胃微生物必须有一定种类和数量的营养物质，并稳定在 pH 6－8 的环境中，才能正常繁殖。粗饲料发酵的碳源由粗饲料本身提供，不足时再加；氮源可用尿素；加入碱性缓冲剂及酸性磷酸盐类，也可用草木灰替代碱。

目前，国内已有机械化或半机械化的发酵装置，每缸一次可制 1 500 kg 的发酵饲料。调制前，先将粗饲料在碱池中浸泡 24 h，发酵过程中的搅拌、出料控制均由机械操作，大大减轻了劳动强度，适宜大、中型牧场利用。

2. 发酵饲料的鉴定

发酵好的饲料，干的浮在上面，稀的沉在下层，表层呈灰黑色，下面呈黄色。原料不同，色泽也不同，如高粱秸呈黄色、黏，呈酱状。若表层变黑，表明漏进了空气。味道有酸臭味，不能有腐臭味，否则为变坏。用手摸，纤维软化。可将滤纸装在用塑料纱窗布做的口袋内，置于缸1/3 处，与饲料一同发酵，经 48 h 后慢慢拉出，将口袋中的饲料冲掉，若滤纸条已断裂说明纤维分解能力强，否则相反。

生物学法操作技术复杂，投入成本较高，但粗饲料经处理后营养价值得到提高，有利于饲料的消化利用，因此，有条件的地区或养殖场可采用此法。

(3) Indoor thermal insulation method. The real buildings are used to build fire walls, stoves, and local heating systems to keep the room temperature at 35－40 ℃. Finally, the roughage can be prepared by stacking or filling cylinder, and compacting the sealing group for 36 hours.

Other nutrients can also be added to the ruminal fermentation roughage. Ruminal microorganisms reproduce normally under certain conditions, such as certain kinds and quantity of nutrients, and stable environment with pH 6－8. The carbon source of roughage fermentation is provided by the roughage itself. The urea is added into roughage as nitrogen source. Alkaline buffer and acid phosphate, are also added into roughage, and alkaline can be replaced with grass ash.

At present, there are mechanized or semi-mechanized fermentation equipment in China, through which each cylinder can produce 1,500 kg of fermentation feed. Before the preparation, the roughage is soaked in an alkali pond for 24 hours, and the stirring and discharging control in the fermentation process are controlled by mechanical operation, which greatly reduces the labor intensity and is suitable for large and medium-sized pastures.

2. Identification of fermented feed

Good quality of fermented feed is colored with gray black on the top and yellow at the bottom, and is classified with dry feed on the top and thin feed at the bottom. Different materials have different colors, such as sorghum straw looks yellow, sticky, and paste shape. If the surface becomes black, it indicates leakage of air. The fermented feed should have a sour taste without a rancid taste, and the fibers should be soft with touch. The filter paper can be put into the bag made of plastic window screen cloth, placed in 1/3 of the cylinder, and fermented together with the feed. After 48 hours, the filter paper is slowly pulled out, and the feed in the bag is washed away. If the filter paper is broken, it means that the fiber decomposition ability is strong, otherwise it is the opposite.

The operation technology of the biological method is complex and the cost is high, but the nutritive value of the roughage is increased after the treatment, which is beneficial to the digestion and utilization of the feed. Therefore, this method can be adopted in well-equipped farms.

四、综合利用
Comprehensive utilization

单一加工处理秸秆饲料往往达不到理想的效果，特别是难以实现产业化规模处理。现在一种新型复合处理技术将不同处理技术合理地组装配套，采用新型调制工艺加工粗饲料。目前，应用较多的是将化学处理与机械成型加工调制相结合，即先对秸秆饲料进行切碎或粗粉碎，进行碱化或氨化等化学预处理，然后添加必要的营养补充剂，再通过机械加工调制成秸秆颗粒饲料或草块，可显著改善秸秆饲料的物理性状和适口性，有利于运输、贮存和利用。要根据当地生产条件、粗饲料的特点、经济投入的大小、饲料营养价值提高的幅度和家畜饲养的经济效益等综合因素，科学地选择加工调制途径。具有一定规模的饲养企业，饲料加工调制要向集约化和工厂化方向发展。广大农村分散饲养的千家万户，要选择简便易行、适合当地条件的加工调制方法，并应向专业加工和建立服务体系方向发展，以便为畜牧业的发展提供可靠的饲料。

饲喂利用时也要注意合理地搭配，因为饲粮的组成对粗饲料的消化影响很大。日粮容积和干物质含量必须保证动物正常消化的需要。消化道未充满或过分充满，均会影响动物的健康和生产性能。粗纤维是保证动物正常消化和代谢过程的重要营养因素，饲粮中粗纤维的含量应占干物质的15%-24%。应注意矿物质营养供应平衡，还要考虑饲草和饲料的适口性。

Single processing of straw feed is often not the ideal strategy, especially is difficult to achieve industrialization scale treatment. Now a new compound processing technology is used to process roughage by assembling different treatment technologies rationally. Currently, the combination of chemical processing and mechanical processing modulation is widely applied, to which the chopped coarse crushed straw is treated by chemical pretreatment using alkaline or ammonia, and then the straw is treated by mechanical processing to create straw pellet feed or grass block after adding the necessary nutritional supplements. The physical properties and palatability of straw can significantly be improved and convenient for transportation, storage and usage. The scientific way of processing and blending should be selected according to the local production conditions, the characteristics of roughage, the size of economic input, the extent of increasing the nutritive value of feed and the economic benefits of raising livestock. For feeding enterprises of a certain scale, feed processing and modulation should be developed towards the intensive and industrial production. In order to provide reliable feed for the development of animal husbandry, thousands of rural households should choose the processing and modulation method which is simple, easy, and suitable for local conditions, and should develop towards professional processing and establish service system.

Proper combination should also be paid attention to when it is used for feeding, because the composition of the diet has a great influence on the digestion of roughage. The ration volume and dry matter content must ensure the normal digestion of animals. Underfilling or overfilling of the digestive tract can affect the health and productivity of animals. Crude fiber is an important nutrient factor to ensure the normal digestion and metabolism of animals. The content of crude fiber in diet should account for 15%-24% of the dry matter. In addition, the balance of minerals and the palatability of forage and feed should be taken into account.

五、粗饲料品质评定
Quality evaluation of roughage

对粗饲料的品质评价,参考卢德勋(2009)的方法并结合中国养殖实际情况提出了粗饲料分级指数评定法(GI),GI 的表示式为:

$$GI = \frac{ME \times DMI \times CP}{NDF(或 ADI)}$$

式中,GI 为粗饲料分级指数(MJ);ME 为粗饲料代谢能(MJ/kg),亦可使用净能(NE)取代 ME,尤其是在对奶牛粗饲料品质的评定上,宜使用 NE;DMI 为粗饲料干物质随意采食量(kg/d);CP 为粗蛋白质占干物质的百分比;NDF 为中性洗涤纤维占干物质的百分比;ADI 为酸性洗涤木质素占干物质的百分比,如有条件,应尽量使用 ADI,不用 NDF。

GI 不仅将粗饲料可利用能量和蛋白质指标联系起来,而且还将粗饲料中难以消化的反映粗饲料物理性质的 ADI 成分包括在内,较为客观地反映了反刍家畜营养利用的规律与粗饲料的营养价值;同时,GI 以系统工程的思想和方法,充分利用粗饲料间的组合效应,来实现粗饲料搭配的最优化;GI 值反映的是粗饲料中可为家畜采食的有效能值,它是一个绝对值,因而可用于指导牧草的种植、确定牧草的最佳刈割期;另外,GI 使用了现在通用的净能(NE)或代谢能(ME)作为描述粗饲料的能值单位,通俗易懂,便于推广。

For the quality evaluation of roughage, a grading index (GI) method was proposed by referring to the method of Lu Dexun (2009) and combining the actual situation of Chinese animal husbandry. The expression of GI is:

$$GI = \frac{ME \times DMI \times CP}{NDF(\text{or } ADI)}$$

GI is roughage grading index (MJ); ME is the metabolic energy of roughage (MJ/kg), and net energy (NE) can also be used to replace ME. NE should be used for evaluation of roughage quality in dairy cows. DMI is the random intake of dry matter of roughage (kg/d). CP is the ratio of crude protein to dry matter; NDF is the ratio of neutral detergent fiber to dry matter; ADI is the ratio of acid detergent lignin to dry matter. If possible, the ADI should be used instead of NDF.

GI not only links the available energy of the roughage with the protein index, but also includes the ADI components in the roughage which are difficult to digest to reflect the physical properties of the roughage. GI objectively reflects the law of nutrient utilization of ruminants and the nutritive value of the roughage. At the same time, GI makes full use of the combination of roughage with the idea and method of system engineering to achieve the optimization of the roughage collocation. GI value reflects the effective energy value in the roughage ingested by livestock. GI is an absolute value, so it can be used to guide the planting of herbage and determine the optimal mowing period of herbage. In addition, GI uses the net energy (NE) or metabolic energy (ME) to describe the energy value of roughage, which is easy to understand and easy to promote.

第四章 青绿饲料
Chapter 4　Green forage

青绿饲料主要指天然水分含量高于60%的青绿多汁饲料,以富含叶绿素而得名。种类繁多,主要包括天然牧草、栽培牧草、青饲作物、非淀粉质根茎瓜类、水生植物及树叶类等。这类饲料种类多、来源广、产量高、营养丰富,对促进动物生长发育、提高畜产品品质和产量等具有重要作用,被誉为"绿色能源"。

Green forage mainly refers to green succulent feed with more than 60% of natural water, which is named for being rich in chlorophyll. They mainly include natural herbage, cultivated herbage, green forage crops, non-starch root gourds, aquatic plants and leaves, etc. This kind of feed has many kinds, wide sources, high yield and rich nutrition, which plays an important role in promoting the growth and development of animals and improving the quality and yield of livestock products. It is known as "green energy".

第一节　青绿饲料的营养特性及影响因素
Section 1　Nutritional characteristics and influencing factors of green forage

青绿饲料是草食家畜的主要饲料之一。近年来由于优质牧草的推广种植,其在农区反刍动物和单胃动物饲养中愈来愈重要。

Green forage is one of the main feeds for herbivores. In recent years, due to the promotion of the cultivation of high-quality herbage, it is becoming more and more important in the feeding of ruminants and monogastric animals in agricultural areas.

一、青绿饲料的营养特性
Nutritional characteristics of green forage

1. 水分含量高,能值较低。陆生植物的水分含量为60%-90%,而水生植物高达90%-95%。因此,其鲜草的干物质含量少,能值较低。陆生植物每千克鲜重的消化能在1.20-2.50 MJ之间。如以干物质为基础计算,由于粗纤维含量较高(15%-30%),故其能值一般低于常规的能量饲料,其消化能值为

1. High moisture content and low energy. The water content of terrestrial plants is 60%-90%, while that of aquatic plants is as high as 90%-95%. Therefore, the dry matter content of fresh grass is less and the energy value is lower. The digestibility of terrestrial plants is between 1.20 and 2.50 MJ/kg of fresh weight. If calculated on the basis of dry matter, due to the high crude fiber content (15%-30%), its energy value is generally lower than that of conventional energy feed, and its digestibility value is 8.37-

8.37－12.55 MJ/kg。

2. 蛋白质含量较高，品质较优。一般新鲜的禾本科牧草和叶菜类饲料的粗蛋白质含量在1.5%-3.0%之间，豆科牧草在3.2%-4.4%之间。若按干物质计算，前者粗蛋白质含量达13%-15%，后者可高达18%-24%。由于青绿饲料含有各种必需氨基酸，尤其以赖氨酸、色氨酸含量较高，故其蛋白质生物学价值较高，一般可达70%以上。

3. 粗纤维含量较低。幼嫩的青绿饲料含粗纤维较少，木质素低，无氮浸出物较高。若以干物质为基础计算，则其中粗纤维为15%-30%，无氮浸出物为40%-50%。随着植物生长期的延长，粗纤维和木质素的含量显著增加。一般来说，植物开花或抽穗之前，粗纤维含量较低。猪对未木质化的纤维素消化率可达78%-90%，对已木质化的纤维素消化率仅为11%-23%。

4. 钙、磷比例适宜。青绿饲料中的矿物质含量因植物种类、土壤与施肥情况而异。以温带草地牧草为例，钙含量为0.25%-0.5%，磷含量为0.20%-0.35%，比例较为适宜，特别是豆科牧草钙的含量较高，因此以青绿饲料为主食的动物不易缺钙。此外，青绿饲料尚含有丰富的铁、锰、锌、铜等微量矿物质元素。但牧草中钠和氯含量一般不足，所以放牧家畜需要补充食盐。

5. 维生素含量丰富。青绿饲料是家畜维生素的良好来源，特别是胡萝卜素含量较高，每千克饲料含50－80 mg。在正常采食情况下，放牧家畜所摄入的胡萝卜素要超过其本身需要量的100倍。此外，青绿饲料中B族维生素、维生

12.55 MJ/kg.

2. High protein content and good quality. The crude protein content of general fresh forage grass and leaf vegetables is between 1.5%-3.0% and that of leguminous forage grass is between 3.2%-4.4%. In terms of dry matter, the crude protein content of the former is up to 13%-15%, while that of the latter is up to 18%-24%. The green forage contains a variety of essential amino acids, especially the higher content of lysine and tryptophan, so its protein biological value is higher, generally up to more than 70%.

3. Low crude fiber content. The young green forage contains less crude fiber, less lignin and higher nitrogen free extract. If the calculation is based on dry matter, the crude fiber is of 15%-30%, and the nitrogen free extract is of 40%-50%. With the prolongation of plant growth period, the content of crude fiber and lignin increases significantly. In general, plants have low crude fiber content before flowering or heading. The digestibility of non-lignified cellulose in pigs is 78%-90%, while that of lignified cellulose is only 11%-23%.

4. The proportion of calcium and phosphorus is suitable. The mineral content in green forage varies with plant species, soil and fertilization. For example, the contents of calcium and phosphorus in herbage of temperate grassland are 0.25%-0.5% and 0.20%-0.35% respectively, which are more suitable. Especially, the content of calcium in leguminous herbage is relatively high, so it is not easy for animals that take green feed as the staple food to be short of calcium. In addition, green forage is rich in iron, manganese, zinc, copper and other trace mineral elements. However, the content of sodium and chlorine in herbage is generally insufficient, so salt needs to be replenished while grazing livestock.

5. Rich in vitamins. Green forage is a good source of vitamins for livestock, especially the high content of carotene, about 50-80 mg/kg of feed. Under normal feeding conditions, grazing animals consume 100 times more carotene than they need. In addition, the content of vitamin B, vitamin E, vitamin C and vitamin K in the green forage is also rich, such as thiamine 1.5 mg/kg, riboflavin 4.6 mg/kg

素E、维生素C和维生素K的含量也较丰富，如青苜蓿中含硫胺素1.5 mg/kg、核黄素4.6 mg/kg、烟酸18 mg/kg，但缺乏维生素D，维生素B6（吡哆醇）的含量也很低。

另外，青绿饲料幼嫩、柔软多汁，适口性好，还含有各种酶、激素和有机酸，易于消化。青绿饲料中有机物质的消化率为：反刍动物75%—85%，马50%—60%，猪40%—50%。

综上所述，青绿饲料是一种营养相对平衡的饲料，但因其水分含量高，干物质中消化能较低，从而限制了其潜在的营养优势。尽管如此，优质的青绿饲料仍可与一些中等的能量饲料相媲美。因此，在动物饲养中，青绿饲料及由其调制的干草可以长期单独组成草食动物饲粮，并可提供一定的动物产品。

对单胃杂食动物（如猪、鸡）来说，由于青绿饲料干物质中含有较多数量的粗纤维，且粗纤维的消化主要在盲肠内进行，因而其对青绿饲料的利用率较差。并且，青绿饲料容积较大，而猪、鸡的胃肠容积有限，使其采食量受到限制。因此，在猪禽饲粮中不宜大量加入青绿饲料，但可作为一种蛋白质与维生素的良好来源适量搭配于饲粮中，以补充其饲料组成的不足，从而满足猪禽对营养的全面需要。

and niacin 18 mg/kg in the green alfalfa. But vitamin D is insufficient, and vitamin B6 (pyridoxine) is also low.

In addition, green forage is young and tender, soft and juicy, and good in palatability. It also contains a variety of enzymes, hormones and organic acids, and is easy to digest. Digestibility of organic matter in green forage is 75%–85% for ruminants, 50%–60% for horses and 40%–50% for pigs.

In summary, green forage is a relatively nutritionally balanced feed, but due to its high moisture content and low digestibility in dry matter, its potential nutritional advantages are limited. Nevertheless, the high quality of green forage can be comparable to some medium energy feeds. Therefore, in animal feeding, the green forage and the hay prepared by it can constitute the forage of herbivore alone for a long time, and can provide certain animal products.

For monogastric omnivores (such as pigs and chickens), the utilization rate of green forage is poor because the dry matter of green forage contains a large amount of crude fiber, and the digestion of crude fiber is mainly carried out in the cecum. Moreover, the volume of green forage is large, and the gastrointestinal volume of pigs and chickens is limited, so that the intake of food is limited. Therefore, it is not suitable to add a large number of green forage in pig and poultry diet, but it can be used as a good source of protein and vitamins in the right amount to supplement the insufficient feed composition, so as to meet the overall nutritional needs of pigs and poultry.

二、影响青绿饲料营养价值的因素
The factors affecting the nutritive value of green forage

青绿饲料的营养价值受多种因素的影响，植物的品种、生长阶段和部位、土壤、肥料、气候条件、草场管理等因素都可能影响青绿饲料的营养价值。

1. 品种、生长阶段和部位。一

The nutritive value of green forage is affected by many factors, such as plant variety, growth stage and position, soil, fertilizer, climatic conditions, grassland management and other factors.

1. The variety, growth stage and position. In general,

一般,豆科牧草和叶菜类的营养价值较高,禾本科次之,水生植物饲料最低。同一种青绿饲料其品种不同,营养价值也有差异。研究者比较了国内外38个和22个紫花苜蓿品种的营养成分含量,各品种的粗蛋白质和粗纤维含量有一定差异,有些苜蓿品种之间上述两项指标差异还比较大,表明青绿饲料的营养成分除受栽培等条件影响外,品种的特质也在起一定的作用。

青绿饲料的生长阶段不同,其营养价值也各异。幼嫩时期水分含量高,干物质中蛋白质含量较多而粗纤维较少,因此在早期生长阶段的各种牧草有较高的消化率,其营养价值也高。随着植物生长期的延长,粗蛋白质等养分含量逐渐降低,而粗纤维特别是木质素的含量则逐渐上升,致使营养价值、适口性和消化率都逐渐降低。

植物体的部位不同,其营养成分差别也很大。例如:苜蓿的上部茎叶中粗蛋白质含量高于下部茎叶,而粗纤维含量则低于下部。一般来讲,茎秆中粗蛋白质含量低而粗纤维含量高,叶片中则恰恰相反。因此,叶片占全株的比例愈大,营养价值就愈高。

2. 土壤与肥料。土壤是植物营养物质的主要来源之一。生长在肥沃和结构良好的土壤上,青绿饲料的营养价值较高;反之,在贫瘠和结构差的土地上收获的青绿饲料其营养价值就较低。特别是青绿饲料中一些矿物质元素的含量在很大程度上受土壤中该元素含量与活性的影响。泥炭土与沼泽土中的钙、磷均较缺乏;干旱的盐碱地中的植物

the nutritive value of legumes and leafy vegetables is relatively higher, followed by gramineae, and the feed of aquatic plants is the lowest. The same kind of green forage has different varieties, and its nutritive value is also different. The researchers compare the nutrient content of 38 and 22 alfalfa varieties at home and abroad. There are some differences in crude protein and crude fiber content among different alfalfa varieties, and there are significant differences in the above two indexes among some alfalfa varieties. It shows that the nutritional composition of green forage is not only affected by cultivation and other conditions, but also by characteristics of varieties to a centain extent.

The growth stage of green forage is different, and its nutritive value is also different. The moisture content is higher in the tender stage, the protein content in the dry matter is rich and the crude fiber is insufficient, therefore, the various herbage in the early growth stage has higher digestibility and its nutritive value is also higher. With the prolongation of plant growth period, the content of crude protein and other nutrients decreases gradually, while the content of crude fiber, especially lignin, increases gradually, resulting in the decrease of nutritive value, palatability and digestibility.

Different parts of the plant body have different nutrient composition. For example, the crude protein content in the upper stems and leaves of alfalfa is higher than that in the lower stems and leaves, while the crude fiber content is lower than that in the lower stems and leaves. Generally speaking, the crude protein content in the stem is low and the crude fiber content is high, while the opposite is true in the leaves. Therefore, the greater the proportion of the leaves in the whole plant, the higher the nutritive value is.

2. Soil and Fertilizer. Soil is one of the main sources of plant nutrients. Growing in fertile and well-structured soil, green forage has higher nutritive value. Conversely, green forage harvested from poor and poorly constructed soils has lower nutritive value. In particular, the content of some mineral elements in the green forage is greatly affected by the content and activity of the elements in the soil. Calcium and phosphorus in peat soil and marsh soil are deficient. Plants in arid saline-alkali lands find it difficult to make use of soil calcium, and plants in calcareous soil are poor

很难利用土壤中的钙;石灰质土壤中的植物对锰和钴吸收不良。

施肥可以显著影响植物中各种营养物质的含量,在土壤缺乏某些元素的地区施以相应的肥料,则可防止这一地区家畜营养性疾病的发生。对植物增施氮肥,不仅可以提高植物的产量,还可增加植物中粗蛋白质的含量,并且施肥后植物生长旺盛,茎叶浓绿,叶绿素含量亦显著增加。

3. 气候条件。气候条件如气温、光照及雨量等对于青绿饲料的营养价值影响也较大。如:在多雨地区或季节,土壤经常被冲刷,土壤中的钙质容易流失,故植物体内钙质积累较少;反之,在干旱地区或季节,植物体内积累的钙质较多。在寒冷地区的植物,其粗纤维含量较温暖地区高,粗蛋白质和粗脂肪的含量则较少。此外,生长在阳光充足的阳坡地的植物,粗蛋白质和六碳糖的含量显著高于阴坡地的植物。

4. 管理因素。牧场放牧制度的健全与否也影响到牧草营养价值。放牧不足,植物变得粗老,营养价值降低;过度放牧则使许多优良牧草如豆科牧草被频繁采食,以致不能恢复生长,逐渐从牧场消失,使牧草营养价值降低。此外,草地经常刈割可打断植物生长发育规律,使其恢复到生理上幼嫩的生长阶段,蛋白质和脂肪的含量可保持在一个较高水平,而粗纤维含量降低。

absorbers of manganese and cobalt.

Fertilization can significantly affect the content of various nutrients in plants, and applying appropriate fertilizers to areas where soil is deficient in certain elements can prevent the occurrence of nutritive diseases of livestock in those areas. Adding nitrogen fertilizer to plants can not only increase the yield of plants, but also increase the crude protein content in plants. Moreover, after fertilization, plants grow vigorously and their stems and leaves are thick green, the chlorophyll content also increases significantly.

3. Climatic Conditions. Climatic conditions such as temperature, light and rainfall have a greater impact on the nutritive value of green forage. For example, in rainy areas or seasons, the soil is often scoured, the calcium in the soil is easy to lose, so the accumulation of calcium in plants is less; Conversely, in dry areas or seasons, plants accumulate more calcium. Plants in cold regions have higher crude fiber content and lower crude protein and fat content than those in warm regions. In addition, the crude protein and six-carbon sugar content of plants growing on sunny slopes are significantly higher than those growing on shady slopes.

4. Management factors. Whether the grazing system is sound or not also affects the total nutritive value of grassland. Lack of grazing makes the plants become coarse and old, and their nutritive value decreases. Overgrazing makes many good forages, such as legumes, to be fed frequently, so that they can not resume growth, gradually disappear from the pasture, and reduce the nutritive value of the forage. In addition, regular mowing of grassland can interrupt the growth and development of plants and restore them to the physiological and tender growth stage. Protein and fat content can be kept at a high level, while crude fiber content can be reduced.

第二节　主要青绿饲料
Section 2　Main green forage

青绿饲料是处于青绿状态的饲料,以天然及栽培的优质牧草为主,青饲作物及叶菜类也占有较大的比例。近年来,青绿树叶的利用日益受到关注。

The green forage is the feed in the green state, mainly consisting of natural and cultivated high quality herbage, and the green feed crops and leaf vegetables also occupy a large proportion. In recent years, more and more attention has been paid to the utilization of green leaves.

一、天然牧草
Natural herbage

我国幅员广大,地域辽阔,在西北、东北、西南地区均有大面积的优良草原、草山和草坡,面积约 4 亿 hm^2,其中可利用草地 3.3 亿 hm^2,约为农业耕地面积的 3 倍。农业地区内还分散有许多小面积的草地,估计约有 1300 万 hm^2。利用这些牧草资源发展畜牧生产有很大的潜力。

天然草地的利用价值受诸如地形地势、草原类型、水源供应以及放牧制度等许多因素的影响。但就草层的营养特性而论,主要取决于牧草的种类和生产阶段。我国天然草地上生长的牧草种类繁多,主要有禾本科、豆科、菊科和莎草科四大类。

这四类牧草干物质中无氮浸出物含量均在 40%-50%。粗蛋白含量有较大差异,其中,豆科牧草的蛋白质含量偏高,在 15%-20%;莎草科中为 13%-20%;菊科与禾本科含量多在 10%-15%,少数可达 20%。粗纤维含量以禾本科草较高,约为 30%,其他三类牧草中约为 25%,个别低于 20%。粗脂肪含量以菊科牧草最高,平均达 5% 左

China has a vast territory, and there are large areas of fine grassland, grassy hills and grassy slopes in northwest, northeast and southwest China, with an area of about 400 million hm^2, in which 330 million hm^2 of grassland can be used, about 3 times of the agricultural arable land area. There are also many small areas of grassland scattered throughout the agricultural area, estimated to be about 13 million hm^2. There is great potential to develop livestock production with these forage resources.

The utilization value of natural grassland is influenced by many factors such as topography, grassland type, water supply and grazing system. But as far as the nutritional characteristics of grass layer are concerned, it mainly depends on the type and production stage of herbage. There are various kinds of herbage growing on natural grassland in China, mainly including gramineae, leguminosae, compositae and sedge family.

The content of nitrogen free extracts in the dry matter of these four kinds of herbage is between 40% and 50%. The crude protein content varies greatly, among which the protein content in legumes is rather high, which is between 15% and 20%, the crude protein in cyperaceae is 13%-20%, and the crude protein in compositae and gramineae is 10%-15%, and for a few, up to 20%. The crude fiber content in gramineae is much more higher, which is about 30%, and in the other three kinds of herbage the content of the crude fibre is about 25%, some less than 20%. The

右，其他类在 2%-4%。矿物质中一般都是钙高于磷，比例适宜。

总的来说，豆科牧草的营养价值较高。禾本科牧草的粗纤维含量较高，对其营养值有一定影响，但由于其适口性较好，特别是在生长早期，幼嫩可口，采食量高，因而也不失为优良的牧草。并且，禾本科牧草的葡萄茎或地下茎再生力很强，比较耐牧，对其他牧草起到保护作用。菊科牧草往往有特殊的气味，除羊以外，一般家畜都不喜采食。

近年来由于全球性气候恶化，加之长期以来超载放牧，使得草原牧草极度稀疏，草原退化严重，畜群的"春乏、夏壮、秋肥、冬死"现象十分普遍。为了有效遏制草原退化，缓解诸如"扬沙""沙尘暴"等恶劣气候现象的发生，草原畜牧业应结束游牧时代，实行围栏放牧及轮牧等放牧技术，有计划扩大人工草场面积，科学确定放牧家畜品种、畜种结构及载畜量，以提高草原的实际畜牧业产值。

crude fat content is the highest in compositae, averaging about 5%, and that in other species is between 2% and 4%. In minerals, calcium is generally higher than phosphorus, and the proportion is appropriate.

In general, legumes have higher nutritive value. The high crude fiber content of gramineae has a certain influence on its nutritive value, but because of its good palatability, especially in the early growth stage, it is tender and delicious, and has high food intake, so it is also an excellent forage. Moreover, the regeneration ability of stolon or underground stem of herbage is very strong, which is more tolerant to grazing and plays a protective role on other herbage. Composite herbage often has a special smell. The livestock do not like to eat except sheep.

In recent years, due to global climate deterioration and overgrazing for a long time, grassland grass is extremely sparse and grassland degradation is serious. The phenomenon of "thin in spring, strong in summer, fat in autumn and dead in winter" of livestock is very common. In order to effectively curb the grassland degradation, lessen the occurences of "flying sand", "dust" and other severe weather phenomena, the nomadic era should be ended in grassland animal husbandry, grazing techniques such as fenced grazing and rotational grazing should be carried out. Measures should be taken to expand the artificial grassland area, and scientifically determine the grazing livestock breeds, breeds structure and grazing capacity to improve the actual output vale of animal husbandry on the grassland.

二、栽培牧草
Cultivated herbage

栽培牧草是指人工播种栽培的各种牧草，其种类很多，但以产量高、营养好的豆科和禾本科牧草占主要地位。栽培牧草是解决青绿饲料来源的重要途径，可为家畜常年提供丰富而均衡的青绿饲料。

Cultivated herbage refers to all kinds of herbage planted artificially. There are many kinds of herbage, but legume and gramineae have high yield and good nutrition. The cultivation of herbage is an important way to solve the problem of green forage source, which can provide abundant and balanced green forage for livestock all year round.

（一）豆科牧草
Leguminous forage

我国栽培豆科牧草有悠久的历

China has a long history of cultivating leguminous

史,2 000年以前紫花苜蓿已在我国西北普遍栽培;草木樨在西北作为水土保持植物也有大面积的种植;其他如紫云英、苕子等既作饲料又是绿肥植物进行种植。

1. 紫花苜蓿。也叫紫苜蓿、苜蓿,是我国最古老、最重要的栽培牧草之一,广泛分布于西北、华北、东北地区,江淮流域也有种植。其特点是产量高、品质好、适应性强,是最经济的栽培牧草,被冠以"牧草之王"。紫花苜蓿的营养价值很高,在初花期刈割的干物质中:粗蛋白质含量为18%-22%,必需氨基酸组成较为合理,赖氨酸可高达1.34%,比玉米高5倍之多;产奶净能5.4-6.3 MJ/kg;含钙1.5%-3.0%;还含有丰富的维生素和微量元素,如胡萝卜素含量可达161.7 mg/kg。紫花苜蓿中含有各种色素,对家畜的生长发育及乳汁、卵黄颜色均有好处。紫花苜蓿的营养价值与刈割时期关系很大,幼嫩时含水多,粗纤维少;刈割过迟,则茎的密度增加,而叶的密度下降,饲用价值降低。

一般认为,紫花苜蓿最适刈割期是在第1朵花出现至1/10开花,根茎上又长出大量新芽的阶段。此时,营养物质含量高,根部养分蓄积多,再生良好。蕾前或现蕾时刈割,蛋白质含量高,饲用价值大,但产量较低,且根部养分蓄积少,影响再生能力。刈割时期还要视饲喂要求来定,青饲宜早,调制干草可在初花期刈割,喂猪禽可早割,喂牛、羊可稍迟。苜蓿为多年生牧草,管理良好时可利用5年以上,第2-4年产草量最高。

forage. Alfalfa was widely cultivated in northwest China 2,000 years ago. Sweet clover is also widely cultivated in northwest China as a soil and water conservation plant. Others such as Chinese milkvetch, vetch and so on are cultivated as feed and green manure.

1. Alfalfa. Alfalfa (*Medicago sativa* L.) is one of the oldest and most important cultivated forages in China. It is widely distributed in northwest, north and northeast of China, and also cultivated in Jianghuai Basin. Characterized by high yield, good quality and strong adaptability, it is the most economical cultivated herbage and is crowned as the "king of herbage". Alfalfa has high nutritive value. In the dry matter cut at the early flowering stage, crude protein content is 18%-22%, and the essential amino acid composition is relatively reasonable. Its content of lysine can be as high as 1.34%, 5 times higher than maize. Its net milk production capacity is 5.4-6.3 MJ/kg, and its calcium content is 1.5%-3.0%, and it is also rich in vitamins and trace elements, such as carotene content is up to 161.7 mg/kg. Alfalfa contains a variety of pigments, which are beneficial to the growth and development of livestock and the color of milk and yolk. The nutritive value of alfalfa is closely related to the cutting period. It contains more water and less crude fiber when it is young. When cutting late, the density of stems increases, while the density of leaves decreases, and the feeding value decreases.

It is generally believed that the optimal cutting period of alfalfa is the stage from the first flower to 1/10 blossom, and a large number of new buds grow on the root stem. At this stage, the nutrient content is high, the root accumulates more nutrient, and the regeneration is good. Cutting before budding or at the time of bud has higher protein content and higher feeding value, but the yield is lower, and the nutrient accumulation in the root is less, which affects the ability of regeneration. Mowing period should be determined according to the feeding requirements such as: for green feeding, early mowing is suitable; for hay, mowing in the first flowering period is suitable; for pigs and poultry, early mowing is necessary while for cattle and sheep, later mowing is suitable. Alfalfa is perennial herbage, which can be used for more than 5 years when well managed, and the highest yield period is from the 2nd to 4th years.

紫花苜蓿的利用方式有多种，可青饲、放牧、调制干草、草粉或青贮，对各类家畜均适宜。用鲜苜蓿喂乳牛，乳牛泌乳量高且乳质好。成年泌乳母牛每日每头可喂 15-20 kg，青年母牛可喂 10 kg 左右。对舍饲的小尾寒羊或大尾寒羊，每只日喂 2-3 kg。用鲜苜蓿喂猪、鸡时，应多利用植株上半部幼嫩枝叶，切碎或打浆饲喂效果较好。

值得注意的是，紫花苜蓿茎叶中含有皂苷，皂苷有降低液体表面张力的作用，牛、羊大量采食鲜嫩苜蓿后，可在瘤胃内形成大量泡沫样物质，引起臌胀病，导致产奶量下降甚至死亡，故饲喂鲜草时应控制喂量，放牧地最好采用无芒雀麦、苇状羊茅等禾本科草与苜蓿混播。但苜蓿皂苷对单胃动物是一种活性成分，对降低动物体内的胆固醇有重要作用。

2. 三叶草。三叶草属植物共有 300 多种，大多数为野生种，少数为重要牧草，目前栽培较多的为红三叶和白三叶。

红三叶又名红车轴草、红菽草、红荷兰翘摇等，为多年生草本植物，生长年限 3-4 年，是江淮流域和灌溉条件良好地区重要的豆科牧草之一。新鲜的红三叶含干物质 13.9%，粗蛋白质 2.2%，产奶净能为 0.88 MJ/kg。以干物质计，其所含可消化粗蛋白质低于苜蓿，但其所含的净能值则较苜蓿略高。红三叶草质柔软，适口性好，各种家畜都喜食。青饲乳牛时有助于提高泌乳量，每日每头最多可喂 40-60 kg。此外也可以放牧，或者调制成干草、青贮利用，放牧时发生臌胀病的概率低于苜蓿，但仍应注意预防。红三叶

Alfalfa can be used in many ways, such as green feeding, grazing, hay making, grass powder or silage, which is suitable for all kinds of livestock. When fresh alfalfa is fed to dairy cattle, the milk yield is high and the milk quality is good. Adult lactating cows can be fed 15-20 kg per day, and young cows can be fed about 10 kg per day. For small-tailed Han sheep or big-tailed Han sheep fed in house, each sheep is fed with 2-3 kg daily. When feeding pigs and chickens with fresh alfalfa, more tender branches and leaves in the upper half of the plant should be used, and the effect of chopping or beating should be better.

It is worth noting that the alfalfa stem leaf contains saponin, which have the effect of reducing the surface tension of liquid. After eating a lot of fresh alfalfa, a large amount of foam-like substances can be formed in the cattle and sheep's rumen, causing bloating disease, resulting in a decline in milk production and even death. Therefore, when feeding fresh grass, the amount of feed should be controlled. The grazing land is best to mix grasses and alfalfa with grasses such as non-mansock brome and tall fescue. But alfalfa saponin is an active ingredient in monogastric animals and plays an important role in lowering cholesterol in animals.

2. Clover. There are more than 300 species of clover in total, most of which are wild species, a few are important herbage, and most of which are red clover and white clover.

Red clover (*Trifolium pratense* L.) is the perennial herb with a growth life of 3-4 years. It is one of the important leguminous grasses in the Jianghuai River basin and areas with good irrigation conditions. Fresh red clover contains 13.9% of dry matter, 2.2% of crude protein and 0.88 MJ/kg of net energy for lactation. In terms of dry matter, its content of digestible crude protein is lower than that of alfalfa, but its net energy value is slightly higher than that of alfalfa. Red clover is soft and has good palatability. It is a favourite for all kinds of livestock. It is helpful to increase the amount of lactation in the green feeding of cows, and each cow can be fed up to 40-60 kg per day. It can also be pastured, made into hay, and used for silage. The incidence of tympanitis is lower than that of alfalfa during pasturing, but prevention should still be observed. There is a high

中有较高含量的黄酮,有提高动物抗氧化性能等作用。

白三叶也叫白车轴草、荷兰翘摇,为多年生草本植物,生长年限可达10年以上,是华南、华北地区的优良草种。由于草丛低矮、耐践踏、再生性好,最适于放牧利用。白三叶鲜草中粗蛋白质含量较红三叶高,而粗纤维含量较红三叶低,因而草质柔嫩,适口性好,饲用价值高,牛、羊喜食。为防止采食过量发生臌胀病,和红三叶一样适宜采取豆、禾草混播。刈割白三叶草也可用来喂猪、禽和兔等动物。此外,白三叶抗逆性强,草丛浓厚,叶色花色美丽,是河堤、公路沿线的良好水土保持植物,也是城市环境美化的草坪植物。

3. 苕子。一年生或越年生豆科植物,在我国栽培的主要有普通苕子和毛苕子两种。

普通苕子,又称春苕子、普通野豌豆、普通箭舌豌豆等。普通苕子的营养价值较高,茎枝柔嫩,生长茂盛,叶片多,适口性好,是各类家畜喜食的优质牧草。但因普通苕子多汁,调制干草时干燥时间较长,故其主要利用方式为青饲,也可青贮或放牧。

毛苕子,又名冬苕子、毛野豌豆等,是水田或棉田的重要绿肥作物。其耐寒力较强,在 -20 ℃以下仍能生存,亦耐碱或耐酸,适宜与麦类物混播以提高产量。毛苕子生长快,茎叶柔嫩,蛋白质和矿物质含量都很丰富,适口性好,营养价值较高。和普通苕子一样,主要利用方式为青饲,以初花期刈割最好,也可青贮或放牧。此外,毛苕子也是良好的绿肥、水土保持和蜜源植物。

普通苕子或毛苕子的籽实中粗蛋白质高达30%,较蚕豆和豌豆稍

content of flavonoids in red clover, which can improve the antioxidant performance of animals.

White clover (*Tri folium repens* L.) is a perennial herb with a growth life of more than 10 years. It is an excellent grass in south and north of China. Because the grass is low, trample-resistant and reproducible, it is most suitable for grazing. The crude protein content of fresh white clover is higher than that of red clover, while the crude fiber content is lower than that of red clover. Therefore, the grass is soft and tender, with good palatability, high feeding value, and cattle and sheep like to eat it. To prevent excessive expansion of food, bean and grasses are suitable for mixed planting as red clover. The white clover is also used to feed pigs, birds, rabbits and other animals. In addition, white clover has strong stress resistance, thick grass and beautiful leaves. It is not only a good soil and water conservation plant along river embankment and highway, but also a lawn plant for urban environment beautification.

3. Vetch. Vetch is an annual or perennial leguminous plant, which is mainly cultivated in China with two kinds of common vetch and hairy vetch.

Common vetch (*Vicia sativa* L.) is of high nutritive value, with tender stems and branches, lush growth, many leaves and good palatability. It is a high-quality forage for all kinds of livestock. However, due to the succulent vetch and long drying time in hay making, the main utilization method is green feeding, silage or grazing.

Hairy vetch (*Vicia villosa Roth.*) is an important green manure crop in paddy fields or cotton fields. It has strong cold resistance and can survive under -20 ℃. It is also alkali or acid resistant, suitable for mixed sowing with wheat to increase yield. Hairy vetch has fast growth, tender stems and leaves, rich protein and mineral content, good palatability and high nutritive value. Like the common vetch, the main way of use is green feeding. It should be cut at the first flowering stage. It is also suitable for silage or grazing. In addition, hairy vetch is also a good green manure, soil and water conservation and honey plant.

The crude protein in the seeds of common vetch or hairy vetch is as high as 30%, which is slightly higher than

高,可作精饲料用。但其中含有生物碱和氰苷,氰苷经水解酶分解后会释放出氢氰酸而使动物中毒,因此饲喂前须浸泡、淘洗、磨碎、蒸煮,同时要避免大量、长期、连续使用。

4. 草木樨。草木樨属植物约有20种,其中最重要的是二年生白花草木樨、黄花草木樨和无味草木樨。其适应性强,分布广,在我国东北、西北、华北以及华南等地均有分布。草木樨既是一种优良的豆科牧草,也是重要的水土保持植物和蜜源植物。草木樨的营养价值较紫花苜蓿稍差,以干物质计:草木樨含有的各种营养成分为:粗蛋白质15% - 19%,粗脂肪1.8%,粗纤维31.6%,无氮浸出物31.9%,钙2.74%,磷0.02%,产奶净能为4.84 MJ/kg。草木樨可青饲、调制干草、放牧或青贮。

草木樨因含有香豆素,具苦味而致适口性差,白花草木樨含1.05% - 1.40%,黄花草木樨含0.84% - 1.22%,无味草木樨仅含0.01% - 0.03%,因而适口性较佳。初喂草木樨家畜不喜食,可与谷草或紫花苜蓿等混喂,最好在现蕾期前或干制后饲用,并且喂量应由少到多,使家畜逐步适应。草木樨因保存不当而发霉腐败时,在霉菌作用下,香豆素会变为双香豆素,其结构式与维生素K相似,二者具有拮抗作用。家畜采食了霉烂草木樨后,遇到内外创伤或手术,血液不易凝固,有时会因出血过多而死亡。减喂、混喂、轮换喂可防止出血症的发生。

that of broad bean and pea, and can be used as refined feed. However, it contains alkaloid and cyanogenetic glycoside, which, after being decomposed by hydrolytic enzymes, can release ahydrogen cyanide (HCN) toxic to animals. Therefore, it must be soaked, washed, ground and steamed before feeding, and at the same time massive, long-term and continuous use should be avoided.

4. Sweet clover. There are about 20 species of sweet clover, of which the most important is the white sweet clover (*Melilotus alba Desr.*), yellow sweet clover [*Melilotus of ficinalis* (*L.*) *Lam.*], toothed sweet clover [*Melilotus dentatus* (W. et K.) Pers.]. It has strong adaptability and wide distribution in northeast, northwest, north and south of China. Sweet clover is not only an excellent leguminous forage, but also an important soil and water conservation plant and honey plant. The nutritive value of sweet clover is slightly lower than that of alfalfa. Measured by dry matter, sweet clover contains various nutrients as follows: 15% - 19% of crude protein, 1.8% of crude fat, 31.6% of crude fiber, 31.9% of nitrogen free extract, 2.74% of calcium, 0.02% of phosphorus, and milk net energy 4.84 MJ/kg. Sweet clover can be used for green feeding, hay making, grazing or silage.

Because it contains coumarin and has bitter taste, sweet clover has poor palatability. White sweet clover contains 1.05% - 1.40% of coumarin and yellow sweet clover contains 0.84% - 1.22% of coumarin. But toothed sweet clover contains only 0.01% - 0.03% of coumarin, so it has good palatability. When feeding sweet clover for the first time, livestock do not like to eat it. It can be mixed with cereal grass or alfalfa, preferably before the bud stage or after drying, and the amount of feeding should be from less to more, so that livestock can gradually adapt. When sweet clover becomes moldy and rotten due to improper preservation, under the action of mold, coumarin will turn into dicoumarin, whose structural formula is similar to vitamin K, and both of them have antagonistic effect. After gathering mildew rotten sweet clover, domestic animals will encounter internal and external trauma or operation, and their blood will not clot easily, sometimes they will die due to excessive bleeding. Reduced feeding, mixed feeding, and rotation feeding can prevent the occurrence of haemorrhage.

5. 紫云英。紫云英又称红花草,在我国长江流域及以南各地广泛栽培。属绿肥、饲料兼用作物,产量较高,鲜嫩多汁,适口性好,尤以猪喜食,也是牛、羊、马、禽、兔的良好青绿饲料。紫云英在现蕾期营养价值最高,以干物质计:含粗蛋白质31.76%,粗脂肪4.14%,粗纤维11.82%,无氮浸出物44.46%,粗灰分7.82%,产奶净能为8.49 MJ/kg。现蕾期产量仅为盛花期的53%,故以盛花期刈割为佳。

在生产中,紫云英可青饲、青贮,也可制成干草或干草粉利用,饲喂牛羊等反刍家畜时也不宜过多,以免引起臌胀病。有些地方直接利用紫云英作绿肥,但从经济效果上看,不如先用紫云英喂猪,后以猪粪肥田,既可保持种植业高产,又可促进养猪业的发展。

6. 沙打旺。又名直立黄芪、苦草,在我国北方各省各地区均有分布。沙打旺适应性强,产量高,具有耐寒、耐旱、耐瘠和抗风沙能力,是饲料、绿肥、固沙保土等方面的优良牧草。沙打旺的茎叶鲜嫩,营养丰富,以干物质计:含粗蛋白质23.5%,粗脂肪3.4%,粗纤维15.4%,无氮浸出物44.3%,钙1.34%,磷0.34%,产奶净能为6.24 MJ/kg。

沙打旺可青饲,放牧,调制青贮、干草或干草粉等,其干草的适口性优于鲜草。沙打旺为黄芪属牧草,含有脂肪族硝基化合物,具苦味,可在动物体内代谢为β-硝基丙酸和β-硝基丙醇等有毒物质。反刍家畜可依靠瘤胃微生物将其分解,因而饲喂较为安全,但最好与其

5. Chinese milkvetch. Chinese milkvetch (*Astragalus sinicus* L.) is widely cultivated in the Yangtze River valley and south of China. It is a green fertilizer and feed crop with high yield, fresh succulency and good palatability. It is especially loved by pigs, and it is also a good green forage for cattle, sheep, horses, birds and rabbits. The nutritive value of Chinese milkvetch is the highest in the budding stage. In terms of dry matter, it contains 31.76% of crude protein, 4.14% of crude fat, 11.82% of crude fiber, 44.46% of nitrogen free extract, 7.82% of crude ash, and 8.49 MJ/kg of milk net energy. The yield in the budding stage is only 53% of that in the full flowering stage, so it is better to cut at the full flowering stage.

Chinese milkvetch can be used as green forage, silage, hay or hay powder, but too much is not suitable for feeding ruminant animals such as cattle and sheep to avoid causing distention. In some places, Chinese milkvetch is directly used as green fertilizer, but from the perspective of economic effect, it is better to feed the pigs with Chinese milkvetch first and then fertilize the fields with pig manure, which can not only maintain the high yield of planting industry, but also promote the development of pig industry.

6. Erect milkvetch. Erect milkvetch (*Astralus adsurgens Pall.*) is distributed in all provinces and regions in the north of China. Erect milkvetch has strong adaptability, high yield, cold tolerance, drought tolerance, barren tolerance and wind sand resistance. It is an excellent forage grass in feed, green manure, sand fixation and soil protection. The stems and leaves of erect milkvetch are fresh and tender and nutritious. In terms of dry matter, it contains 23.5% of crude protein, 3.4% of crude fat, 15.4% of crude fiber, 44.3% of nitrogen free extract, 1.34% of calcium, 0.34% of phosphorus, and 6.24 MJ/kg of milk net energy.

Erect milkvetch can be used for green feeding, grazing, preparing silage, hay or hay powder, etc., and the palatability of hay is better than that of fresh grass. Erect milkvetch is a kind of forage grass belonging to *Astragalus membranaceus*, which contains aliphatic nitro compound (aliphatic nitrocompounds), with bitter taste and can be metabolized into toxic substances such as β-nitropropionic acid and β-nitropropanol in animals. Ruminants can rely on

他牧草搭配使用。对单胃动物而言,沙打旺属低毒牧草,但仍可在饲粮中占有一定比例。用沙打旺草粉喂猪时可占饲粮的10%-20%,在鸡饲粮中可占5%-7%。沙打旺可与青刈玉米或禾本科牧草混合青贮,经青贮后,其有毒成分减少,饲喂安全性提高。

7. 小冠花。也称多变小冠花,原产于欧洲南部和东地中海地区,我国从20世纪70年代开始引进,在南京、北京、陕西、山西、辽宁等地生长良好。小冠花根系发达、耐寒、耐瘠,繁殖力强,覆盖度大,花色多、鲜艳,既可作为牧草、保土、蜜源植物,又可作为美化庭院净化环境的观赏植物。小冠花茎叶繁茂柔软,叶量丰富,营养价值与紫花苜蓿接近,以干物质计:含粗蛋白质20.0%,粗脂肪3.0%,粗纤维21.0%,无氮浸出物46.0%,钙1.55%,磷0.30%,产奶净能为6.30 MJ/kg。

小冠花茎叶有苦味,适口性比紫花苜蓿差,牛、羊喜食,特别是羊更喜食,除青饲和青贮外,也可调制干草或草粉。但小冠花草地耐牧性差,应在连续放牧之后围栏割草,待草地恢复生机后再行放牧。小冠花含有毒物质β-硝基丙酸,单独或大量饲喂单胃畜禽时易引起中毒,尤以幼兔危害为大,因此应限量或与其他牧草搭配饲喂。

8. 红豆草。也叫驴食豆、驴喜豆,原产于欧洲,是豆科红豆草属多年生草本植物,在山西、甘肃、内蒙古、陕西、青海等地种植较多。红豆草是我国干旱和半干旱地区有价值的牧草之一,其花色粉红艳丽,气味

rumen microorganisms to decompose erect milkvetch, so it is safer to feed, but it is best to use it with other forages. For monogastric animals, erect milkvetch is a low toxic forage, but it can still account for a certain proportion in the feed. When fed with erect milkvetch powder, it can account for 10%-20% of the pig feed and 5%-7% of the chicken feed. Erect milkvetch can be mixed with green corn or gramineous forage. After silage, the toxic components are reduced and the feeding safety is improved.

7. Crownvetch. Crownvetch (*Coronilla varia* L.) is native to southern Europe and eastern Mediterranean region. Introduced into China since the 1970s, it grows well in Nanjing, Beijing, Shaanxi, Shanxi, Liaoning and other places. It has developed root system, cold tolerance, barren tolerance, strong fecundity, large coverage, multi-color and bright flowers, which can be used not only as forage grass, soil conservation, honey plants, but also as ornamental plants to beautify the courtyard and purify the environment. The crownvetch has luxuriant and soft stems and rich leaves, and its nutritional value is similar to that of alfalfa. In terms of dry matter, it contains 20.0% of crude protein, 3.0% of crude fat, 21.0% of crude fiber, 46.0% of nitrogen free extract, 1.55% of calcium, 0.30% of phosphorus and 6.30 MJ/kg of milk net energy.

The stems and leaves of crownvetch have bitter taste, and the palatability of it is worse than alfalfa. Cattle and sheep like to eat it, especially sheep. In addition to green feeding and silage, hay or grass powder can also be mixed. However, the grazing tolerance of the crownvetch is poor, so the grass should be mowed by fence after continuous grazing and grazed again after the grassland is restored to life. Crownvetch contains toxic substance β-nitropropionic acid (beta-nitropropionicacid), which is easy to cause poisoning when feeding monogastric livestock and poultry alone or in large quantities, especially in young rabbits, so it should be fed in limited quantity or in combination with other forages.

8. Sainfoin. Sainfoin (*Onobrychis viciifolia*) is original to Europe. It is a perennial herb of Leguminosae, which is widely planted in Shanxi, Gansu, Inner Mongolia, Shaanxi, Qinghai and other places of China. Sainfoin is one of the valuable forage grasses in arid and semi-arid areas of China. It has bright pink flowers, fragrant smell and rich

芳香,营养丰富,除蛋白质外,还含有丰富的维生素和矿物质,是家畜的优质饲草,饲用价值可与紫花苜蓿相媲美,被称为"牧草皇后"。开花期干物质中营养成分为:粗蛋白质15.1%,粗脂肪2.0%,粗纤维31.5%,无氮浸出物43.0%,钙2.09%,磷0.24%,产奶净能6.01 MJ/kg。

红豆草适口性很好,各种家畜均喜食,除了青草粉喂食外,还可以用鲜草打浆喂猪。收种子后的茎秆也可作为牛、羊等草食家畜的良好粗饲料。无论单播,还是和禾本科牧草混播,其干草和种子产量均较高,而且易调制干草,其最大优点是在调制干草过程中叶片损失少、易晾干。其抗病虫害能力强,而且返青较早,是提供早期青饲料的牧草之一,在早春缺乏青饲料的地区栽培尤为重要。红豆草各个生育阶段的茎叶均含有很高的浓缩单宁,可沉淀能在瘤胃中形成大量持久性泡沫的可溶性蛋白质,故可使反刍家畜在青饲、放牧利用时不致发生膨胀病。

nutrition. In addition to protein, it is also rich in vitamins and minerals. It is a high-quality forage grass for livestock. Its feeding value is comparable to that of alfalfa and is called "Herbage Queen". The nutrients in dry matter during flowering are as follows: 15.1% of crude protein, 2.0% of crude fat, 31.5% of crude fiber, 43.0% of nitrogen free extract, 2.09% of calcium, 0.24% of phosphorus, and 6.01 MJ/kg of milk net energy.

Sainfoin has excellent palatability, and all kinds of livestock like to eat it. In addition to green feeding of grass powder, it can also be pulped with fresh grass to feed pigs. The stalk after harvesting seeds can also be used as a good roughage for cattle, sheep and other herbivorous livestock. No matter single sowing or mixed sowing with grasses, the hay and seed yields are higher, and it is easy to prepare hay, and its biggest advantage is that the leaves lose less and are easy to dry in the process of preparing hay. It has strong resistance to diseases and insect pests and returns to green earlier, so it is one of the forages to provide early green fodder, and it is especially important to cultivate in the areas where green fodder is lacking in early spring. The stems and leaves of sainfoin at all growth stages contain very high concentrated tannin (condensed tannins), which can precipitate a large number of soluble proteins that can form persistent foam in the rumen, so that ruminants will not have bloating disease during green feeding, grazing and utilization.

(二)禾本科牧草
Gramineae forage

重要的禾本科牧草包括黑麦草、无芒雀麦、羊草、苏丹草、鸭茅、象草等。

Important gramineae forage includes perennial ryegrass (*Lolium perenne* L.), smooth bromegrass (*Bromus inermis* Leyss.), Chinese wildrye [*Aneurolepidium chinense* (Trin.) Kitag.], sudan grass [*Sorghum sudanense* (Piper) Stapf.], orchard grass (*Dactylis glomerate* L.), elephant grass (*Pennisetum purpureum* Schumach.), etc.

1. 黑麦草。本属有20多种,其中最有饲用价值的是多年生黑麦草(*Lolium perenne* L.)和一年生黑麦草(*Lolium multi florum* Lam.),我国南北方都有种植。黑麦草生长

1. Ryegrass. There are more than 20 species of this genus, among which perennial ryegrass (*Lolium perenne* L.) and Italian ryegrass (*Lolium multi florum* Lam.) are the most valuable forage species, which are planted in the south and north of China. Ryegrass grows fast, has many tillers,

快,分蘖多,一年可多次刈割,产量高,茎叶柔嫩光滑,适口性好,以开花前期的营养价值最高。新鲜黑麦草约含干物质17%,粗蛋白质2.0%,产奶净能为1.26 MJ/kg。

黑麦草干物质的营养组成随其刈割时期及生长阶段而不同。随生长期的延长,黑麦草的粗蛋白质、粗脂肪、粗灰分含量逐渐减少,粗纤维明显增加,尤其是难以消化的木质素增加显著,故刈割时期要适宜。

黑麦草可青饲、调制干草、放牧利用或青贮,各类家畜都喜食。特别是多年生黑麦草,由于分蘖多、再生性强、耐践踏,是很好的放牧草,与白三叶、红三叶、百脉根等混播,能建成高产优质的刈牧兼用草地。黑麦草制成干草或干草粉再与精料配合,是肉牛育肥的好饲料。试验证明,周岁阉牛在黑麦草地上放牧,日增重为700 g;喂黑麦草颗粒料(占饲粮40%、60%、80%),日增重分别为994 g、1 000 g、908 g,而且肉质较好。

2. 无芒雀麦。又名无芒草、禾萱草,原产于欧洲、西伯利亚和中亚,在我国东北、西北、华北等地均有分布。无芒雀麦适应性广,生活力强,适口性好,茎少叶多,营养价值高,是世界重要的栽培牧草之一。幼嫩的无芒雀麦干物质中所含粗蛋白质不亚于豆科牧草,到种子成熟时,其营养价值明显下降。

无芒雀麦刈割后用于青饲、制备干草或青贮。同时,无芒雀麦有深且强固的地下根茎,能形成絮结草皮,耐践踏,再生力强。但单播时在3-4年后会迅速衰退,和苜蓿等豆科牧草混播因补充氮源可防止草地早衰。无芒雀麦春天早发,秋天

can be cut many times a year, has high yield, tender and smooth stems and leaves, good palatability, and has the highest nutritive value in the early stage of flowering. Fresh ryegrass contains about 17% of dry matter, 2.0% of crude protein and 1.26 MJ/kg of milk net energy.

The nutrient composition of the dry matter of ryegrass varies with the cutting stage and growth stage. The crude protein, crude fat and crude ash content of ryegrass decreases gradually with the prolongation of the growing period, while the crude fiber increases significantly, especially the indigestible lignin increases significantly, so the cutting period should be appropriate.

Ryegrass can be green feeding, hay making, grazing or silage, and all kinds of livestock like to eat it. In particular, perennial ryegrass is a good grazing grass because of its many tillers, strong regeneration and trampling resistance. It can be mixed with white clover, red clover and hundred vein root to build a cutting grassland with high yield and high quality. Ryegrass is made into hay or hay powder and then combined with concentrate. It is a good feed for beef cattle fattening. The experiment shows that the one-year-old eunuchs grazed on ryegrass, the daily gain is 700 g, and the daily gain is 994 g, 1,000 g and 908 g respectively when fed with ryegrass pellets (accounting for 40%, 60% and 80% of the diet), and the meat quality is fine.

2. Smooth bromegrass. Smooth bromegrass (*Bromus inermis Leyss.*) is native to Europe, Siberia and Central Asia, and is distributed in northeast, northwest and north of China. Smooth bromegrass is one of the most important cultivated forages in the world because of its wide adaptability, strong vitality, good palatability, less stems and more leaves and high nutritive value. The crude protein in the dry matter of young brome is no less than that of leguminous forage, and its nutritive value decreases significantly when the seeds mature.

Smooth bromegrass is used for green feeding, hay preparation or silage after cutting. At the same time, smooth bromegrass has deep and strong underground rhizomes, which can form flocculated turf, and is resistant to trampling and strong regeneration. However, single sowing will decline rapidly after 3-4 years, and mixed sowing with leguminous forages such as alfalfa can prevent premature

晚枯,直到深秋,再生草适口性仍较好,所以能延长青草放牧期。据报道,无芒雀麦在160天左右放牧期内肉牛可获增重45 kg;用来放牧早期断奶的肥育羔羊,每公顷牧地可获增重32.5 kg,平均日增重0.11 kg。

3. 羊草。又名碱草,是欧亚大陆草原区东部草甸草原及干旱草原上的重要建群种之一,在我国东北、华北、西北等地均有大面积分布。羊草为多年生禾本科牧草,在北方草原区多为群落的优势种或建群种,以羊草为主构成的各种类型草原草场的面积达333.3万 hm² 以上。近年来,经过人工驯化栽培已成为北方地区的优良栽培草种。羊草叶量丰富,适口性好,马、牛、羊等家畜都喜食。羊草鲜草中成分为:干物质28.64%,粗蛋白质3.49%,粗脂肪0.82%,粗纤维8.23%,无氮浸出物14.66%,粗灰分1.44%。

羊草生长期长,种子成熟后茎叶仍可保持绿色,可放牧或青饲。羊草主要供调制干草用,其干草产量高,营养丰富,但刈割时间要适当,过早或过迟都会影响其质量。抽穗期刈割调制成干草,颜色浓绿,气味芳香,是各种家畜的上等青干草,也是我国出口的主要草产品之一。优质羊草干草,1头奶牛日喂量可达15-20 kg,切短喂或整喂效果均好。放牧利用时宜从5月下旬开始至10月上旬结束,也可在冬季利用枯草放牧牛、羊、马。幼嫩时切碎或打浆后也可喂猪。

4. 苏丹草。也称野高粱,属一

senescence of grassland by supplementing nitrogen sources. Smooth bromegrass develops early in spring and withers late in autumn. Until late autumn, the palatability of regenerated grass is still good, so it can prolong the grazing period of grass. It is reported that the weight gain of beef cattle can be increased by 45 kg during the grazing period of about 160 days, and the weight gain for fattening lambs can be 32.5 kg per hectare, with an average daily gain of 0.11 kg.

3. Leymus chinensis. Leymus chinensis [*Aneurolepidium chinense* (Trin.) Kitag.] is one of the important constructive species on the eastern meadow grassland and arid grassland in the Eurasian steppe region. It is widely distributed in northeast, north, and northwest of China. Leymus chinensis is a perennial gramineous forage. It is mostly the dominant or constructive species of the community in the northern grasslands. The area of various types of grasslands mainly composed of Chinese wildrye is over 3.333 million hm^2. In recent years, after artificial domestication and cultivation, it has become an excellent cultivated grass species in the north. Leymus chinensis has abundant leaves and good palatability. Horses, cattle, sheep and other domestic animals like to eat it. The ingredients of Leymus chinensis are: 28.64% of dry matter, 3.49% of crude protein, 0.82% of crude fat, 8.23% of crude fiber, 14.66% of nitrogen free extract, and 1.44% of crude ash.

The vegetative growth period of Leymus chinensis is long, and the stems and leaves of the seeds can remain green after maturity, and it can be grazing or green feeding. Leymus chinensis is mainly used for hay making. Its hay yield is high and nutrition is abundant, but the cutting time should be appropriate, and its quality will be affected if it is cut too early or too late. The hay is cut and modulated during the heading period. The color is dark green and the smell is aromatic. It is the finest green hay for various livestock and one of the main grass products exported from China. High quality Leymus chinensis hay can be fed with a daily feed rate of 15-20 kg per cow, and the effect is good when fed short or whole. Grazing should start in late May and end in early October, and hay can also be used to graze cattle, sheep, and horses in winter. It can also be fed to pigs after being chopped or beaten when it is young.

4. Sudan grass. It's also known as wild sorghum which

年生禾本科牧草，原产于非洲苏丹，现遍布我国各地，尤以西北和华北干旱地区栽培最多。苏丹草具有高度的适应性，抗旱能力特强，在夏季炎热干旱地区，苏丹草也能旺盛生长。苏丹草的营养价值取决于其刈割时期，抽穗期刈割要比开花期和结实期刈割营养价值高，适口性也好，草食家畜均喜食，同时也是草食性鱼类的优质饲料。

苏丹草品质佳、产量高，可青饲、青贮或调制干草。由于苏丹草的茎叶比玉米、高粱柔软，更适宜晒制干草。此外，苏丹草的再生力强，也可放牧利用。具体利用时，第一茬适于刈割鲜喂或晒制干草，第二茬以后可用于牛、羊放牧。由于苏丹草幼嫩茎叶含少量氢氰酸，为防止发生中毒，要等到株高达 50-60 cm 以后才可以刈割、放牧。苏丹草喂乳牛时，每日每头可喂 30-40 kg 鲜草或 3-5 kg 干草。

5. 高丹草。高丹草是由饲用高粱和苏丹草自然杂交形成的一年生禾本科牧草，由第三届全国牧草品种审定委员会第二次会议于 1998 年 12 月 10 日审定通过。高丹草综合了高粱茎粗、叶宽和苏丹草分蘖力、再生力强的优点，能耐受频繁的刈割，并能多次再生。其特点是：产量高，抗倒伏和再生能力出色，抗病抗旱性好，茎秆更为柔软纤细，可消化的纤维素和半纤维素含量高而难以消化的木质素低，消化率高，适口性好，营养价值高。经测定，高丹草在拔节期的营养成分为：水分 83%，粗蛋白质 3%，粗脂肪 0.8%，无氮浸出物 8.3%，粗纤维 3.2%，粗灰分 1.7%，是饲喂草食家畜的一种优良青绿饲料，适于饲喂牛、羊、兔、鹅等多数畜禽和鱼类。

is an annual grass native to Sudan, Africa. It is now widely cultivated in all parts of China, especially in the arid areas of northwest and north China. Sudan grass has abilities of high adaptability and drought resistance. In hot and dry areas in summer, Sudan grass can also flourish and grow. The nutritive value of Sudan grass depends on its cutting period. The value of cutting at heading stage is higher than that at flowering and fruiting stage, which is good for palatability. Herbivorous livestock all like to eat it, and it is also a high-quality feed for herbivorous fish.

Sudan grass has great quality and high yield. It can be processed into succulence, silage or hay making. Because its stems and leaves are softer than maize and sorghum, it is more suitable for hay making. In addition, sudan grass has strong regeneration ability and can also be used for grazing. The first crop is suitable for fresh cutting and hay making, and the second crop can be used for cattle and sheep grazing. As the young stems and leaves of sudan grass contain a small amount of hydrocyanic acid, it is necessary to wait until the plant grows up to 50-60 cm high before cutting and grazing for the sake of avoiding being poisoned. When sudan grass is fed to dairy cattle, 30-40 kg of fresh grass or 3-5 kg of hay can be fed to per head per day.

5. Gaodan grass. Gaodan grass (sordan, hybrid *sorghum-sudangrass*) is a kind of annual grass which is formed by the natural cross between forage sorghum and Sudan grass. It was approved by the 2nd meeting of the 3rd national grass variety examination committee on December 10, 1998. Gaodan grass has the advantages of thick stem and wide leaf of sorghum and strong tiller and regeneration ability of sudan grass. It can endure frequent cutting and can regenerate for many times. Its characteristics are high yield, excellent lodging resistance, egeneration ability, good disease resistance and drought resistance, softer and slimer stalks, high content of digestible cellulose and hemicellulose but low lignin indigestible, high digestibility, good palatability and high nutritive value. It is found that the nutritive composition of the grass in the jointing stage is as follows: 83% of water, 3% of crude protein, 0.8% of crude fat, 8.3% of nitrogen free extract, 3.2% of crude fiber and 1.7% of crude ash. It is an excellent green feed for herbivorous livestock and suitable for feeding most livestock, poultry and

高丹草的主要利用方式是调制干草和青贮,也可直接用于放牧。干草生产适宜刈割期为抽穗期,即播种6-8周后,植株高度达到1.5-2.0 m时可开始第1次刈割,此时干物质中蛋白质含量较高,粗纤维含量较低,留茬高度应不低于15 cm,过低的刈割会影响再生。再次刈割的时间以3-5周以后为宜,间隔过短会引起产量降低。高丹草青贮前应将含水量由80%-85%降到70%左右。高丹草放牧时也应注意预防氢氰酸中毒,适宜放牧的时间是播种后5-6周,株高达45-80 cm时,此时的消化率可达到60%以上,粗蛋白质含量高于15%。过早放牧会影响牧草的再生,放牧可一直持续到初霜前。

6. 黑麦。黑麦是禾本科黑麦属一年或越年生草本植物,原产于中东地区及地中海。于1979年由美国引入我国的黑麦品种为冬牧-70,在我国南北方推广面积均较大。此草株高1.7 m左右,适应性广,耐旱、抗寒、耐瘠薄,分蘖再生能力强,生长速度快,产量高。冬牧-70具有营养丰富全面、适口性好、饲用价值高等优点,干物质中粗蛋白质占18%,尤其是赖氨酸含量较高,是玉米、小麦的4-6倍,脂肪含量也高,并含有丰富的铁、铜、锌等微量元素和胡萝卜素,是各类家畜冬春季节的良好青绿饲料,同时也是鱼类的好饲料。

冬牧-70以秋播为主,一般冬前不刈割,待翌年3月初进入旺盛生长期开始刈割,直到夏播前还可刈割2-3次,每次刈割留茬7-10 cm,最后一次麦收时刈割,但不留茬。随着黑麦物候期的延长,植株逐渐老化,粗蛋白质含量逐渐下降,

fish such as cattle, sheep, rabbits and geese.

The main use of Gaodan grass is hay making and silage, but it can also be used directly for grazing. The suitable cutting period for hay production is heading stage, that is, the first cutting can be started 6-8 weeks after sowing and when the plant height reaches 1.5-2.0 m, the content of crude protein in the dry matter is high, the crude fiber content is low, and the stubble height should be no less than 15 cm. Cutting too much will affect regeneration. The span of cutting should be 3-5 weeks. Too short interval will lead to a decrease in yield. The water content should be reduced from 80%-85% to 70% before silage. Attention should also be paid to the prevention of hydrocyanic acid poisoning in Gaodan grass. The suitable grazing time is 5-6 weeks after sowing, when the plant is as high as 45-80 cm, the digestibility can reach more than 60% and the crude protein content is higher than 15%. Early grazing affects pasture regeneration and can continue until the first frost.

6. Rye. Rye (rye; secale *cereale* L.) is a species of grass of the genus Gramineae, native to the Middle East and Mediterranean Sea. Winter grazing-70, a rye variety introduced into China from the United States in 1979, has been widely spread in both north and south China. The plant height of this grass is about 1.7 m, with wide adaptability, drought resistance, cold resistance and barren resistance, strong tillering regeneration ability, fast growth rate and high yield. Winter grazing-70 has abundant nutrition, good palatability, high forage value, etc. The crude protein of dry matter is 18%, especially the lysine content is high, 4-6 times that of corn of wheat, and fat content is high too. The dry matter is rich in iron, copper, zinc and other trace elements and carotene. So it is a good green forage for all kinds of livestock and fish feed in winter and spring.

Winter grazing-70 is mainly planted in autumn. Generally, it will not be cut before winter. It will start to be cut at the beginning of march of the next year when it enters the vigorous growing period. It can be cut for 2-3 times until summer sowing, with 7-10 cm of stubble left each time, and there is no stubble left during the last wheat harvest. With the prolongation of the phenological phase of rye, the

头茬饲草粗蛋白质含量高,可以作为蛋白质饲料使用。冬牧-70青饲时,乳牛可日喂30－40 kg,羊可日喂7 kg。除了利用其青饲外,也可制作青贮或晒制青干草。

7. 鸭茅。又称鸡脚草、果园草,为多年生草本植物,原产于欧洲西部,我国湖北、湖南、四川、江苏等省有较大面积栽培。鸭茅草质柔嫩,叶量多,营养丰富,适口性好,是牛、羊、马、兔等草食家畜和草食性鱼类的优良牧草,幼嫩时也可以喂猪禽。抽穗期茎叶干物质中含营养成分为:粗蛋白质12.7%,粗脂肪4.7%,粗纤维29.5%,无氮浸出物45.1%,粗灰分8%。鸭茅植株繁茂,产草量高,再生性强,适于放牧或调制干草,也可刈割后青饲或制作青贮料。由于鸭茅的营养成分随其生长期延长而下降,茎秆也因木质化而变得粗硬,故其利用上一定要注意适时刈割。青饲宜在抽穗前或抽穗期进行刈割,晒制干草时收获期不迟于抽穗盛期,放牧时以拔节中后期至孕穗期为好。

8. 象草。又称紫狼尾草,为多年生草本植物,原产于热带非洲,在我国南方各省区有大面积栽培。象草具有产量高、管理粗放、利用期长等特点,已成为南方青绿饲料的重要来源。象草营养价值较高,茎叶干物质中含营养成分为:粗蛋白质10.58%,粗脂肪1.97%,粗纤维33.14%,无氮浸出物44.70%,粗灰分9.61%。

象草质地柔软,叶量丰富,主要用于青饲和青贮,也可以调制成干草备用。适时刈割的象草,柔软多

plant gradually ages, and the crude protein content decreases gradually. The crude protein content of the first stubble forage is higher, so it can be used as protein feed. During feeding the winter grazing-70 green, dairy cattle can be fed 30–40 kg and sheep can be fed 7 kg per day. In addition to the use of its green feeding, it can also be used to make silage or drying green hay.

7. Orchardgrass. It is a perennial herb native to Western Europe and cultivated in a large area in Hubei, Hunan, Sichuan, Jiangsu and other provinces of China. Orchardgrass (*Dactylis glomerate* L.) has soft and many leaves, which is nutrient-rich and has good palatability. It's the excellent grass for cattle, sheep, horses, rabbits and other herbivorous livestock and herbivorous fish. It can also be fed to pigs and poultry when it is young. The nutrients in dry matter of stem and leaf at heading stage are: 12.7% of crude protein, 4.7% of crude fat, 29.5% of crude fiber, 45.1% of nitrogen free extract, 8% of crude ash. Orchardgrass has flourish plant, high yield, strong regeneration. It is suitable for grazing or hay making and can also be mowed for green feeding or silage. Because the nutrient composition of orchard grass decreases with the extension of its growing period, the stem also becomes thick and hard because of lignification, it must be cut timely when using. It is advisable for green forage to be cut before heading or at heading stage, and the harvest time for hay making should be no later than peak heading stage. When grazing, it is better to take the middle and late jointing stage to booting stage.

8. Elephant grass. It's also known as purple wolfsbane, which is a perennial herb, native to tropical Africa and cultivated in a large area in southern China. Elephant grass (*Pennisetum purpureum Schumach.*), characterized by high yield, extensive management and long utilization period, has become an important source of green forage in the south. Elephant grass is of high nutritional value. The nutrients in dry matter of stem and leaf are: 10.58% of crude protein, 1.97% of crude fat, 33.14% of crude fiber, 44.70% of nitrogen free extract, 9.61% of crude ash.

Elephant grass is soft in texture and abundant in leaves. It is mainly used for green feeding and silage. It can also be made into hay for use. The elephant grass that is mowed at

汁,适口性好,利用率高,是牛、羊、马、兔、鹅的良好饲草。幼嫩时也可以喂猪、禽,亦可作为养鱼饲料。象草为上繁草,再生性能好,耐践踏,故也可放牧利用。

the right time is soft and juicy, with good palatability and high utilization rate. It is a good forage for cattle, sheep, horses, rabbits, geese. It can also be fed to pigs and poultry when it is young, and it can be used as feed for fish farming. Elephant grass belongs to the finest grass, which has great ability of regeneration and resistance to trampling, so it can also be used for grazing.

三、青饲作物
Green forage crops

青饲作物是指农田栽培的农作物或饲料作物,在结实前或结实期刈割作为青绿饲料用。青饲作物涉及的种类很多,除禾本科、豆科外,也有十字花科及菊科等作物。和牧草类相比,青饲作物具有生长周期短、生长发育旺盛和短期产量高等优点,但在栽培和管理上费时费力。对于同一种作物,与采种利用相比,青饲利用往往能获得更大的单位面积营养物质产量,同时栽培时间也可缩短1个月以上,并含有更丰富的维生素和矿物质元素。不同作物相比,豆科比禾本科含有更多的粗蛋白质和钙,营养价值也高,禾本科有较高的产量。

青饲作物,必须掌握最佳的刈割时间。通常认为在开花期到乳熟期利用为宜,用于直接饲喂时可适当提前,调制青干草或作青贮时可适当推迟。特别是近年来培育的一些青贮专用品种,青贮时既利用作物的茎叶,也利用子实,称之为整株青贮,刈割时以乳熟期至蜡熟期为宜。在农区推广青饲作物的栽培和利用,对调整农业种植结构,实现"粮食-饲料-经济作物"三元结构的有机结合,保证青绿饲料的正常供应,推动畜牧业的稳步发展等具有重要意义。以下概要介绍青刈玉米、青刈大麦、青刈燕麦、大豆苗、豌

Green forage crops refer to the crops or fodder crops cultivated in farmland, which are cut before or during the period of fruiting and used as green fodder. There are many kinds of green forage crops, including gramineae, legumes, cruciferae, compositae and so on. Compared with herbage, green forage crops have the advantages of short growth cycle, vigorous growth and high short-term yield, but they are time-consuming and laborious in cultivation and management. For the same crop, compared with seed collection and utilization, the utilization of green forage can obtain larger unit area of nutrient yield, shorten the cultivation time by more than 1 month, and contain more vitamins and minerals. Compared with different crops, legume contains more crude protein and calcium than graminaria, and has higher nutritive value. However, gramineae has a higher yield.

It is necessary for green forage crops to master the best cutting time. It is generally considered that it is advisable to use it from the flowering period to the milk-ripe period. It can be appropriately advanced for direct feeding and postponed for making hay or silage. In particular, some special silage varieties cultivated in recent years use both the stems and leaves of the crops and the seeds during silage, which is called whole crop silage. It is appropriate to cut from milk-ripe stage to dough stage. Popularizing the cultivation and utilization of green forage crops in agricultural areas is of great significance to the adjustment of agricultural planting structure, the realization of the organic combination of the ternary structure of "grain-feed-cash crops", the guarantee of the normal supply of green forage and the promotion of the steady development of animal husbandry. The following

豆苗和蚕豆苗等常见的青饲作物。

is a brief introduction of green corn, green barley, green oat, soybean seedling, pea seedling and broad bean seedling.

(一) 青刈玉米
Green corn

玉米是重要的粮食和饲料兼用作物,在全世界广泛栽培,其植株高大、生长迅速、产量高,茎中糖分含量高,胡萝卜素及其他维生素丰富,饲用价值高。玉米的种类很多,作青刈用的通常为传统农田种植的马齿型玉米,其味甜多汁,适口性好,消化率高,营养价值远远高于收获子实后剩余的秸秆,是牛、羊、猪的良好青绿饲料。青刈玉米用作猪的饲料时,宜在株50-60 cm拔节以后开始刈割,到抽雄前后割完;作为牛、羊饲料时,可从吐丝到蜡熟期分批刈割。

近年来,由于养殖业的发展和育种技术的进步,涌现了一批高产优质青饲青贮玉米品种,种植这些专用或兼用品种正逐渐成为玉米种植业的一个主导方向。这类品种大体分为两种不同的类型:一是以利用茎叶为主的分蘖多穗型;二是茎叶和果穗同时利用的单秆大穗型。墨西哥玉米,也叫大刍草,属于玉米的野生近缘种,具有植株高大、分蘖多穗、根系发达、晚熟和生物产量高等特点,是许多新品种培育的良好的基本材料,但通常不宜作为青刈玉米种植。借其培育出的诸如"京多1号""新多2号""科多8号"等均属于多茎多穗型,既适合于青饲也可行青贮,草质优良,每公顷鲜草产量可达45-135 t。"中原单32号""科青1号"则属于单秆大穗的粮饲兼用型品种,茎秆粗,果穗大,蛋白质含量高,适口性好,即使子实成熟后茎叶仍保持鲜绿,因此,也可以在收获子实后用秸秆青饲或青贮。

Corn is an important food and feed crop, which is widely cultivated all over the world. It has large plants, rapid growth, high yield, high sugar content in the stem, rich carotene and other vitamins, and high feeding value. There are many kinds of corns. Horse tooth type corn, which is usually used for clipping in traditional farmland, has sweet and juicy taste, good palatability and high digestibility. Its nutritive value is far higher than the remaining straw after harvest. It is a good green forage for cattle, sheep and pigs. When the green corn is used as pig feed, it is advisable to start cutting after the plant is jointing's 50-60 cm and finish cutting before or after tasseling. When fed to cattle and sheep, it can be cut in batches from spinning to dough stage.

In recent years, due to the development of breeding industry and the progress of breeding technology, some high-yield and high-quality green silage varieties have sprung up, and planting these special-purpose or mixed-use varieties is gradually becoming a leading direction of corn planting industry. This kind of species can be roughly divided into two different types: one is the multi-panicle type of tiller which mainly uses stems and leaves; the other is the big-panicle type of single stalk which uses stems and fruit cluster simultaneously. Mexican corn (*Euchlaena Mexicana Schrad.*), also known as teosinte, is a wild relative of corn that is characterized by tall plants, tillering, well-developed roots, late maturation, and high biological yield. It is a good base material for many new varieties, but is generally not suitable for the cultivation of green corn. It has cultivated such as "Jingduo No. 1", "Xinduo No. 2" and "Keduo No. 8", all of which belong to the multi-stem and multi-spike type, which are suitable for green feeding and feasible silage. The grass quality is excellent, and the yield of fresh grass per hectare can reach 45-135 t. "Zhongyuandan No. 32" and "Keqing No. 1" are grain and feed varieties which belong to the big-panicle type of single stalk. They have thick stem, large ear, high protein content and good palatability, and their stems and leaves remain fresh green even after the seeds are mature. Therefore, straw can also be used for green feeding or silage after harvesting the seeds.

(二) 青刈大麦
Green barley

大麦也是重要的粮食和饲料兼用作物之一，有冬大麦和春大麦之分。大麦有较强的再生性，分蘖能力强，及时刈割后可收到再生草，因此是一种很好的青饲作物。青刈大麦可根据畜禽的要求，在拔节至开花时分期刈割，随割随喂。延迟收获则品质迅速下降。早期收获的青刈大麦质地鲜嫩，适口性好，可粉碎或打浆喂给猪禽，稍晚刈割则可以作为牛、羊、马的饲料，或者供调制成干草或青贮利用。

Barely is also one of the most important crops for both grain and fodder, including winter barley and spring barley. Barley is a good green forage crop because of its strong regeneration and tillering ability, and can receive regenerated grass after cutting in time. Green barley can be cut in stages from jointing to flowering according to the requirements of livestock and poultry, and fed as needed. Delayed harvesting results in rapid deterioration of quality. The early harvested green barley has a tender texture and good palatability. It can be crushed or pulped for feed to pigs and poultry, and the older barley can be used as fodder for cattle, sheep and horses, or for making hay or silage.

(三) 青刈燕麦
Green oat

燕麦叶多茎少，叶片宽长，柔嫩多汁，适口性强，是一种比较好的青刈饲料。青刈燕麦可在拔节至开花时刈割，各种禽类、兔、鱼都喜食，喂猪可粉碎或打浆再饲用。若抽穗后刈割，产量高，可以饲喂牛、羊、马等草食畜。青刈燕麦营养丰富，干物质中营养成分为：粗蛋白质14.7%，粗脂肪4.6%，粗纤维27.4%，无氮浸出物45.7%，粗灰分7.6%，钙0.56%，磷0.36%，产奶净能为6.40 MJ/kg。

Oats have more leaves and fewer stems. Its wide and long leaves are tender and juicy, and have a strong palatability. It is a relatively good green mowing feed. Green oats can be cut from jointing to flowering. All kinds of poultry, rabbits, and fish like to eat it. It can be crushed or beaten to feed pigs. If cutting after earing, it can result in high yield and can feed cattle, sheep, horses and other herbivore livestock. Green oats are rich in nutrition, and the nutrients in dry matter are as follows: 14.7% of crude protein, 4.6% of crude fat, 27.4% of crude fiber, 45.7% of nitrogen-free extract, 7.6% of crude ash, 0.56% of calcium, 0.36% of phosphorus, and 6.40 MJ/kg of net energy of milk production.

(四) 青刈豆苗
Green beans

青刈豆苗包括青刈大豆、青刈秣食豆、青刈豌豆、青刈蚕豆等，也是很好的一类青饲作物。与青饲禾本科作物相比，其蛋白质含量高，品质好，营养丰富，家畜喜食，但大量

Green beans include green soybeans, green fodder soybeans, green peas, green broad beans and so on. They are also good green forage crops. Compared with gramineae crops, it has high protein content, good quality, and rich nutrition, so livestock like to eat it. However, ruminants

饲喂反刍家畜时易发生臌胀病。刈割时间因饲喂目的及对象不同而异,早期作为猪、禽、鱼饲料时,可在现蕾至开花初期株高 40－60 cm 时刈割,刈割越早品质越好,但产量也低。通常在开花至荚果形成时期刈割,此时茎叶生长繁茂,干物质产量最高,品质也好。适时刈割的豆苗茎叶鲜嫩柔软,适口性好,富含蛋白质和各种氨基酸、胡萝卜素、维生素 B_1、维生素 B_2、维生素 C 和各种矿物质,是各种畜禽的优质青绿饲料。幼嫩的青刈豆苗是猪、鸡、鹅、兔、鱼的良好饲料,粉碎或打浆后拌入精料饲喂,效果很好。稍晚刈割的可用作牛、羊的饲料,可整喂或切短饲喂,但多量采食易患臌胀病,故应与其他饲料搭配饲喂。除供青饲外,在开花结荚时期刈割的豆苗,还可供调制干草用。秋季调制的干草颜色深、品质佳,是牛、羊、马的优良越冬饲料。也可制成草粉,作为畜禽配合饲料的原料。

are easy to have distention when fed in large quantities. The cutting time varies with different feeding purposes and objects. When used as feed for pigs, poultry and fish in the early stage, it can be cut from budding to early flowering stage when the plant height is 40－60 cm. The earlier the cutting, the better the quality is, but the lower the yield is. Cutting is usually done between flowering and pod formation when the stems and leaves are flourishing, and dry matter yield is the highest, so is the quality. Timely cutting pea seedlings can make the stems and leaves tender and soft with good palatability, rich in protein and various amino acids, and high content of carotene, vitamin B_1, vitamin B_2, vitamin C and various minerals, making them the high-quality green forage for all kinds of livestock and poultry. The young green bean sprouts are a good feed for pigs, chickens, geese, rabbits and fish, which is effective for mixing with concentrate after grinding or beating. The later mowed can be used as the feed for cattle and sheep. It can be whole or cut short, but large amounts of feed are prone to bloating disease, so it should be fed with other feeds. In addition to green feeding, in the period of flowering and pod, clipping bean seedlings can also be used to make hay. The hay prepared in autumn is dark in color and good in quality. It is an excellent overwintering feed for cattle, sheep and horses. It can also be made into grass powder and used as a raw material for livestock and poultry coordination feedstuff.

四、叶菜类
Leaf vegetables

叶菜类,主要是利用其宽大而浓密的叶片部分,其来源广泛,涉及的种类也很多,包括菊科、紫草科、藜科、蓼科、苋科等,通常作为饲料栽培的有苦荬菜、聚合草、甘蓝、牛皮菜、猪苋菜、串叶松香草、菊苣、杂交酸模等几种,此外也包括食用蔬菜、根茎瓜类的茎叶及野草野菜等。

叶菜类在生长旺盛时期多汁而鲜绿,能保持较高的营养价值,富含

For leaf vegetables, their broad and dense leaves are mainly made use of. They come from a wide range of sources and involve many species, including the Indian lettuce, comfrey, chenopodiaceae, polygonaceae, amaranthaceae, etc. Usually cultivated as feed, there are bitter thistle, comfrey, cabbage, leaf beet, pig amaranth, pine vanilla, chicory, hybrid acid mold and so on, as well as edible vegetables, stems and leaves of root melons, wild weeds and so on.

Leaf vegetables are juicy and bright green in the period of vigorous growth, which can maintain high nutritive value

胡萝卜素、维生素 C、B 族维生素等维生素和一些矿物质元素。粗蛋白质含量因种类不同而有差异,有些甚至高于紫花苜蓿等豆科牧草,但其中有部分属于非蛋白含氮物。粗纤维含量较低,钙磷比例也较适宜。这类饲料的利用方式多为刈割后直接饲喂,或切短打浆后饲喂。因其水分含量高,一般很少用来制备青干草或青贮。

叶菜类含有一些抗营养因子或毒素,主要是草酸和硝酸盐。草酸易和钙结合形成不溶物从而影响钙的利用,所以,在叶菜类饲喂量大时应注意钙的补充。硝酸盐本身无毒,但经酶或细菌的作用可被还原成亚硝酸盐而呈毒性,因此利用叶菜类饲料时要避免长时间堆放或加热焖煮。

1. 苦荬菜

苦荬菜又叫苦麻菜或山莴苣等,是菊科莴苣属一年生或越年生草本植物。苦荬菜生长快、再生力强、利用率高,南方一年可刈割 5 - 8 次,北方一年 3 - 5 次,一般每公顷产鲜草 75 - 100 t,高者可达 150 t。苦荬菜茎叶嫩绿多汁,易消化,粗蛋白质含量较高,粗纤维含量较少,富含维生素,营养价值较高。其味稍苦,性甘凉,适口性好,猪、牛、鸡、鹅、兔均喜食,也是喂鱼的好饲料。

苦荬菜主要用于青饲,也可制作青贮。喂猪时通常切碎或打浆后拌糠麸饲喂,喂给母猪可防止便秘,改善食欲和促进泌乳。一头成年母猪,日喂量可达 9 - 10 kg,既节省精料又有利于繁殖。青贮利用时可在现蕾至开花期刈割,含水分过高时,要晒半天到一天后再青贮,也可和玉米、苏丹草等混贮。

and are rich in vitamins such as carotene, vitamin C, B vitamins and some mineral elements. The content of crude protein varies with different species, some of which are even higher than those of leguminous forages such as alfalfa, but some of them belong to non-protein nitrogen. The utilization way of this kind of feed is mostly direct feeding after cutting, or feeding after cutting short and beating. Because of its high moisture content, it is generally seldom used to prepare green hay or silage.

Leaf vegetables contain some antinutritional factors or toxins, mainly oxalic acid and nitrate. Oxalic acid is easy to combine with calcium to form insoluble substance and affect the utilization of calcium. Therefore, attention should be paid to the supplement of calcium when feeding large amount of leaf vegetables. Nitrite is not toxic by itself, but can be reduced to nitrite by enzymes or bacteria to make it toxic. Therefore, the use of leaf vegetables should avoid prolonged stacking or heating and braising.

1. Indian lettuce

Indian lettuce (*Lactuca indica* L.) is an annual or perennial herb of the genus *Asteraceae lettuce*. It has the advantages of rapid growth, strong regeneration and high utilization. It can be cut 5 - 8 times a year in the south and 3 - 5 times a year in the north. It generally produces 75 - 100 tons of fresh grass per hectare, and the highest yield can reach 150 tons. It is green, juicy, easy to digest with more crude protein and less crude fiber. It is rich in vitamins and high in nutritive value. Its taste is slightly bitter, sweet and cool in nature, and has a good palatability. Pigs, cows, chickens, geese and rabbits all like it. It is also a good feed for fish.

It is mainly used in green feed and may also be made into silage. When feeding pigs, they are usually chopped or beaten and then mixed with bran. Feeding to sows can prevent constipation, improve appetite and promote lactation. An adult sow can be fed up to 9 - 10 kg per day, which saves concentrate and is beneficial to reproduction. When silage is used, it can be cut from budding to flowering, and when the moisture content is too high, it should be dried for half a day to 1 day before silage, or it can be mixed with corn, Sudan grass and so on.

2. 聚合草

聚合草又称饲用紫草、爱国草等,为紫草科多年生草本植物。聚合草产量高、营养丰富、利用期长、适应性广,全国各地均可栽培,是畜、禽、鱼的优质青绿多汁饲料。聚合草再生性很强,南方一年可刈割5-6次,北方为3-4次,第一年每公顷可产鲜草75-90 t,第二年以后每公顷产112.5-150 t。聚合草营养价值较高,其干草的粗蛋白质含量与苜蓿接近,高的可达24%,而粗纤维含量则比苜蓿低。风干聚合草茎叶的营养成分为:粗蛋白质21.09%,粗脂肪4.46%,粗纤维7.85%,无氮浸出物36.55%,粗灰分15.69%,钙1.21%,磷0.65%,胡萝卜素200.0 mg/kg,核黄素13.80 mg/kg。

聚合草有粗硬刚毛,畜禽不喜食,可在饲喂前先经切碎或打浆,则具有黄瓜香味,或与粉状精料拌和,适口性提高,饲喂效果较好。聚合草也可调制成青贮料或干草。如晒制干草,须选择晴天刈割,就地摊成薄层晾晒,宜快干,以免日久颜色变黑,品质下降。值得注意的是,聚合草茎叶中含吡咯双烷类生物碱,是一类损害动物肝脏的毒素,含量可达0.2%-0.3%,因此要限饲,畜禽日粮中所占比例不应超过干物质的20%,最好与其他饲草搭配饲喂。

3. 牛皮菜

牛皮菜又称根达菜、叶用甜菜,为藜科甜菜属二年生草本植物,我国各地均有栽培。牛皮菜为喜温作物,生长的适温为15-25℃,温度过低,则生长缓慢或停止。牛皮菜产量高、叶量大、利用期长、易于种植,既可食用又可饲用。

牛皮菜叶厚柔嫩多汁,适口性

2. Comfrey

Comfrey is a perennial herbaceous plant of the Comfrey family. Comfrey (*Symphytum peregrinum* Ledeb.) has high yield, rich nutrition, long utilization period and wide adaptability. It can be cultivated all over the country and is a good green and juicy feed for livestock, poultry and fish. Comfrey has a strong reproducibility. It can be cut 5-6 times a year in the south and 3-4 times in the north. Fresh grass can yield 75-90 t per hectare in the first year and 112.5-150 t per hectare after the second year. The nutritive value of comfrey is higher. The crude protein content of hay is close to that of alfalfa, up to 24%, while the crude fiber content is lower than that of alfalfa. The nutritional components of the stems and leaves of air-dried polygonum are as follows: 21.09% of crude protein, 4.46% of crude fat, 7.85% of crude fiber, 36.55% of nitrogen free extract, 15.69% of crude ash, 1.21% of calcium, 0.65% of phosphorus, 200.0 mg/kg of carotene, 13.80 mg/kg of riboflavin.

Comfrey has rough bristles. Livestock and poultry do not like to eat it. It can be cut or beaten before feeding, then it has cucumber fragrance. Mixed with powdery fine material, its palatability is improved, and its feeding effect is better. Comfrey can also be made into silage or hay. If the hay is made in the sun, it must be cut on sunny days and spread into thin layers to air in the sun. It should be dried quickly so as not to turn black and reduce the quality. It should be noted that pyrrolizidine alkaloids, a class of toxins that damage the liver of animals, are found in the stems and leaves of polygrasses, and their content can be up to 0.2% to 0.3%. Therefore, diet of livestock should not exceed 20% of the dry matter, and it is best to combine with other forages.

3. Leaf beet

Leaf beet (*Beta Vulgaris* L. var. *cicla* L.) is a chenopodiaceae vulgaris biennial herbaceous plant and is cultivated all over China. It is a thermophilic crop and the suitable temperature for growth is 15-25 ℃. If the temperature is too low, the growth will slow or stop. With high yield, large amount of leaves, long utilization period and easy growing, it can be both edible and feedable.

Leaf beet leaves are thick, tender and juicy, with good

好,营养价值也较高,是猪喜食的一种青绿饲料。投喂时应当生喂,投喂量逐渐增加,如果一次投喂量过多会导致动物排稀粪。不能煮熟投喂,因为在煮熟放置时,会产生亚硝酸盐从而导致动物中毒。除喂猪外,还可喂牛、兔、鸭、鹅等,也可打浆喂鱼。

4. 杂交酸模

杂交酸模也叫酸模菠菜、高秆菠菜、鲁梅克斯 k-1 等。该品种是 1974—1982 年苏联乌克兰国家科学院中央植物园以巴天酸模为母本、天山酸模为父本远缘杂交育成。我国于 1995 年开始引进,并在新疆、黑龙江、山东、江西等地推广利用。

杂交酸模为蓼科酸模属多年生草本植物,既具有生长快、产量高和品质优良等特性,又有极强的耐寒性,可耐 -40 ℃ 的低温。此外,它还具耐旱涝、耐盐碱、喜水肥、适应性广、抗逆性强等特性,适于在盐渍土上种植。但它易感白粉病,也易发生虫害。在水肥条件较好的情况下,每公顷产量可达 150 - 225 t,折合干草为 15.0 - 22.5 t。

杂交酸模蛋白质含量高,干物质中粗蛋白质含量在叶簇期达 30% - 34%,还含有较高的胡萝卜素、维生素 C 等维生素。鲜喂时将茎叶切碎直接喂或打浆拌入糠麸等饲料后再喂,可整株喂牛。青贮时可加 20% - 30% 的禾本科干草粉或秸秆,效果很好。因其水分很高,干物质含量低,故不适宜调制青干草。该草在利用方式和选择饲喂的畜种方面有一定的限制,宜先少量引种试验后再推广。杂交酸模抗热性差,七八月份高温季节生长缓慢或停止生长,因而夏季产量很低且因

palatability and high nutritive value. It is a kind of green forage that pigs like to eat. It is suitable for raw feeding, and the amount of feeding increases gradually, while too much feeding at one time can easily lead to sparse defecation. It cannot be cooked before feeding, because when cooked and stored, nitrite is likely to form and will cause poisoning for animals. In addition to feeding pigs, it can be fed to cows, rabbits, ducks, geese, etc.. It can also be beaten to feed fish.

4. Hybrid acid mold

Hybrid acid mold is also known as sorbic spinach, tall stalk spinach, Rumax K-1, etc. The cultivar was bred in the Central Botanical Garden of the National Academy of Sciences of Ukraine from 1974 - 1982, using Rumex Batianum as female parent and Rumex Tianshan as male parent. It was introduced in China in 1995 and popularized in Xinjiang, Heilongjiang, Shandong, Jiangxi and other places.

Hybrid acid mold (*Rumex Patientia* × R. *talshanicus*) is a perennial herb of polygonaceae. It has characteristics of fast growth, high yield and excellent quality, as well as strong cold resistance and can resist low temperature of -40 ℃. In addition, it is also drought and waterlogging-resistant, saline-alkali resistant, water-fertilizer-loving, widely adaptable, and strong stress resistant, and so is suitable for planting in saline soil. But it is susceptible to powdery mildew and insect pests. Under good water and fertilizer conditions, the yield per hectare can reach 150 - 225 t, equivalent to 15.0 - 22.5 t hay.

The crude protein content in dry matter is up to 30% - 34% at leaf cluster stage, and it also contains higher carotene, vitamin C and other vitamins. When feeding fresh, the stems and leaves are chopped and fed directly or beaten and mixed with bran and other feeds before feeding. The whole plant can be fed to cattle. During Silage 20% - 30% of grass straw powder or straw can be added, and the effect is very good. Because of its high water content and low dry matter content, it is not suitable for making green hay. This grass has certain limitation in the utilization mode and the selection of feeding breeds, so it is advisable to introduce a small amount of the grass first and then popularize it. The hybrid acid mold has poor heat resistance and grows slowly or stops growing in high temperature in July and August, so

单宁含量高,对猪、牛、羊等适口性差,饲喂量不宜过多。

5. 菊苣

菊苣原产欧洲,用于蔬菜、饲料或制糖,1988年山西农业科学院畜牧兽医研究所从新西兰引入饲用型普那菊苣,现已在山西、陕西、浙江、河南、河北、山东、四川等地推广种植。菊苣为菊科多年生草本植物,喜温暖湿润气候,抗旱、耐寒、耐盐碱、喜水肥,一年可刈割3-4次,每公顷产鲜草120-150 t。

菊苣产量高,叶片大,叶量多,营养丰富。莲座期干物质中营养成分为:粗蛋白质21.4%,粗脂肪3.2%,粗纤维22.9%,无氮浸出物37.0%,粗灰分15.5%;开花期干物质中营养成分为:粗蛋白质17.1%,粗脂肪2.4%,粗纤维42.2%,无氮浸出物28.9%,粗灰分9.4%。动物必需氨基酸含量高而且齐全,茎叶柔嫩,适口性良好,牛、羊、猪、兔、鸡、鹅均喜食。一般多用于青饲,还可与无芒雀麦、紫花苜蓿等混合青贮,以备冬春饲喂奶牛。

6. 菜叶、蔓秧和蔬菜类

菜叶是指菜用瓜果、豆类的叶子及一般蔬菜副产品,人们通常不食用而作废料遗弃。菜叶种类多、来源广、数量大,是值得重视的一类青绿饲料。以干物质计,其能量较高、易消化,畜禽都能利用。尤其是豆类叶子营养价值很高,蛋白质含量也较丰富。

蔓秧是指作物的藤蔓和幼苗,一般粗纤维含量较高,不适于喂鸡、

the yield in summer is very low. Besides, due to the high tannin content, it has poor palatability for pigs, cattle and sheep, so the feeding quantity should not be limited.

5. Chicory

Chicory (*Cichorium intybus* L.) was originally planted in Europe and used in vegetables, feed or sugar production, The Animal husbandry and Veterinary Institute of Shanxi Academy of Agricultural Sciences introduced feed-type Puna Radicchio from New Zealand in 1988. Now it is widely cultivated in Shanxi, Shaanxi, Zhejiang, Henan, Hebei, Shandong, Sichuan and other places. Chicory is a perennial herbaceous plant of compositae, which likes warm and humid climate. It has characteristics of drought resistance, cold resistance, salt resistance, and it likes water and fertilizer. It can be cut 3-4 times a year, and produces 120-150 t of fresh grass per hectare.

Chicory has high yield, large and a lot of leaves and rich nutrition. Nutritional components of dry matter in indus period are:21.4% of crude protein, 3.2% of crude fat, 22.9% of crude fiber, 37.0% of nitrogen free extract, 15.5% of crude ash. The nutritional components of the dry matter in flowering period are:17.1% of crude protein, 2.4% of crude fat, 42.2% of crude fiber, 28.9% of nitrogen free extract, and 9.4% of crude ash. Its essential amino acid content for animals is high and complete. Its stems and leaves are soft; its palatability is good. Cattle, sheep, pigs, rabbits, chickens, geese are fond of it. Commonly used in green feeding, it can also be mixed with smooth bromegrass, alfalfa and other silage to prepare for feeding cows in winter and spring.

6. Vegetable leaves, trailing plants and vegetables

Vegetable leaves refer to the leaves of fruits and vegetables, legumes and general vegetable byproducts. People usually don't eat them and discard them as waste. Vegetable leaves are rich in variety, wide in source and large in quantity. It is a kind of green feed which deserves notice. In terms of dry matter, it has high energy and is easy to digest, and it can be fed to livestock and poultry. Especially, its leaves of legumes have high nutritive value and rich protein content.

Trailing plants refer to the vines and crop seedlings. Generally, they have high crude fiber content and are not

可作猪饲料,但老化后只能饲喂反刍家畜。

白菜、甘蓝和菠菜等食用蔬菜也可用作饲料。在蔬菜旺季,大量剩余的蔬菜、次菜及菜帮等均可饲喂家畜。为了均衡全年的青绿饲料供应,还可适时栽种些蔬菜。

7. 野草野菜类

野草野菜类是指人们在山林、野地、渠旁、田边、屋前房后挖掘的喂猪、兔等家畜的饲料。种类繁多,有豆科、菊科、旋花科、蓼科、苋科、十字花科等。这类饲料多数是在幼嫩生长阶段用作饲料,故营养价值较高,蛋白质含量较多,粗纤维含量较低,钙磷比例适当,均具有青绿饲料营养相对平衡的特点。但采集饲料的工作费时费力,采集时要注意鉴别毒草及是否喷洒过农药,以防中毒。

suitable for feeding chickens. They can be used as pig feed, but they can only be fed to ruminant livestock after aging.

Eating vegetables such as cabbage, kale and spinach also can be used as feed. In the vegetable season, a lot of leftover vegetables, secondary vegetables, vegetable outer leaves can be fed to livestock. In order to balance the supply of green feed throughout the year, some vegetables can also be planted in due course.

7. Wild weeds and wild herbs

Wild weeds and wild herbs refer to the feed for pigs, rabbits and other livestock dug by people in mountain forests, fields, ditches, fields and houses. There are many kinds, such as legumes, compositae, convolvulaceae, Polygonaceae, Amaranaceae, cruciferae, etc. This kind of feed is mostly used as feed in the tender growth stage, so it has higher nutritional value, more protein content, lower crude fiber content, appropriate proportion of calcium and phosphorus, which are the characteristics of green feed nutrition balance. However, the work of collecting feed is time-consuming and laborious, so attention should be paid to identify poisonous weeds and whether they have been sprayed with pesticides in order to prevent poisoning.

五、非淀粉质根茎瓜类
Non-starchy rhizome melons

非淀粉质根茎瓜类包括胡萝卜、芜菁甘蓝、甜菜及南瓜等,是家畜冬季的主要青绿多汁饲料。这类饲料天然水分含量很高,可达70%–90%,粗纤维较低,矿物质中钙和磷的含量都低,而无氮浸出物较高,且多为易消化的淀粉或糖分,所以能量较高。维生素含量因饲料种类不同而异,一般维生素C和B族维生素中硫胺素、核黄素和尼克酸含量高,胡萝卜和南瓜中含有丰富的胡萝卜素。至于马铃薯、甘薯、木薯等块根块茎类,因其富含淀粉,在生产中多被干制成粉后用作饲料原料,因此放在能量饲料部分介绍。

Non-starchy rhizome melons include carrots, rutabagas, beets and pumpkins, which are the main green and succulent fodder for livestock in winter. This kind of feed has a high natural moisture content which is up to 70%–90% and low crude fiber, low calcium and phosphorus content in minerals, and high nitrogen free extracts, and most of them are easily digestible starch or sugar, so its energy is high. Its vitamin content varies with different feed types. Generally, vitamin C and B vitamins contain high levels of thiamine, riboflavin and niacin. Carrots and pumpkins are rich in carotene. As for potato, sweet potato, cassava and other root tubers, they are often made into dry powder and used as feed raw materials in production for the rich starch, so they are introduced in the energy feed section.

1. 胡萝卜

胡萝卜是伞形科胡萝卜属二年生草本植物，以肉质根作饲料用。其产量高、易栽培、耐贮藏、营养丰富，是家畜冬春季重要的多汁饲料。胡萝卜的营养价值很高，大部分营养物质为无氮浸出物，含有蔗糖和果糖，故具甜味。其中胡萝卜素含量尤其丰富，为一般牧草饲料所不及。胡萝卜还含有大量的钾盐、磷酸盐和铁盐等。一般来说，颜色愈深，胡萝卜素或铁盐含量愈高，红色的比黄色的高，黄色的又比白色的高。胡萝卜按干物质计产奶净能为 7.65－8.02 MJ/kg，可列入能量饲料，但由于其鲜样中水分含量高、容积大，故在生产实践中并不依赖它来供给能量。它的重要作用是冬春季节作为多汁饲料并供给胡萝卜素等维生素。

在青绿饲料缺乏季节，在干草或秸秆含量较高的饲粮中添加一些胡萝卜，可改善饲粮口味，调节消化机能。乳牛饲料中若有胡萝卜作为多汁饲料，则有利于提高产奶量和乳的品质，所制得的黄油呈红黄色。对于种畜，饲喂胡萝卜可供给丰富的胡萝卜素，对于公畜精子的正常生成及母畜的正常发情、排卵、受孕与怀胎都有良好作用。胡萝卜熟喂，其所含的胡萝卜素、维生素C及维生素E会遭到破坏，因此最好生喂，一般奶牛日喂 25－30 kg，成年猪日喂 5－7.5 kg，家禽可日喂 20－30 g。

2. 芜菁甘蓝

芜菁在我国较少用作饲料，但芜菁甘蓝（也称灰萝卜）在我国已有近百年的栽培历史，两者均为十字花科芸薹属二年生草本植物。这两种块根饲料性质基本相似，水分含量都很高（约90%），干物质中无

1. Carrot

Carrot (carrot; *Daucus carota* L. *var. sativa* DC.) is a biennial herb of the genus Carrot in the umbelliferae, which is fed with succulent roots. It has high yield, easy cultivation, storage tolerance and rich nutrition, and is an important succulent feed for livestock in winter and spring. Carrots have high nutritive value. Most of the nutrients are nitrogen-free extracts, containing sucrose and fructose, so they are sweet. The content of carotene is especially rich, which is much more than that of the general forage. Carrots also contain a lot of potassium, phosphate and iron salts. In general, the darker the color, the higher the carotene or iron content. The red is higher in carotene than the yellow, and the yellow is higher in carotene than the white. The net milk yield of carrot is 7.65－8.02 MJ/kg according to dry matter, which can be included in energy feed. However, due to its high moisture content and large volume in fresh sample, people do not depend on it to supply energy in production practice. Its important role in winter and spring season is juicy feed and for supplying carotene and other vitamins.

In the season of lack of green forage, adding some carrots to the forage with high hay can improve the food taste and adjust the digestive function. If carrots are used as succulent feed for dairy cattle, it is beneficial to improve milk production and quality. Moreover, the butter made of milk is red and yellow. For breeding stock, feeding carrot can provide abundant carotene, which has a good effect on the normal production of male sperm and normal estrus, ovulation, conception and pregnancy of female animals. When carrots are fed cooked, the carotene, vitamin C and vitamin E in it will be destroyed, so it's best for raw feed. Generally, dairy cows are fed with 25－30 kg per day, and adult pigs, with 5－7.5 kg per day and poultry, with 20－30 g per day.

2. Rutabaga

Turnip is less used as forage in China, but rutabaga (rutabaga; *Brassica napobrassica* Mill.) (also known as Grey radish) has been cultivated for nearly a hundred years in China. Both of them are Brassica biennial herbaceous plants of the family Brassica. The two kinds of root tuber feed are basically similar in nature, with high water content

氮浸出物含量相当高,大约为70%,因而能量较高,每千克消化能可达14.02 MJ,鲜样由于水分含量高,只有1.34 MJ/kg。

芜菁与芜菁甘蓝含有某种挥发性物质,在饲喂奶牛时,可通过空气扩散波及牛乳,使乳沾染某种特殊气味。奶牛采食后可立即由乳腺排出,因此,注意不在挤乳前饲喂,减少牛乳在空气中的暴露机会,就可以避免牛乳异味的产生。

这两种块根饲料在国外多用于喂牛、羊,在我国多用来喂猪。由于其不仅能量价值高,而且块根在田里存留时间可以延长,即使抽薹也不空心,因而可以解决块根类饲料在部分地区夏初难以贮藏的问题。

3. 甜菜

甜菜又名糖萝卜,属藜科甜菜属二年生草本植物,原产于欧洲中南部,我国主要分布在东北、华北、西北地区,其他地区种植较少。甜菜的品种较多,按其块根中干物质与糖分含量多少,可大致分为糖甜菜、半糖甜菜和饲用甜菜三种。

各类甜菜的无氮浸出物主要是糖分(蔗糖),但也含有少量淀粉与果胶物质。由于糖用与半糖用甜菜中含有大量蔗糖,故其块根一般不用作饲料,而是先用以制糖,然后用其副产品甜菜渣作为饲料。关于甜菜渣的有关内容将在第八章中阐述。

根据不同畜种对甜菜消化率的差异,饲用甜菜喂牛、糖用甜菜喂猪最为适宜。用甜菜喂奶牛,产奶量与乳脂率无不良影响,且有所提高。甜菜尤适于饲喂生长肥育猪,但不宜长期饲喂种公羊和去势公羊,以免引起尿道结石。喂乳牛时,饲用甜菜可日喂40 kg,糖用甜菜日喂

(about 90%). The nitrogen free extract content in dry matter is quite high, which is about 70%. Therefore, they have high energy, up to about 14.02 MJ per kilogram of digestion energy, and only 1.34 MJ/kg of fresh sample due to their high water content.

Turnips and rutabagas contain a volatile substance. When fed to cows, it can spread through the air and infect their milk, giving it a special smell. Cows can be expelled from their mammary glands immediately after feeding. Therefore, pay attention not to feed them before milking and reduce the exposure of milk in the air, so as to avoid the milk odor.

These two kinds of root feed in foreign countries are mostly used to feed cattle and sheep, and to feed pigs in our country. Because of its high energy value and the prolonged retention time of root tubers in the field, and it is not hollow even when licked, it can solve the problem that root tubers are difficult to store in early summer in some areas.

3. Beet

Beet (beet; *Beta Vulgaris* L.) is also known as beetroot. It belongs to the chenopodiaceae beet biennial herb, and is native to central and southern Europe, mainly distributed in northeast China, North China and northwest China, and is less cultivated in other areas. There are many varieties of beet. According to the amount of dry matter and sugar content in root tubers, it can be roughly divided into sugar beet, half sugar beet and fodder beet.

The nitrogen-free extracts of various beets are mainly sugar (sucrose), but also contain a small amount of starch and pectin. As sugar beets and semi-sugar beets contain a large amount of sucrose, their root tubers are generally not used as feed, but firstly used to make sugar, and then the by-product beet residue is used as feed. The content of beet pulp will be discussed in Chapter 8.

According to the difference of digestibility of beet among different breeds, feeding fodder beet to cattle and sugar beet to pigs is reasonable. When beet is fed to cows, it has no adverse effects but can improve the ratio of milk yield and milk fat. Beet is especially suitable for feeding growth fattening pigs, but should not be fed for a long time to breeding rams and castrated rams, so as not to cause urethral stones. For dairy cattle, 40 kg of feed beets and 25 kg

25 kg；喂成年猪时，饲用甜菜可日喂5-7.5 kg，糖用甜菜日喂4-6 kg；幼猪喂量应酌减，切碎或打浆喂给效果较好。刚收获的甜菜不可立即饲喂家畜，否则易引起腹泻，这可能与块根中硝酸盐含量有关，当经过一个时期贮藏以后，大部分硝酸盐即可能转化为天门冬酰胺而变为无害。用甜菜喂动物时，宜生喂，不可熟喂。蒸煮不仅破坏甜菜中的维生素，而且会生成较多量亚硝酸盐。

甜菜叶富含草酸，为避免其在动物体内积累，可在甜菜茎叶汁液中加适量的0.2%石灰乳，以形成草酸钙。草酸钙不能被动物肠壁吸收，随粪便排出。未脱除草酸的甜菜不能长期用作公羊等动物的饲料，否则会使尿道结石发病率提高。

4. 南瓜

南瓜又名倭瓜，属葫芦科南瓜属一年生植物，既是蔬菜，又是优质高产的饲料作物。南瓜营养丰富、耐贮藏，运输方便，是猪、牛、羊及鸡的良好饲料，尤适于猪的育肥。

南瓜中无氮浸出物含量高，且其中多为淀粉和糖类。南瓜含淀粉量多，而饲料南瓜含果糖和葡萄糖较多。南瓜中还含有很多的胡萝卜素和核黄素，喂各类畜禽都适宜，尤适宜饲喂繁殖和泌乳家畜。南瓜含水分在90%左右，不宜单喂。喂奶牛时，10 kg南瓜（带子）饲用价值约与1.5-1.8 kg混合干草或3.65 kg玉米青贮料相当；喂猪时，10 kg南瓜的饲用价值约相当于1 kg谷物。南瓜喂鸡效果也很好，有促进换羽、提前产蛋的作用。

of sugar beets can be fed daily. When fed to adult pigs, the daily feeding of feed beets can be 5-7.5 kg, and the daily feeding of sugar beets can be 4-6 kg. The quantity of feeding to young pigs should be reduced, and it is better chopped or beaten. Fresh beets should not be fed to livestock immediately, or they are prone to cause diarrhea, which may be related to the nitrate content in root tubers. After a period of storage, most of the nitrate may be converted into asparagine and become harmless. When feeding animals with beets, they should be fed raw but not cooked. Steaming not only destroys the vitamins in beets, but also produces more nitrites.

Beet leaves are rich in oxalic acid. In order to avoid its accumulation in animals, an appropriate amount of 0.2% lime milk can be added to the leaf juice of beet stem, aiming to form calcium oxalate. Calcium oxalate cannot be absorbed by the intestinal walls of animals and is excreted in the faeces. Beet without removing oxalic acid cannot be used as fodder for rams and other animals for a long time, otherwise it will increase the incidence of urinary calculi.

4. Pumpkin

Pumpkin (cushaw; *Cucurbita Maschata* Duch.) is also known as squash. It is an annual plant of Cucurbita, which is not only a vegetable, but also a forage crop with high yield and high quality. Pumpkin is nutritious, storable, and convenient to transport. It is a good feed for pigs, cattle, sheep and chickens, especially suitable for fattening pigs.

The content of nitrogen free extract in pumpkin is high, and the content of starch and sugar are the main components. Pumpkin contains more starch, while feed pumpkin contains more fructose and glucose. Pumpkin also contains a lot of carotene and riboflavin, which is suitable for feeding all kinds of livestock and poultry, especially suitable for feeding breeding and lactation livestock. Pumpkin contains moisture at around 90%, so it is not suitable to be fed alone. The feeding value of 10 kg of pumpkin (tape) for cows is equivalent to 1.5-1.8 kg of mixed hay or 3.65 kg of corn silage. When fed to pigs, 10 kg of pumpkin is worth about 1 kg of grain. The effect of feeding chicken is also very good, which has the effect of promoting moulting and laying eggs in advance.

六、水生植物
Aquatic plants

用作饲料的水生植物是指生长在水中或潮湿土壤中的草本植物，有水浮莲、水葫芦、水花生、绿萍、水芹菜和水竹叶等。这类植物具有生命力强、生长繁殖快、适应性广、利用时间长和产量高等特点。但正是由于这些特点导致其易过度繁殖，形成单一的优势群落，影响或抑制其他物种的生长，破坏生态多样性，并进而堵塞航道，影响水运。特别是外来引进的物种如水葫芦、水花生在南方一些水域已经达到了泛滥成灾的地步。因此，将其作为饲料加以合理利用，既可以扩大青绿饲料来源，又可以减轻其对环境生态的危害。

水生植物茎叶柔软、细嫩多汁，富含胡萝卜素及微量矿物质元素，水分含量高达90%－95%，因而干物质含量很低，营养价值也比陆生植物饲料低，饲喂时应与其他饲料搭配使用，以满足家畜的营养需要。此外，水生饲料最易带来寄生虫病如猪蛔虫、姜片吸虫、肝片吸虫等，利用不当往往得不偿失。解决的办法除了注意水塘的消毒、灭螺工作外，最好将水生饲料青贮发酵或煮熟后饲喂，有的也可制成干草粉。熟喂时宜随煮随喂，不宜过夜，以防产生亚硝酸盐。

Aquatic plants used as feed refer to the herbs growing in water or moist soil, such as water lettuce, water hyacinth, water peanut, azolla, water celery and water bamboo leaves, etc. This kind of plant has the characteristics of strong vitality, fast growth and reproduction, wide adaptability, long utilization time and high yield. However, because of these characteristics, it tends to over reproduce, forming a single dominant community, affecting or inhibiting the growth of otherspecies, destroying ecological diversity, and then blocking the waterway, affecting water transport. In particular, introduced species such as water hyacinth and water peanut have reached the point of inundation in some waters of the south. Therefore, the rational use of it as feed can not only expand the source of green forage, but also reduce its harm to the environment and ecology.

The stems and leaves of aquatic plants are soft, tender and juicy, rich in carotene and trace mineral elements, and the moisture content is as high as 90%－95%, so content of dry matter is very low, and their nutritive value is lower than that of terrestrial plant feeds. Feeding should be used in combination with other feeds to meet the nutritional needs of livestock. In addition, aquatic feed is most likely to bring parasitic diseases such as swine roundworm, ginger trematode, fasciola hepatica, etc. Improper use often outweighs the gain. The solution is to pay attention to the disinfection of ponds and snail control work. The best way is aquatic silage fermentation or cooked before feeding, and some can also be made into hay powder. It should be cooked and fed when it is cooked, and should not be kept overnight to prevent the production of nitrite.

七、树叶类
Leaves

我国有丰富的树木资源，除少数不能饲用外，大多数树木的叶子、嫩枝及果实都可用作畜禽的饲料。合理采集可食树叶，多途径、多渠道

China is rich in tree resources, and most of the leaves, including twigs and fruit of trees can be used as feeds for livestock and poultry except a few kinds. Reasonable collection of edible leaves, multi-channel and multi-channel

开发和占领"空中牧场",是扩大饲料资源,促进林、牧业并进的重要一环。用作饲料的树叶较多,有苹果叶、杏树叶、桃树叶、桑叶、梨树叶、榆树叶、柳树叶、紫穗槐叶、刺槐叶、泡桐叶、橘树叶及松针叶等。

1. 紫穗槐叶、刺槐叶

紫穗槐又名紫花槐,是很有价值的饲用灌木类。刺槐又名洋槐,为豆科乔木。两者叶中蛋白质含量都很高,以干物质计可达20%以上,且粗纤维含量又较低,因此鲜叶制成的青干叶粉又属于蛋白质饲料。槐叶中的氨基酸也十分丰富,如刺槐叶中含有的氨基酸为:赖氨酸1.29%-1.68%,苏氨酸0.56%-0.93%,精氨酸1.27%-1.48%等。此外,维生素(尤以胡萝卜素和B族维生素含量高)和矿物质含量丰富,如紫穗槐青干叶中胡萝卜素含量可达270 mg/kg,与优质苜蓿相当。就营养特性而言,两者除具相同的共性外,刺槐叶尚有口味香甜、适口性好等特点。试验表明,用新鲜槐叶或槐叶粉饲喂猪、鸡,可取得增长快、饲料利用率高和节粮等效果。

采集季节不同,槐叶质量也不同。一般春季质量较好,夏季次之,秋季较差。但过早采集会影响林木生长,因此,科学采集时间宜在不影响林木生长的前提下尽量提前。北方可在7月底、8月初开采,最迟不超过9月上旬。采集过迟,则绿叶变黄,营养价值大幅度下降。采集部位一般为叶柄和叶片。槐叶叶片薄、易晒,一般两天水分可降至10%左右,即可粉碎贮藏。暂不加工的,可装入麻袋或化纤袋内,置于

development and occupation of "air pasture" is an important link to expand feed resources and promote forestry and animal husbandry. There are many kinds of leaves used as fodder, such as apple leaves, almond leaves, peach leaves, mulberry leaves, pear leaves, elm leaves, willow leaves, amorpha fruticosa leaves, black locust leaves, paulownia leaves, orange leaves and pine needles.

1. Amorpha fruticosa leaves, black locust leaves

Amorpha fruticose (indiobush amorpha falseindigo; *Amorpha fruticosa* L.), also known as purple flower locust, is a valuable feedstock shrub. Black locust (black locust; *Robinia Pseudoacacia* L.), also known as pseudoacacia, is a legume tree. The protein content in both leaves is high, up to more than 20% in dry matter, and the crude fiber content is low. Therefore, the green and dry leaf powder made from fresh leaves also belongs to protein feed. The amino acids in locust leaves are also very rich. For example, the amino acids in locust leaves are: 1.29%-1.68% of lysine, 0.56%-0.93% of threonine, 1.27%-1.48% of arginine, etc. In addition, it is rich in vitamins (especially high in carotene and B vitamins) and minerals. For example, the carotene content in the dry green leaves of Amorpha fruticose can reach 270 mg/kg, which is comparable to that of high-quality alfalfa. In terms of nutritional characteristics, the leaves of black locust have sweet taste and good palatability, in addition to having the same commonness with Amorpha fruticosa. The experimental results show that fresh locust leaves or locust leaves powder could be used to feed pigs and chickens to achieve effects of rapid growth, high feed utilization and grain saving.

The quality of locust leaves is different with the difference of collection season. Generally, its quality is better if collected in spring, followed by that of summer and autumn. However, early collection will affect the growth of trees. Therefore, scientific collection time should be advanced as far as possible without affecting the growth of trees. The north can be mined in late July and early August, no later than early September at the latest. If collected too late, the green leaves will turn yellow and the nutritive value will greatly decrease. The collection sites are generally petioles and leaves. Locust leaves are thin and easy to sun. Generally water can be reduced to about 10% after 2 days,

通风、阴凉、干燥处保存。鲜槐叶可直接青饲,叶粉可作为配合饲料原料。添加量为:一般鸡饲料中加3%~5%,猪饲料中加8%~15%。

2. 泡桐叶

泡桐又名白花泡桐、大果泡桐,为玄参科泡桐属落叶乔木,分布于我国中部及南部各省、区,在河南、山东、河北、陕西等地都生长良好。泡桐生长快,管理得当时5~6年即可成材,故有"3年成檩、5年成梁"之说。据测定,一株10年轮伐期的泡桐,年产鲜叶100 kg,折干叶28 kg左右;一株生长期泡桐,年可得干叶10 kg左右。按此计算,全国泡桐叶若能被充分利用,其量亦十分可观。

用泡桐叶、花、果作饲料由来已久。早在《博物志》中就记载有"桐花、叶饲猪极肥大且易养",《本草纲目》中也记载"泡桐花傅猪疮司猪肥大三倍",可见泡桐叶、花、果均可作为猪、羊、兔等动物的饲料。鲜泡桐叶家畜不喜食,干制可改善适口性。泡桐叶干物质中主要营养成分含量为:粗蛋白质19.3%,粗纤维11.1%,粗脂肪5.82%,无氮浸出物54.8%,钙1.93%,磷0.21%。试验表明,6~8 kg断奶仔猪日喂30 g泡桐叶粉,连喂60天,其增重较对照组提高24%;20~25 kg架子猪隔日每头每天喂60 g,连喂30天,平均每头猪增重较对照组多14%。但也有报道认为,泡桐叶、花适于饲喂40 kg以上肥育猪。饲喂50 kg左右肥育猪3个月,其体重可达115~125 kg,有缩短饲养

then they can be crushed and stored. If not processed temporarily, they can be put into gunny bags or chemical fiber bags and stored at a ventilated, cool and dry place. Fresh locust leaves could be fed directly and the leaves powder could be used as feedstuff. The addition amount is: 3%-5% in general chicken feed, and 8%-15% in pig feed.

2. Paulownia leaves

Paulownia [fortune paulownia, foxglove tree; *Paulowinia fortunei* (seem.) Hemsl.] is also known as fortune paulownia or big fruit paulownia. It is a deciduous tree of paulownia genus in the family of Radix Aucklandiae, distributed in central and southern provinces and regions of China, growing well in Henan, Shandong, Hebei, Shaanxi and other places. Paulownia grows fast. It can grow to yield within 5-6 years if managed properly, so there is a saying that goes "3 years into purlin, 5 years into beam". According to the measurement, the annual output of a paulownia plant in the 10-year ring cutting period is 100 kg of fresh leaves and 28 kg of dry leaves. A paulownia plant in its growing period can obtain about 10 kg of dry leaves per year. According to this calculation, if the whole country's paulownia leaves can be fully utilized, the quantity is also very considerable.

Paulownia leaves, flowers and fruits have long been used as fodder. As early as in the *Annals of Natural History*, it was recorded that "pigs fed with tung flowers and leaves are extremely fat and easy to raise", and it was also recorded in *Compendium of Materia Medica* that "pigs fed with paulownia flowers and sores are fat 3 times". It can be seen that paulownia leaves, while flowers and fruits can all be used as feed for pigs, sheep, rabbits and other animals. Domestic animals do not like fresh paulownia leaves and drying can improve its palatability. The main nutrients in the dry matter of paulownia leaves are as follows: 19.3% of crude protein, 11.1% of crude fiber, 5.82% of crude fat, 54.8% of nitrogen free extract, 1.93% of calcium, 0.21% of phosphorus. The results show that the weight gain of 6-8 kg weaned piglets is 24% higher than that of control group when they are fed 30 g paulownia leaf powder for 60 days. Every 20-25 kg piglet is fed 60 g for 30 days, and the average weight gain of each pig is 14% more than that of the control group. However, it has also been reported that paulownia leaves and flowers are suitable for feeding fattening

期、提高经济效益之效。还有人认为泡桐花含生长激素,连续喂用可促进猪健康生长。

3. 桑叶

桑也称桑树,原产于中国,已有3 000多年的栽培历史,除高寒地区外,全国都有种植。桑叶的产量高,生长季节内可采4-6次。桑叶不仅是蚕的基本饲料,也可作猪、兔、羊、禽的饲料。鲜桑叶含有:粗蛋白质4%,粗纤维6.5%,钙0.65%,磷0.85%,还含有丰富的维生素E、维生素B2、维生素C及各种矿物质。桑树枝、叶营养价值接近,均为畜禽的优质饲料。桑叶、枝采集可结合整枝进行,宜鲜用,否则营养价值下降。枝叶量大时,可阴干贮藏供冬季饲用。

4. 苹果叶、橘树叶

苹果枝叶来源广、价值高。据分析,苹果叶中一般含营养成分:粗蛋白质9.8%,粗脂肪7%,粗纤维8%,无氮浸出物59.8%,钙0.29%,磷0.13%。鲜苹果枝条兔喜食,喂时可将枝条折成小段(每段相连),放兔笼一侧,且枝条越粗、皮越厚越好。

橘树叶粗蛋白质含量较高,其量比稻草高3倍。每千克橘树叶含维生素C约151 mg,并含单糖、双糖、淀粉和挥发油,故其具舒肝、通气、化痰、消肿解毒等药效。试验表明,鸡饲料中添加橘叶粉可提高产蛋量和产肉量。长期给兔喂橘叶或橘叶粉,可使兔健壮生长,并可有效

pigs of more than 40 kg. Feeding fattening pigs weighing about 50 kg for 3 months can make them weigh up to 115-125 kg, which can shorten the feeding period and increase economic benefits. Others believe paulownia flowers contain growth hormones and that continuous feeding can promote healthy growth in pigs.

3. Mulberry leaves

Mulberry (white mulberry, mulberry; *Morus Alba* L.), also known as mulberry trees, is native to China. It has been cultivated for more than 3,000 years, and can be found all over the country except in alpine regions. Yield of mulberry leaves is high. It can be picked 4-6 times in the growing season. Mulberry leaves are not only the basic feed for silkworms, but also the feed for pigs, rabbits, sheep and poultry. Fresh mulberry leaves contain 4% of crude protein, 6.5% of crude fiber, 0.65% of calcium, 0.85% of phosphorus. It is also rich in vitamin E, vitamin B2, vitamin C and various minerals. Mulberry branches and leaves have similar nutritive value, and they are both high-quality feeds for livestock and poultry. Mulberry leaves and branches can be collected in combination with pruning, and should be fresh, otherwise the nutritive value will decline. When the amount of branches and leaves is large, they can be stored in the shade and used for winter feeding.

4. Apple leaves, orange leaves

Apple (apple; *Malus Pumila* Mill.) has wide sources of branches and leaves and high value. According to the analysis, the general nutrients in apple leaves are: 9.8% of crude protein, 7% of crude fat, 8% of crude fiber, 59.8% of nitrogen free extract, 0.29% of calcium, 0.13% of phosphorus. Rabbit likes fresh apple branches. We can fold the branches into small sections on one side of the rabbit cage when feeding (each section is connected). The thicker the branches and skin, the better.

The crude protein of orange leaves is higher, and its amount is 3 times higher than that of straw. Each kilogram of orange leaves contains 151 mg of vitamin C, and contains simple sugar, disaccharide, starch and volatile oil, so it has a pesticide effect of soothing liver, ventilation, phlegm, swelling and detoxification. The experiment shows that adding orange leaves powder to chicken feed could increase the egg and meat yield. Feeding the rabbit orange leaves or

地预防球虫病及咳喘等症发生,使毛兔产毛量提高。据《日本农业新闻》报道,将整枝剪下的橘树枝叶加工成2-3 cm的碎条青贮一个月后喂役牛或肉牛,可促进牛健康生长且生长速度加快。橘叶采集宜结合秋末冬初修剪整枝时进行。

5. 松叶

松叶主要是指马尾松、黄山松、油松以及桧、云杉等树的针叶。据分析,马尾松针叶干物质为53.1%-53.4%,含总能9.66-10.37 MJ/kg,粗蛋白质6.5%-9.6%,粗纤维14.6%-17.6%,钙0.45%-0.62%,磷0.02%-0.04%。松叶富含维生素、微量元素、氨基酸、激素和抗生素等,对多种畜禽均具抗病、促生长之效,在提高产蛋量、产奶量、节省精料、改善蛋黄色泽和提高瘦肉率等方面有明显效果。针叶一般在每年11月至翌年3月采集较好,其他时间采集针叶则含脂肪和挥发性物质较多,易对畜禽胃肠和泌尿器官产生不良影响。采集时应选嫩绿肥壮松针,采集后避免阳光曝晒,从采集到加工要求不应超过3天。

据报道,用鲜松针喂兔,应将鲜针叶切成细段,按每千克活重1 g计混入粉料中饲喂。用干叶饲喂时,宜将干叶与大麦、统糠按1:1混合使用。喂用时应坚持由少到多的原则。针叶在畜禽日粮中的配比如下:猪料中5%-8%,肉鸡料中3%,产蛋鸡料中5%,奶牛、鹅料中10%。

orange leaves powder for a long time can make the rabbits grow steadily, and can effectively prevent coccidiosis, cough and asthma and other diseases, resulting in the increase of hair production. According to the *Nihon Agri News*, turning the branches and leaves of orange trees into 2-3 cm pieces of silage for a month and then feeding them to serving cattle or beef cattle can promote healthy growth and speed up the growth of cattle. The collection of orange leaves should be combined with pruning in late autumn and early winter.

5. Pine leaves

Pine leaves (pine; *Pinus massoniana* Lamb.) mainly refer to the needle leaves of pinus massoniana, pinus taiwanensis, Chinese pinus, cypress and pruce, etc. According to the analysis, the dry matter of pinus massoniana's needle leaves is 53.1%-53.4%, 9.66-10.37 MJ/kg of total energy, 6.5%-9.6% of crude protein, 14.6%-17.6% of crude fiber, 0.45%-0.62% of calcium, 0.02%-0.04% of phosphorus. It's rich in vitamins, trace elements, amino acids, hormones and antibiotics, etc. and has the effect on a variety of livestock and poultry of disease resistance and growth promotion. It also has obvious effects in improving egg yield, milk yield, saving concentrate, improving egg yolk color and increasing lean meat rate. Needles are generally collected from November to March of the following year, while needles at other times contain more fat and volatile substances, which tend to have adverse effects on the gastrointestinal and urinary organs of livestock and poultry. Tender green and fat pine needles should be selected for collection. After collection, sunlight exposure should be avoided. The time interval from collection to processing should not exceed 3 days.

It is reported that when feeding rabbits with fresh pine needles, the fresh needles should be cut into fine pieces and fed into powder according to 1g of live weight per kilogram. When feeding with dry leaves, it is advisable to mix the dry leaves with barley and mixture of rice chaff and husk by 1:1. The principle of feeding from less to more should be adhered to when feeding. The ratio of needles in livestock and poultry diet is as follows: 5%-8% in pig feed, 3% in broiler feed, 5% in laying hens feed, 10% in cows and geese feed.

第五章 青贮饲料
Chapter 5　Silage

　　青贮饲料是指将新鲜的青刈饲料作物置于厌氧条件下，经乳酸菌等有益微生物发酵而成的具有特殊芳香气味、营养丰富的多汁饲料。青贮饲料能够长期保存青绿饲料的营养特性，并能扩大饲料资源。高品质青贮饲料的主要营养品质与其青贮原料相接近。对于反刍动物，青贮料青贮饲料的采食量、有机质消化率和有效能值与青贮原料相似。

　　Silage is a type of fodder with special aromatic odor and rich nutrition, which has been preserved by acidification from green forage crops, achieved through anaerobic fermentation by anaerobic microorganisms such as lactic acid bacteria. Silage can preserve the nutritional features of green forages for long periods and expand feed resources. The high-quality silage has a similar nutritional quality with silage raw materials. For ruminants, silages' feed intake, organic matter digestibility and effective energy value are similar to those of silage raw materials.

第一节　青贮饲料的优点及其发展
Section 1　Advantages and development of silage

　　青贮饲料的营养价值因原料种类的不同而有所差异，但其共同的特点是：富含水分、粗蛋白质、维生素和矿物质营养成分，非蛋白氮中以酰胺和氨基酸的比例最高。青贮饲料气味酸香，柔软多汁，因此适口性好，易于消化。

　　青贮饲料的发展源于传统农业生产中对剩余资源利用不足以及秋冬季反刍动物对青绿饲料的需求。在传统农业生产中，农作物收割后，大量的农作物秸秆被废弃，一定程度上影响了经济社会的可持续发展。而通过将牧草、饲料作物、秸秆等进行青贮后饲养牲畜，既可以长期保存和供应饲料，提高饲料品质，节省饲料成本，又可以减少病虫危

　　The nutritive value of silage varies with types of raw materials, but its common characteristics are: rich in water, crude protein, vitamins and mineral nutrients, and non-protein nitrogen has the highest ratio of amides and amino acids. Silage smells sour, soft and juicy, so it has good palatability and is easy to digest.

　　The development of silage stems from the insufficient utilization of surplus resources in traditional agricultural production, as well as the ruminant's requirement for green forages in autumn and winter. In traditional agricultural production, a large amount of crop residues are discarded after harvesting, which affects the sustainable development of the economy and society to a certain extent. By ensiling pastures, forage crops, straws and so on, we can not only preserve and supply feed for a long time, improve the quality of feed, save the cost of feed, but also reduce pests and dis-

害,实现过腹还田,促进农牧业良性循环。

一、青贮饲料的优点
The advantages of silage

1. 有利于青绿饲料长期保存和供应

青贮饲料可调剂青绿饲料供应的不平衡。由于青绿饲料生长期短,老化快,受季节影响较大,很难做到一年四季均衡供应。我国西北、东北、华北地区气候寒冷,作物生长期短,青绿饲料生产受到限制,整个秋冬季节都缺乏青绿饲料,而青贮饲料一旦做成,可以把夏季多余的青绿饲料保存起来供秋冬使用,保存年限可达2-3年或更长,从而可以弥补青绿饲料利用的时差之缺,解决冷季时家畜缺乏青绿饲料的问题。以此做到营养物质的全年均衡供应,从而使家畜始终保持高水平的营养状态和生产水平。特别是对乳牛饲养业,青贮饲料已经成为维持和创造高产水平不可缺少的重要饲料之一。

2. 保存青绿饲料的营养特性

(1) 最大保持青绿饲料营养物质。实验表明,青绿饲料在晒制过程中,植物细胞并未立即死亡,仍在继续呼吸,需消耗和分解营养物质,当达到风干状态时,养分损失一般达到20%-40%。若在风干过程中遇到雨雪淋洗或发霉变质,则损失更大。但在青贮过程中,青绿饲料在密封厌氧条件下保存,由于不受日晒、雨淋、机械损失的影响,物质的氧化分解作用微弱,养分损失仅为3%-10%,从而保存了大部分养分,特别是在保存蛋白质和维生素(胡萝卜素)方面要远远优于其他

eases, realize the return to the field, and promote the virtuous circle of agriculture and animal husbandry.

1. Conducive to the long-term preservation and supply of green forage

Silage can adjust the imbalance in the supply of green forage. Due to the short growing period and fast aging, the green forage is greatly influenced by the season, and it is difficult to achieve a balanced supply throughout the year. The climate in Northwest, Northeast, and North of China is cold, with short growth period and restricted production of green forage, which is suffering from green forage shortage throughout the autumn and winter. The silage makes it possible for the excess green forage in summer to be preserved for autumn and winter. It can be preserved for 2-3 years or even longer, which can solve the problem for lacking green forage in cold weather. In this way, a balanced supply of nutrients can be achieved by silage throughout the year and keep livestock always maintain a high level of nutrition and production levels. Silage has become one of the most important feed resources to maintain a healthy body and high production level, especially for dairy farming.

2. Preserving the nutritional characteristics of green forage

(1) Maximum retention of nutrients in the green forage. Experiments show that during the drying process of the green forage, the plant cells do not die immediately, but continue to breathe, and need to consume and decompose nutrients. When it reaches the air-dried state, the nutrient loss generally reaches 20% to 40%. In case of rain and snow leaching or mildew deterioration during air drying, the loss will be even greater. However, during the silage process, the green forage is stored under sealed anaerobic conditions. It is not affected by the sun, rain and mechanical factors so that oxidative decomposition of the substance is weak, with the nutrient loss only 3%-10%, which makes it possible for the preservation of most nutrients, especially in the preservation of protein and vitamins (carotene),

保存方法。

（2）青贮饲料鲜嫩多汁，使水分得以保存。青贮饲料含水量可达70%。同时，在青贮过程中由于微生物的发酵作用，产生大量乳酸和芳香族化合物，具有酸香味，柔软多汁，增强了其适口性和消化率。向日葵、菊芋、蒿草及玉米秸秆等，有的在新鲜时有臭味，有的质地较粗硬，一般家畜不喜食或利用率很低，若将其制成青贮饲料，不但可以软化秸秆，提高适口性，还能充分发挥此类饲料的作用，增加可食饲料的数量和种类。用同类青草制成的青贮饲料和干草，青贮饲料的消化率较干草有所提高。

3. 利于反刍动物集约化生产经营

青贮饲料调制方便，可以扩大饲料资源，并且青贮饲料调制方法简单，易于掌握。修建青贮设备的费用较少，且青贮饲料一次调制可长久使用。调制过程受天气条件的限制较少，在阴雨天或天气不好时，晒制干草困难，调制青贮料影响较小。例如：饲喂鲜草需要每天刈取，不仅增加劳动成本、延长占地时间，而且对饲料产量和质量均有影响；调制干草虽可一次收割，但晾晒干草需要时间，且因翻动、日晒、风吹、雨淋等因素影响，造成营养物质大量流失。而利用人工种植的牧草或饲料作物调制青贮饲料，可在产量最高、营养价值最好的时期，一次收割储存起来，比饲喂鲜草或调制干草缩短占地时间，也可及时播种下茬作物。

另外，青贮饲料储藏空间比干草小，可节约存放场地，提高饲喂效率。每立方米青贮饲料质量为450－700公斤，其中干物质含量为150公斤；而每立方米干草质量仅为70

which is far superior to other preservation methods.

(2) Silage is tender and juicy, and silage preserves moisture. The water content of silage can reach 70%. At the same time, due to the fermentation of microorganisms during the silage process, a large amount of lactic acid and aromatic compounds are produced, which have a sour flavor, soft and juicy, good palatability, and enhance its palatability and digestibility. Sunflower, jerusalem artichoke, wormwood, corn stalks, etc., are smelly when fresh, or have a rough texture. Most livestock do not like to eat so that utilization rate is low. If they are made into silage, they can not only soften straw and improve palatability, but also give full play to the role of this type of feed and increase the amount and type of the edible feed. The digestibility of silage and hay made from grass of the same kind is improved.

3. Conducive to intensive management of ruminants

Convenient preparation of silage can expand feed resources, and the silage preparation method is simple and easy to master. The cost of building silage equipment is less, and the silage can be used for a long time after being prepared. The preparation process is less restricted by weather conditions. In rainy days or bad weather, it is difficult to dry hay, and the influence of the preparation of silage is less. For example, feeding fresh grass needs to be harvested every day, which not only increases labor costs and prolongs the land occupation time, but also affects the output and quality of feed. Although the hay can be harvested at the right time, it takes time to dry the licorice. Flip, sun, wind, rain and other factors have caused a large loss of nutrients. Artificially planted forage or fodder crops used to modulate silage can be harvested and stored at one time during the period when the yield is the highest and the nutritive value is the best. It takes less time than feeding fresh grass or modulating hay, and can be planted in time for the next stubble.

In addition, the storage space of silage is smaller than that of hay, which can save storage space and improve feeding efficiency. The mass of 1 m^3 silage is 450－700 kg, of which the dry matter content is 150 kg; while the mass of 1 m^3 hay is only 70 kg, which contains about 60 kg dry

公斤,约含干物质60公斤。例如:每吨青贮苜蓿体积为1.25立方米,而每吨苜蓿干草体积则为13.3-13.5立方米。

4. 有利于净化饲料、保护环境

很多害虫寄生在收割后的秸秆上越冬。这时将秸秆铡碎青贮,由于青贮窖缺乏氧气,并且酸度较高,能够杀死青贮饲料中的病菌、虫卵,且能够破坏杂草种子的再生能力,从而减少对畜禽和农作物的危害。例如:玉米螟的幼虫大多潜伏在玉米秸秆中越冬,经过青贮的玉米秸,玉米螟会全部失去活力。另外,秸秆青贮使长期以来焚烧秸秆的现象大为改观,使这一资源变废为宝,减少了对环境的污染。

matter. For example, 1 t silage alfalfa occupies 1.25 m^3, while 1 t alfalfa hay occupies 13.3 – 13.5 m^3.

4. Conducive to purifying feed and protecting the environment

Many pests parasitize the harvested straw for overwintering. After the straw is chopped and ensilaged, the lack of oxygen in the silage and the high acidity can kill the germs and insect eggs in the green fodder and destroy the regeneration ability of weed seeds, thereby reducing the harm to livestock, poultry and crops. For example, most of the larvae of the corn borer live in the corn stalks for overwintering. After the silage of the corn stock, the corn borer will lose all vitality. In addition, straw silage has greatly changed the phenomenon of straw burning for a long time, turning this resource into treasure and reducing environmental pollution.

二、青贮饲料的发展
The development of silage

1. 我国青贮饲料的生产现状

我国青贮秸秆技术的推广和研究已有几十年历史,但其发展速度并不理想,且时起时落。我国玉米种植面积仅次于美国,是世界上第二大玉米生产国。丰富的玉米及玉米秸秆成为我国畜牧养殖业的重要饲料来源,尤其是我国奶业进入快速发展期,急需大量优质青贮饲料,以提高饲养水平,提高奶的产量和品质。

而目前青贮设备多采用地下或半地下红砖水泥窖,还有部分青贮塔,农村以地下土窖为多见。这无法保证青贮饲料的厌氧发酵水平,而且经常的分次取用更加剧了好氧微生物的侵入,常常造成青贮饲料的有氧腐败。生产青贮饲料的工艺多采用自然发酵方式,但秸秆本身

1. The production status of silage in China

The promotion and research of silage straw technology in China has a history of several decades, but its development speed is not ideal, and it fluctuates from time to time. China's corn planting area is second only to America, and it is the second largest corn producer in the world. Abundant corn and corn stalks have become an important source of feed for my country's animal husbandry industry. In particular, China's dairy industry has entered a period of rapid development, and so there is an urgent need for a large amount of high-quality silage to improve feeding levels and increase milk production and quality.

At present, silage equipment mostly uses underground or semi-underground red brick cement cellars, and some silage towers. Underground crypts are more common in rural areas. This cannot guarantee the level of anaerobic fermentation of the silage, and the frequent use of fractions increases the intrusion of aerobic microorganisms and often causes the aerobic corruption of the silage. The process of producing silage mostly uses natural fermentation, but the straw

乳酸菌非常少，自身的营养成分也有限，无法保证乳酸菌在发酵过程中成为优势菌群，而青贮饲料中最有益的成分应该是L-乳酸，所以生产出的青贮饲料质量不佳。但需要肯定的是，在教学研一体的模式下，校企合作频次和质量显著增加，现在较多中大规模的养殖企业也能够生产优质青贮饲料。

2. 国外青贮饲料的研究进展

目前，世界大多数国家都把青贮饲料作为反刍动物日粮中的主要粗饲料，因世界近三分之一国家缺粮，且谷物和优质蛋白质饲料的价格不断上涨。秸秆畜牧业已经成为发展趋势，越来越多国家的肉牛、奶牛等草食家畜生产已经形成集约化、系列化经营。在欧美畜牧业发达国家，特别是反刍动物生产发达的国家，都广泛使用青贮饲料，大量种植青贮用玉米。如：法国、加拿大、英国、荷兰等国家已培育出优质高产的青贮饲料专用玉米进行全株青贮饲喂，并进行大面积推广种植。英国、荷兰等国种植的玉米几乎全部用于制作青贮饲料。

itself has very few lactic acid bacteria and its own nutritional content is also limited. It cannot be guaranteed that lactic acid bacteria will become the dominant flora during the fermentation process. The most beneficial component in silage should be L-lactic acid. Therefore, the silage produced is of poor quality. However, it is necessary to be confirmed that under the model of integrated teaching and research, the frequency and quality of school-enterprise cooperation have significantly increased, and now many medium and large-scale breeding enterprises are also able to produce high-quality silage.

2. Research progress of silage abroad

At present, most countries in the world use silage as the main roughage in ruminant diets. As nearly one third of the countries in the world lack food and grains and high-class protein feed prices continue to rise. Straw animal husbandry has become a developing trend. In more and more countries, the production of beef cattle, dairy cattle and other herbivorous livestock has formed intensive and series operation. In animal developed countries of husbandry such as European countries and the United States, especially the developed countries in ruminant production, large quantities of silage corn are planted and silage is extensivly used. For example, in France, Canada, the United Kingdom, the Netherlands and other countries, high-quality and high-yield special maize for silage feeding have been developed, and a large area of planting is promoted. Almost all corn grown in the United Kingdom, the Netherlands and other countries is used to make silage.

第二节　青贮发酵原理及基本过程

Section 2　Principle and basic process of silage fermentation

一、青贮发酵原理
The principle of silage fermentation

青贮发酵是一个复杂的微生物活动和生物化学变化过程。当青贮原料铡碎入窖并压实密封后，植物

Silage fermentation is a complex process of microbial activities and biochemical changes. When silage raw materials are cut into the cellars and compact-sealed, plant cells

细胞仍具有活性，持续呼吸，进行有机物的氧化分解，产生二氧化碳、水和热量，这一过程使得在密闭的环境内氧气逐渐减少，故一些好氧性微生物逐渐死亡，而乳酸菌在厌氧环境下迅速繁殖扩大。青贮过程为青贮原料中乳酸菌的生长繁殖创造了有利条件，使乳酸菌大量繁殖，并将青贮原料中可溶性糖类变成乳酸，当达到一定浓度和无氧状态时，抑制了有害细菌的生长和繁殖，保证青贮饲料能够长期保存。对青贮饲料来说，成败的关键在于能否创造一定条件保证乳酸菌的迅速繁殖，形成有利于乳酸菌发酵的大环境。

are still active and continue to breathe and carry out the oxidative decomposition of organic matter, producing carbon dioxide, water and heat. This process results in a gradual loss of oxygen in a confined surroundings, so that some aerobic microorganisms gradually die, and lactic acid bacteria in the anaerobic environment rapidly expand. The process of silage creates favorable conditions for the growth and reproduction of lactic acid bacteria in silage raw materials, which makes the lactic acid bacteria multiply in large numbers and turn the soluble sugars in silage raw materials into lactic acid. When the concentration and oxygen state reach a certain concentration and anaerobic state, the growth and reproduction of aerobic bacteria are inhibited to ensure the long-term preservation of silage. For silage, the key to success or failure is to create certain conditions to ensure the rapid reproduction of lactic acid bacteria and to form an environment conducive to the fermentation of lactic acid bacteria.

二、青贮发酵的基本过程
The basic process of silage fermentation

青贮发酵过程大致可以分为好氧菌活动期、乳酸发酵期和发酵稳定期三个阶段。

1. 好氧菌活动期

该阶段是其他杂菌与乳酸菌进行竞争优势菌群的时期，这个阶段大概持续一周左右，与原料的性质和储藏条件有关。新鲜青贮原料在青贮容器中压实密封后，植物细胞并未立即死亡，在1-3天内仍进行呼吸作用，植物的蛋白酶和需氧微生物都具有活性，分解有机物质，直至青贮饲料内氧气消耗尽、呈厌氧状态时才停止呼吸。这个阶段，青贮窖内的pH值在6.0-6.5之间。在青贮开始时，附着在原料上的酵母菌、腐败菌、霉菌和醋酸菌等好氧微生物，利用植物细胞因受机械压榨而排出的富含可溶性碳水化合物的液汁，迅速繁殖。腐败菌、霉菌等繁殖最为强烈，破坏青贮料中的蛋

The process of silage fermentation can be divided into three stages: aerobic bacteria activity stage, lactic acid fermentation stage and fermentation stable stage.

1. Aerobic bacteria activity stage

This stage is the period when other miscellaneous bacteria compete with lactic acid bacteria for dominance. This stage lasts about a week and is related to the nature and storage conditions of raw materials. After the fresh silage raw material is compacted and sealed in the silage container, the plant cells do not die immediately, but continue to breathe within 1-3 days to decompose the organic material, and then stop breathing until the oxygen in the silage is exhausted and anaerobic. At this stage, the pH value in the silage cellar is between 6.0 and 6.5, plant respiration continues, and both protease and aerobic microorganism of the plant are active, in which respiration and enzyme reaction play a major role. At the beginning of silage, the air-friendly microorganisms attach to the raw materials, such as yeast, rot fungi, hail bacteria, and acetic acid bacteria, taking advantage of the liquid rich in soluble carbohydrates that the plant cells excrete as a result of mechanical pressing, and

白质,形成大量吲哚、气体和少量醋酸等。好氧微生物的活动以及植物细胞的呼吸,使得青贮原料间存在的少量氧气很快消耗殆尽,形成厌氧环境。另外,植物细胞的呼吸作用及微生物的活动还放出热量。厌氧和温暖的环境为乳酸菌发酵创造了条件。

如果青贮原料中氧气过多,植物呼吸时间过长,好氧微生物活动旺盛,则会使原料内温度升高,有时高达60℃左右,从而削弱乳酸菌与其他微生物的竞争能力,导致青贮饲料营养成分损失过多,青贮饲料品质下降。一旦达到厌氧环境后,第一阶段就结束了。如果青贮原料为糖分高的理想作物,而且管理措施比较完善,那么这个过程仅需要几个小时;如果青贮原料糖分低或管理不好,这个过程会持续几周。因此,青贮技术的关键是尽可能缩短第一阶段的时间,通过及时青贮和切短压紧密封从而减少植物呼吸作用和好氧微生物的繁殖,以降低养分损失,提高青贮饲料质量。

2. 乳酸发酵期

厌氧条件及青贮原料中的其他条件形成后,乳酸菌迅速繁殖,经糖酵解途径产生大量乳酸,同时乳酸可部分转化为醋酸、丙酸及丁酸,酸度增大,pH值下降,使腐败菌、醋酸菌等活动停止,甚至绝迹。当pH值下降到4.2以下时,各种好氧微生物都不能生存,就连乳酸链球菌的活动也受到抑制,只有乳酸杆菌存在。当pH值降到3时,乳酸杆菌也停止活动,乳酸发酵即基本结束。

一般情况下,糖分适宜的原料发酵5-7天后微生物总数达到高

reproducing rapidly. Putrid bacteria, molds, etc. multiply the most intensely, destroying the protein in silage, forming a large amount of indole, gas and a small amount of acetic acid, etc. The activities of aerophilic microorganisms and respiration of plant cells make the small amount of oxygen in the silage raw materials quickly exhausted, forming an anaerobic environment. In addition, plant cell respiration, enzyme oxidation and microbial activity also give off heat. Anaerobic and warm conditions create conditions for lactic acid bacteria to ferment.

If the oxygen in silage raw material is too much, the plant's respiration time is too long, and the activity of good aerative microorganism is vigorous, the temperature in the raw material will rise, sometimes as high as 60 ℃, thus weakening the competitiveness of lactic acid bacteria and other microorganisms, leading to excessive loss of nutrients in silage and deterioration of the quality of silage. Once the anaerobic environment is reached, the first stage is over. If silage is an ideal crop with high sugar content and is well managed, the process takes only a few hours; if silage is low in sugar or poorly managed, this process can take several weeks. Therefore, the key of silage technology is to shorten the time of the first stage as much as possible, and to reduce the respiration and the reproduction of aerobic harmful microorganisms by timely silage and cutting and sealing, so as to reduce nutrient loss and improve the quality of silage.

2. Lactic acid fermentation stage

After anaerobic conditions and other conditions in silage raw materials are formed, lactic acid bacteria multiply rapidly to form a large amount of lactic acid, the acidity increases, and the pH value drops, which promotes the suppression of spoilage bacteria, butyric acid bacteria and other activities, or even extinction. When the pH value drops below 4.2, all kinds of aerobic microorganisms cannot survive, and even the activities of Streptococcus lactis are inhibited, and only Lactobacillus exists. When the pH value is 3, Lactobacillus also stops its activity, and the lactic acid fermentation basically stops.

In general, the total number of microorganisms reaches a peak after 5-7 days fermentation of sugar suitable raw ma-

峰,其中以乳酸菌为主。乳酸菌是第二阶段中的主要发酵菌,它们利用可溶性碳水化合物产生大量的乳酸。乳酸是青贮中最好的发酵酸,其含量应该占青贮中总有机酸含量的 60% 以上。饲喂青贮饲料时,乳酸会成为反刍动物的一种重要能量来源。玉米青贮过程中,玉米青贮半天后,乳酸菌数量即达到最高峰,每克饲料中达 16.0 亿个。第 4 天时下降到 8.0 亿个,pH 值达 4.5,而其他微生物则已全部停止繁殖而绝迹。因此,玉米青贮过程比豆科牧草快,青贮品质也好,是最优良的青贮作物。

青贮第二阶段是青贮发酵过程中最长的阶段,它会一直持续到所有的微生物停止活动,一般为 20 天左右。糖分含量较高的玉米、高粱等青贮后 20-30 天就可以进入稳定阶段,豆科牧草需 3 个月以上。当达到这个条件后,青贮就处于稳定状态。只要氧气不进入到青贮窖中,就不会有进一步的破坏过程发生。

3. 发酵稳定期

在青贮设备良好的情况下,由于青贮饲料中有大量的乳酸菌和乳酸,形成了一个稳定的平衡状态,乳酸可以保持饲料不受其他杂菌影响而变质,同时又限制了乳酸菌的过量繁殖,乳酸菌在乳酸不足时又继续生长分泌乳酸,微生态如此循环,从而能够使饲料保存很长时间。在此期间,如果青贮管理不当,也会引发二次发酵。例如:青贮窖开封提取青贮饲料时,空气随之进入,青贮料的表面因与空气接触,好氧微生物在这样的条件下大量繁殖,青贮饲料中的养分遭受大量损失,出现好氧性腐败,并产生大量的热。二次发酵的微生物包括酵母菌和霉

terials, and lactic acid bacteria are the main ones. Lactic acid bacteria are the main fermentation bacteria in the second stage. They use soluble carbohydrates to produce large amounts of lactic acid. Lactic acid is the best fermented acid in silage, and its content should account for more than 60% of the total organic acid content in silage. Lactic acid can be an important source of energy for ruminants when fed with silage. During the silage of maize, after half a day of silage, the number of lactic acid bacteria reaches its peak, reaching 1.60 billion per gram of feed. On the fourth day, it drops to 800 million, with a pH value of 4.5, while all other microorganisms stop reproducing and become extinct. Therefore, corn silage is the best silage crop with faster fermentation process and better silage quality than legumes.

The second stage of silage is the longest stage in the silage fermentation process, which lasts until all the microorganisms stop their activities, generally around 20 days. Maize and sorghum with high sugar content can enter into the stable stage after 20-30 days of silage, and legumes need more than 3 months. When this condition is reached, silage is in a stable state. As long as the oxygen does not enter the silo, no further destruction will occur.

3. Fermentation stable stage

In case of good silage equipment, due to the large amounts of lactic acid bacteria and lactic acid in the silage, a stable equilibrium state is formed. Lactic acid can keep the feed from deterioration by other miscellaneous bacterium, and limit the excessive breeding of lactic acid bacteria. When the lactic acid is insufficient, lactic acid bacteria continue to grow, and the micro-ecology circulates in this way, so that the feed can be stored for a long time. During this period, secondary fermentation can also occur if the silage is not properly managed. For example, when the silo is opened to extract the silage, the air enters then the surface of the silage is in contact with the air, and the aerobic microorganisms multiply in large numbers under such conditions, resulting in a large loss of nutrients in the silage and aerobic corruption, and much heat is generated. The microorganisms of secondary fermentation include yeast and

菌，一般情况下，首先是 1-2 天酵母繁殖，产生热量，然后是霉菌繁殖，使饲料腐败。随着发热，饲料的乳酸含量迅速下降，pH 值迅速上升，pH 值超过 5 时，由饲料中蛋白质和氨基酸分解产生的挥发性氨态氮急剧增加。二次发酵后的青贮饲料重量损失很大，由 700 kg/m³ 降为 400 kg/m³，且二次发酵使青贮饲料腐烂变质，不能再饲喂家畜。饲料中引起腐烂的霉菌还产生毒素，引起家畜中毒、下痢、流产等，造成畜牧生产的损失。二次发酵的主要原因有：一是发酵过程中可溶性碳水化合物缺乏；二是乳酸产生速率过慢，未能抑制像梭状芽孢杆菌这样的孢子的生长。梭状芽孢杆菌的生长会引起青贮饲料的 pH 值升高，这样的青贮饲料在厌氧条件下也是不稳定的，而且干物质损失较大，饲喂价值降低；三是饲喂青贮饲料时多次少量取出。为了防止青贮饲料在发酵稳定期的二次发酵，应进行针对性的预防，如提高制作青贮饲料的质量，掌握取料技术，喷洒微生物菌剂或有机酸等。

mold. Generally, after 1-2 days, yeast propagates and produces heat, and then mold propagates, which putrefies the feed. With the heat, the lactic acid content of the feed decreases rapidly, and the pH value increases rapidly. When the pH value is more than 5, the volatile ammonia nitrogen produced by the decomposition of proteins and amino acids in the feed increases sharply. As a result, the weight loss of silage after secondary fermentation is very large, which decreases from 700 kg/m³ to 400 kg/m³, and the secondary fermentation makes the silage rot and deteriorate and they can not be fed to livestock, and the mold that causes corruption in the feed also produces toxins, causing poisoning, dysentery, abortion and so on in livestock, resulting in the loss of animal husbandry production. Secondary fermentation is mainly due to the following reasons: one is the lack of soluble carbohydrates during fermentation; the second is that lactic acid production rate is too slow to inhibit the growth of spores such as Clostridium difficile. The growth of such bacteria causes a rise in the pH value of silage, which is also unstable under anaerobic conditions and has a greater dry matter loss and reduces feeding value; the third is that people take out a small amount of silage many times when feeding silage. In order to prevent the secondary fermentation of silage in the fermentation stable period, targeted prevention should be carried out, such as improving the quality of silage, improving the technology of taking silage, and spraying microbial agents or organic acids.

第三节　青贮饲料的原料及调制

Section 3　Raw materials and preparation of silage

一、青贮原料和添加剂
Silage raw materials and additives

1. 青贮原料

凡无毒的新鲜植物均可作为青贮原料，如青刈带穗玉米、农作物秸秆、优质禾本科牧草、饲料作物等。青贮原料含水量根据制作工艺不同

1. Silage raw materials

Any non-toxic fresh plants can be used as silage raw materials, such as green clipping ear corn, crop straw, high quality grass, feed crops, etc. The water content of silage raw materials varies according to the different production

而要求不同,一般应为60%-70%。青贮原料要有一定的含糖量,一般不低于1%-5%,禾本科含糖量一般较高,青贮容易成功,而豆科牧草蛋白质含量较高,单独青贮难以成功。另外,青贮前须保证青贮原料干净,无泥土和其他杂质。

2. 青贮微生物

从20世纪早期开始,人们就开始探索青贮饲料发酵过程中的微生物。目前知晓,在青贮饲料发酵过程中,起重要作用的是乳酸菌,同时其他细菌或者真菌也扮演着正面或负面的角色。刚刈割的青饲料中带有各种细菌、霉菌、酵母等微生物,其中腐败菌最多,乳酸菌很少。新鲜青草饲料上腐败菌的数量远远超过乳酸菌的数量。青饲料如不及时青贮,在田间堆放2-3天后,腐败菌大量繁殖,每克青饲料中往往达数亿以上。因此,为了促进青贮过程中乳酸菌的繁殖活动,必须了解各种微生物的活动规律和对环境的要求,以便采取措施,抑制各种不利于青贮的微生物活动,尽可能消除一切妨碍乳酸菌生长的条件。

(1) 乳酸菌 是能够将碳水化合物发酵产生大量乳酸的一类细菌。乳酸菌的种类很多,其中对青贮有益的主要是乳酸链球菌、德氏乳酸杆菌。它们均为同型发酵的乳酸杆菌,发酵后只产生乳酸。此外,还有异型发酵的乳酸杆菌,除产生乳酸外,还产生大量的乙醇、醋酸、甘油和二氧化碳等。乳酸链球菌属兼性厌氧菌,耐酸能力较低,青贮饲料中酸量达0.5%-0.8%,pH值达4.2时即停止活动。乳酸杆菌为厌氧菌,只在厌氧条件下生长繁殖,耐酸力强,青贮中酸量达1.5%-2.4%,pH值为3时才停止活动。

process, and generally should be within 60%-70%. Silage raw materials should contain a certain amount of sugar, generally no less than 1%-5%. Generally, grass with high carbohydrate content is easy to succeed, while legume forage with high protein is difficult to succeed alone. Before silage, make sure that the silage material is clean and free of soil and other impurities.

2. Silage microorganisms

From the early 20th century, people began to explore the microorganisms in the fermentation process of silage. It is known that lactic acid bacteria play an important role in silage fermentation, while other bacteria or fungi also play a positive or negative role. There are various bacteria, mold, yeast and other microorganisms in the green feed of freshly cut, among which spoilage bacteria are the most, lactobacillus is the least. The amount of spoilage bacteria on the fresh grass feed far exceeds the amount of lactic acid bacteria. If the green fodder is not ensilaged in time, after 2-3 days in the field, the spoilage bacteria will multiply in large quantities, often to hundreds of millions of per gram of green feed. Thus, in order to promote the normal reproduction of lactic acid bacteria in the silage process, it is necessary to understand the activity rules and environmental requirements of all kinds of microorganisms, so as to take measures to inhibit all kinds of microbial activities that are not conducive to silage and eliminate all conditions that hinder the formation of lactic acid bacteria.

(1) Lactic acid bacteria. Lactic acid bacteria are a kind of bacteria that ferment carbohydrates to produce large amounts of lactic acid. There are many kinds of lactic acid bacteria, among which the main ones beneficial to silage are Streptococcus lactis and Lactobacillus destilis. They are all homozygous fermented Lactobacillus, which produces only lactic acid after fermentation. Moreover, the heterogeneous fermentation of Lactobacillus, in addition to producing lactic acid, also produces a large number of ethanol, acetic acid, glycerin and carbon dioxide, etc. Streptococcus lactis is facultative anaerobe with low acid resistance. The acid content in silage reaches 0.5%-0.8%, and the activity stops when the pH value reaches 4.2. Lactobacillus is an anaerobic bacterium, which can only grow and reproduce under anaerobic conditions, and has strong acid resistance. It stops its activi-

各类乳酸菌在含有适量的水分和碳水化合物、缺氧条件下,生长繁殖快,可使单糖和双糖分解成大量乳酸。

在青贮过程中,同型发酵乳酸菌是最有益的,因为它可以快速地产生乳酸,使青贮的 pH 值下降。而且,与异型发酵乳酸菌相比,同型发酵乳酸菌发酵的青贮饲料的干物质损失比较低,主要是可溶性碳水化合物转化成乳酸的效率比较高。然而,乳酸菌的发酵类型取决于底物的组成。己糖、葡萄糖、果糖和多糖是乳酸菌利用的主要底物,它们在青贮作物中的比例和可利用程度经常会影响到乳酸菌的发酵类型。在作物的可溶性碳水化合物比较低的情况下,如果葡萄糖、果糖的比例比较低,就会促进乳酸菌的异型发酵。当果糖含量不足时,某些乳酸菌会利用乳酸作为底物生成乙酸。

据乳酸菌对温度的要求不同,可分为嗜冷性乳酸菌和嗜热性乳酸菌两类。嗜冷性乳酸菌在 25－35 ℃条件下繁殖最快,正常青贮时,主要是嗜冷性乳酸菌活动。嗜热性乳酸菌发酵,结果可使温度达到 52－54 ℃,如超过这个温度,则意味着还有其他好氧性腐败菌等微生物参与发酵。高温青贮养分损失大,青贮饲料品质差,应当避免。乳酸的大量形成,一方面为乳酸菌本身生长繁殖创造了条件,另一方面产生的乳酸使其他微生物如腐败菌、酪酸菌等死亡。乳酸积累导致酸度增强,乳酸菌自身也受抑制而停止活动。在良好的青贮饲料中,乳酸含量一般占青贮饲料重量的 1%－2%,pH 值下降到 4.2 以下时,只有少量的乳酸菌存活。

ties only when the acid content in silage reaches 1.5%-2.4% and pH value is 3. Lactic acid bacteria can grow and multiply rapidly under the condition of adequate water, carbohydrate and hypoxia, and can decompose monosaccharides and disosaccharides into large amounts of lactic acid.

In the process of silage, the homozygous lactic acid bacteria are most beneficial because they rapidly produce lactic acid, lowering the pH value of silage. Moreover, compared with heterozygous lactic acid bacteria, homozygous lactic acid bacteria fermentation silage has lower dry matter loss, mainly because of higher efficiency of soluble carbohydrate conversion into lactic acid. However, the type of fermentation of lactic acid bacteria depends on the composition of the substrate. Hexose, glucose, fructose and polysaccharides are the main substrate used by lactic acid bacteria. Their proportion and availability in silage crops often affect the fermentation type of lactic acid bacteria. When the soluble carbohydrates of crops are relatively low, if the proportion of glucose and fructose is low, it will promote the fermentation of heterolactic acid bacteria. When fructose content is insufficient, some lactic acid bacteria use lactic acid as a substrate to produce acetic acid.

According to the different requirements of lactic acid bacteria to temperature, it can be divided into two types: psychrotrophic lactic acid bacteria and thermophilic lactic acid bacteria. Psychrotrophic lactic acid bacteria proliferate faster at the temperature of 25-35 ℃. In normal silage, psychrotrophic lactic acid bacteria are mainly active; in thermophilic lactic acid bacteria fermentation, the temperature can reach 52-54 ℃. If the temperature exceeds this, it means that there are other good air spoilage bacteria and other microorganisms involved in the fermentation. High temperature silage has great nutrient loss and poor quality of silage, so it should be avoided. On the one hand, the formation of lactic acid creates conditions for the growth and reproduction of Lactobacillus. On the other hand, the formation of lactic acid causes the death of other microorganisms, such as spoilage bacteria and tyrosine bacteria. The accumulation of lactic acid leads to the increase of acidity and the inhibition of lactic acid bacteria. In good silage, the lactic acid content generally accounts for 1-2% of the weight of the green feed, and when the pH value drops below 4.2,

（2）酪酸菌（丁酸菌） 是一种厌氧、不耐酸的有害细菌，主要有丁酸梭菌、蚀果胶梭菌、巴氏固氮梭菌等。其在 pH 值 4.7 以下时不能繁殖，青贮原料上本来数量不多，只在温度较高时才能繁殖。酪酸菌活动的结果使葡萄糖和乳酸分解产生具有挥发性臭味的丁酸，也能将蛋白质分解为挥发性脂肪酸，使原料发臭变黏，降低青贮饲料的品质。丁酸发酵的程度是鉴定青贮饲料好坏的重要指标，丁酸含量越多，青贮饲料的品质越差。另外，丁酸菌还能利用各种有机氮化合物，从而破坏青贮饲料中的蛋白质，使营养流失。当青贮饲料中丁酸含量达到万分之几时，即影响青贮饲料的品质。青贮原料幼嫩，碳水化合物含量不足，含水量过高，装压过紧等均易促进酪酸菌的活动和大量繁殖。

（3）腐败菌 凡能强烈分解蛋白质的细菌统称为腐败菌。此类细菌很多，有嗜高温的，也有嗜中温或低温的；有好氧的如枯草杆菌、马铃薯杆菌，也有厌氧的如腐败梭菌和兼性厌氧菌如普通变形杆菌。它们能使蛋白质、脂肪、碳水化合物等分解产生氨、硫化氢、二氧化碳、甲烷和氢气等，使青贮原料变臭变苦，养分损失大，导致青贮失败，不能饲喂家畜。青贮原料中腐败菌在数量上占主要地位，但它们不耐酸，正常青贮条件下，当乳酸逐渐形成、pH 下降、氧气耗尽后，腐败菌活动即迅速抑制，以致死亡。只有在青贮料装压不紧、残存空气较多或密封不好时才大量繁殖。

（4）酵母菌 酵母菌是好氧性真菌，喜潮湿，不耐酸。在青饲料切

only a small amount of lactic acid bacteria exist.

（2）Butyric acid bacteria. Butyric acid bacteria are anaerobic and acid intolerant harmful bacteria, mainly including Clostridium butyricum, Clostridium pectin, Clostridium pasteuri, etc. It cannot reproduce when the pH value is below 4.7, and its quantity in raw materials is less. It can only reproduce when the temperature is high. As a result of the activity of tyrosine bacteria, glucose and lactic acid can be decomposed into butyric acid with volatile odor, and protein can be decomposed into volatile fatty acid, which makes raw materials smelly and sticky and reduces the quality of silage. The degree of butyric acid fermentation is an important indicator to judge the quality of silage. The more butyric acid in the content it has, the worse the quality of silage is. In addition, butyrate bacteria can also use a variety of organic nitrogen compounds, so as to destroy the protein in silage, resulting in nutrient loss.

（3）Spoilage bacteria. Any bacteria that can strongly decompose proteins are collectively called Spoilage bacteria. There are many such bacteria, some of which are thermophilic, some of which are mesophilic or hypothermic. There are aerobic bacteria such as Bacillus subtilis and Bacillus potato, and anaerobic bacteria such as Clostridium putrifolium and facultative anaerobe such as Proteus. They can decompose proteins, fats and carbohydrates to produce ammonia, hydrogen sulfide, carbon dioxide, methane and hydrogen, etc., making silage raw materials smelly and bitter, leading to great loss of nutrients and failure of feeding livestock. Spoilage bacteria are predominant in quantity in silage materials, but they are not acid resistant. Under normal silage conditions, when lactic acid is gradually formed, its pH value drops, and oxygen is exhausted, spoilage bacterial activity is rapidly suppressed, and will die. Only when the silage is not tightly packed, the residual air is too much, or the seal is not good, can spoilage bacteria reproduce in large quantities.

（4）Yeast. Yeast is an aerobic fungus, which likes humidity and is not acid tolerant. Yeasts reproduce only on the

碎尚未装贮完毕之前，酵母菌只在青贮原料表层繁殖，分解可溶性糖，产生乙醇及其他芳香类物质。待封窖后，空气越来越少，其作用随即减弱。在正常青贮条件下，青贮饲料装压较紧，原料间残存氧气少，酵母菌只能在最初几天内繁殖，可进行乙醇发酵，随着氧气的耗尽和乳酸的积累而很快受到抑制，所产生的少量乙醇等芳香物质使青贮具有特殊气味。但是在糖分不足的青贮原料中，由酵母菌引起的乙醇发酵可造成糖分减少，影响乳酸生成。

（5）醋酸菌　醋酸菌属于好氧性细菌。在青贮初期有空气存在的条件下，可大量繁殖。酵母或乳酸发酵产生的乙醇，可被醋酸菌发酵产生醋酸。醋酸的产生可抑制各种不耐酸的微生物如腐败菌、霉菌、酪酸菌的活动与繁殖。但是在不正常情况下，青贮窖内氧气残存过多、产生大量醋酸时，因醋酸有刺鼻气味，会影响家畜的适口性并使饲料品质降低。

（6）霉菌　它是导致青贮饲料变质的主要好氧性微生物，通常仅存在于青贮饲料的表层或边缘等易接触空气的部分。正常青贮情况下，霉菌仅生存于青贮初期，青贮过程中的酸性环境和厌氧条件足以抑制霉菌的生长。霉菌可破坏有机物质，分解蛋白质产生氨，使青贮饲料发霉变质并产生酸败味，降低其品质，甚至失去饲用价值。

（7）肠细菌　在青贮发酵初期与乳酸菌竞争发酵底物，生成乙酸，并且部分具有蛋白水解活性。肠细菌将硝酸还原为亚硝酸，再还原为氮和一氧化二氮，一氧化二氮还原为一氧化氮和硝酸。在有氧条件

surface of the silage material to decompose soluble sugars to produce ethanol and other aromatic substances before the shredding and storage is completed. After the cellar is sealed, there is less and less air, and its effect then weakens. Under normal silage conditions, the silage is tightly packed, and there is little residual oxygen between the raw materials. Yeast can only reproduce in the first few days, and alcohol fermentation can be carried out. With the depletion of oxygen and the accumulation of lactic acid, it is quickly inhibited, and the small amount of ethanol and other aromatic substances thus produced make silage have special smell. However, in silage materials with insufficient sugar, alcohol fermentation caused by yeast can lead to sugar reduction and affect the formation of lactic acid.

(5) Acetic acid bacteria. Acetic acid bacteria belong to aerobic bacteria. In the early stage of silage, it can reproduce in large numbers under the condition of much air. Ethanol produced by yeast or lactic acid fermentation can be fermented by acetic acid bacteria to produce acetic acid. The production of acetic acid can inhibit the activity and reproduction of various harmful and acid-free microbe such as spoilage bacteria, molds and butyric acid bacteria. However, under abnormal conditions, excessive residual oxygen in the silo and large amount of acetic acid will affect the palatability of livestock and reduce the quality of feed due to the irritating odor of acetic acid.

(6) Mould. It is the main aerobic microorganism that causes the deterioration of silage. It usually only exists in the surface layer or the edge of silage and other parts that are easy to contact with air. Under normal silage conditions, mold only exists in the early stage of silage, and the acidic environment and anaerobic conditions during silage are enough to inhibit the growth of mold. Mold can destroy organic material, decompose protein to produce ammonia, make silage moldy and produce rancid taste, reduce its quality, and even lose feeding value.

(7) Enteric bacteria. It competes with lactic acid bacteria to produce acetic acid at the initial stage of silage fermentation, and some of them have proteolytic activity. Enteric bacteria reduce nitric acid to nitrite, and then to nitrogen and nitrous oxide, which are reduced to nitric oxide and nitric acid. Under aerobic conditions, nitric oxide is oxi-

下，一氧化氮氧化成二氧化氮、三氧化二氮、四氧化二氮的混合气体，呈黄褐色。气态一氧化氮和二氧化氮与水接触生成硝酸和亚硝酸，对肺细胞有损伤。同时生成的亚硝酸和一氧化氮能够抑制梭菌生长。

（8）芽孢杆菌 可利用有机酸、乙醇、2,3-丁二醇、甘油作为碳源。而且部分芽孢杆菌的孢子可通过牛消化道、粪便污染牛奶，引起牛奶和乳制品腐败变质，甚至食物中毒。

（9）梭菌 在无氧条件下进行丁酸发酵，分解糖和乳酸形成丁酸，引起糖分的损失，并且造成青贮饲料 pH 的升高。同时，梭菌还能分解蛋白质形成氨基酸、胺、硫化氢等。

3．青贮添加剂

对于一些较难调制为优质青贮饲料的原料，或对产品的不同需求，在青贮调制时往往人为添加某些添加剂来调控青贮发酵过程，其主要目的是促进乳酸发酵、抑制不良发酵、提高青贮饲料的营养价值等，例如：乳酸菌制剂、酶制剂、糖类等发酵促进剂，甲酸等发酵抑制剂，尿素等营养性添加剂，丙酸等防腐剂。

（1）发酵促进剂 主要包括微生物添加剂、碳水化合物、纤维素酶制剂。在青贮原料中可以直接添加乳酸菌菌种，这是为了增加乳酸菌起始状态的比例，促进乳酸菌尽快繁殖，保证短时间内发酵产生大量乳酸，降低 pH 值，从而达到抑制有害微生物，减少干物质的损失，提高青贮品质的目的。至于其中的纤维素酶制剂，对于秸秆饲料来说，纤维素含量很高，添加纤维素酶可以把纤维物质分解为单糖和双糖，为乳酸菌发酵提供充足的碳源，并且还

dized to a mixture of nitrogen dioxide, nitrogen trioxide and nitrogen tetroxide, showing a yellow-brown color. Gaseous nitric oxide and nitrogen dioxide contact with water to produce nitric acid and nitrite, which can damage lung cells. The nitrite and nitric oxide thus produced simultaneously can inhibit the growth of clostridium.

(8) Bacillus. Bacillus can use organic acid, ethanol, 2- and 3-butanediol and glycerol as carbon sources. And the spores of partial Bacillus can pollute milk through milk alimentary canal and feces, causing milk and dairy products to spoil and deteriorate, and even food poisoning.

(9) Clostridium. Butyric acid fermentation is carried out under anaerobic conditions, which decomposes sugar and lactic acid to form butyric acid, causing the loss of sugar and increasing the pH value of silage. Meanwhile, Clostridium can also break down proteins to form amino acids, amines, hydrogen sulfide, etc.

3. Silage additives

For some stubborn materials which are difficult to be made into high-class silage, or for different requirements of products, some additives are often added to regulate the fermentation process of silage during silage. Its main purpose is to promote lactic acid fermentation, restrain harmful fermentation and improve the nutritive value of silage. The main additives are: lactobacillus preparations, enzyme preparations, carbohydrate and other fermentation promoters, fermentation inhibitors such as formic acid, nutritional additives such as uric acid, and preservatives such as propionic acid.

(1) Fermentation promoters. They mainly include microbial additives, carbohydrates and cellulase preparations. Lactic acid bacteria can be added directly in the silage raw materials to increase the proportion of the initial state of lactic acid bacteria, promote the reproduction of lactic acid bacteria as soon as possible, ensure a short period of fermentation to produce a large amount of lactic acid, reduce its pH value, thus achieving the purpose of inhibiting aerobic microorganisms, reducing the loss of dry matter, and improving the quality of silage. As for the cellulase preparation, the content of cellulose is very high for straw feed. Adding cellulase can decompose the fiber into monosaccharides and disaccharides, providing sufficient carbon source

可以提高饲料的消化率。

（2）发酵抑制剂　是最早使用的一类添加剂，这类添加剂主要是无机酸，后来使用的是有机酸。加入这类添加剂能有效降低青贮料的pH值，抑制杂菌繁殖，直接形成适合乳酸菌生长的环境。

（3）营养性添加剂　这类添加剂主要有尿素、糖蜜、矿物质等，用来补充青贮饲料营养成分不足，改善发酵品质。

（4）防腐剂　常用的防腐剂有丙酸、山梨酸、乙酸、氨等。防腐剂不能改善发酵过程，但可以防止饲料变质。

(2) Fermentation inhibitors. They are inorganic acids firstly used, and later on organic acids are used. Adding such additives can effectively reduce the pH value of silage, inhibit the proliferation of heterozygous bacteria, and directly form the environment suitable for the growth of lactic acid bacteria.

(3) Nutritional additives. They mainly include urea, carbohydrate, minerals, etc., which are used to supplement the nutrient deficiency of silage and improve the fermentation quality.

(4) Preservative additives. Commonly used preservative additives are propionic acid, sorbic acid, acetic acid, ammonia, etc. Preservative additives do not improve the fermentation process, but they can prevent feed deterioration.

二、青贮类型
Silage type

青贮类型多种多样，根据青贮原料的类型、气候以及具体环境差异，青贮饲料被分成很多类别。按照原料含水量的高低可以分为低水分青贮（半干青贮）和高水分青贮（普通青贮）。按照发酵难易和复杂程度分为一般青贮、混合青贮和添加剂青贮。

1. 一般青贮

也称普通青贮，即对常规青绿饲料（如青刈玉米），按照一般的青贮原理和步骤使之在厌氧条件下进行乳酸菌发酵而制作的青贮。是将原料切碎、压实、密封，在厌氧环境下使乳酸菌大量繁殖，从而将饲料中的淀粉和可溶性糖变成乳酸。当乳酸积累到一定浓度后，便抑制腐败菌的生长，将青绿饲料中的养分保存下来。

2. 半干青贮

也称作低水分青贮，具有干草和青贮料两者的优点，是近20年来

Silage comes in many varieties and is divided into many categories depending on the types of raw materials for silage, climate, and specific environment. According to the water content of the raw material, it can be divided into low water silage (semi-dry silage) and high water silage (general silage). According to the difficulty and complexity of fermentation, it can be divided into general silage, mixed silage and additive silage.

1. General silage

General silage is also called common silage, that is, silage made by lactic acid bacteria fermentation under anaerobic conditions for conventional green feed (such as green mowing corn) according to the general principles and steps of silage. The raw materials are chopped, compacted, and sealed to make lactic acid bacteria multiply in an anaerobic environment, so as to turn the starch and soluble sugar in the feed into lactic acid. When lactic acid accumulates to a certain concentration, it inhibits the growth of putrid bacteria and preserves the nutrients in the green feed.

2. Semi-dry silage

Semi-dry silage is also called low water silage. It has the advantages of both hay and silage and has been a popular

在国外盛行的方法。它将青贮原料风干到含水量40%-55%时，植物细胞渗透压达到55×100 000-60×100 000 Pa。这样便于使某些腐败菌、酪酸菌及乳酸菌的生命活动接近于生理干燥状态，因受水分限制而被抑制。这样，不仅使青贮品质提高，而且还克服了高水分青贮由于排汁所造成的营养流失。原料水分含量低，使微生物处于生理干燥状态，生长繁殖受到抑制，饲料中微生物发酵弱，养分不被分解，从而达到保存养分的目的。该类青贮由于水分含量低，其他条件要求不严格，故较一般青贮而言扩大了原料的范围。

3. 混合青贮

有些饲草单独青贮不易成功，若把两种或两种以上的原料进行混合青贮，则可以调制成品质优良的青贮饲料。混合青贮一般有三种类型：含水量大的与含水量少的原料混合青贮，如甜菜叶、块根、块茎类、瓜类、蔬菜副产物等与秸秆麸皮混合青贮，不仅提高了青贮质量，而且省去了加水工序；含糖量少的与含糖量高的原料混合青贮，例如豆科牧草与禾本科牧草混合青贮；为提高青贮饲料的营养价值而调制的配合饲料，如用玉米、向日葵与其他饲料混合青贮，豌豆、燕麦与其他作物混合青贮等。

4. 添加剂青贮

是在青贮中加进一些添加剂来影响青贮的发酵作用。如：添加各种可溶性碳水化合物、接种乳酸菌、加入酶制剂等，可促进乳酸发酵，迅速产生大量的乳酸，使 pH 很快达到要求（3.8-4.2）；加入各种酸类、抑菌剂等可抑制腐败菌等不利于青贮的微生物的生长，例如黑麦

method abroad in recent 20 years. When the silage material is air-dried to 40%-55% of water content, the osmotic pressure of plant cells reaches 55×100 000-60×100 000 Pa. This causes the life activities of some spoilage bacteria, butyric acid bacteria and lactic acid bacteria to be close to physiological dry state, which is inhibited due to water restrictions. In this way, not only the quality of silage is improved, but also the nutrient loss caused by juice drainage in high moisture silage is overcome. The low moisture content of raw materials keeps the microorganisms in a physiological dry state, the growth and reproduction are inhibited, the fermentation of microorganisms in the feed is weak, the nutrients are not decomposed, so as to achieve the purpose of nutrient preservation. Due to its low moisture content and less stringent requirements for other conditions, this kind of silage expands the range of raw materials compared with general silage.

3. Mixed silage

Some forage silage alone is not easy to succeed. If two or more kinds of raw materials are mixed for silage, it can be made into high-quality silage. There are generally three types of mixed silage: mixed silage with high water content and low water content raw materials, such as beet leaves, tubers, melons, vegetable by-products and straw and bran, which not only improves the quality of silage, but also saves the water adding process; mixed silage with low sugar content and high sugar content raw materials, such as legume and gramineous forage mixed silage; mixed silage with high sugar content raw materials, such as legume and gramineous forage mixed silage; compound feed such as corn, sunflower mixed with other feed silage, pea, oats mixed with other crops silage, etc. for improving the nutritive value of silage.

4. Additive silage

Additives are added to silage to affect the fermentation of silage, for example: adding various soluble carbohydrates, inoculating lactobacillus or adding enzyme preparation, etc., can promote lactic acid fermentation, quickly produce a large amount of lactic acid, so that its pH value quickly meets the requirements (3.8-4.2); or the addition of various acids and bacteriostatic agents can inhibit the growth of spoilage bacteria and other microorganisms that

草青贮可按10克/公斤比例加入甲醛/甲酸(3∶1)的混合物;加入尿素、氨化物等可提高青贮饲料的养分含量。通过添加剂青贮,不仅可提高青贮效果,还可扩大青贮原料的范围。

are not conducive to silage. For instance, the mixture of formaldehyde/formic acid (3∶1) can be added to the proportion of 10 grams/kg for ryegrass silage; the nutrient content of silage can be increased by adding urea and ammoniate. In this way, the silage effect can be improved and the range of silage materials can be expanded.

三、调制过程
The modulation process

1. 原料采收

原料要适时收割,确保产量,同时要保证原料中含有较高含量的干物质、蛋白质和可发酵碳水化合物等营养物质。收割过早,原料含水多,可消化营养物质少,铡碎过程中营养物质随流出液损失的量也增多,导致饲料整体营养价值不高;收割过晚,纤维素含量增加,适口性差,消化率降低。

玉米秸的采收:全株玉米秸青贮,一般在玉米籽乳熟期采收。收果穗后的玉米秸,一般在玉米棒子蜡熟至70%完熟时,叶片尚未枯黄或玉米茎基部1-2片叶开始枯黄时立即采摘玉米棒,在采摘玉米棒的当日,最迟次日将玉米茎秆采收制作青贮。

牧草的采收:豆科牧草一般在现蕾至开花始期刈割青贮;禾本科牧草一般在孕穗至刚抽穗时刈割青贮;甘薯藤和马铃薯茎叶等一般在收薯前1-2日或霜前收割青贮。幼嫩牧草或杂草收割后可风干3-4小时(南方)或1-2小时(北方)后青贮,或与玉米秸等混贮。

2. 原料切碎

为了便于装袋、贮藏和动物采食,原料须经过切碎。玉米秸、串叶松香草秸秆或菊苣的秸秆青贮前均

1. Harvesting of raw materials

Raw materials should be harvested timely to ensure yield. At the same time, the raw materials must contain high levels of nutrients such as dry matter, protein and fermentable carbohydrates. Raw materials harvested too early will contain more water, less digestible nutrients and the amount of nutrients lost with the outflow of fluid in the process of chaff cutting is also increased, resulting in the overall nutritive value of feed is not high. Late harvesting results in increased cellulose content, poor palatability, and decreased digestibility.

Corn stalk harvesting. For silage of whole corn stalk, it is usually harvested at the milk stage of corn seed. Corn stalks after ear harvest are generally picked when the corn cobs are waxed to 70% of the time, when the leaves are not yellow or when 1 or 2 leaves at the base of the corn stem begin to yellow. On the day of picking the corn cobs, the corn stalk is harvested for silage, the next day at the latest.

Herbage harvesting. Leguminous herbage are generally mowed and made into silage from budding to flowering. Grasses are generally cut and made into silage from booting to just heading. Sweet potato vine and potato stem and leaf are generally harvested and made into silage 1-2 days before the harvest or before frost. After harvesting, young grass or weeds can be dried for 3-4 hours (in the south) or 1-2 hours (in the north) and then silaged, or mixed with corn stalks.

2. Chopping the raw materials

In order to facilitate bagging, storage and animal feeding, raw materials must be chopped. The corn straw, rosin straw or chicory straw must be shredded to about 1-2 cm

须切碎到长约 1-2 厘米,青贮时才能压实。牧草和藤蔓柔软,易压实,切短至 3-5 厘米青贮,效果较好。

3. 加添加剂

原料切碎后立即加入添加物,目的是让原料快速发酵。可添加 2%-3% 的糖、甲酸(每吨青贮原料加入 3-4 kg 85% 的甲酸)、淀粉酶和纤维素酶、尿素、硫酸铵、氯化铵等铵化物等。

4. 装填储存

青贮场地应地势高并且干燥,土质坚硬,地下水位低,易排水、不积水,靠近畜舍,远离水源,远离圈厕和垃圾堆,防止污染。通常可以用塑料袋和窖藏等方法。装窖前,底部铺 10-15 厘米厚的秸秆,以便吸收液汁。窖四壁铺塑料薄膜,以防漏水透气,装时要踏实,可用推土机碾压,人力夯实,一直装到高出窖沿 60 厘米左右,即可封顶。封顶时先铺一层切短的秸秆,再加一层塑料薄膜,然后覆土拍实。四周距窖 1 米处挖排水沟,防止雨水流入。窖顶有裂缝时,及时覆土压实,防止漏气漏水。袋装法须将袋口张开,将青贮原料每袋装入专用塑料袋,用手压和用脚踩实压紧,直至装填至距袋口 30 厘米左右时,抽气、封口、扎紧袋口。

(1)青贮塔 可分为全塔式和半塔式两种。一般为圆筒形,直径 3-6 米,高 10-15 米。可青贮水分含量 40%-80% 的青贮饲料,装填原料时,较干的原料应在底部。青贮塔由于取料出口小,深度大,青贮原料自重压实程度大,空气含量少,贮存质量好。但造价高,仅大型牧场采用。

long before silage, with which the products can be compassed. Forage and vines are soft and easy to compact, and can be cut short to 3-5 cm long, which is good for silage.

3. Adding additives

The additives are added immediately after the raw materials are chopped to allow rapid fermentation of the raw materials. 2-3% of sugar, formic acid (3-4 kg 85% of formic acid for it silage materials), amylase and cellulase, urea, ammonium sulfate, ammonium chloride and other ammonium compounds can be added.

4. Loading and storage

The silage site should be in high dry terrain with hard soil, low groundwater level. The site should be easy to drain, close to the livestock shed, away from water source and the toilet and garbage to prevent pollution. Plastic bags and cellaring are often used. Before filling the cellar, lay 10-15 cm thick straw at the bottom to absorb liquid. The four walls of the cellar are covered with plastic film to prevent water leakage and ventilation. The installation should be steadfast. It can be rolled by a bulldozer and compacted by manpower. After installing to approximate 60 cm height, the bag can be sealed. To seal the top, the first layer is short-cutted straw covered by a layer of plastic film, and then soil and pat. Drainage ditch around 1 meter away from the cellar should be digged to prevent rainwater from flowing in. When there are cracks in the cellar roof, timely earth compaction is necessary to prevent leakage of air and water. With the bag packing method, the mouth shall be opened, and each bag of silage raw materials shall be put into the special plastic bag, which can be pressed with the hands and feet to compact, until the products are located around 30 cm of the bag mouth, which can be used to extract, seal and bind the bag mouth.

(1) Silage towers. It can be divided into full tower type and half tower type. Generally, it is cylindrical, with a diameter of about 3-6 m and a height of about 10-15 m. Silage with 40%-80% of moisture content can be silaged, with the drier ingredients at the bottom when filling. The silage tower has a small outlet, a large depth, a large compaction degree of silage raw materials, less air content and good storage quality. But the cost is high, and it is only used in large pastures.

（2）青贮窖 青贮窖分地下式、半地下式和地上式三种，圆形或方型，直径或宽2～3米，深2.5～3.5米。通常用砖和水泥做材料，窖底预留排水口。一般根据地下水位高低、当地习惯及操作方便决定采用哪一种。但窖底必须高出地下水位0.5米以上，以防止水渗入窖。青贮窖结构简单，成本低，易推广。

（3）堆贮 分地表堆贮和半地表堆贮。

① 地表堆贮：选择干燥、利水、平坦、地表坚实并带倾斜的地面，将青贮原料堆放压实后，再用较厚的黑色塑料膜封严，上面覆盖一层杂草之后，再盖上厚约20～30厘米的泥土，四周挖出排水沟排水。地表堆贮简单易学，成本低，但应注意防止家畜踩破塑料膜而进气、进水造成腐烂。

② 半地表堆贮：选择干燥、利水、带倾斜度的地面，挖60厘米左右的浅坑，坑底及四周要抹平，将塑料膜铺入坑内，再将青贮原料置于塑料膜内，压实后，将塑料膜提起封口，再盖上杂草和泥土，四周开排水沟深约30～60厘米。地表青贮的缺点是取料后，与空气接触面大，不及时利用，青贮质量会变差，造成损失。

（4）塑料袋青贮 除大型牧场采用青贮圆捆机和圆捆包膜机外，农村普遍推广塑料袋青贮。青贮塑料袋只能用聚乙烯塑料袋，严禁用装化肥和农药的塑料袋，也不能用聚苯乙烯等有毒塑料袋。青贮原料装袋后，应整齐摆放在地面平坦光洁的地方，或分层存放在棚架上，最

(2) Silage pits. Silage pits can be divided into underground, semi-underground and aboveground types. They are round or square, with the diameter or width of 2 - 3 m, and the depth of 2.5 - 3.5 m. It is usually made of brick and cement, and a drain is reserved at the bottom of the cellar. Generally, the type of pit is chosen according to the level of the water table, local customs and convenience of operation. But the bottom of the cellar must be more than 0.5 m above the water table to prevent water from seeping into the cellar. The silage pit is easy to promote for its simple structure, and low cost.

(3) Surface storage and semi-surface storage.

① Surface storage. A dry, water-friendly, flat, solid and inclined ground is firstly chosen. After the silage raw materials are stacked and compacted, they are sealed with a thick black plastic film and covered with a layer of weeds, and then covered with a thickness of about 20 - 30 centimeters of soil. Drainage ditches are dug all around for drainage. Surface storage is easy to promote for its low cost, but attention should be paid to prevent livestock from stepping on the plastic film which will cause water or air leakage to rot.

② Semi-surface storage: Choose a dry, conducive, and inclined ground, and dig a shallow hole of about 60 cm, and its bottom and surrounding should be balanced. Spread the plastic film into the pits, and then put the silage raw materials into the plastic film. After compacting, carry the plastic film to seal, and then cover it with weeds and soil. Open drainage ditch around about 30 - 60 cm deep. The disadvantage of surface silage is that it has a large contact area with the air after taking the material, and the quality of silage becomes worse if it is not used in time, resulting in loss.

(4) Plastic bag silage. It is widely promoted in rural areas, except for the use of silage round baling machine and round baling envelop machine in large pastures. Only polyethylene plastic bags can be used for silage plastic bags. Plastic bags containing chemical fertilizers and pesticides are strictly prohibited, and toxic plastic bags such as polystyrene are not allowed. After the silage material is bagged, it shall be neatly placed in a smooth and clean place on the ground,

上层袋的封口处用重物压上。在常温条件下,青贮1个月左右,低温2个月左右,即青贮完成,可饲喂家畜。在较好环境条件下,存放一年以上仍保持较好质量。塑料袋青贮的优点:投资少,操作简便;贮藏地点灵活,青贮省工,不浪费,节约饲养成本。

or stored in layers on the shelf, and the seal of the top bag shall be pressed on with heavy objects. Under normal temperature, silage for about one month or two months under low temperature means complete silage, which can be fed to livestock. Under better environmental conditions, storage for more than one year can still maintain good quality. The advantages of plastic bag silage is less investment, easy operation, flexible storage, labor saving, no waste, and saving of feeding costs.

第四节 青贮饲料品质鉴定及利用
Section 4　Quality identification and utilization of silage

一、感官评定
Sensory evaluation

感官评定一般在开启青贮容器时,根据青贮饲料的色泽、质地、气味和酸度四方面进行评定。色泽优质的青贮饲料若青贮前作物颜色为绿色,青贮后仍为绿色或黄色的为最佳。气味优良的青贮饲料通常具有轻微的香甜的酒酸味(乙醇)和酸味(乳酸);霉味则说明压得不实,空气进入了青贮窖,引起饲料霉变;若有刺鼻的酸味,则说明醋酸较多,品质较差。质地优质的青贮饲料中植物的茎叶等结构应当能够清晰辨认,结构破坏呈黏滑状态是腐败的标志,具体表现为:质地上乘的青贮饲料经用力攥握后仍保持松散、柔软;劣质青贮饲料握有黏腻感,松手后成团;若质地干燥、粗硬,则说明原料含水量比较低。

Sensory evaluation is generally based on silage color, texture, odor and acidity when the silage container is opened. For high-quality silage, if the crop color is green before silage and green or yellow after silage, it is the best. Well-smelling silage usually has a slight sweet acidity (ethanol) and acidity (lactic acid); the musty smell indicates that the pressure is not solid, and the air enters the silo, causing the feed mildew. If the silage has a pungent sour taste, it means that there is more acetic acid and the quality is poor. The stems, leaves and other structures of plants in high-quality silage should be clearly identifiable. The slimy state of structural failure is a sign of corruption, which is specifically manifested as follows: excellent silage remains loose and soft after being firmly grasped; poor-quality silage has a sticky feel, forming a lump after letting go; if the texture is dry, coarse and hard, it means that the raw material moisture content is relatively low.

二、实验室评定
Laboratory evaluation

实验室评定是以化学分析为主。根据青贮饲料的不同对青贮饲料质量进行评定,包括青贮饲料的有机酸含量、pH 值、氨态氮/总氮。测定氨态氮与总氮的比值,主要是评价蛋白质被破坏的程度;测定总有机酸含量,用来判断发酵情况。

1. pH 值(酸碱度) pH 值是衡量青贮饲料品质好坏的重要指标之一。实验室测定 pH 值可用精密酸度计测定,生产现场可用精密石蕊试纸测定。优良青贮饲料 pH 值在 2 以下,超过 4.2(低水分青贮除外)则说明青贮发酵过程中腐败菌、酪酸菌等活动较为强烈。劣质青贮饲料 pH 值在 5.5 - 6.0,中等青贮饲料的 pH 值介于优良与劣等之间。

2. 氨态氮 氨态氮与总氮的比值反映青贮饲料中蛋白质及氨基酸分解的程度,比值越大,说明蛋白质分解越多,青贮质量不佳。由于豆科牧草蛋白质含量高,蛋白降解程度严重,氨态氮含量远高于禾本科等其他青贮牧草含量,对青贮发酵品质影响较大,故在豆科牧草评价体系中需要将氨态氮作为一种评价指标。

3. 有机酸含量 有机酸总量及其构成可以反映青贮发酵过程的好坏,其中最重要的是乳酸、乙酸和丁酸,乳酸所占比例越大越好。优良的青贮饲料,含有较多的乳酸和少量乙酸,而不含丁酸;品质差的青贮饲料,含丁酸多而乳酸少。

Laboratory evaluation is based on chemical analysis. The quality of silage is evaluated according to different measurement indexes of silage, including organic acid content, pH value, ammonium nitrogen/total nitrogen of silage. The ratio of ammonium nitrogen to total nitrogen is mainly used to evaluate the degree of protein destruction, and the content of total organic acids is used to judge the fermentation condition.

1. pH value. pH value is one of the important indexes to measure the quality of silage. The pH value can be determined in the laboratory with precision pH meter and in the production field with precision litmus paper. A good silage with a pH value below 2 and over 4.2 (except for low water silage) indicates that the activities of spoilage bacteria and butyric acid bacteria in the silage fermentation process are relatively strong. The pH value of inferior silage is between 5.5 and 6.0, and that of medium silage is between good and inferior.

2. Ammonium nitrogen. The ratio of ammonium nitrogen to total nitrogen reflects the degree of protein and amino acid decomposition in silage. The higher the ratio is, the more proteins are decomposed, and the quality of silage is poor. Due to the high protein content and serious protein degradation degree of leguminous herbage, ammonium nitrogen content is much higher than that of gramineae and other silage herbage, which has a great influence on the quality of silage fermentation. Therefore, ammonium nitrogen should be used as an evaluation index in the evaluation system of leguminous herbage.

3. Organic acid content. The total amount of organic acids and their composition can reflect the quality of the silage fermentation process. The most important ones are lactic acid, acetic acid and butyric acid. The greater the proportion of lactic acid in the content, the better the silage. Excellent silage contains more lactic acid and a little acetic acid, but does not contain butyric acid; silage with poor quality contains more butyric acid and less lactic acid.

三、饲用价值评定
Evaluation of feeding value

上述评定方法对以青贮饲料为基础的家畜营养配方指导性不强,青贮饲料的质量最终应以饲用价值来体现。目前较为常用的评定方法主要是测定能量、总可消化养分和中性洗涤纤维的消化率,并进一步计算干物质采食量、消化能、奶牛泌乳净能、肉牛产肉净能和生长净能等,这些指标在奶牛生产中应用较为成熟,得到普遍的认可。

Although the above methods can be used to evaluate the quality of silage to some extent, they are not very instructive for the nutrient-formulation of livestock based on silage, and the quality of silage should be reflected by feeding value. More commonly used evaluation methods at present are mainly by measuring energy, total digestible nutrient digestibility and neutral detergent fiber digestibility (NDFD), and furthermore, by calculating dry matter intake, digestion energy, milk cow next energy, beef net meat production and growth net energy, etc. These parameters, widely used in dairy production, have been generally recognized.

四、青贮饲料的利用
The utilization of silage

1. 取用方法

青贮过程进入稳定阶段,一般糖分含量较高的玉米秸秆等需要20天以上即可发酵成熟,开窖取用,或待秋冬季节开窖饲喂家畜。

开窖取用时,如发现表层呈黑褐色并有腐败臭味时,应把表层弃掉。对于直径较小的圆形窖,应由上到下逐层取用,保持表面平整。对于长方形窖,自一端开始分段取用,不要挖窝掏取,取后最好覆盖,以尽量减少与空气的接触面;每次用多少取多少,不能一次大量取用后堆放在畜舍慢慢饲用,要用新鲜青贮饲料。青贮饲料只有在厌氧条件下才能保持良好品质,如果堆放在畜舍里和空气接触,很快就会感染霉菌和杂菌,导致发霉变质。特别是夏季,正是各种细菌繁殖最旺盛的时候,青贮饲料也最易坏。

当青贮窖被打开,青贮饲料被

1. Access method

When the silage process enters the stable stage, corn stalks with high sugar content can be fermented and matured for more than 20 days, and the pits can be opened for use, or the pits can be opened for livestock feeding in autumn and winter.

When opening the cellar for use, if the surface layer is found to be dark brown and rotten, the surface layer should be discarded. For circular pits with smaller diameter, they should be taken layer by layer from top to bottom to keep the surface smooth and flat. For the rectangular cellar, they should be taken from one end of the section but not be digged out. After taking the best cover, be sure to minimize the contact with the air surface: take as much as you use each time, and you can not take a large amount at one time and pile it up in the barn for upcoming feeding. Fresh silage should be used. Silage can be kept in good quality only under anaerobic conditions. If it is piled up in the barn and exposed to the air, it will quickly become infected with mold and bacteria, which will make the silage moldy. Especially in summer, when all kinds of bacteria multiply most vigorously, silage is also the most vulnerable.

When the silage cellar is opened and the silage is ex-

暴露在空气中时,氧气的存在会激活好氧微生物的活性,这些微生物会利用剩余的发酵底物和发酵产物,从而导致营养物质的损失显著增加。有氧腐败的主要表现为产生大量的二氧化碳和释放出大量的热,导致乳酸浓度降低,pH 值增加。有研究发现,在这个过程中发生腐败的青贮饲料,每天大概有 1.5%~4.5% 的干物质损失,几乎与密封保存几个月的损失一样。只要青贮饲料暴露在空气中,此阶段就会发生。因此,为了降低这些损失和改善青贮饲料的好氧稳定性,必须加强青贮饲料的管理和科学利用。

2. 饲喂技术

青贮饲料可以作为草食家畜牛羊的主要粗饲料,一般占日粮干物质的 50% 以下。青贮饲料虽然是一种优质粗饲料,但不能作为家畜的单一日粮,否则不利于家畜的生长发育。饲喂时应根据牛羊的实际需要与精饲料、优质干草搭配使用,以提高瘤胃微生物对氮素和饲料的利用率,以及动物干物质采食量。刚开始喂时家畜不喜食,喂量应由少到多,逐渐适应后即可习惯采食。训练方法是:先空腹饲喂青贮饲料,再饲喂其他草料;先将青贮饲料拌入精料喂,再喂其他草料;先少喂,后逐渐增加;或将青贮饲料与其他料拌在一起饲喂,在饲喂初期或青贮饲料酸度较高时,可以添加适量的小苏打饲喂,以降低酸度,提高适口性,促进消化吸收,避免酸中毒现象的发生。由于青贮饲料含有大量有机酸,具有轻泻作用,故患有胃肠炎的家畜要少喂或不喂。母畜妊娠后期不宜多喂,产前 15 天停喂,产后 10-15 天在饲粮中重新加入青贮饲料。劣质的青贮饲料有害畜体健康,易造成流产,不能饲喂。冰冻

posed to the air, the presence of oxygen activates aerobic microorganisms that use the remaining fermentation substrates and fermentation products, causing a significant increase in nutrient loss. The main phenomenon of aerobic decay is the production of large amounts of carbon dioxide and the release of large amounts of heat, leading to a decrease in lactic acid concentration and an increase in pH value. Studies have found that during this process, about 1.5% to 4.5% of dry matter loss occurs per day, almost as much as the loss of sealed storage for several months. This stage occurs whenever silage is exposed to air. Therefore, in order to reduce these losses and improve the aerobic stability of silage, it is necessary to strengthen the management and scientific utilization of silage.

2. Feeding techniques

Silage can be used as the main roughage for herbivores, such as cattle and sheep, generally accounting for less than 50% of the dry matter of the diet. Although silage is a kind of high-quality roughage, it cannot be used as a single feed for livestock, otherwise it is not conducive to the growth and development of livestock. Feeding should be used in combination with concentrate and high-quality hay according to the actual needs of cattle and sheep to improve the utilization rate of rumen microorganisms to nitrogen and feed, as well as dry matter intake of animals. At the beginning of feeding, livestock do not like the silage, which should be added into the feed gradually from less to more, and then they can get used to the feeding after adaptation. The training method is: feed silage on an empty stomach first followed by other forage; feed the silage mixture with the concentrate first, and then feed the other forage; feed a small amount of silage at first, and then increase gradually; or silage feeding, together with other material on early feeding with right amount of baking soda to reduce the acidity, improve the palatability, promote the digestion and absorption, and avoid the acidosis phenomenon. As silage contains a lot of organic acids and has laxation effect, livestock with gastroenteritis should be fed less or not. Female animals should not be fed more in the late pregnancy, and feeding should be stopped 15 days before delivery. The silage is readded to the diet 10-15 days of postpartum. Poor-quality

的青贮饲料也易引起母畜流产,应待冰融化后再喂。

3. 饲喂量

不同家畜在不同阶段的饲喂量也不同。成年牛每100公斤体重日喂青贮量为:泌乳牛5-7公斤,肥育牛4-5公斤,役牛4-4.5公斤,种公牛1.5-2.0公斤。小母牛每50公斤体重饲喂1.25-1.5公斤。绵羊每100公斤体重日喂量为:成年羊4-5公斤,羔羊0.4-0.6公斤。奶山羊每100公斤体重日喂量为:泌乳母羊1.5-3.0公斤,青年母羊1.0-1.5公斤,公羊1.0-1.5公斤。马的日喂量为:役马每匹每天可喂12-15公斤,种母马和1岁以上的幼驹每天可喂6-10公斤。繁殖母猪每天饲喂2-3公斤。

silage is harmful to the health of livestock and can easily cause abortion and cannot be fed. Frozen silage is also easy to cause abortion for female animals, and should be fed after the ice melts.

3. Amount of feeding

Different livestock need different feeding amounts at different stages. The daily silage of adult cattle (per 100 kg of body weight) is: 5-7 kg for lactating cattle, 4-5 kg for fattening cattle, 4-4.5 kg for working cattle, and 1.5-2.0 kg for breeding bulls. Heifers are fed with 1.25-1.5 kg per 50 kg body weight. The daily feeding rate of sheep (per 100 kg of body weight) is: adult sheep 4-5 kg, lamb 0.4-0.6 kg. Dairy goats are fed 1.5-3.0 kg of lactating ewes, 1.0-1.5 kg of young ewes and 1.0-1.5 kg of rams per 100 kg of body weight per day. Daily feeding of horses is: each serving horse can be fed 12-15 kg per day, and the breeding mares and foals over 1 year old can be fed 6 to 10 kg per day. Breeding sows are fed 2-3 kg per day.

第六章 能量饲料
Chapter 6 Energy feed

在国际饲料分类法中,能量饲料属于第四类,是指饲料干物质中粗纤维含量低于18%,粗蛋白质含量低于20%,消化能含量大于10.46 MJ/kg 的饲料,其中如消化能在12.55 MJ 以上的称为高能饲料。这类饲料包括谷实类、糠麸类、块根块茎瓜果类、糖蜜类、动植物油脂类和乳糖等。在饲料生产中,饲料成本约占生产成本的70%,而能量成本在饲料费用中约占75%。能量饲料在畜禽饲粮中所占比例较大,一般在50%以上,是畜禽重要的能量来源。

该类饲料能量含量高,除小麦麸外,通常每千克的代谢能含量在11 MJ 以上。蛋白质含量低(8.6%-15.7%),且品质差,必需氨基酸含量不足,尤其是赖氨酸和蛋氨酸缺乏,难以满足珍禽的蛋白质要求;该类饲料矿物质含量不平衡,钙少(一般低于0.1%),磷多(可达0.3%-0.5%),且主要为植酸磷,利用率低,并可影响其他矿物质元素的利用;维生素含量不平衡,缺乏维生素A 和维生素D,只含有少量β-胡萝卜素,除了维生素 B_2 含量较低外(只有1-2.2 mg/kg),其余B 族维生素含量较丰富,不同种类的能量饲料因养分组成不同,饲用价值亦不同。

Energy feed belongs to the fourth category in the International Feed Classification, which refers to the feed with less than 18% of crude fiber content, less than 20% of crude protein content, and more than 10.46 MJ/kg of digestible energy content in the dry matter. The feed with digestible energy above 12.55 MJ is called high-energy feed. This type of feed includes grain, bran, roots, tubers, melons, fruits, molasses, animal and vegetable fats, and lactose, etc. During feed production, feed costs account for about 70% costs of production, and energy costs account for about 75% of feed costs. Energy feed occupies a large proportion in livestock and poultry diet, generally more than 50%, which is an important energy source for livestock and poultry.

Energy feed has high energy content. Except for wheat bran, the metabolizable energy content of energy feed is usually above 11 MJ/kg. Energy feed has low protein content (8.6%-15.7%), poor quality, insufficient essential amino acid content, especially lack of lysine and methionine, which is difficult to meet the protein requirements of rare birds. Energy feed has unbalanced mineral content, low calcium (generally less than 0.1%), high phosphorus (up to 0.3%-0.5%), and mainly phytate phosphorus, which has low utilization and affects the utilization of other mineral elements. The vitamin content of energy feed is unbalanced. It lacks vitamin A and vitamin D, and only contains a small amount of β-carotene. The content of B vitamins is rich, but vitamin B_2 is low (only 1-2.2 mg/kg). Different types of energy feed have different feeding values due to their different nutrient composition.

第一节 谷实类饲料
Section 1　Cereal grain

谷实类饲料是禾本科植物的成熟种子,常用的有玉米、小麦、大麦、稻谷、高粱和燕麦等,在动物饲粮组成中占有重要位置,一般占饲粮组成的50%以上,是畜禽重要的能量来源。

Cereal grains are mature seeds of gramineous plants. The commonly used are corn, wheat, barley, rice, sorghum, and oats, which occupy an important position of animal feeds. Cereal grains are important sources of energy for livestock and poultry, generally account for more than 50% of the feed composition.

一、营养特点
Nutritional characteristics

(一) 禾谷类籽实的结构
The structure of cereal grains

禾谷类籽实,即通常所称的"种子",其外形随种类不同而异,但其结构由外向里被分为四个部分,即种皮、糊粉层、胚乳和胚。

1 种皮　禾谷类籽实的种皮占种子组成的比例不同。种皮为种子的保护组织,粗纤维、维生素和矿物质元素含量高。种子的粗纤维绝大部分集中在种皮中,如小麦种皮和糊粉层中粗纤维含量占89%。

2 糊粉层　粗蛋白质含量较丰富,如在小麦中约占全部蛋白的20%,维生素含量也丰富。

3 胚乳　为作物养分的储存器官,占籽实的70%-82%,主要为淀粉,其次为二糖和极少量单糖,蛋白质含量较少,主要为醇溶蛋白。

4 胚　是禾谷类种子最重要的部分,占籽实的3%-5%,为植物新个体的生长组织。脂肪含量高,有的高达30%以上;蛋白质含量丰富;矿物质和维生素含量高,尤其维生素E含量高。

Cereal grains, which are usually called "seeds", vary in shape with different species. Their structure is divided into four parts from the outside to the inside: ① testa; ② aleurone layer; ③ endosperm; ④ embryo.

1 Testa. The ratio of testa to cereal grains is different. The seed coat is the protective tissue of the seed, with a high content of crude fiber (CF), vitamins and mineral elements. Most of the CF of grains is concentrated in the testa, e.g., the CF in wheat testa and aleurone layer accounts for 89%.

2 Aleurone layer. The crude protein and vitamin content are relatively rich, e.g., accounts for about 20% of the total protein in wheat.

3 Endosperm. It is the organ that stores nutrients of crops, accounting for 70% - 82% of the grains, mainly starch, followed by disaccharides and very few monosaccharides. There's less protein in endosperm, mainly gliadin.

4 Embryo. It is the most important part of cereal seeds, accounting for 3% to 5% of the seeds, and is the growth tissue of new plants. Its fat content is high, some up to 30% or more; its protein content is rich; its mineral and vitamin content are high, especially the vitamin E content.

（二）禾谷类籽实的营养价值
The nutritive value of cereal grains

1. 无氮浸出物含量丰富、有效能值高　禾谷类籽实无氮浸出物含量高，一般占籽实干物质的63%-75%，主要为淀粉，且粗纤维含量较低，一般不超过5%，带颖壳的大麦、燕麦、稻谷和粟等粗纤维含量则在6.8%-8.2%。因此，谷实类饲料的干物质消化率很高，有效能值高，在鸡饲料中每千克含代谢能11.00-14.00 MJ，如玉米无氮浸出物的消化率为90%（牛）和93%（猪），而猪对玉米的能量代谢率为85%，故该类饲料是畜禽饲粮中的主要组成部分和能量来源。

2. 蛋白质含量低且品质差　蛋白质含量在7.9%-13.9%，主要为醇溶蛋白和谷蛋白质，约占蛋白质组成的80%以上。而清蛋白和球蛋白含量低，约占蛋白质组成的20%左右，这就导致赖氨酸、蛋氨酸、苏氨酸和色氨酸含量较低，氨基酸平衡性差。因此，禾谷类籽实蛋白质的生物学效价较低，处于50%-70%。

3. 粗脂肪　禾谷类籽实脂肪含量为3.5%左右（1.6%-5.2%），构成脂肪的脂肪酸主要为不饱和脂肪酸，其中亚油酸和亚麻酸含量较高，如玉米中约含45%的亚油酸，这对满足畜禽对必需脂肪酸的需要有好处。

4. 矿物质含量低且不平衡　钙少磷多，钙含量低，仅为0.02%-0.09%，磷含量较高为0.25%-0.36%，但主要为利用率较低的植酸磷（鸡对植酸磷的利用率为0%-50%，猪为10%-40%）为主，约占总磷的60%左右，故钙、磷比例极不平衡（1:6.4），如果仅以谷实类

1. The nitrogen free extract is rich in content and high effective energy value. Nitrogen free extracts of cereal grains are high, generally accounting for 63%-75% of the dry matter of the seed, mainly starch, and the crude fiber content is low, generally not exceeding 5%. Only barley with glumes, oats, rice and millet is between 6.8%-8.2%. Therefore, the dry matter digestibility and effective energy of cereal grains are high. In chicken feed, it contains 11.00-14.00 MJ of metabolizable energy per kilogram. The digestibility of nitrogen free extracts of corn is 90% (cattle) and 93% (swine). Moreover, the energy metabolism rate of corn to swine is 85%, so that cereal grains are the main component and energy source of livestock and poultry feed.

2. Poor quality and low protein content. The protein content of cereal grains is between 7.9%-13.9%, mainly gliadin and gluten, accounting for more than 80% of the total protein. The albumin and globulin content are low, accounting for about 20% of the total protein, which leads to the content of lysine, methionine, threonine and tryptophan is low, and the amino acid balance is poor. Therefore, the biological potency of cereal grains protein is low, between 50%-70%.

3. Crude fat. The fat content of cereal grains is about 3.5% (1.6%-5.2%). The fatty acids that make up the fat are mainly unsaturated fatty acids, among which linoleic acid and linolenic acid are relatively high. For example, corn contains about 45% of linoleic acid which is the main source of essential fatty acids for livestock and poultry.

4. Low and unbalanced mineral content. Cereal grains have low calcium and high phosphorus in content. There are only 0.02%-0.09% of calcium, and relatively high 0.25%-0.36% of phosphorus. But it is mainly phytate phosphorus with low utilization rate (the utilization rate of chicken for phytate phosphorus is 0%-50%, of swine is 10%-40%), which accounts for about 60% of the total phosphorus. Therefore, the ratio of calcium to phosphorus

为唯一饲料来源,钙、磷不仅不能满足动物的需要,而且钙、磷比例也不恰当。

5. 维生素含量低且不平衡 一般维生素 B_1 和维生素 E 较为丰富,但缺乏维生素 B_2、维生素 A 和维生素 D。

谷实类饲料由于其养分组成不同,其消化性不同,因此,用其饲喂同一种家畜其饲用价值也不同;同时,动物种类不同,其消化力不同,故用同一种籽实饲喂不同家畜其饲用价值也不同。

is extremely unbalanced at 1∶6.4. If cereal grains are used as the only source of feed, the unbalanced calcium and phosphorus ratio cannot meet the needs of livestock.

5. Low and unbalanced vitamin content. Generally, vitamin B_1 and vitamin E are abundant, but vitamin B_2, vitamin A and vitamin D are lacking.

Cereal grains have different digestibility due to their different nutrient composition. Therefore, feeding the livestock with several kinds of cereal grains has different feeding values. At the same time, different species have different digestibility. Feeding different livestock with the same feed, the effects are also different.

二、主要谷实类饲料的特点
The characteristics of main cereal grains

(一) 玉米
Corn

我国粮食作物生产中玉米产量占第三位,仅次于水稻和小麦,近几年产量徘徊在 1.06 亿–1.25 亿吨。据统计,2003 年我国玉米产量为 1.14 亿吨,比 2002 年下降了 6%。全世界玉米的 70%–75% 作为饲料,我国 2002 年饲料玉米消费占总消费量的 75% 以上。除此之外,玉米在食品和酿造工业应用的副产品,如酒糟、玉米蛋白粉、玉米胚芽饼等也作为饲料被利用。因此,玉米是主要的饲料来源之一,被称为"饲料之王"。

1. 玉米的营养价值 玉米籽实含生理有效能量高,鸡饲料中每千克含代谢能 13.56 MJ,是禾谷类籽实中生理有效能值最高的。玉米粗纤维含量低,仅为 2% 左右,易消化,适口性好,对畜禽而言,饲用价值高于其他禾谷类籽实(对猪而言,小麦除外)。因此,通常在配合饲料中的用量在 50% 以上。但玉

In China's food crop production, corn output ranks third, preceded only by rice and wheat. In recent years, the output is between 106 million and 125 million tons. According to statistics, corn output of China in 2003 was 114 million tons, a decrease of 6% from 2002. 70%–75% of the corn is used as feed all over the world, while feed corn consumption accounted for more than 75% of the total consumption of China in 2002. In addition, the by-products of corn used in food and brewing industries, e. g., distillers grains, corn gluten meal, corn germ cake, etc., are also used as feed. Therefore, corn is one of the main feed sources and is called "the king of feed".

1. Nutritive value of corn. Corn seeds contain high physiological effective energy, chicken feed contains 13.56 MJ of metabolizable energy per kilogram, which is the highest energy value of cereal grains. Corn has a low crude fiber content, only about 2%, that is digestible. For livestock and poultry, the feeding value of corn is higher than that of other cereal grains (except for wheat for pigs). Therefore, it is usually used in compound feed whose amount is above 50%. However, the corn's protein content is low (7%–

米蛋白质含量较低（7%－9%），且品质较差，缺乏赖氨酸和色氨酸；钙含量极低，仅为0.02%，磷含量约0.27%，但有效磷仅为0.12%，钙、磷比例不平衡（Ca∶P＝1∶6，有效磷）；黄玉米中含β-胡萝卜素和叶黄素较高，有利于禽类蛋黄着色，但缺乏维生素D和维生素K，水溶性维生素中维生素B_1含量较丰富，而维生素B_2和烟酸缺乏。

2. 饲用价值　玉米适口性好，能值高，是猪、禽饲粮中主要的饲料原料，最大用量可达70%。玉米的饲用价值（除小麦对猪的价值外）高于其他谷实类饲料，而且黄玉米有助于鸡皮肤和脚胫的着色。但育肥后期猪大量使用玉米会导致胴体变软，背膘变厚而影响加工。另外，在应用玉米时，除注意补充缺乏的营养素外，还应注意防止黄曲霉污染而造成黄曲霉毒素中毒。目前，通过玉米育种，已培育出了高赖氨酸玉米和高油脂玉米，这两个品种玉米在饲粮中的应用将改善动物的生产性能。

3. 我国饲料玉米的质量标准　我国将饲料玉米的质量标准分为三级。

（二）小麦
Wheat

小麦是全世界主要粮食作物之一，我国产量居全世界第二位，年产约9 600万吨。小麦在我国主要作为粮食，很少用作饲料。北欧国家的能量饲料主要为麦类，其中小麦的用量较大；亚洲国家中，日本饲料用小麦的比重约占总进口量的1/4。小麦作为畜禽饲料，除能量比玉米低外，其他营养指标均优于玉米。

1. 营养价值　小麦籽实粗纤

9%) and the quality of protein is poor, lacking lysine and tryptophan; the calcium content is extremely low, only 0.02%. The phosphorus content of corn is about 0.27%, but the available phosphorus is only 0.12%, the ratio of calcium to phosphorus is not balanced (Ca∶P＝1∶6, available phosphorus). Yellow corn contains high β-carotene and lutein, which is conducive to the coloring of poultry egg yolk, but lacks vitamin D and vitamin K. Vitamin B_1 is concentrated in water soluble vitamins, while vitamin B_2 and niacin are lacking.

2. Feeding value of corn. Corn has good palatability and high energy value. It is the main feed material in pig and poultry diets, among which the maximum amount can reach 70%. The feeding value of corn (except for the value of wheat to pigs) is higher than that of other cereal grains, and yellow corn helps color the chicken skin and shin. However, the abuse of corn in pigs' late fattening period will cause the carcass softer and the back fat thicker, which affects the processing. In addition to covering the lack of nutrients when using corn, attention should also be paid to preventing aspergillus flavus pollution, which may cause aflatoxicosis. At present, high-lysine corn and high-fat corn have been bred through corn breeding. The application of these two varieties of corn in diets will improve animal production performance.

3. The quality standards of feed corn in China are divided into three levels.

Wheat is one of the main food crops in the world. Its output in China ranks second in the world, with an annual output of about 96 million tons. Wheat is mainly used as food in China and is rarely used as feed. The energy feed in the Nordic countries is mainly wheat, of which the amount of wheat is relatively large. Among Asian countries, the proportion of wheat for feed in Japan accounts for about 1/4 of the total import volume. As a feed for livestock and poultry, wheat is superior to corn in other nutritional indicators except that it has lower energy than corn.

1. Nutritive value of wheat. Wheat seeds and corn

维含量和玉米相当,为2%左右;粗脂肪含量较玉米低,为1.7%;但蛋白质含量高于玉米,为13.9%(产地不同粗蛋白质含量不同,高的可达19%,低的不到12%),是谷实类饲料中蛋白质含量最高的,但蛋白质的品质仍然不好,第一限制性氨基酸为赖氨酸、苏氨酸和异亮氨酸,与豆粕相比可消化赖氨酸仅为11%;生理有效能值较高,仅低于玉米,为每千克含代谢能12.7 MJ(鸡);矿物质元素方面,小麦仍然是钙少(0.17%)磷(0.31%)多,且磷仍然以植酸磷(0.18%)为主,约占总磷的58.1%,微量元素铁、铜、锰、锌和硒含量较少;小麦含B族维生素和维生素E较多,而维生素A、维生素D和维生素C含量极低。

2. 饲用价值 小麦作为畜禽饲料,除能量比玉米低外,其他营养指标均优于玉米,对各种动物都有较高的饲用价值。然而小麦中所含有的阿拉伯木聚糖、β-葡聚糖等抗营养因子限制了小麦在猪和家禽等动物饲粮中的大量(超过饲粮原料组成的30%)应用。阿拉伯木聚糖和β-葡聚糖的主要作用为增加畜禽消化道的黏度,降低饲料利用率,使非特异性结肠炎的发病率增高,导致动物生产性能降低。对此可采用在饲粮中添加酶、抗生素和增加一定燕麦壳等粗纤维含量高的原料,来消除或降低其抗营养作用,提高动物的生产性能。用小麦喂禽类,饲用价值为玉米的90%左右,而用来喂猪其饲用价值与玉米相当,甚至略高于玉米。用小麦饲喂猪、鸡时将其粉碎(但不宜粉碎太细)饲喂效果较好,其在猪饲粮中的最大用量为70%,而在仔鸡、后备母鸡、产蛋母鸡、育肥期肉鸡饲粮中的最大用量分别为20%、40%、

seeds have similar crude fiber content, which is about 2%. Corn has a relatively lower fat content of 1.7% and higher protein content of 13.9% (crude protein content differ with different origins, from 12% to 19%). Wheat has the highest protein content in cereal grains, but the quality of protein is not good. Its first limiting amino acids are lysine, threonine and isoleucine, and digestible lysine content is only 11% compared to soybean meal. Physiologically effective value of wheat is high, only lower than corn, containing 12.7 MJ/kg (chicken) of metabolizable energy. To minerals, wheat has few calcium (0.17%) and more phosphorus (0.31%). And phytate phosphorus (0.18%) still dominates, accounting for 58.1% of total phosphorus. Trace elements, e. g., iron, copper, manganese, zinc and selenium content are few. Wheat contains lots of B vitamins and vitamin E, while vitamin A, vitamin D and vitamin C are extremely low.

2. Feeding value of wheat. In addition to lower energy, wheat has more nutrients than corn as feed for livestock and poultry. Wheat has good feeding value for various animals. However, the antinutritional factors such as arabinoxylan, β-glucan contained in wheat restrict the massive use (more than 30% of raw material) of wheat in swine and poultry feeds. The main function of arabinoxylan and β-glucan is to increase the viscosity of the digestive tract of livestock and poultry, reduce the feed utilization, increase the incidence of non-specific colitis and reduce the performance of animal production. In this regard, add enzymes, antibiotics and some raw materials with high crude fiber like oat husks can eliminate or reduce their antinutritional effects, improve animal production. Feeding poultry with wheat has about 90% feeding value compared with corn, while feeding pigs has similar or even slightly higher feeding value. It is better to crush wheat (but not too finely) before feeding pigs and chickens. The maximum dosage of wheat in pig feed is 70%. However, in chickens, hens, laying hens, broiler fattening feeds, the maximum dosage of wheat are respectively 20%, 40%, 40% and 60%. In ruminant feed, less than 50% in the total feed is appropriate.

40%和60%,在反刍动物饲粮中用量以不超过50%为宜。

3. 我国饲料用小麦的质量标准　饲料用小麦的国家标准是 NY/T 117—1989。

(三) 大麦
Barley

大麦有两种,即普通(皮)大麦和裸大麦。在我国,大麦栽种地区很广,年产量约为310万吨,其中皮大麦约占总产量的2/3。裸大麦是我国藏族人民的主要粮食,也是酿制青稞酒、啤酒的重要原料。作为饲料用的大麦在美国约占总产量的60%,在西欧国家占70%－90%。我国用作饲料的大麦数量不确定。

1. 营养价值　裸大麦的粗纤维含量为2.0%,与玉米相当,皮大麦的粗纤维含量较高,为4.8%;无氮浸出物含量为67%以上,主要成分为淀粉,其他糖约占10%,主要是非淀粉多糖,即阿拉伯木聚糖和β-葡聚糖,含量分别占干物质的6.25%－6.93%和3.85%－4.51%,故大麦的生理有效能值低于玉米和小麦,每千克鸡饲料含代谢能11.01－11.30 MJ;蛋白质含量为11%－113%,蛋白质的品质是能量饲料中较好的,氨基酸组成中赖氨酸、蛋氨酸、色氨酸、异亮氨酸等含量高于玉米,尤其是赖氨酸含量为0.43%,比玉米几乎高出一倍;矿物质含量仍为钙少(0.04%－0.09%)磷多(0.33%－0.39%),钙、磷比例不恰当。相对于小麦和玉米,植酸磷占总磷的比例较少,约占总磷的46.15%－48.49%;大麦含微量元素铁较高,但含铜较低。

2. 饲用价值　总体而言,大麦的饲用价值较玉米低,用来饲喂猪、

3. Quality standard of feed wheat in China. Chinese national standard quality indicators of wheat feed is NY/T 117—1989.

There are two types of barley, i. e. ordinary barley and naked barley. Barley is planted widely in China, with an annual output of about 3.1 million tons, of which ordinary barley accounts for about 2/3 of the total output. Naked barley is the main food for the Tibetan people in China, and it is also an important raw material for brewing highland barley wine and beer. Barley as feed accounts for about 60% of the total feed yield in the United States and 70%－90% in Western European countries. The amount of barley used as feed in China is uncertain.

1. Nutritive value of barley. Similar to corn, the crude fiber content of naked barley is 2.0%. Ordinary barley has a higher crude fiber content of 4.8%. The nitrogen free extract content is more than 67%, mainly composed of starch, and other sugars account for 10%. Starch polysaccharides, namely arabinoxylan and β-glucan, respectively accounts for 6.25%－6.93% and 3.85%－4.51% of the dry matter. Therefore, the physiologically effective energy value of barley is lower than that of corn and wheat, metabolizable energy per kilogram of chicken feed is 11.01－11.30 MJ. Protein content of barley is 11%－113%, which is the best in quality in energy feed. Amino acid, including lysine, methionine, tryptophan and isoleucine, are higher than corn, especially the lysine content is almost double that of corn, is 0.43%. The mineral content is still low in calcium (0.04%-0.09%) and high in phosphorus (0.33%-0.39%), and the calcium-phosphorus ratio is not appropriate. Relative to wheat and corn, phytate accounts for less among total phosphorus, which is about 46.15%－48.49%. When it comes to microelement, barley contains higher iron content but lower copper content.

2. Feeding value of barley. In general, the feeding value of barley is lower than that of corn, e. g., the feeding

家禽和牛的饲用价值为玉米的88%（猪）、80%-85%（鸡）和90%（牛）。通常情况下，大麦不宜用来配制仔猪饲粮，但有研究表明，将大麦进行加热处理后，能显著改善仔猪的生产性能（相较于生大麦）。用大麦饲喂育肥猪，以不超过25%为宜，并可获得质量较高的、有利于加工的硬脂胴体。我国举世闻名的金华火腿产区，曾将大麦作为养猪必备精料之一。大麦对鸡的饲用价值明显不如玉米，在肉仔鸡饲粮中的用量以不超过20%为宜，育肥鸡最好控制在10%以下，后备母鸡和产蛋母鸡用量可高点，但应控制在45%以内，应注意消除非淀粉多糖的不利影响。大麦是反刍动物优良的精料，在奶牛精料补充料用量上以控制在40%以内为宜。

3. 我国饲料用大麦的质量标准

我国制定了饲料用皮大麦的国家标准(NY/T 118—1989)及饲料用裸大麦国家标准(NY/T 210—1992)

value of barley for pigs, poultry and cattle respectively is 88%, 80%-85% and 90% compared to corn. Barley should not be added in piglet diets, but studies have shown that heat-treated barley can significantly improve the performance of piglets (compared with raw barley). Feed fattening pigs with no more than 25% barley can get a high-quality stearin carcasses. Barley was one of the essential ingredients for pig raising in China's world-famous Jinhua ham producing area. The feeding value of barley to chicken is significantly lower than that of corn. The usage of barley for broiler chicken feed should not exceed 20%, and for fattening chickens, 10% or less, while for pullets and laying hens, dosage can be higher (within 45%), and attention should be paid to eliminate the adverse effects of non-starch polysaccharides. Barley is an excellent concentrate for ruminants, and the amount of concentrate supplements for dairy cows should be controlled within 40%.

3. Quality standard of feed barley in China. China has formulated national standards for feed barley (NY/T 118—1989) and feed naked barley (NY/T 210—1992).

（四）高粱
Sorghum

高粱，又称蜀黍、荻子，原产于热带。我国高粱总产量约占全世界总产量的1/10，其种植面积和产量位于世界第五位，主产区为辽宁和黑龙江两省。

1. 营养价值 高粱籽粒的结构和养分含量与玉米相似。粗纤维含量为1.7%，粗脂肪含量3.4%，能量含量略低于玉米，在鸡饲料中，每千克含代谢能12.30 MJ；而蛋白质含量略高于玉米，为9.0%，但蛋白质品质仍然较差，缺乏赖氨酸、组氨酸和蛋氨酸；维生素中烟酸和生物素的含量较玉米高，但利用率较低，如烟酸含量为41 mg/kg（玉米为24 mg/kg），但能被利用的仅为

Sorghum is native to the tropic area. The planting area and output of sorghum in China rank fifth in the world, and sorghum output accounts for about 10% of the world's total production. The main planting areas are Liaoning and Heilongjiang provinces.

1. Nutritive value of sorghum. The structure and nutrient content of sorghum grains are similar to those of corn. Its crude fiber content is 1.7%, its crude fat content is 3.4%, and its energy content is slightly lower than that of corn. In chicken feed, it contains 12.30 MJ of metabolizable energy per kilogram. The protein content is slightly higher than that of corn at 9.0%, but the protein quality is poor, lacking of lysine, histidine and methionine. Nicotinic acid and biotin content are higher than that of corn, but utilization is lower. For example, the nicotinic acid content of sorghum is 41 mg/kg (while corn is 24 mg/kg), but only

14 mg/kg,利用率不到35%,而维生素 B_2、维生素 B_6 的含量与玉米相当;钙(0.13%)、磷(0.36%)含量高于玉米,但仍然为钙少磷多比例失衡。

2. 饲用价值 高粱对猪和鸡的饲用价值为玉米的95%。由于含单宁较高(黄谷高粱和白谷高粱单宁含量一般为0.2%-0.4%,而褐高粱单宁含量高,为0.6%-3.6%),导致适口性明显变差,并影响养分利用率,从而使动物的生产性能降低。因此,应控制高粱尤其是含单宁高的高粱在畜禽饲粮中的用量,在猪饲粮中以不超过30%为宜,在鸡饲粮中的用量应控制在10%以内。

3. 我国饲料用高粱的质量标准 我国制定了饲料用高粱的国家标准(NY/T 115—1989)。

(五) 稻谷
Rice

在我国,稻谷栽培的历史悠久,种植区域广泛,总产量居世界第一位,占国内粮食生产总量的40%左右,居粮食作物之首。稻谷为带外壳的水稻种子,外壳的比重约占稻谷的25%,脱去外壳的稻谷即为糙米,约占75%,而糙米又由5%-6%的种皮、2%-3%的胚芽和90%-92%的胚乳组成。糙米生产成白米即胚乳,为人类食用,而脱去的种皮、胚芽和少量的胚乳即为细米糠。稻谷、糙米和细米糠均可作为饲料用。

1. 营养价值 稻谷粗蛋白质含量比玉米低,为7.8%,但品质优于玉米,其赖氨酸(0.29%)、蛋氨酸(0.19%)和色氨酸(0.10%)含量均高于玉米。粗纤维含量较高,为8.2%,因而有效能值低于玉米

14 mg/kg can be used, the utilization rate is less than 35%. Its content of Vitamin B_2 and Vitamin B_6 is equivalent to corn. Its content of calcium (0.13%) and phosphate (0.36%) is higher than that of the corn, but the ratio of calcium and phosphorus is still out of balance.

2. Feed value of sorghum. The feeding value of sorghum for pigs and chickens is 95% of that of the corn. Due to the high tannin content (The tannin content of yellow valley sorghum and white valley sorghum is generally 0.2%-0.4%, while that of brown sorghum is 0.6%-3.6%.), the palatability is significantly worsened and the utilization of nutrients is influenced, thereby reducing the animal production performance. Thus, the amount of sorghum, especially sorghum with high tannin content, should be controlled in the diets of livestock and poultry. The amount in pig diet should not exceed 30%, and the amount in chicken diet should be controlled within 10%.

3. Quality standard of feed sorghum in China. National standard of feed sorghum in China is NY/T 115—1989.

China has a long history of planting rice with extensive planting areas, accounting for about 40% of China's total domestic food production, and the total output ranks first in the world. Rice grain is rice seed with husk, of which the husk accounts for 25% of the whole rice grain, and the rice without the husk is unpolished rice, accounting for 75% of the whole rice grain. Unpolished rice is composed of 5%-6% of testa, 2%-3% of germ and 90%-92% of endosperm. Brown rice is produced into white rice, or endosperm for human consumption, while the removed testa, germ and a small amount of endosperm are fine rice bran. Rice, brown rice and rice bran can all be used as feed.

1. Nutritive value of rice. The crude protein content of rice is 7.8%, lower than that of the corn. But the protein quality of rice is better. In it, lysine (0.29%), methionine (0.19%) and tryptophan (0.10%) are higher than that of the corn. Crude fiber content of rice is relatively high at 8.2%. Therefore, the effective energy value is lower than that of

和小麦,对鸡每千克代谢能含量为11.00 MJ。而糙米的粗纤维较低,为2.7%,因而,有效能值比稻谷有所提高,在鸡饲料中,每千克代谢能为14.06 MJ,与玉米相当。稻谷灰分含量为4.6%,且硅酸盐含量高,钙少(0.03%)磷多(0.36%),钙、磷比例不平衡。

2. 饲用价值 稻谷由于粗纤维含量较高,对猪和禽类的饲用价值仅为玉米的85%。糙米可完全取代玉米,有研究表明,在仔猪饲粮中利用熟稻米的采食量、平均日增重、干物质消化率与熟玉米相比得到极显著改善,粗蛋白质消化率也得到显著改善。用糙米饲喂家禽,可能对禽类皮肤、胫和蛋黄着色不利。

3. 我国饲料用稻谷的质量标准 我国制定了饲料用稻谷的国家标准(NY/T 116—1989)。

（六）其他谷实类
Other grains

燕麦和粟在西北、内蒙古、东北有少量种植。这两种籽实均含有纤维颖壳,如燕麦颖壳占整粒籽实重量的23%－35%,粗纤维含量高,约为10.5%,有效能值低,在鸡饲料中,每千克含代谢能10.7 MJ,营养价值与大麦相同。脱壳后则为优良的能量饲料。鸡对燕麦的饲用价值为玉米的70%－80%,一般不宜大量用于猪、鸡饲料,如用量多,将影响采食量,猪、鸡饲粮中用量宜控制在20%以内。

the corn and the wheat, and the metabolizable energy content per kilogram of chicken is 11.00 MJ. The crude fiber of brown rice is low (2.7%). Therefore, the effective energy value is higher than that of the rice. In chicken feed, its metabolizable energy per kilogram is 14.06 MJ, which is equivalent to the corn. The ash content of rice is 4.6%, and the content of silicate is high, calcium is less (0.03%) and phosphorus is more (0.36%), and the calcium-phosphorus ratio is unbalanced.

2. Feeding value of rice. Due to the high crude fiber content of the rice, its feeding value to pigs and poultry is only 85% of the corn. Brown rice can completely replace the corn. Studies have shown the feed intake, average daily gain, and dry matter digestibility of cooked rice in piglet diets are significantly improved compared with the cooked corn, and crude protein digestibility has also been significantly improved. Feeding poultry with brown rice may be harmful to the skin, shin and egg yolk of poultry.

3. Quality standard of feed rice in China. National standard for feed rice formulated in China is NY/T 116—1989.

The oat and the millet are planted in a small amount in the Northwest, Inner Mongolia, and Northeast. Both grains contain fiber husks, e.g., oat glume accounts for 23%－35% of the weight of the whole grain, which has high crude fiber content, about 10.5%, low effective energy value. In chicken feed, it contains 10.7 MJ of metabolizable energy per kilogram, the nutritive value is the same as that of barley. After shelling, the oat is an excellent energy feed. The feeding value of the oat for chicken is 70%－80% of corn. Generally, it is not suitable to be added a lot in swine and chicken feed. Otherwise, it will affect the feed intake. Dosage should be controlled within 20% (in swine and chicken feed).

第二节　谷实类加工副产物饲料
Section 2　By-product feed of the grains

谷实类加工副产物饲料是禾谷类籽实的加工副产物。禾谷类籽实加工成人类食品,如大米、面粉和玉米粉后,剩余的种皮、糊粉层、胚及少量胚乳等则构成了糠麸类饲料。制米的副产物称为糠,制面粉的副产物为麸。糠麸类饲料主要有米糠、麦麸、高粱糖和玉米皮等,不能作为人类食物,主要用作畜禽饲料和酿造业的原料。

The by-product feed of the grains is the by-products of grain processing. Cereal grains can be processed into human food, such as rice, flour and corn meal. The remaining seed coat, aleurone layer, embryo and a small amount of endosperm, etc., constitute the bran feed. The by-product of rice-making is called bran, and the by-product of flour-making is bran. Milling by-product feeds mainly include rice bran, wheat bran, sorghum sugar, corn husk and so on, which cannot be used as human food, and are mainly used as raw materials for feed and brewing.

一、营养特点
Nutritional characteristics

糠麸类饲料的营养价值与其所含的种皮、糊粉层和胚的比例有关,即与加工成粮食的程度有关,如面粉有精粉和普粉、大米有中米和上米之分等。一般是种皮比例越大营养越差,而糊粉层和胚的比例越高,其营养价值也高。

The nutritive value of bran feed is related to the ratio of seed coat, aleurone layer and embryo, that is, related to the processing degree. For example, flour has refined flour and general flour, and rice includes medium rice and top rice. Generally, the thicker the seed coat, the poorer the nutrition, while the higher the ratio of aleurone layer and embryo, the higher its nutritive value.

1. 粗蛋白质　糠麸类饲料的粗蛋白质含量比其原料籽实有提高,粗蛋白质含量在9.3%-15.7%,蛋白质的品质有所改善,尤其是赖氨酸的含量(0.29%-0.74%)比相应原料籽实(0.18%-0.44%)有较大幅度提高,如米糠赖氨酸含量为0.74%,比稻谷提高1.5倍,小麦麸比小麦提高了0.9倍。

1. Crude protein. The crude protein content of bran feed is higher than that of its raw material grains, the crude protein content is between $9.3\%-15.7\%$. The protein quality is improved, especially lysine content ($0.29\%-0.74\%$) grows greatly than corresponding grain material ($0.18\%-0.44\%$). For example, the lysine content of rice bran is 0.74%, 1.5 times higher than that of rice, and the lysine content of wheat bran is 0.9 times higher than that of wheat.

2. 粗纤维　糠麸类饲料由于为籽实的种皮、糊粉层、胚和少量胚乳组成,粗纤维含量比原籽实高,为3.9%-9.1%,有效能值较低,每千克代谢能含量为6.82 MJ-11.21 MJ。

2. Crude fiber. The bran feed is composed of seed coat, aleurone layer, embryo and a small amount of endosperm. Therefore, the crude fiber content is higher than that of the original grain feed, which is $3.9\%-9.1\%$. And the effective energy value of bran feed is lower. The metabolizable energy content per kilogram is 6.82 MJ-11.21 MJ.

3. 粗脂肪　糠麸类饲料中粗脂肪含量高,为3.4%－16.5%(大麦麸),尤其是米糠粗脂肪含量为16.5%,最高的可达22.4%,并且脂肪酸主要为不饱和脂肪酸。因此,在利用时应注意防止氧化酸败,在育肥猪后期利用米糠等时应控制用量,以免产生软猪肉而影响加工品质。

4. 矿物质　糠麸类饲料粗灰分含量较高,通常为2.6%－8.7%,但仍然是钙少(0.02%－0.30%)磷多(0.26%－1.69%),且磷以植酸磷为主,平均约占总磷的85%,钙、磷比例极度不平衡(钙与总磷比为1∶13),将严重影响钙、磷的吸收利用。微量元素锰和铁含量高。

5. 维生素　糠麸类饲料中水溶性维生素B族含量丰富,尤其是维生素B_1含量高,另外未脱脂的糠麸类饲料中维生素E的含量也高。

3. Crude fat. The crude fat content of bran feed is high, ranging from 3.4% to 16.5% (barley bran), especially the crude fat content of rice bran is 16.5%, and the highest can be 22.4%, and the fatty acid is mainly unsaturated fatty acids. Therefore, attention should be paid to prevent oxidative rancidity. In the late grwoing period of finishing pigs, the dosage should be limited when feeding rice bran, so as to avoid producing soft pork to affect the processing quality.

4. Minerals. The crude ash content of bran feeds is high, usually 2.6%－8.7%, but it is still low in calcium (0.02%－0.30%) and high in phosphorus (0.26%－1.69%), and phosphorus is dominated by phytate phosphorus, accounting for about 85% of total phosphorus on average. Calcium-phosphorus ratio is extremely uneven (calcium-phosphorus ratio is 1∶13), which will seriously affect the absorption and utilization of calcium and phosphorus. The trace elements are high in manganese and iron.

5. Vitamin. Bran feed is rich in water-soluble vitamin B group, especially vitamin B_1. And vitamin E is also high in bran feed before defatted.

二、主要加工副产物饲料
Main processing by-product feed

(一) 小麦麸和次粉
Wheat bran and wheat middling

小麦麸和次粉是小麦加工成面粉后的副产品。小麦麸是由小麦的种皮、糊粉层、少量的胚和胚乳组成,即为通常所称的麸皮;而次粉则由糊粉层、胚乳和少量细麸组成。小麦精制过程可得到25%左右的小麦麸,3%－5%次粉,而生产普粉可获得15%左右的小麦麸。有资料表明,我国每年次粉和小麦麸的产量分别为100万吨和1 000万吨以上,主要作为饲料被利用。

1. 营养价值　小麦麸和次粉

Wheat bran and wheat middling are the by-products of wheat processed into flour. Wheat bran is composed of seed coat, aleurone layer, a small amount of embryo and endosperm, which is commonly called bran. Wheat middling is composed of the aleurone layer, endosperm and a small amount of fine bran. The wheat refining process can obtain about 25% of wheat bran and 3%－5% of wheat middling, while the production of common wheat flour can obtain about 15% of wheat bran. Data indicate that, annually wheat middling and wheat bran production can be up to 1 million tons and 10 million tons respectively in China, and they are mainly used as animal feed.

1. The nutritive value. The nutritive value of wheat

的营养价值与加工工艺和出粉率有关，以干物质计，粗纤维含量较小麦高，为8.9%，因而，在能量饲料中其生理有效能含量最低，在鸡饲料中每千克仅含代谢能6.82 MJ，在猪饲料中每千克仅含消化能9.37 MJ；粗蛋白质含量比小麦高，是能量饲料中含量最高的，为15.7%，且赖氨酸含量也较丰富，为0.58%，故蛋白质品质比小麦高；富含B族维生素尤其是维生素B_1和维生素E，但烟酸利用率低，仅为34%，并且缺乏维生素B_{12}；灰分含量为4.9%，矿物质元素含量高，尤其是铁、锰、锌等微量元素含量高，钙少（0.11%）磷多（0.92%），钙、磷比例不平衡（1∶8.4），磷仍然以植酸磷为主，占总磷的74%左右。

2. 饲用价值 小麦麸由于其容重小，常用来调节饲粮的能量浓度；同时，由于小麦麸物理结构疏松、吸水性强、可刺激胃肠道的蠕动，因而具有轻泻作用。小麦麸对所有家畜都是较好的饲料，对奶牛和马属动物用量可大一些；而对鸡其饲用价值为玉米的75%，因此在鸡饲粮中的最大用量应控制在10%-15%；猪饲粮中最大用量以不超过15%为宜。

3. 我国饲料用小麦麸和次粉的质量标准 我国制定了饲料用小麦麸的国家标准（NY/T 119—1989）及饲料用次粉的国家标准（NY/T 216—1992）。

（二）米糠和米糠饼
Rice bran and rice bran meal

米糠是稻谷加工成白米后的副产品。稻谷脱去外壳得到糙米的副产物为砻糠，其营养价值低，不作为

bran and wheat middling is related to processing technology and flour yield. In terms of dry matter, the crude fiber content of them is about 8.9%, higher than that of wheat. Thus, the physiological effective energy of wheat bran and wheat middling is the lowest among the energy feeds. It only contains 6.82 MJ of metabolizable energy per kilogram in chicken feed, and only 9.37 MJ of digestible energy per kilogram in pig feed. The crude protein content is also higher than wheat, which is the highest in the energy feed (15.7%). And lysine content is also rich (0.58%), so their protein quality is higher than wheat. Wheat bran and wheat middling are rich in B vitamins, especially vitamin B_1 and vitamin E, but the utilization rate of niacin is low (only 34%), and lack in vitamin B_{12}. The ash content is 4.9%, and the mineral element content is high, especially the high content of trace elements such as iron, manganese, and zinc. There is little calcium (0.11%) and more phosphorus (0.92%). The ratio of calcium to phosphorus is not balanced (1∶8.4). Phosphorus is still dominated by phytate phosphorus, accounting for about 74% of total phosphorus.

2. The feeding value. Because of its small unit weight, wheat bran is commonly used to adjust the energy level of dietary. At the same time, wheat bran has a loose physical structure, strong water absorption, and can stimulate the peristalsis of gastrointestinal tract, so it has a laxative effect. Wheat bran is a good feed for all livestock. For horses and cows, the dosage can be increased. But for chicken, the feeding value can only reach 75% of that of corn. So, the maximum usage of wheat bran in chicken diets should be controlled at 10%-15%, and the maximum usage in pig feed should not exceed 15%.

3. Quality standards of wheat bran and wheat middling for feed in China. China has formulated a national standard for feed wheat bran (NY/T 119—1989) and wheat middling (NY/T 216—1992).

Rice bran is a by-product of rice processed into white rice. Rice hull is the by-product of rice removing husk into brown rice, which has low nutritive value and is not used as

畜禽饲料用。由糙米生产成白米的副产物即为米糠，包括种皮、糊粉层和少量胚和胚乳，即通常所称的细米糠或洗米糠，约占糙米重量的8%–11%。一般100 kg稻谷可得到75–80 kg糙米，25–20 kg砻糠，或65–70 kg白米，30–35 kg精米和统糠。

生产上通常将砻糠和洗米糠按一定比例混合。而统糠，如二八糠和三七糠等，其营养价值与砻糠所占比例有关。

洗米糠经压榨或有机溶剂浸提脱去脂肪后，被称为米糠饼或米糠粕。目前我国米糠脱脂有80%采用压榨法，20%采用有机溶剂浸提法。

1. 营养价值　米糠的蛋白质、赖氨酸、粗纤维、脂肪含量均高于玉米，尤其是脂肪含量可达16.5%，最高达22.4%，脂肪的脂肪酸组成中以不饱和脂肪酸为主，油酸和亚油酸占79.2%，故米糠的生理有效能值较高，在鸡饲料中每千克含代谢能11.21 MJ，在猪饲料中每千克含消化能12.64 MJ；钙少（0.07%）磷多（1.43%），钙、磷比例严重失调为1:20左右，而且磷以植酸磷为主，占总磷的93%左右，因而利用率较低；富含微量元素铁和锰；维生素B族和维生素E含量高，但缺乏维生素A、维生素C和维生素D。

与米糠相比，脱脂米糠饼或米糠粕的粗脂肪含量降低，尤其米糠粕粗脂肪含量仅为2.0%，另外粗纤维含量增加到7.4%左右，因此有效能值降低，在鸡饲料中每千克含代谢能8.28 MJ–10.17 MJ，而

第六章　能量饲料

feed for livestock and poultry. Rice bran is a by-product of brown ice processed into white rice, including the seed coat, aleurone layer and a small amount of embryo and endosperm, commonly known as fine rice bran or rice skin, accounts for about 8%–11% of the weight of brown rice. Generally, 100 kg rice can produce 75–80 kg brown rice and 25–20 kg rice hull, or 65–70 kg white rice, 30–35 kg polished rice and rice mill by-product.

In production, rice hull and rice skin are usually mixed in a certain proportion. But for rice mill by-product, such as twenty percent of rice hull and eighty percent of rice skin, and thirty percent of rice hull and seventy percent of rice skin, etc., the nutritive value is related to the proportion of rice hull.

The washed rice bran is squeezed or extracted with organic solvents to remove the fat of rice skin, and it is called rice bran cake or rice bran meal. Currently in China, 80% of rice skin is defatted by pressing method, and the rest is extracted by organic solvent extration method.

1. The nutritive value. The content of protein, lysine, crude fiber, fat in rice bran is higher than that of corn, especially the fat content can reach 16.5%, even up to 22.4%. The fatty acid composition of the fat is mainly unsaturated fatty acids, of which oleic acid and linoleic acid account for 79.2%. Therefore, the physiologically effective energy value of rice bran is relatively high. It contains 11.21 MJ of metabolizable energy per kilogram in chicken feed and 12.64 MJ per kilogram of digestible energy in pig feed. It has less calcium (0.07%) and more phosphorus (1.43%), with unbalanced calcium-phosphorus rate at about 1:20, and phosphorus is dominated by phytate phosphorus, accounting for about 93% of the total phosphorus, so the utilization rate of rice bran is low. It is rich in trace elements, such as iron and manganese. B vitamins group and vitamin E content are high, but vitamin A, vitamin C and vitamin D are insufficient.

Compared with rice bran, the crude fat content of defatted rice bran cake or defatted rice meal is reduced, especially the crude fat content of rice bran meal is only 2.0%. And the crude fiber content increases to about 7.4%, so the effective energy is reduced, and metabolizable energy per kilogram in chicken feed is 8.28 MJ–10.17 MJ. The crude

粗蛋白质、氨基酸和微量元素含量有所提高。

2. 饲用价值　应用米糠时应注意消除胰蛋白酶抑制剂、植酸磷、非淀粉多糖（NSP）等抗营养因子的不利影响，应防止脂肪酸氧化酸败而影响适口性和营养价值。米糠是糠麸类饲料中能值最高的，新鲜米糠适口性较好，对猪是较好的能量饲料。新鲜米糠在生长猪饲粮中的用量应控制在12%以内，育肥猪最大用量应控制在15%。米糠对鸡饲用价值较低，为玉米的50%，细米糠饲用价值较高，为玉米的85%-90%，因此，在鸡饲粮中的最大用量应控制在5%-10%。

3. 我国饲料用米糠的质量标准　我国制定了饲料用米糠（NY/T 122—1989）、饲料用米糠饼（NY/T 123—2019）和饲料用米糠粕（NY/T 124—2019）的国家标准。

（三）其他糠麸类
Other brans

主要包括高粱糠、玉米皮和大麦麸等。猪饲料中高粱糠的消化能和鸡饲料中高粱糠的代谢能均比小麦麸高，由于其含单宁较多，适口性差，应注意限量用于猪禽饲粮，用于奶牛、肉牛效果较好。玉米皮是玉米制粉过程中的副产物，粗纤维含量较高，为9.1%，营养价值与麦麸相当。对猪、鸡有效能值低，一般不宜作为仔猪的饲料原料，而可作为奶牛和肉牛的饲料原料。

protein, amino acid and microelement content has increased.

2. The feeding value. When rice bran is fed, the adverse effects of antinutritional factors, such as trypsin inhibitor, phytate phosphorus, and non-starch polysaccharide (NSP), etc. should be eliminated to prevent oxidative rancidity from affecting palatability and nutritive value. Rice bran has the highest energy value of bran feed, and fresh rice bran is palatable. It is a preferable energy feed for pigs. Fresh rice bran in the growing swine dietary should be controlled within 12% or less, and in fattening swine dietary maximum amount should be controlled at 15%. The feeding value of rice bran to chicken is low, in line with 50% of that of corn. The feeding value of fine rice bran is high, in accord with 85%-90% of that of corn, thus, the maximum amount of fine rice bran in chicken diets should be controlled at 5%-10%.

3. Quality standards of feed rice bran, rice bran cake and rice bran meal in China. The national standard of feed rice bran, feed rice bran cake and feed rice bran meal is NY/T 122—1989, NY/T 123—2019 and NY/T 124—2019 respectively.

They mainly include sorghum bran, corn husk and barley bran, etc. The digestible energy of sorghum bran in pig feed and the metabolizable energy of sorghum bran in chicken feed are higher than those of wheat bran. Wheat bran contains more tannin and has poor palatability, so it should be limited for use in swine and poultry diets, and the effect is better when used in dairy cows and beef cattle. Corn husk is a by-product in the process of corn flour milling, with a high crude fiber content of 9.1%, and its nutritive value is equivalent to that of wheat bran. The effective energy value for pigs and chickens is low, and it is generally not suitable as a feed material for piglets, but can be used as a feed material for dairy cows and beef cattle.

第三节 块根、块茎及瓜果类饲料
Section 3 Root, tuber, melon and fruit feed

块根、块茎和瓜果类饲料主要包括甘薯、马铃薯、木薯、萝卜、胡萝卜、饲用甜菜、菊芋和南瓜等,这些在我国均有广泛种植。该类饲料如以鲜样计,容积大、水分含量高,一般为75%-90%,干物质含量低为10%-25%,能值低,每千克含消化能仅为1.8 MJ-4.69 MJ,故被称为稀释的能量饲料;如以干物质计,粗纤维和蛋白质含量低,但无氮浸出物含量高在76%左右,因此,有效能值高,与谷类籽实相当,属能量饲料。

Root tuber, tuber, melon and fruit feed include sweet potato, potato, cassava, radish, carrot, fodder beet, jerusalem artichoke and pumpkin, etc., which are widely planted in China. If the feeds are fresh, they have large volume, generally 75%-90% of high water content, 10%-25% of low dry matter content, low energy value, 1.8 MJ/kg-4.69 MJ/kg of digestible energy, so they are also called diluted energy feed. If calculated by dry matter, the crude fiber content and protein content of the feeds are low, but the nitrogen free extract content is high at about 76%. Therefore, the high effective energy value is as high as that of cereal seeds, which is an energy feed.

一、营养特点
Nutritional characteristics

1. 粗蛋白质

以干物质计,该类饲料粗蛋白质含量为2.5%-10%,蛋白质品质差,赖氨酸(平均为0.15%)和蛋氨酸(平均为0.06%)含量低,满足不了动物的需要,并且非蛋白氮含量较高。

2. 无氮浸出物

无氮浸出物含量高达60%-88%,其中主要为淀粉,粗纤维含量低,约为3%-10%,因而消化率较高,有效能值高,猪饲料中每千克含消化能12.55 MJ-14.46 MJ,含代谢能12.13 MJ-12.55 MJ,与谷类籽实的能值相当,故属于高能饲料之一。

3. 矿物质

以干物质计,粗灰分含量为

1. The crude protein (CP)

The crude protein content of this kind of feed based on dry matter is 2.5%-10%, and the protein quality is poor. The content of lysine (average 0.15%) and methionine (average 0.06%) is low, which can not meet the needs of animals. And the content of non-protein nitrogen is relatively high.

2. Nitrogen free extract (NFE)

The content of nitrogen free extract is as high as 60%-88%, of which starch is mainly used, and the crude fiber content is low about 3%-10%, and thus the digestibility is high. Effective energy value of root tuber, tuber, melon and fruit feed is high. Pig feed contains 12.55 MJ-14.46 MJ of digestible energy per kilogram and 12.13 MJ/kg-12.55 MJ/kg of metabolic energy, which is equivalent to the energy value of grains, so it is also one of the high-energy feeds.

3. Minerals

In terms of dry matter, the crude ash content is 1.9%-

1.9%-3.0%,钙含量比谷类籽实高,且钙多(0.19%-0.27%)磷少(0.02%-0.09%),钙、磷比例与动物需要也不匹配,为3-9.5:1。

4. 维生素

该类饲料中的胡萝卜、黄南瓜和红心甘薯含有较丰富的β-胡萝卜素,其他的则缺乏β-胡萝卜素,该类饲料普遍缺乏B族维生素。

块根、块茎及瓜果类饲料以干物质计,是重要的能量饲料。鲜用时是反刍动物冬季不可缺少的多汁饲料,某些则为补充β-胡萝卜素的重要原料。

3.0%, the calcium content is higher than that of cereal grains. The calcium content (0.19%-0.27%) is high and the phosphorus content (0.02%-0.09%) is low, and the calcium-phosphorus ratio is not appropriate to the needs of the animals, which is 3-9.5:1.

4. Vitamins

This type of feed generally lacks B vitamins. Carrots, yellow squash and red-core sweet potatoes in this type of feed are rich in β-carotene, while they are insufficient in the others. This type of feed lacks B vitamins.

Root tuber, tuber, melon and fruit feed are important energy feeds in terms of dry matter. Fresh feeds are indispensable juicy fodder for ruminants in winter, and some are important raw materials to supplement the β-carotene.

二、常见的块根、块茎类饲料的特点
The characteristics of main root tuber, tuber feed

(一) 甘薯
Sweet potato

甘薯,也称红苕、红薯、地瓜等,是我国四大粮食作物之一,以其易种植、产量高,而广为栽种,年产量约1亿吨。

1. 营养价值 鲜甘薯水分含量高达60%-80%,干物质为20%-40%,其中淀粉约为88%,故鲜甘薯有效能值低,鸡饲料中每千克含代谢能3.31 MJ,猪饲料中每千克含消化能3.85 MJ,红心甘薯中β-胡萝卜素含量高。

甘薯干干物质含量为87%、粗蛋白质含量为4%,其中蛋白质氮占40%-60%,氨基酸氮占20%-30%,另外含有10%-30%的酰胺、氨和生物碱等。必需氨基酸如赖氨酸(0.16%)和蛋氨酸(0.06%)等含量均很低,满足不了动物需要;粗脂肪含量低,为0.8%,粗纤维含量

Sweet potato is also called *camote*, *batata*, etc. It is easy to be planted and has a high yield, so is widely planted, which is one of the four major food crops in China. The annual output is about 100 million tons.

1. The nutritive value. The water content of fresh potatoes is high (60%-80%), and the dry matter content is 20%-40%. Among them, starch is about 88%. Therefore, the effective energy value of fresh sweet potatoes is low. Chicken feed contains 3.31 MJ of metabolizable energy per kilogram, and pig feed contains 3.85 MJ of digestible energy per kilogram. The red-core sweet potato contains high β-carotene.

The dry matter content of sweet potato tuber flakes is 87%, the crude protein content is 4%, among which the protein nitrogen accounts for 40%-60%, the amino acid nitrogen accounts for 20%-30% and amides, ammonia and other alkaloids account for 10%-30%. The content of essential amino acids such as lysine (0.16%) and methionine (0.06%) are very low, which can not meet the needs of animals. The crude fat content is low (only 0.8%), the

为 2.8%，无氮浸出物含量高，为 76.4%，因而，有效能值较高，与玉米相当，鸡饲料中每千克含代谢能 9.79 MJ，猪饲料中每千克含消化能 11.80 MJ；粗灰分含量低为 3.0%，钙含量为 0.19%，高于磷 0.02%；红心甘薯中含有较高的 β-胡萝卜素，但 B 族维生素含量低。

2. 饲用价值　新鲜甘薯多汁、味甜、适口性好，畜禽喜食，尤其对肥育猪和泌乳奶牛具有促进消化、储积脂肪和增加产奶的作用。生喂或熟喂，动物都爱吃。但熟喂可改善能量和干物质的消化率，尤其能显著改善蛋白质的消化率。但如保存不当，甘薯会发芽、腐烂或出现黑斑，黑斑中含有黑芭霉酮毒素，动物误食将会中毒。

甘薯干可用于配合饲料，且由于有黏性利于制粒，但可能增加饲料厂粉尘。甘薯干喂猪的饲用价值仅为玉米的 59%。有研究表明，生甘薯或甘薯粉由于含有一定量的胰蛋白酶抑制剂且易腐烂，直接喂猪会降低猪日增重，增加猪每千克增重的饲料成本。因此，甘薯（风干基础）生喂时，生长猪不宜超过日粮的 15%，肥育猪不宜超过日粮的 20%。

（二）马铃薯
Potato

马铃薯又称为土豆、洋芋、地蛋和山药蛋等。我国种植马铃薯的地域广阔，其主要作为人类食物和生产淀粉用，较少用于饲料。

1. 营养价值　鲜马铃薯的水分含量高为 70%－80%，干物质为 20%－30%，其中粗蛋白质为 1.6%，

crude fiber content is 2.8%, and the nitrogen-free extract content is high (76.4%). Therefore, the effective energy value is high, equivalent to corn, and chicken feed contains 9.79 MJ of metabolizable energy per kilogram, and pig feed contains 11.80 MJ of digestible energy per kilogram. The crude ash content is low (3.0%), the calcium content is 0.19%, 0.02% higher than phosphorus. Red-code sweet potato contains higher β-carotene, but lower B vitamins.

2. The feeding value. Fresh sweet potato is juicy, sweet and palatable, suitable for livestock and poultry, especially for fattening pigs and lactation cows. Sweet potato has the effect of promoting digestion, accumulating fat and increasing milk production. Animals love to eat—whether it is raw or cooked. But cooked feeding can improve the digestibility of energy and dry matter, especially digestibility of protein. However, if stored improperly, sweet potatoes will sprout, rot and dark spots will appear. Dark spots contain nigrotoxin, which can cause animal poisoning if ingested by mistake.

Dried sweet potatoes are available to formulated feed, and it is conducive to granulation because of its viscosity, but it may increase the dust in feed plant. The feeding value of dried sweet potato to pigs is only 59% of that of corn. Researches have shown that raw sweet potato or sweet potato flour contains a certain amount of trypsin inhibitor and is perishable. Feeding pigs directly will reduce the daily weight gain of pigs and increase the feed cost per kilogram of weight gain. Therefore, when feeding raw sweet potatoes, for growing pigs the amount should not exceed 15% in the diet (DM basis), and for finishing pigs, the amount should not exceed 20% in the diet (DM basis).

The potato is also called *common potatoes*, *bastard potatoes*, *Irish potatoes* and *white potatoes*, and it is widely planted in China. Potato mainly acts as human food or production of starch, and is rarely used in the production of feed.

1. The nutritive value. Fresh potatoes have 70%-80% of moisture content, 20%-30% of dry matter, of which the crude protein accounts for 1.6%, crude fat accounts for

粗脂肪0.1%,淀粉含量占80%左右,粗纤维含量为0.7%。马铃薯的营养成分因品种不同而异。

以干物质计,粗蛋白质含量较甘薯高,与玉米接近,蛋白质的品质比小麦好,用于20 kg的仔猪,蛋白质的生物学效价为71%-75%。脂肪含量少,干物质中主要为淀粉,含量高达80%。另外,含有一定量的蔗糖、葡萄糖和果糖,以及一定量的柠檬酸和苹果酸等,而粗纤维含量低,因此有效能值含量高,与玉米相当,鸡饲料中每千克含代谢能13.14 MJ,猪饲料中每千克含消化能14.77 MJ。

2. 饲用价值 马铃薯在畜禽饲粮中的应用效果与喂时的状态有关,对草食家畜,生喂和熟喂的效果基本一致;而用于猪时,熟喂效果优于生喂,生喂消化率低,还易引起腹泻。在利用马铃薯时应防止龙葵素中毒,尤其当发芽或皮变为青绿色时应特别注意。

0.1%, starch accounts for about 80%, and crude fiber accounts for 0.7%. The nutrients of potatoes differ with varieties.

In terms of dry matter, the crude protein content of potato is higher than that of sweet potato, which is close to corn. And the quality of protein is better than that of wheat. When used for 20 kg piglets, the biological value of protein of potato is 71%-75%. The fat content is low, and starch content in dry matter is up to 80%. Furthermore, it contains a certain amount of sucrose, glucose, fructose, citric acid and malic acid, etc. But the crude fiber content is low, so the effective energy value is high, equivalent to corn. Chicken feed contains 13.14 MJ of metabolizable energy per kilogram, and pig feed contains 14.77 MJ of digestible energy per kilogram.

2. The feeding value. The feeding effect of potato in livestock and poultry is related to the condition of feeding. To herbivore, raw feed and cooked feed basically have the same effect. But to swine, cooked feed has better effect, raw feed has low digestibility and can easily cause diarrhea. When feeding potatoes, it is necessary to prevent solanine poisoning, especially when it germinates or the skin turns blue green.

(三) 木薯
Cassava

木薯,又称树薯,为热带多年生灌木,其茎干基部形成块茎的部分可用来生产淀粉和饲喂动物。木薯分为苦味种和甜味种两大类,都含有氢氰酸(HCN),尤其皮中含量多。鲜样中,苦木薯氢氰酸含量为每千克250 mg以上,甜木薯含量较低,仅为苦木薯的1/5,因而不需脱毒即可饲喂。

1. 营养价值 鲜木薯水分含量为70%左右,干物质为25%-30%。以干物质计,木薯干粗蛋白质含量为2.5%,且蛋白质品质差,赖氨酸含量为0.13%,蛋氨酸含量仅为0.05%,满足不了动物的需要;粗脂肪含量低,为0.7%;粗纤

Cassava, also known as *Manihot esculenta*, manioc, *yuca* and *mandioca*, is a tropical perennial shrub. The tuber at the base of its stem can be used to produce starch and feed animals. Cassava is divided into two categories, bitter and sweet, both of which contain hydrocyanic acid (HCN), especially eriched in the skin. Fresh bitter cassava contains HCN 250 mg/kg or more, and fresh sweet cassava only contains 1/5 of that of bitter cassava, and thus can be fed without detoxification.

1. The nutritive value. The water content of fresh cassava is 70% or so, and the dry matter content is 25%-30%. In terms of dry matter, the crude protein content of cassava tuber flake is 2.5%, and the protein quality is poor. The content of lysine is 0.13% and the content of methionine is only 0.05%, which can not meet the needs of animals. The contents of crude fat and crude fiber are low

维含量低,与玉米相近,为 2.5%;而无氮浸出物含量高,为 79.4%;生理有效能值高于甘薯,猪饲料中每千克含消化能 13.10 MJ,鸡饲料中每千克含代谢能 12.38 MJ;灰分含量低,为 1.9%;而钙的含量为 0.27%,比玉米高,但木薯缺乏磷,磷含量是该类饲料中最低的,仅含 0.09%。

2. 饲用价值 木薯皮中含有较高的氢氰酸,尤其是苦木薯,因此一般不宜鲜喂,通常需经脱毒后如晒干或脱皮干燥后以木薯干的形式利用。同时,木薯黏性强,有利于制粒,但用量太高,对制粒又不利。木薯干可作为配制畜禽平衡饲粮的原料。木薯对鸡的饲用价值为玉米的 70%,通常鸡饲粮木薯干的用量以不超过 20% 为宜。

3. 我国饲料用甘薯和木薯的质量标准 我国制定了块根块茎类饲料干制品的国家质量标准。

(0.7% and 2.5% respectively), while the content of nitrogen-free extract is high (79.4%). The physiologically effective energy value of cassava is higher than that of sweet potato, digestible energy per kilogram in pig feed is 13.10 MJ, and metabolizable energy per kilogram in chicken feed is 12.38 MJ. The ash content is as low as 1.9%, and the calcium content is 0.27%, higher than that of corn. But cassava lacks phosphorus, and has the lowest phosphorus content of such feed, only which is 0.09%.

2. The feeding value. Cassava skin, especially bitter cassava's skin contains high hydrocyanic acid, so fresh cassava are generally not suitable to be fed directly. It usually needs to be used in the form of dried cassava after detoxification such as sun drying or peeling and drying. At the same time, cassava has strong stickiness, which is good for granulation, but too high dosage is bad for granulation. Dried cassava can be used as a raw material for producing a balanced diet for livestock and poultry. The feeding value of cassava to chicken is 70% of that of maize. The amount of cassava in chicken feed usually should not exceed 20%.

3. Quality standards for sweet potato and cassava for feed. China has formulated national quality standards for dry feed products of tubers, roots and tubers.

(四) 其他块根、块茎及瓜果类饲料
Other root tuber, tuber, melon and fruit feed

1. 饲用甜菜 饲用甜菜产量高,但干物质含量低,只有 8%-11%,含糖量也只有 5%-11%。以干物质计,饲用甜菜生理有效能含量与高粱和大麦相近,猪饲料中每千克含消化能 13.39 MJ。新鲜甜菜不宜马上用来饲喂家畜,否则易引起腹泻;同时应注意补充钙。

2. 菊芋 又称洋姜和姜不辣,是菊科植物,其地下块茎部分富含果聚糖——菊糖。其营养价值不如马铃薯和甘薯。

1. Fodder beet. The yield of fodder beet is high, but the dry matter content is low, only 8%-11%, and the sugar content is only 5%-11%. In terms of dry matter, the physiologically effective energy content of fodder beet is similar to that of sorghum and barley. Pig feed contains 13.39 MJ of digestible energy per kilogram. Fresh beet should not be used to feed livestock immediately, otherwise it will easily cause diarrhea. Calcium should be supplemented at the same time.

2. Helianthus tuberosus. It is a plant of the Asteraceae family. Its under ground tubers are rich in fructan—inulin. Its nutritive value is not as good as that of potato and sweet potato.

第四节 其他能量饲料
Section 4 Other energy feeds

这类能量饲料包括动物脂肪、植物油、制糖工业的副产品糖蜜、乳制品加工的副产物乳清粉以及人类食品工业的副产物、富含碳水化合物的液态产品如液态小麦淀粉、马铃薯蒸煮脱皮后的副产品和干酪乳清等。

This type of energy feed includes animal fats, vegetable oils, molasses—a by-product of the sugar industry, whey powder—a by-product of dairy processing, and by-products of the human food industry, and liquid products rich in carbohydrates such as liquid wheat starch (LWS), potato steam peeling (PSP), and cheese whey (CW), etc.

一、脂肪
Fat

脂肪作为一种高能量饲料,以单位重量来计算,它含有约 2.25 倍纯淀粉及 3 倍于谷物的能量。在畜禽饲粮中利用,能提高饲粮的能量浓度,并且对提高平均日增重和改善饲料转化率均有积极作用。作为饲料脂肪来源很广,主要包括:植物油(如大豆油、菜籽油、棉籽油、玉米油和椰子油等)、动物脂肪(牛脂、猪油等)和工业生产的粉末油脂(又称为固体脂肪)等三大类。

As a kind of high-energy feed, fat contains about 2.25 times the pure starch and 3 times the energy of cereals calculated by unit weight. For livestock, the utilization of energy feed can improve the energy level of dietary, and has a positive effect on improving average daily gain and feed conversion rate. There are a wide range of feed fat sources, mainly including three categories, i.e. vegetable oils (such as soybean oil, rapeseed oil, cottonseed oil, corn oil, coconut oil, etc.), animal fats (tallow, lard, etc.) and industrially produced powdered oils (also known as solid fat), etc.

(一)营养价值
Nutritive value

1. 动物脂肪 屠宰厂将卫生检验不合格的胴体、内脏和皮脂等经高温处理而得到动物脂肪,它不仅可作为工业原料,而且也是一种高能量饲料。常温条件下,动物脂肪呈固态,加热熔化为液态,因而使用不是很方便。鸡饲料中动物脂肪每千克含代谢能 33.1 MJ。

2. 植物油 植物油在常温下通常呈液态。常见的植物油有大豆油、菜籽油、棉籽油、玉米油、花生油

1. Animal fats. Animal fats are obtained from high-temperature treated of carcass, viscera and sebum from slaughtering plant that did not pass the sanitary inspection, which can not only be used as industrial raw materials, but also a high energy feed. At room temperature, the animal fat is solid, and becomes liquid when heated, so it is not very convenient to be used. Animal fat in chicken feed contains 33.1 MJ of metabolizable energy per kilogram.

2. Vegetable oils. Vegetable oils are usually liquid at room temperature. Common vegetable oils are soybean oil, rapeseed oil, cottonseed oil, corn oil, peanut oil and coco-

和椰子油等。植物油中不饱和脂肪酸含量高,约占油脂的30%-70%,因而植物油中必需脂肪酸亚油酸的含量一般高于动物脂肪,但该类油脂易发生氧化酸败。其有效能值含量略高于动物脂肪,鸡饲料中每千克含代谢能36.8 MJ,使用方便。

3. 工业生产的粉末油脂　粉末油脂又称固体脂肪,它是将动物或植物油脂经特殊处理,即将油脂与酪蛋白、乳糖和淀粉等赋形剂混合加工而成的小颗粒状油脂。其脂肪含量在90%以上,不易被氧化,而且使用方便。

（二）饲用价值
Feeding value

1. 营养作用　在肉鸡饲粮和商品猪饲粮中添加油脂可以提供低成本的高能量饲料,增加饲料中必需脂肪酸的含量,促进营养素的吸收和利用,提高蛋白质和干物质的表观利用率,提高动物生产性能和产品质量,降低生产成本。在蛋鸡饲粮中添加油脂可以提高幼母鸡产蛋的蛋重及产蛋量,改善整个产蛋期的饲料报酬,提高饲料营养素,尤其是淀粉和蛋白质的利用率,在热应激时有助于产蛋期的持续。在反刍动物精料补充料中添加油脂可以提高日粮的能量浓度,它是脂溶性维生素的溶剂,能改变奶和储存油脂的脂肪酸组成,改进饲料的适口性。

2. 非营养作用
（1）降低饲料粉尘:在微量成分预混料中添加1%-2%的油脂可减少粉尘20%-50%,降低微量成分损失30%-50%;在粉料中添加5%的油脂,能减少空气中粉尘达50%;猪采食添加油脂的饲粮,呼吸道发病率呈下降趋势。

nut oil, etc. The content of unsaturated fatty acids in vegetable oil is high, accounting for about 30%-70% of the fat. Therefore, the content of essential fatty acid linoleic acid in vegetable oil is generally higher than that of animal fat, but such fat is prone to oxidative rancidity. Its effective energy content is slightly higher than that of animal fat, chicken feed contains 36.8 MJ of metabolizable energy per kilogram, which is easy to be used.

3. Industrially produced powdered oils. Powdered oils are also called solid fats. They are small granular fats that are processed by special treatment of animal fats or vegetable oils. That is, mix the fats and excipients, casein, lactose and starch. The fat content of powdered oils is above 90%, and is not easy to be oxidized and is convenient to use.

1. Nutritive effect. Adding fats in broiler diets and commercial pig diets can decrease the costs of high energy feed, increase the content of essential fatty acids, promote the absorption and utilization of nutrients, and improve the apparent utilization of protein and dry matter, and thus improve the production performance and product quality of animals, reducing the production costs. Adding fats in laying hens' diets can increase the egg weight and egg production, improve the feed conversion throughout the laying period, and improve the utilization of nutrients, especially starch and protein, which helps to sustain the laying period during heat stress. Adding fats in concentrate supplements of ruminants can increase the energy concentration of the feed. Fat is a solvent for fat-soluble vitamins. It can change the fatty acid composition of milk and stored oils, and improve the palatability of the feed.

2. Non-nutritive effect
（1）Reducing feed dust. Adding 1%-2% fat to premix of the trace components can reduce dust by 20%-50% and the loss of micro amount components by 30%-50%. Adding 5% fat to powder can reduce dust in the air by 50%. Adding fat in swine feed can decrease the incidence of respiratory illness.

（2）润滑作用：油脂在制粒过程中起润滑作用，从而减少饲料机械的磨损，延长使用寿命，降低生产成本。

（3）改善饲料外观，有利于销售。

3. 饲粮中油脂的添加量

（1）猪饲料：在保育期第2—3阶段日粮脂肪含量可添加到8%，但应注意脂肪类型的选择，一般脂肪酸链的长度和饱和度对脂肪消化率影响较大。研究表明，含高比例短链饱和脂肪酸和长链不饱和脂肪酸的脂肪饲料比含长链饱和脂肪酸的脂肪饲料利用率高。仔猪在断奶后第一周，对含不饱和脂肪酸较高的植物油比动物油有更高的消化率。通常，椰子油、乳脂和猪油能被仔猪很好地利用，大豆油和玉米的饲喂效果也很好，牛脂的效果最差。仔猪饲粮中添加量为5%—10%；生长肥育猪为3%—5%；妊娠泌乳母猪为10%—15%。

（2）鸡饲料：肉鸡饲粮适宜添加量为5%—8%，产蛋鸡饲粮中适宜添加量为3%—5%。

（3）奶牛饲料：在奶牛饲粮中脂肪添加量并不是越多越好，添加量过多，特别是未经包被的脂肪，容易导致瘤胃负担过重，降低微生物的活性和粗纤维的消化率，使干物质采食量下降，产奶量随之降低。一般奶牛的适宜添加量为3%—4%，即每头牛每天补充0.45 kg—1.36 kg 即可。但若添加量超过4%，超过部分应由包被的脂肪代替。同时，要注意奶牛的产奶量，若平均泌乳低于25.5 kg，不必添加油脂；泌乳量超过此值，则应添加脂肪以提高日粮能量浓度，满足其生产需要。在奶牛饲粮中添加的脂肪最好是经包被处理的脂肪，添加未经

(2) Lubrication: Fat plays a lubricating role in the process of granulation, thereby reducing the abrasion of feed machinery, prolonging the service life and reducing the production cost.

(3) Improving the appearance of feed, which is conducive to sale.

3. Dosage of fat added in feed

(1) Pig feed. The fat content of pig feed can be added to 8% in the second to third stages of the nursery period, but attention should be paid to the choice of the type of fat. Generally, the length and saturation of fatty acid chains on the general have great influence on the digestibility of fats. Studies have shown that fat feed containing a high proportion of short-chain saturated fatty acids and long-chain unsaturated fatty acids has higher utilization rate than that of containing long-chain saturated fatty acids. In the first week after weaning, piglets have a higher digestibility of vegetable oils with higher unsaturated fatty acids than animal oils. Generally, coconut oil, milk fat and lard can be used well by piglets. Soybean oil and corn are also very effective in feeding, and beef tallow is the worst. The addition amount in the piglet diet is 5%—10%, 3%—5% for growing-finishing pigs and 10%—15% for pregnant-lactation sows.

(2) Chicken feed. The appropriate dietary supplemental level of fat for broilers is 5%—8%, and that for laying hens is 3%—5%.

(3) Dairy cow feed. The feeding effect of fat added in the dairy cow feed is not the more the better. The excessive addition, particularly uncoated fat, can easily lead to overloading of the rumen, reduce the activity of microorganisms and the digestibility of crude fiber, and thus reducing dry matter intake (DMI) and milk production. In general, the appropriate dosage for dairy cows is 3% to 4%, that is, 0.45 kg to 1.36 kg per cow per day. However, if the added amount exceeds 4%, the excess should be replaced by coated fat. At the same time, attention should be paid to the cow milk production. If the average milk output is less than 25.5 kg, no fat is necessary. If the milk output exceeds 25.5 kg/d, fat should be added to increase the energy level of feed to meet their production needs. Coated fat is the best choice of fat that added in dairy feed since uncoated fat has a strong inhibition on rumen. In addition, the type of fat is

包被处理的脂肪,对瘤胃具有强烈的抑制作用;除此以外,还应注意脂肪类型,研究表明,奶牛日粮中添加脂肪,以长链饱和脂肪酸为好。

4. 使用油脂的注意事项

(1) 饲粮添加油脂后,能量浓度增加,应相应增加饲粮其他养分的水平。

(2) 添加油脂时针对不同动物应注意选择脂肪的类型,尤其是饱和与不饱和脂肪酸的比例和中短链脂肪酸的含量。

(3) 脂肪容易氧化酸败,应避免使用已发生氧化酸败的脂肪,在高温高湿季节使用脂肪时应添加抗氧化剂。

(4) 避免使用劣质油脂,如高熔点油脂(椰子油和棉籽油)和含毒素油脂(棉籽油、蓖麻油和桐子油等)以及被二噁英污染的油。

(5) 生长肥育猪后期饲喂含脂肪高的饲粮,会增加背膘厚度,胴体质量变差。在屠宰前14天,在饲粮中添加较高水平的维生素E或维生素C,可延长富含不饱和脂肪酸的猪肉、禽肉的货架期。

(三) 饲用脂肪的品质
Quality of feed fat

饲用脂肪的品质要求

(1) 总脂肪酸:总脂肪酸含量要在90%以上,水分少于1.0%,不溶性杂质小于1.5%。

(2) 动植物脂肪的不可皂化物分别小于1.0%和4.0%。

(3) 稳定性(AOM值):70 h试验中过氧化值小于20 meq/kg。

(4) 农药残毒及聚氯联苯含量符合饲用规范。因此,使用脂肪时,要注意脂肪的品质,特别是不能使用酸败的脂肪饲喂。鱼油质量的行业标准是SC/T 3502—2016。

also important. Researches have shown that long-chain saturated fatty acids are the best fats to be added in dairy feeds.

4. Precautions for using fats

(1) The energy concentration increases after adding fat to the diet, and the level of other nutrients should be increased accordingly.

(2) The type of fat should be selected according to the varieties of animals, especially the ratio of saturated fatty acids and unsaturated fatty acids and the content of medium and short-chain fatty acids.

(3) Fats are prone to be oxidatively rancid. Avoid using fats that have already been oxidizely rancid. Antioxidants should be added when using fats in the hot and humid season.

(4) Avoid using low-grade fats, such as high melting point fats (coconut oil and cottonseed oil) and the toxin-containing oils (cottonseed oil, castor oil, and jatropha oil, etc.) and oil contaminated with dioxins.

(5) Feeding high-fat diets to growing-finishing pigs at the late period will increase the backfat thickness and worsen the carcass. Adding high levels of vitamin E or vitamin C in dietary 14 days before slaughter can extend the shelf life of pork and chicken which are rich in unsaturated fatty acids.

Qualitly stipulations for feed fat

(1) Total fatty acids. The total fatty acid content should be above 90%, the water content should be less than 1.0%, and the insoluble impurities should be less than 1.5%.

(2) The unsaponifiable matter of animal and vegetable fats should be less than 1.0% and 4.0% respectively.

(3) Stability (AOM value). The peroxide value should be less than 20 meq/kg in 70 h test.

(4) Pesticide residues and polychlorinated biphenyls content is in line with specifications for feeding. Therefore, pay attention to the quality of fat when adding, especially not to use rancid fat for feeding fish. The industry standard for fish oil is SC/T 3502—2016.

二、糖蜜
Molasses

糖蜜主要是以甜菜或甘蔗为原料制成的副产品,柑橘类的植物、玉米淀粉、高粱属的植物等也可用作制糖的原料。根据生产糖蜜的具体条件不同,糖蜜可分为原糖蜜、A、B型糖蜜。原糖蜜产品是指未经精制的甜菜或甘蔗液汁,A、B型糖蜜是废糖蜜的中间体,而在畜禽饲粮中运用的糖蜜通常为废糖蜜。

1. 营养价值

糖蜜中的干物质含量为74%-79%(水分含量21%-26%),粗蛋白质含量为2.9%-7.6%,蔗糖以及游离葡萄糖和果糖含量约为50%,在猪饲料中每千克代谢能9.7 MJ。因此,糖蜜是一种低热量、低蛋白质产品,回肠碳水化合物消化率较高,干物质消化率为80%左右,其他成分消化率较低,为40%-70%。常见糖蜜灰分含量较高,为8.0%(8.1%-10.5%),其中钙含量为0.1%-0.81%,总磷含量为0.02%-0.08%,钾含量很高,为2.4%-4.8%。在禽饲粮中添加糖蜜可解决低豆粕饲粮中钾含量较低的问题。

2. 饲用价值

饲粮中添加适量的糖蜜可减少饲料粉尘和改善动物的生产性能。但糖蜜与脂肪或其他液态物质一样,在平衡饲粮中的添加量受到限制(在商品饲粮中用量为1%-9%)。由于添加到立式混合机较困难,添加量不易控制,且用量受到限制,所以一般以不超过5%为宜;同时,添加糖蜜可能会导致在饲粮配制和储存过程中运送机、谷物料斗和料箱被阻塞,装袋困难,产品堆放后变硬等问题。因

Molasses is a by-product of sugar beet or sugar cane. Citrus plants, corn starch, sorghum plants, etc. can also be used as raw materials for molasses production. According to the specific conditions for production, molasses can be divided into raw molasses, A and B type molasses. Raw molasses products refer to unrefined beets or sweet juice. A, B type molasses are the intermediates of waste molasses, which are typically used in livestock and poultry feed.

1. Nutritive value

The dry matter content of molasses is 74%-79% (water content is 21%-26%), the crude protein content is 2.9%-7.6%, and the content of sucrose, free glucose and fructose is about 50%, 9.7 MJ of metabolizable energy per kilogram in pig feed. Thus, molasses is kind of low calorie, low protein products. The digestibility of ileal carbohydrates is relatively high, and the digestibility of dry matter is 80% or so. The digestibility of other components is 40%-70%. Common molasses has a relatively high ash content of 8.0% (8.1%-10.5%), of which the calcium content is 0.1%-0.81%, the total phosphorus content is 0.02%-0.08%, and the potassium content is very high (2.4%-4.8%). The problem of low potassium content in low soybean feed can be dealt with by adding molasses.

2. Feeding value

Adding appropriate molasses in feed can reduce feed dust and improve animal performance. But like fat or other liquid substances, the dosage added in balanced diets is limited (1%-9% is good in commercial feed). Since it's hard to add molasses to vertical mixers, the amount of addition is difficult to control, and the amount is limited, generally not more than 5%. At the same time, adding molasses may cause problems in the process of feed mixing and storage, such as blocking the conveyor, grain hopper and feed bin, and more troubles in bagging, and feed tends to harden when stacked. Therefore, the amount should be appropriately

此，在利用时可适当控制用量，或添加脂肪（可有利于糖蜜混合），或添加适当水将其稀释而降低黏结问题，但会降低混合率。目前，某些猪场采用液态养猪的生产体系，可解决在平衡日粮中（以干物质计）添加40%-50%糖蜜时遇到的问题。研究表明，在猪日粮中糖蜜添加量在30%以内，随添加量增加猪的生产性能得到改善，家禽平衡日粮中以不超过20%为宜，如用量达到24%以上，则使家禽饮水增加，有轻泻作用，而猪饲料添加量超过25%时，也会出现轻泻作用。

controlled while feeding. We can add much fat to facilitate the mixing of molasses, or add appropriate water to dilute feed to reduce the bonding problem, but it will reduce the mixing rate at the same time. At present, some pig farms adopt a liquid pig production system, which can solve the problems encountered when adding 40%-50% (calculated by dry matter) molasses in the balanced diet. Studies have shown that when the amount of molasses added in the pig diet is within 30%, the performance of pigs will be improved with the increase in the amount of addition. If the dosage of molasses is up to 24% or more, the poultry will drink more water, and will have mild diarrhea. And it will also cause mild diarrhea if the dosage of molasses in pig feed is more than 25%.

三、乳清粉
Whey powder

乳清粉是全乳除去乳脂和酪蛋白后干燥而成的乳制品之一。其含有牛奶的大部分水溶性成分，包括乳蛋白、乳糖、水溶性维生素和矿物质元素，是近年来开发的一种在哺乳期动物开食料和早期断奶动物高档平衡饲粮中使用的高档饲料原料。据估计，美国乳清年产量的50%用于配制动物饲粮，其中一半用于养猪业。

Whey powder is one of the dairy products dried after removing milk fat and casein from whole milk, containing most of water-soluble components of milk, i.e. milk protein, lactose, water soluble vitamins and minerals. It's a high-grade feed material developed in recent years used in lactating animal starter's and early weaning animal's high-grade balanced diet. It's estimated that 50% of the annual wheay production in the United States is used for the formulation of animal diets, half of which is used for the pork industry.

（一）营养价值
Nutritive value

乳清粉能提供给幼龄动物非常容易利用的乳糖、优质乳蛋白和生物学效价很高的矿物质和维生素。据测定，乳清粉中粗蛋白质平均含量为12%，蛋白质品质好，赖氨酸含量为1.1%，灰分含量较高，为9.7%，其中钙含量为0.87%，磷含量为0.79%，钙磷比例恰当，但其钠含量较高为2.5%；也有一种乳清粉，其部分乳糖被结晶，粗蛋白含量大约为17%。

Whey powder can provide lactose, high-quality milk protein, minerals and vitamins with high biological value that are very easy to be utilized to young animals. According to experiments, the average crude protein content in whey powder is 12%, the protein quality is good, the lysine content is 1.1%. The ash content is 9.7%, of which the calcium content is 0.87% and the phosphorus content is 0.79%, the ratio of calcium to phosphorus is appropriate, but sodium content is high at 2.5%. There is also a kind of whey powder in which part of lactose is crystallized, and the crude protein content is about 17%.

（二）饲用价值
Feeding value

早期断奶仔猪的日粮中需要简单的碳水化合物，如乳糖，而像淀粉这种复杂的碳水化合物很少被利用。因此，乳清粉或乳糖被广泛地应用在早期断奶仔猪的日粮设计中。乳清粉在2.2-5.0 kg的猪日粮中占15%-30%，在5.0-7.0 kg的猪日粮中占10%-20%，在7.0-11.0 kg的猪日粮中占10%。由于干乳清含盐量很高，在早期断奶仔猪日粮中使用时必须注意食盐的添加量。乳糖也被应用在断奶仔猪的日粮中，而且经常比乳清价格低。通常乳糖的推荐量是：在2.2-5.0 kg仔猪的日粮中占18%-25%，在5.0-7.0 kg仔猪的日粮中占15%-20%，在7.0-11.0 kg仔猪的日粮中占10%。

The diets of early weaned piglets require simple carbohydrates, such as lactose, while complex carbohydrates can seldom be used, such as starch. Therefore, whey powder and lactose are widely used in the feed of early weaned piglets. The dosage of whey powder in daily feed of 2.2-5.0 kg piglets is 15%-30%, in feed of 5.0-7.0 kg piglets, is 10%-20% and in feed of 7.0-11.0 kg piglets, is 10%. Due to the high salt content of dry whey, it is necessary to pay attention to the addition of salt in the early weaned piglet diet. Lactose is also used in the diets of weaned piglets and is usually cheaper than whey. Typically, the recommended addition of lactose in daily diet of 2.2-5.0 kg piglets is 18%-25%, in daily diet of 5.0-7.0 kg piglets, is 15%-20%, and in daily diet of 7.0-11.0 kg piglets, is 10%.

四、其他
Others

富含碳水化合物的液态副产品是来源于人类食品工业的液态副产品。从营养学的角度将其分为三大类，即富含碳水化合物类、富含脂肪类和富含蛋白质类。其中以富含碳水化合物类在动物生产中应用最多，在荷兰养猪业中每年大约利用230万吨来自人类食品工业的液态副产品，其中有70%为富含碳水化合物类。

添加液态副产品于猪饲料中对改善生产性能和动物健康、减少死亡率和降低养猪成本有积极作用。富含碳水化合物的液态副产品，储存时易发酵，产生乳酸和乙酸使pH值降低，从而抑制饲料和消化道中的有害微生物。富含碳水化合物的

The liquid by-products rich in carbohydrates are liquid by-products from the human food industry. It is divided into three categories from the perspective of nutrition: rich in carbohydrates, rich in fat and rich in protein. The liquid by-products rich in carbohydrates are most commonly used in animal production. Approximately 2.3 million tons of liquid by-products from the human food industry are used every year in the pig industries in the Netherlands, of which 70% is rich in carbohydrates.

The addition of liquid by-products to pig feed has a positive effect on improving the production and animal health, reducing the mortality and cost of raising pigs. Liquid by-products rich in carbohydrates tend to ferment when stored, produce lactic acid and acetic acid that decrease pH value, and inhibit harmful microorganisms in feed and digestive tract. Liquid by-products rich in carbohydrates have

液态副产品水分含量高,为74.9%-94.8%;干物质含量相应较低,为25.1%-5.2%。因此,如以鲜样计,很难将其归为能量饲料类,但如以干物质计则为能量饲料。

1. 种类

富含碳水化合物的液态副产品有三种,即液态小麦淀粉、马铃薯蒸煮脱皮后的副产品和干酪乳清。

2. 营养价值

如以干物质计,富含碳水化合物的液态副产品的粗蛋白质含量为11.67%-15.51%,粗脂肪含量为1.30%-2.86%,粗纤维含量为0%-6.39%,无氮浸出物含量为50.22%-65.17%。

3. 饲用价值

在生长猪和肥育猪饲粮中用液态副产品分别取代日粮干物质采食量的35%和55%,试验组和对照组的能量、粗蛋白质、氨基酸等营养水平相同,结果表明利用液态副产品能明显改善生长肥育猪的平均日增重和饲料利用率,但瘦肉率比对照组明显下降。在断奶仔猪日粮中以液态副产品取代22.5%干物质(15% LWS、7.5% CW),分3次喂给,每天每头采食量控制在2.5 L,结果表明液态副产品组的采食量、平均日增重与对照组基本一致,饲料利用率有所改善,但瘦肉率显著降低。

high water content, ranging from 74.9% to 94.8%. The dry matter content is correspondingly low, ranging from 25.1% to 5.2%. Thus, it can not be classified as energy feed by the proportion of fresh sample, but can be classified as the energy feed by the dry matter.

1. Categories

Liquid by-products rich in carbohydrates have three categories, i.e. liquid wheat starch, by-product of peeling potato after cooking, and whey cheese.

2. Nutritive value

In terms of dry matter, the crude protein, crude fat, crude fiber, and nitrogen-free extract content of liquid by-products rich in carbohydrates is 11.67%-15.51%, 1.30%-2.86%, 0%-6.39%, 50.22%-65.17% respectively.

3. Feeding value

There was an experiment that substituted liquid by-products for 35% and 55% of dry matter intake in growing and finishing pig diets, respectively. The test group and the control group had the same nutritional levels of energy, crude protein and amino acids. The results showed that the addition of liquid by-products could significantly improve the average daily gain and feed utilization of growing and finishing pigs, but the lean meat rate was significantly lower than that of the control group. If replace 22.5% of dry matter (15% LWS, 7.5% CW) with liquid by-products in the weaned piglets' diet, and feed them 3 times and control the daily feed intake at 2.5 L per pig, the results showed that the feed intake, average daily gain of the liquid by-product group were basically the same as those of the control group. The feed utilization rate was improved, but the lean meat rate was significantly reduced.

第七章 蛋白质饲料
Chapter 7　Protein feed

蛋白质饲料是指干物质中粗蛋白质含量大于或等于20%、粗纤维含量低于18%的饲料原料。可分为植物性蛋白质饲料、动物性蛋白质饲料、单细胞蛋白质饲料和非蛋白氮饲料。蛋白质饲料是动物配合饲料中重要且比较缺乏的饲料原料之一，应深入开发利用。

Protein feed refers to the feed materials that contain more than 20% of crude protein and less than 18% of crude fiber in dry matter. Protein feed can be grouped into plant protein feed, animal protein feed, single-cell protein feed and non-protein nitrogen feed. Protein feed is one of the most important components of animal compound feeds, which should be further studied and utilized.

第一节　植物性蛋白饲料
Section 1　Plant protein feed

植物性蛋白饲料包括豆类籽实、饼粕类和其他植物性蛋白饲料，是动物生产中使用最多、最常用的蛋白质饲料。该类饲料具有以下共同特点。

1. 蛋白质含量高且质量好。

一般来说，植物性蛋白质饲料粗蛋白质含量在20%-50%，因种类不同所以差异较大。其蛋白质主要由球蛋白和清蛋白组成，且必需氨基酸含量和平衡程度明显优于谷蛋白和醇溶蛋白，因此蛋白质品质高于谷物类蛋白，蛋白质利用率是谷类的3-6倍。但植物性蛋白质的消化率一般仅有80%左右，原因在于大量蛋白质与细胞壁多糖结合，有明显抗蛋白酶水解的作用，存在蛋白酶抑制剂，可阻止蛋白酶消化蛋白质含胱氨酸丰富的清蛋白，

The plant protein feed mainly includes bean seeds, bean cake, bean meal, and other plant proteins. It is one of the mostly used protein feeds in animal production. This kind of feed has the common characteristics as follows.

1. The protein content is high and the quality is good.

Generally, the crude protein content of plant protein feed is between 20%-50%, which varies greatly due to different types. Its protein is mainly consisted of globulin and albumin, and the content and balance of essential amino acids are better than that of gluten and gliadin. Therefore, the protein quality is higher than that of cereals, and the protein utilization rate is 3-6 times that of cereals. However, the digestibility of plant protein is generally only about 80%. The reason is that a large amount of protein is combined with cell wall polysaccharide, which has obvious resistance to protease hydrolysis. The existence of protease inhibitors can prevent protease from digesting the protein which are rich in cystine, possibly resulting in a core residue

可能产生一种核心残基,能对抗蛋白酶的消化。此类饲料经适当加工调制,可提高蛋白质利用率。

2. 粗脂肪含量变化大。

油料籽实脂肪含量在15%-30%,非油料籽实只有1%左右。饼粕类脂肪含量因加工工艺不同差异较大,高的可达10%,低的仅1%左右。

3. 粗纤维含量一般不高,基本上与谷类籽实近似,饼粕类稍高些。

4. 矿物质含量中钙少磷多,且主要是植酸磷。

5. 维生素含量与谷实类相似,B族维生素较为丰富,而维生素A、维生素D较缺乏。

6. 大多数含有一些抗营养因子,影响其饲用价值。

that resists digestion by proteases. Such feed can be properly processed and modulated to improve protein utilization.

2. The crude fat content is changeable.

The concentration of fat in oil seeds is approximately 15% to 30%, while the fat concentration in non-oil seeds is just about 1%. The concentration of fat in cake or meal feed changes from 1% to 10% due to the different processing to them.

3. The concentration of crude fibre is relatively low, which is basically similar to cereal seeds, while the concentration of crude fibre in cakes are slightly higher.

4. It contains less calcium and more phosphorus which is mainly consisted of phytate phosphorus.

5. The concentration of vitamins is similar to that of cereals. B vitamins are more abundant, while vitamin A and vitamin D are relatively deficient.

6. Most of the plant protein feed contains anti-nutritional factors which have negative effects on its feeding values.

一、豆类籽实
Bean seeds

豆类籽实包括大豆、豌豆、蚕豆等,曾作为我国主要役畜的蛋白质饲料,现在一般以食用为主。全脂大豆经加热或膨化后常用在高热能饲料和颗粒饲料中。

Bean seeds, including soybeans, peas and broad beans, etc., were once used as the protein feed for the main draught animals in China. The full-fat soybeans can be used for the production of high-calorie feed and pellet feed after being heated or extruded.

(一) 大豆
Soybean

1. 概述 大豆为双子叶植物纲豆科大豆属一年生草本植物,原产于中国。2020年,全世界大豆总产量3.62亿t,其中巴西产量最高,占全世界总产量的36.6%,第二、三位分别为美国、阿根廷,中国的总产量1 960万t,约占全世界总产量的5.4%。我国大豆主要产区为黑龙江、河北、安徽、江苏、河南及山西等省。按种皮颜色,大豆可分为黄

1. Summary. Soybean [*Glycine max* (*L*) *Merr.*] is a species of dicotyledonous, leguminosae, Glycine L., and it is native to China. In 2020, the total output of soybeans in the world is 362 million tons, of which Brazil has the highest output, accounting for 36.6% of the world's total output, and the output of the United States, Argentina ranks second and third respectively. China's total output is 19.6 million tons, accounting for about 5.4% of the world's total output. The main producing areas of soybeans in China are Heilongjiang, Hebei, Anhui, Jiangsu, Henan, and Shanxi

色大豆、黑色大豆、青色大豆、其他大豆和饲用豆五类。

2. 营养特性　大豆粗蛋白质含量为32%－40%，生大豆中蛋白质多属水溶性蛋白质（90%），加热后即溶于水。氨基酸组成良好，植物蛋白质中普遍缺乏的赖氨酸含量较高（如黄豆和黑豆分别为2.30%和2.18%），但含硫氨基酸含量低。大豆中脂肪含量高，达17%－20%，其中不饱和脂肪酸较多，亚油酸和亚麻酸可占其脂肪总量的55%。脂肪的代谢能约比牛油高出29%，油脂中存在磷脂质，占1.8%－3.2%。大豆中碳水化合物含量不高，无氮浸出物仅26%左右，其中蔗糖占无氮浸出物总量的27%，水苏糖、阿拉伯木聚糖、半乳糖分别占16%、18%和22%。淀粉在大豆中含量甚少，仅0.4%－0.9%；纤维素占18%。阿拉伯木聚糖、半乳聚糖及半乳糖酸结合呈黏性的半纤维素，存在于大豆细胞膜中，有碍消化。矿物质中钾、磷、钠较多，但60%的磷为不能利用的植酸磷，铁含量高。维生素含量略高于谷实类，B族维生素含量较多，而维生素A、维生素D少。

生大豆中存在多种抗营养因子，其中加热可破坏胰蛋白酶抑制因子、血细胞凝集素、抗维生素因子、植酸十二钠、脲酶等；加热无法破坏皂苷、胃肠胀气因子等。此外，大豆中还含有大豆抗原蛋白，该物质能够引起仔猪肠道过敏、损伤，进而导致腹泻。

3. 大豆的加工　生大豆含有许多抗营养因子，直接饲喂会造成动物下痢和生长抑制，饲喂价值较低，因此，生产中一般不直接使用生

provinces. Soybeans can be classified to yellow soybeans, black soybeans, cyan soybeans, forage soybeans, and the other soybeans.

2. Nutritional characteristics. Soybean contains 32%-40% of crude protein. The protein in raw soybean is mostly water-soluble protein (approximately 90%), which can be dissolved in water after heating. The composition of amino acids is good, and the content of lysine, which is generally insufficient in plant protein, is relatively high (2.3% in yellow soybeans and 2.18% in black soybeans), but the content of sulphur-containing amino acids is low. Soybean has a high fat content of 17%-20%, which is mainly unsaturated fatty acids. The linoleic acid and linolenic acid can account for 55% of the total fat. The metabolic energy of fat is about 29% higher than that of tallow, and there are phospholipids in oil, accounting for 1.8%-3.2%. The carbohydrate content in soybean is not high, and the nitrogen-free extract is only about 26%, of which sucrose accounts for 27% of the total nitrogen-free extract, and stachyose, arabinoxylan, and galactose account for 16%, 18%, and 22% respectively. The content of starch in soybean is only 0.4%-0.9%, and the content of cellulose is 18%. Arabinoxylan, galactan and galactonic acid combine to form viscous hemicellulose, which exists in soybean cell membranes and hinders digestion. There are many potassium, phosphorus and sodium in minerals, but 60% of the phosphorus is phytate phosphorus that cannot be used, and the iron content is high. The vitamin content is slightly higher than that of cereals, with more B vitamins, but less vitamin A and vitamin D.

There are a variety of anti-nutritional factors in raw soybean. Among them, those that can be destroyed by heating include trypsin inhibitor, hemagglutinin, anti-vitamin factor, sodium phytate, urease, etc. Heating cannot destroy saponins, flatulence factors, etc. In addition, soybean also contains soybean antigen protein, which can cause intestinal allergy and injury in piglets, which in turn leads to diarrhea.

3. Processing. Raw soybean cannot be used for animal production directly, as it contains lots of antinutritional factors which may cause intestinal allergies, diarrhoea and growth inhibition in young animals. The most common way

大豆。加工大豆最常见的方法是加热。热处理使生大豆中不耐热的抗营养因子变性和失活,提高了大豆的利用率和饲料价值。大豆的加工方法主要有:焙炒、干式挤压法、湿式挤压法等方法。

4. 加工大豆的品质判定　大豆的加工方法不同,饲用价值也不同。用炒制方法加工的大豆有烤豆的特殊香气,但是容易出现加热不均的情况,过生和过熟都会影响饲用价值。挤压法产品脂肪消化率高,代谢能较高。大豆湿法膨化处理有利于破坏全脂大豆的抗原活性。在加热过程中,大豆蛋白质中一些不耐热的氨基酸会分解,主要是还原糖与氨基酸之间发生的美拉德反应,导致大多数氨基酸尤其是赖氨酸利用率下降,降低了大豆的营养价值。因此,大豆的适宜加工非常重要。经过加工生产的全脂大豆与生大豆相比,水分含量较低,其他养分含量相对较高,抗营养因子含量低,使用安全,因此在畜禽饲粮中应用较多。

5. 原料标准　我国国家标准《饲料用大豆》(GB/T 20416—2006)中规定:饲料用大豆应色泽、气味正常,杂质含量≤1.0%,生霉粒≤2.0%,水分≤13%。以不完善粒、粗蛋白质为定级指标,分为3级。

6. 大豆的饲用价值　生大豆饲喂畜禽可导致腹泻和生产性能下降,但经过加热处理的全脂大豆对各种畜禽均有良好的饲喂效果。在肉鸡饲粮中,因加工全脂大豆密度低,故用于粉状料宜在10%以下,否则会影响采食量,造成增重降低,而颗粒料则无此虑。以颗粒料饲喂时,添加全脂大豆与豆粕+豆油相比,可提高肉鸡的代谢能和对饲料脂肪的消化率。饲喂全脂大豆的肉

to process soybean is heating. Heating processing denatures and inactivates the heat-labile anti-nutritional factors in raw soybean, increasing its utilisation rate and feeding value. The processing like stir-frying, dry pressing and wet pressing are the main methods used to pre-treat raw soybean.

4. The characterization of processed soybean. The feeding value of soybean is different due to different processing methods. For example, the soybeans processed with rostering have special flavour of aroma and roasted beans. However, the uneven heating may cause some soybeans to be overcooked while some beans are still raw, which have negative effects on feeding value. The soybeans processed with pressing have higher digestibility for fat and higher metabolic energy. Moreover, wet pressing is beneficial to inhibit the activity of antigen in full-fat soybeans. During the processing of heating, the thermolabile amino acids resolves, and the reducing sugar reacts with amino acids (Maillard reaction) to reduce the utilization rate of most of the amino acids, especially the lysine, and so reduce the nutrition value of soybeans. The processed full-fat soybeans contain less water and antinutritional factors, and higher content of other nutrients. Thus, processing of raw soybeans has been widely applied to animal feed production.

5. Material standard. According to the national standard GB/T 20416—2006 Soybean for Feed, the color and smell of feed soybeans should be normal. The content of impurity, moldy kernel, and water should be lower than 1.0%, 2.0% and 13% respectively. Feed soybeans can be classified into 3 different levels according to the content of crude protein and unsound kernel.

6. Feeding values. Feeding raw soybeans to livestock and poultry may cause diarrhea and decrease production performance, but the full-fat soybeans treated with heating have good feeding effects on various livestock and poultry. The density of soybeans is low, thus the addition of full-fat soybeans should be lower than 10% in powered broiler feed, while there is no limitation for the addition of soybeans to pellet feed. When feeding pellets, adding full-fat soybeans can better improve the metabolic energy and digestibility of feed fats by broilers, compared with soybean pulp plus soybean oil. Moreover, the broiler fed with full-fat soybeans

鸡,胴体和脂肪组织中亚油酸和ω-3脂肪酸含量较高。全脂大豆可替代蛋鸡饲粮中的豆粕,提高蛋重并显著提高亚麻酸和亚油酸的含量。

在猪饲粮中应用生大豆作为唯一蛋白质来源,对其生产性能有很大影响,与大豆饼相比,饲喂大豆粕会增加仔猪的腹泻率、降低生长育肥猪的增重和饲料转化率、降低母猪的生产性能。全脂大豆因其蛋白质和能量水平都较高,是配置仔猪全价料的理想原料。

牛饲料中使用生大豆,但不宜超过精饲料的50%,且需配合胡萝卜素含量高的粗饲料,否则会降低维生素A的利用率,造成牛奶中维生素A含量剧减。生大豆也不宜与尿素同用。肉牛饲料中生大豆用量过大会影响采食量,且有软脂倾向。经热处理的全脂大豆适口性比生大豆好,并且具有较高的瘤胃蛋白质降解率。

全脂大豆无论从化学组成上还是从养分的利用效率上,都是饲用价值较高的反刍动物和水产动物饲料原料,在鱼饲料中可以部分代替鱼粉,较豆粕营养价值更高。全脂大豆中的高油脂含量减少了鱼类自身能量的分解,这对冷水鱼很有意义。全脂大豆中含有的亚油酸和亚麻酸,为如鲑、鲤、罗非等鱼类提供了所必需的大量不饱和脂肪酸。

(二) 豌豆
Pea

1. 概述 豌豆又名毕豆、小寒豆、淮豆、麦豆。豌豆适应性强,喜阴凉而湿润的气候。我国豌豆种植面积约为200万hm^2,总产量约150万t,在四川省种植最多。豌豆除作

has higher content of linoleic acid, ω-3 unsaturated fatty acid in the carcass and fat. The replacement of soybean pulp with processed full-fat soybeans in layer feed can increase the weight, and the content of linolenic acid and linoleic acid in eggs.

The application of raw soybean as the only protein source in the diet of pigs has a great influence on their production performance. Compared with soybean cake, feeding soybean meal can increase diarrhea rate of piglets, decrease weight gain and feed conversion rate of growing-finishing pigs, and decrease reproduction performance for sows. Full fat soybean is an ideal raw material for full-price feeding of piglets due to its high protein and energy levels.

The dosage of raw soybeans cannot exceed 50% in concentrated feed of cattle. The dairy cattle should be fed with the fodder which contains high concentration of carotene when fed with raw soybeans. Otherwise, the absorption rate of vitamin A will decrease, causing the low content in milk in dairy cows. Moreover, raw soybeans should not used to fed cattle with urea at the same time. Otherwise, the adding of too much raw soybeans has negative effects on feed intake and the cattle tend to produce soft lipid. The full-fat soybeans processed with heating has better palatability, and higher digestibility of ruminal protein, compared with the raw soybeans.

The nutritional composition and utilization rate of full-fat soybeans are suitable for making feeds for ruminants and aquatic livestock. Feeding with full-fat soybean which contains high concentration of oil can reduce the metabolism of energy in cold water fish. It has better feeding values than soybean meal and can be used to replace fish meal in fish feed. Moreover, the high content of linoleic acid, linolenic acid in full-fat soybeans provides plenty unsaturated fatty acids to fish like trout, carp and tilapia mossambica.

1. Summary. Pea, which is also known as Bi beans, Xiaohan beans, Huai beans, and Wheat beans. It has good adaptability and prefers cool and humid climates. The planting area of peas in China is about 2 million hm^2, and the total output is about 1.5 million tons, and is planted most in

食用外,也供作饲料。

2. 营养特性　豌豆风干物中粗蛋白质含量约为24%,蛋白质中含有丰富的赖氨酸,而其他必需氨基酸含量都较低,特别是含硫氨基酸与色氨酸。豌豆中粗纤维含量约为7%,粗脂肪约为2%,各种微量矿物质元素含量都偏低。另外,豌豆中也含有胰蛋白酶抑制因子、外源植物凝集素、致胃肠胀气因子,故不宜生喂。

3. 原料标准　我国农业行业标准《饲料用豌豆》(NY/T 136—1989)中规定:以粗蛋白质、粗纤维、粗灰分为质量控制指标,按含量可分为3级。

4. 饲用价值　豌豆在鸡饲料中可使用10%-20%。粉碎后在生长育肥猪饲粮中可用至12%,但需补充蛋氨酸,对生长及屠体品质无不良影响;种猪亦可用之,煮熟后可用到20%-30%。奶牛、肉牛和肉羊精饲料中可分别用20%、12%和25%以下。

Sichuan province. Pea is used as fodder as well as food.

2. Nutritional characteristics. The content of crude protein in air-dry peas is about 24%. The protein is rich in lysine, while the content of other essential amino acids is low, especially sulfur-containing amino acid and tryptophan. Peas contain approximate 7% of crude fibre, 2% of crude fat, and little minerals. In addition, peas also contain tryptase inhibitors, exogenous plant lectins, and gastrointestinal flatulence factors, so they should not be used to feed animals directly.

3. Material standards. According to the national agricultural standard NY/T 136—1989 Pea for Feed, feed peas can be classified into 3 different levels according to the content of crude protein, crude fiber and crude ash.

4. Feeding values. The recommended adding range of peas in chicken feed changes from 10% to 20%. The crushed peas can be added to feed (12%) for fattening pigs, but need to be used with the addition of methionine. The crushed peas can also be used for feed in breeding pigs, and the addition range can be raised to 20%-30% when peas are cooked. The adding proportions of peas in dairy cattle, beef and mutton sheep should be lower than 20%, 12%, and 25% respectively.

二、饼粕类
Cake and meal

(一) 大豆饼粕
Soybean cake and meal

1. 概述　大豆饼粕是以大豆为原料取油后的副产物。由于制油工艺不同,通常将压榨法取油后所得的饼状产品称为大豆饼,而用溶剂法提取油脂后得到的产品称为豆粕。2020年,全国饲料用豆粕约为7 000万t。

大豆饼粕的加工方法有四种:液压压榨法、旋压压榨法、溶剂浸提法和预压后浸提法。压榨法的取油工艺主要分为两个过程:第一过程

1. Summary. Soybean cake and meal is the by-products of soybeans after extracting oil. According to the different oil production processes, the cake-like products obtained after oilextraction by pressing method are called soybean cakes, while the products obtained after oil extraction by solvent method are called soybean meals. The national soybean meals for feed was 70 million tons in 2020.

There are four processing methods for soybean meal: hydraulic pressing, spinning pressing, solvent extraction and pre-pressing plus solvent extraction. The pressing extraction methods mainly have 2 steps: the first step is the cleaning,

为油料的清洗、破碎、软化、轧胚,油料温度保持在60-80 ℃;第二过程为料胚蒸炒(100-125 ℃)后再加机械压力,使油分离出来。浸提法取油工艺为:利用有机溶剂在55-65 ℃下浸泡料胚,提取油脂后将其残余烘干(105-120 ℃)而得到粕。浸提法比压榨法可多取油4%-5%,且粕中残脂少,易保存,为目前生产中主要采用的工艺。

大豆饼粕是使用最广泛、用量最多的植物性蛋白质原料,世界各国普遍使用,一般其他饼粕类的使用与否以及使用量都以与大豆饼粕的比价来决定。

2. 营养特性 大豆饼粕粗蛋白质含量高,一般在40%-50%,必需氨基酸含量高,且组成合理。赖氨酸含量在饼粕类中最高,为2.4%-2.8%,赖氨酸与精氨酸比约100:130,比例适当,若配合大量玉米和少量的鱼粉,很适合家禽氨基酸营养需求。异亮氨酸含量是饼粕饲料中最高者,约1.8%。色氨酸、苏氨酸含量也很高,与谷食类饲料配合可起到互补作用。蛋氨酸含量不足,在以玉米—大豆饼粕为主的饲粮中,一般要额外添加蛋氨酸才能满足畜禽营养需求。粗纤维含量较低,主要来自大豆皮。无氮浸出物主要是蔗糖、棉籽糖、水苏糖和多糖类,淀粉含量低。胡萝卜素、核黄素和硫胺素含量少,烟酸和泛酸含量较多,胆碱含量丰富,维生素E在高脂肪残渣和贮存期短的饼粕中含量较高。矿物质中钙少磷多,磷多为植酸磷(约占61%),硒含量低。此外,大豆饼粕色泽佳,风味好,加工适当的大豆饼粕仅含有微量抗营养因子,不易变质,使用上无用量限制。

crushing, softening, and rolling of raw materials at 60-80 ℃; the second step is that the oil materials are fried with hot steam at 100-125 ℃, and processed with mechanical pressure to separate oil. The processes of solvent extraction method are as follows: soak the material embryos in the organic solvent at 55-65 ℃. The soybean meal is the dried residue after oil extraction (105-120 ℃). The solvent extraction method produces 4%-5% of more oil than the pressing method, and produces residue which contains less fat and is easy to store. It is the main process currently used in production.

Soybean cake and meal is the most widely used plant-based protein material around the world. Generally, the use of other types of cake and meal is determined by the price ratio of soybean cake and meal.

2. Nutritional characteristics. Soybean cake and meal has a high content and a reasonable ratio of crude protein (40%-50%), and a high content of essential amino acids. The content of lysine is the highest among the cakes and meals (2.4%-2.8%). The ratio of lysine to arginine is appropriate (approximately 100:130). Soybean cake and meal is very suitable for chicken feed if combined with a large amount of corn and a small amount of fish meal. The content of isoleucine in soybean cake and meal is the highest (1.8%). Moreover, the content of tryptophan and threonine in soybean cake and meal is also high. The mixing use of soybean cake and meal, and cereal feed can play a complementary role. Extra methionine needs to be added to the feed based on corn-soybean cake and meal, to meet the nutritional needs of livestock and poultry. The content of crude fibre which are mainly from soybean bulls is low. The nitrogen free extracts are mainly consisted of sucrose, raffinose, stachyose, polysaccharides, with low content of starch. The content of carotene, riboflavin and thiamine is low, the content of niacin and pantothenic acid is high, the content of choline is rich, and the content of vitamin E in cakes with high fat residue and short storage is high. The minerals in soybean cake and meal are consisted of high content of phosphorus which is mainly phytate phosphorus (about 61%) but low content of calcium and selenium. In addition, soybean meal has good color and good flavor, and

大豆粕和大豆饼相比,脂肪含量较低,而蛋白含量较高,且质量稳定,是目前市场上的主要产品。去皮豆粕是用去皮大豆榨油后得到的,近10多年来此产品有所增加,其与大豆粕相比,粗纤维含量低,一般在3.3%以下,蛋白质含量高,一般为48%-50%,营养价值较高。

3. 原料标准 我国国家标准《饲料用大豆粕》(GB/T 19546—2004)中规定,饲料用大豆粕呈浅黄褐色或浅黄色不规则的碎片状或粗粉状,色泽一致,无发酵、霉变、结块、虫蛀及异味异臭;不得掺入饲料用大豆粕以外的物质,若加入抗氧化剂、防霉剂、抗结块剂等添加剂,需要具体说明加入的品种和数量。

4. 质量评定方法 大豆饼粕是大豆加工后的产品,也含有一些抗营养因子。评定大豆饼粕质量的指标主要为抗胰蛋白酶活性、脲酶活性、水溶性氮指数、维生素 B_1 含量、蛋白质溶解度等。许多研究结果表明,当大豆饼粕中的脲酶活性在0.03-0.4范围时,饲喂效果最佳,而对家禽来说,在0.02-0.2时最佳。大豆饼粕最适宜的水溶性氮指数值标准不一,一般在15%-30%。生产实践中也可根据饼粕的颜色来判定大豆饼粕加热程度适宜与否。正常加热时为黄褐色;加热不足或未加热时,颜色较浅或呈灰白色;加热过度则呈暗褐色。

5. 饲用价值 大豆饼粕适当

soybean meal that is properly processed only contains trace anti-nutritional factors, which is not easy to deteriorate, and there is no limit on the addition ratios to animal feeds.

Soybean meal has lower fat content and higher protein content, and the quality is stable compared with soybean cake. It is the main product on the market at present. The dehulled soybean meal is the meal obtained after the oil extraction with dehulled soybeans. The yield of dehulled soybean meal has increased in the past 10 years. The dehulled soybean meal has low crude fibre content (lower than 3.3%), higher protein content (48%-50%), and high nutritive value compared with soybean meal.

3. Material standard. According to the national standard GB/T 19546—2004 Soybean Meal for Feed, soybean meal in good quality is in irregular fragments or coarse power, with consistent colour (light yellow or light yellow brown). It should be free of fermentation, mildew, agglomeration, moth-eaten, and peculiar smell. Substances other than soybean meal for feed shall not be added. The additives like antioxidants, anti-fungal agents, anti-caking agents should be specified with the types and quantities.

4. Quality evaluation method. Soybean cake and meal is the processed product of soybeans, and it also contains some antinutritional factors. The main indicators for evaluating the quality of soybean cake and meal are the activities of antitrypsin and urease, water-soluble nitrogen index, vitamin B_1 content, and protein solubility, etc. Many studies have shown that the feeding values of soybean cake and meal are best when the activity of urease changes within 0.03-0.4. For poultry feed, the feeding values of soybean cake and meal are best when the activity of urease changes within 0.02-0.2. There is no recommendation for water-soluble nitrogen index in soybean cake and meal. The index usually ranges from 15% to 30%. The colour of soybean cake and meal can also be used to determine whether the heating treatment is appropriate in production practice. The colour of soybean cake and meal is yellowish brown when they are heated normally. The colour is lighter or off-white when they are underheated or not heated. The colour is dark brown when they are overheated.

5. Feeding value. The soybean cake and meal which is

加热后添加蛋氨酸，即为养鸡最好的蛋白质来源，适用于任何阶段的家禽，对幼雏效果更好，其他饼粕原料不及大豆饼粕。加热不足的大豆饼粕能引起家禽胰脏肿大，发育受阻，添加蛋氨酸也无法改善，对雏鸡影响尤甚，这种影响随着动物年龄的增长而下降。适当处理后的大豆饼粕也是猪的优质蛋白质饲料，适用于任何种类、任何阶段的猪。因大豆饼粕中粗纤维含量较多，多糖和低聚糖类含量较高，幼畜体内无相应消化酶，故在人工代乳料中，应对大豆饼粕的用量加以限制，以小于10%为宜，否则易引起下痢。乳猪宜饲喂熟化的脱皮大豆粕，育肥猪无用量限制。以豆粕为唯一的蛋白源的饲粮中，添加蛋氨酸可提高猪生产性能。

大豆饼粕也是奶牛、肉牛的优质蛋白质原料，各阶段牛饲料中均可使用，适口性好，长期饲喂也不会厌食。采食过多会有软便现象，但不会下痢。牛可有效利用未经加热处理的大豆饼粕，含油脂较多的豆饼对奶牛有催乳效果，在人工代乳料和开食料中应加以限制。羊、马也可使用，效果优于生大豆。目前，我国大豆饼粕用于反刍动物的量逐渐下降，代之以 NPN 和其他粗纤维含量高而价格低的饼粕类。

在水产动物中，草食鱼及杂食鱼对大豆粕中的蛋白质的利用率很好，可达90%左右，大豆粕能够取代部分鱼粉作为蛋白质的主要来源。肉食鱼对大豆粕利用率低，应尽量少用。

properly heated and added with methionine can be used as the best protein sources for raising chickens. It is suitable for poultry at any stage and has a better effect on young chicks. Others are not as good as soybean meal. Feeding soybean cake and meal which is processed with insufficient heating may cause pancreas swelling and growth retardation in poultry, especially in young poultry. This effect decrease with the age of the animals. The soybean cake and meal which is processed properly is also a high-quality protein feed for pigs. It is suitable for pigs of any type and at any stage. Soybean cake and meal contains high concentration of crude fibre, polysaccharides and oligosaccharides which cannot be digested by young animals. The addition amount of soybean cake and meal should be limited to less than 10% in artificial milk replacers. Otherwise, it is easy to induce diarrhea. Suckling pigs should be fed with mature dehulled soybean cake and meal, and there is no restriction on the dosage for fattening pigs. In feed which uses soybean cake and meal as the only protein source, the addition of methionine can improve growth performance in pigs.

Soybean cake and meal is also a high-quality protein source for dairy cows and beef cattle. It has good palatability and can be used for feed at all stages. It will not be anorexia after long-term feeding. Excessive intake of soybean cake and meal will induce soft faeces, but will not cause diarrhea. The soybean cake which contains more fat has a lacto-promoting effect on dairy cows, so it needs to be restricted in artificial milk replacers and starters. Soybean cake and meal is better than raw soybean when used for feed in sheep and horse. At present, the amount of soybean cake and meal used in ruminants is gradually decreasing, and replaced by non-protein nitrogen and other meals which contain high crude fibre content.

In aquatic animals, herbivorous and omnivorous fish have a good utilization of protein in soybean cake and meal (up to about 90%). Soybean cake and meal can be used to replace part of fish meal as the main source of protein. Carnivorous fish has a low utilization of soybean cake and should be used as little as possible.

(二) 菜籽饼粕
Rapeseed cake and meal

1. 概述 油菜是我国主要的油料作物,我国油菜籽总产量约为1 210万t,主产区在四川、湖北、湖南、江苏、浙江、安徽等省。除作种用外,95%用于生产食用油。菜籽饼是以油菜籽为原料,经压榨法取油后所得的饼状产品;菜籽粕是以油菜籽为原料,经预榨—溶剂浸提或直接浸提法取油、脱溶剂、干燥后得到的产品。

油菜品种可分为四大类:甘蓝型、白菜型、芥菜型和其他型油菜,不同品种含油量和有毒物质含量不同。为解决菜籽的毒性问题,改善菜籽饼粕的饲用价值,植物育种学家一直致力于"双低"油菜品种的培育。第一个"双低"油菜品种于1974年在加拿大诞生,之后许多"双低"油菜品种陆续育种成功并得到迅速推广,到20世纪80年代末,欧洲一些国家基本实现了油菜品种双低化。我国双低油菜品种的研究始于20世纪70年代中后期,但发展迅速,已选育出多个双低油菜品种,推广面积也迅速扩大,达到目前油菜种植总面积的30%以上。榨油工艺主要为动力螺旋压榨法和预压浸提法,目前生产上以后者占主导地位。

2. 营养特性 菜籽饼粕含有较高的粗蛋白质,为34%-38%。氨基酸组成平衡,含硫氨基酸较多,精氨酸含量低,精氨酸与赖氨酸的比例适宜,是一种氨基酸平衡良好的饲料。粗纤维含量较高,为12%-13%,有效能较低。碳水化合物为不易消化的淀粉,且含有8%的戊聚糖,雏鸡不能利用。菜籽外壳几

1. Summary. Rape (*Brassica compestris* L. var. *oleifera* DC.) is one of the main oil-bearing crops in China. The total output of rapeseed is about 12.1 million tons, and the main planting areas are Sichuan, Hubei, Hunan, Jiangsu, Zhejiang and Anhui provinces. Except for seed use, 95% is used to produce edible extraction. The cake-like products which are obtained after oil extraction by pressing method are rapeseed cakes. Rapeseed meal (rape seed meal) is a product obtained from rapeseed as raw material, which is obtained by pre-pressing-solvent extraction or direct extraction to extract oil, desolventize and dry.

There are 4 main categories of rapes which are wild cabbage, Chinese cabbage, senvy and other types. The content of oil and toxics in different rapeseeds is different. To solve the toxic issues and improve the feeding value of rapeseed, plant breeders have been committed to the cultivation of "double-low" rapeseed varieties. The first breed of double-low rape was cultivated in Canada in 1974. After that, many breeds of double-low rape were bred in some European countries by the end of the 1980s. The research on double-low rape breeding in our country started in the middle and late 1970s, and developed rapidly. Several double-low breeds of rape have been bred, and widely planted in China. The promotion area of double-low breed occupies more than 30% in the total planting area of rape. The rapeseed oil extraction process is mainly the power screw pressing method and the pre-pressing leaching method, and the latter is the dominant method in the current production.

2. Nutritional characteristics. Rapeseed cake and meal has high content of crude protein (34%-38%), and the composition of amino acids is balanced. The content of sulfur-containing amino acids is high, the content of arginine is low, and the ratio of lysine to arginine is appropriate. Rapeseed cake and meal has higher crude fibre content (12%-13%), and lower effective energy. Carbohydrates in rapeseed cake and meal are mainly consisted of non-digestible starches and 8% of pentosans which cannot be used by

乎无利用价值,是造成菜籽饼粕代谢能值低的根本原因。矿物质中钙、磷含量均高,但大部分为植酸磷,富含铁、锰、锌、硒,尤其硒含量远高于豆粕。维生素中胆碱、叶酸、烟酸、核黄素、硫胺素含量均比豆粕高,但胆碱与芥子碱呈结合状态,不易被肠道吸收。

此外,菜籽饼粕含有硫葡萄糖苷、芥子碱、植酸、单宁等抗营养因子,影响其适口性。"双低"菜籽饼粕与普通菜籽饼粕相比,粗蛋白质、粗纤维、粗灰分、钙、磷等常规成分含量差异不大,但有效能略高,赖氨酸含量和消化率显著高于普通菜籽饼粕,蛋氨酸、精氨酸含量略高。

3. 原料标准 我国国家标准《饲料用菜籽粕》(GB/T 23736—2009)中规定:饲料用菜籽粕呈褐色、黄褐色或金黄色小碎片或粗粉状,有时夹杂小颗粒,色泽均匀一致,无虫蛀、霉变、结块及异味臭味;不得掺入菜籽粕以外的物质(非蛋白氮等),若加入抗氧化剂、防霉剂、抗结块剂等添加剂时,要具体说明加入的品种和数量。

4. 饲用价值 菜籽饼粕因含有多种抗营养因子,饲喂价值明显低于大豆粕,并可引起动物甲状腺肿大,采食量下降,生产性能下降。国内外培育的"双低"(低芥酸和低硫葡萄糖苷)品种已在我国部分地区推广,并获得较好效果。

在鸡配合饲料中,菜籽饼粕应限量使用,一般幼雏饲粮中应避免使用。品质优良的菜籽饼粕,肉鸡成长后期可用至10%-15%,但为防止肉鸡风味变劣,用量宜低于

chicks. Rapeseed shells are almost useless, which is the root cause of the low metabolizable energy value of rapeseed cake and meal. For the minerals in rapeseed cake and meal, the content of calcium and phosphorus which are mainly consisted of phytate phosphorus is high, the content of iron, manganese, zinc, and selenium is high, especially the content of selenium is much higher than that of soybean meal. For vitamins, the content of choline, folic acid, niacin, riboflavin, and thiamine is higher than that of soybean meal. However, choline and sinapine are in combined state and are not easily absorbed by the intestine.

Moreover, rapeseed cake and meal contains antinutritional factors like glucosinolate, sinapine, phytic acid and tannin, which have negative effects on its palatability. Compared with normal rapeseed cake and meal, the double-low breeds contain similar content of crude protein, crude fibre, crude ash, calcium, phosphorus, and higher content of lysine, methionine, and arginine. The digestibility rate and effective energy are also higher in double-low breeds.

3. Material standard. According to the national standard GB/T 23736—2009 Rapeseed Cake and Meal for Feed, rapeseed meal in good quality is in irregular fragments or coarse power, with consistent colour (brown, yellow-brown or golden yellow). It should be free of mildew, agglomeration, moth-eaten, and peculiar smell. Substances other than rapeseed meal (non-protein nitrogen, etc.) shall not be added. The using of additives like antioxidants, anti-fungal agents, anti-caking agents should be specified with the types and quantities.

4. Feeding value. Since rapeseed cake and meal contains a variety of antinutritional factors, its feeding value is significantly lower than that of soybean meal, and can cause animal goiters, decrease feed intake and production performance. "Double-low" (low erucic acid and low glucosinolate) varieties cultivated at home and abroad have been promoted in some areas of my country and achieved good results.

In chicken compound feed, the rapeseed cake and meal should be used in a limited amount, and it should be restricted in young chick feed. The rapeseed cake and meal with good quality can be used up to 10%-15% in broiler's later growing stage, but in order to prevent the risk of dete-

10%。蛋鸡、种鸡可用至8%,超过12%即可引起蛋重和孵化率下降。褐壳蛋鸡采食多时,鸡蛋有鱼腥味,应谨慎使用。

毒物含量高的菜籽饼粕,对猪的适口性差,在饲料中过量使用会引起不良反应,如甲状腺肿大、肝肾肿大等,生长率下降30%以上,显著影响母猪繁殖性能。因此,菜籽饼粕在肉猪饲粮中的用量应限制在5%以下,母猪饲粮中低于3%。经处理后的菜籽饼粕或"双低"品种的菜籽饼粕在肉猪饲粮中可用至15%,但为防止软脂现象,用量应低于10%;对种猪用至12%对繁殖性能并无不良影响,奶牛精饲料中使用10%以下,产奶量及乳脂率正常。低毒品种菜籽饼粕饲养效果明显优于普通品种,可提高使用量,奶牛最高可用至25%。

riorating flavour, the dosage should be lower than 10%. In feeds of layer and breeding hens, rapeseed cake and meal can be used up to 8%, and more than 12% can lead to drop of egg weight and hatchability. When brown-shell hens eat for a long time, the eggs have a fishy smell and should be used with caution.

Rapeseed meal with high toxic content has poor palatability to pigs, and excessive use of rapeseed cake and meal will cause adverse reactions like goiter, liver and kidney enlargement, drop in growth rate (more than 30%), and reduction in reproductive performance. Thus, the use of rapeseed cake and meal should be limited to less than 5% in the diets of pigs, and less than 3% in the diets of sows. The cake and meal of rapeseed or the double-low rapeseed can be used up to 15% in the diets of pigs, while the dosage should be less than 10% to reduce risk of producing soft fat. The use of processed rapeseed cake and meal to 12% in breeding pigs feed has no negative effects on reproduction. Less than 10% of processed rapeseed cake and meal in dairy cows feed has no negative effects on milk production and milk fat rate. The feeding effect of low-toxic varieties of rapeseed meal is significantly better than that of ordinary varieties, which can increase the amount of use, and the maximum dosage for dairy cows is 25%.

(三) 棉籽饼粕
Cottonseed cake and meal

1. 概述 以棉籽为原料,经脱壳或部分脱壳后再以压榨法取油后所得的饼状产品称为棉籽饼;以棉籽为原料,经脱壳或部分脱壳后再以预榨—溶剂浸提或直接浸提法取油、脱溶剂、干燥后得到的产品称为棉籽粕。2018年,我国棉籽饼粕年产量为800万t,主产区在新疆、河南、山东等省(自治区)。

2. 营养特性 棉籽饼粕粗纤维含量主要取决于制油过程中棉籽的脱壳程度。国产棉籽饼粕粗纤维含量较高,达13%以上,有效能值低于大豆饼粕。而脱壳完全的棉仁饼粕粗纤维含量约为12%,代谢能

1. Summary. The cake-like products obtained after oil extraction with cotton seed by pressing method are cottonseed cakes. Using cottonseed as raw material, after shelling or partial shelling, the product obtained by pre-pressing-solvent extraction or direct extraction for oil extraction, desolvation and drying is called cottonseed meal. The total production of cottonseed cake and meal in China was 8 million tons in 2018, and the main producing areas are Xinjiang, Henan and Shandong, etc.

2. Nutritional characteristics. The content of crude fibre in cottonseed cake and meal mainly depends on the dehulling degree in oil extraction. The domestic cottonseed cake and meal has higher crude fibre content (13%), and lower effective energy than soybean cake and meal. The fully dehulled cotton kernel cake and meal has 12% of crude

水平较高。棉籽饼粕粗蛋白质含量较高，达 34% 以上，棉仁饼粕粗蛋白质可达 44%－46%。氨基酸中赖氨酸含量较低，仅相当于大豆饼粕的 50%－60%，蛋氨酸亦低，精氨酸含量较高，赖氨酸与精氨酸之比在 100∶270 以上。矿物质中钙少磷多，其中 71% 左右为植酸磷，含硒少。维生素 B_1 含量较多，维生素 A、维生素 D 含量少。棉籽饼粕中的抗营养因子主要为棉酚、环丙烯脂肪酸、单宁和植酸。

3. 原料标准　我国国家标准《饲料用棉籽粕》(GB/T 21264—2007) 中规定：饲料用棉籽粕呈黄褐色或金黄色小碎片或粗粉状，有时夹杂小颗粒，色泽均匀一致，无发酵、霉变、结块及异味异臭。

4. 饲用价值　对鸡的饲用价值主要取决于游离棉酚和粗纤维的含量。含壳多的棉籽饼粕，粗纤维含量高，热能低，应避免在肉鸡饲料中使用。用量依游离棉酚含量而定，含量在 0.05% 以下的棉籽饼粕，在肉鸡饲粮中可占 10%－20%，在产蛋鸡饲料中可占 5%－15%。未经脱毒处理的饼粕，饲粮中用量一般不得超过 5%。蛋鸡饲粮中棉酚含量在 200 mg/kg 以下，不影响产蛋率；若要防止"桃红蛋"，应限制在 50 mg/kg 以下。亚铁盐的添加可增加鸡对棉酚的耐受力。鉴于棉籽饼粕中的环丙烯脂肪酸对动物的不良影响，棉籽饼粕中的脂肪含量越低越安全。

品质好的棉籽饼粕是猪良好的蛋白质饲料原料，可代替猪饲料中 50% 的大豆饼粕，但需补充赖氨酸、钙、磷和胡萝卜素等。品质差的棉籽饼粕用量过大则会影响适口性，

fiber content and high metabolizable energy. Cottonseed cake and meal has high content of crude protein (more than 34%). Cotton kernel cake and meal has 44%－46% of crude protein content. The content of lysine is relatively low, which is only 50%－60% of that in soybean cake and meal. The content of methionine is also low, but the content of arginine is high, the ratio of lysine to arginine is above 100∶270. For the minerals in cottonseed cake and meal, the content of calcium and selenium is low, and the content of phosphorus is high, but 71% of phosphorus are phytates. The content of vitamin B_1 is high, but the content of vitamin A and D is low. The main antinutritional factors in cottonseed cake and meal are gossypol, cyclopropene fatty acid, tannin and phytic acid.

3. Material standard. According to the national standard GB/T 21264—2007 Cottonseed Cake and Meal for Feed, cottonseed meal in good quality is in irregular fragments or coarse power, with consistent colour (yellow-brown or golden yellow). It should be free of fermentation, mildew, agglomeration, and peculiar smell.

4. Feeding value. The feeding value for chickens mainly depends on the content of free gossypol and crude fibre. Cottonseed meal with high husk content, high crude fiber content and low heat energy should be avoided in broilers. Usually, the cottonseed cake and meal containing lower than 0.05% of free gossypol can be added to 10%－20% in broiler feed and 5%–15% in layer feed. The cottonseed meal without detoxified processing cannot be added over 5% in animal feeds. The content of 200 mg/kg free gossypol in layer feed has no effects on egg production rate, while the free gossypol content should be lower than 50 mg/kg in layer feed to avoid producing pink eggs. The supplementation of ferrous salt can increase the tolerance of chickens to gossypol. Due to the adverse effects of cyclopropylene fatty acids in cottonseed cake and meal, the lower the fat content, the safer it is.

Cottonseed cake and meal with high quality is a good protein feed material for pigs, which can be used to replace 50% of soybean cake and meal in pig feed, but needs to be added with lysine, calcium, phosphorus, and carotene. Excessive use of low-quality cottonseed cake and meal has

并有中毒可能。棉仁饼粕是猪良好的色氨酸来源,但其蛋氨酸含量低,一般乳猪、仔猪不宜使用。游离棉酚含量低于0.05%的棉籽饼粕,在肉猪饲粮中可用至10%-20%,母猪可用至3%-5%;若游离棉酚高于0.05%,则应谨慎使用。

棉籽饼粕对于反刍动物不存在中毒问题,是反刍家畜良好的蛋白质来源。奶牛饲料中适当添加棉籽饼粕可提高乳脂率,若用量超过精饲料的50%则影响适口性,同时乳脂变硬。棉籽饼粕属便秘性饲料原料,必须搭配芝麻饼粕等软便性饲料原料使用,一般用量以精饲料中占20%-35%为宜。喂幼牛时,以低于精饲料的20%为宜,且需搭配含胡萝卜素高的优质粗饲料。肉牛可以棉籽粕为主要的蛋白质饲料,但应供应优质粗饲料,再补充胡萝卜素和钙,方能获得良好的增重效果,一般在精饲料中可占30%-40%。

游离棉酚可使种用动物尤其是雄性动物生殖细胞发生障碍,因此种用雄性动物应禁止使用棉籽饼粕,雌性种畜应尽量少用。

(四)花生仁饼粕
Peanut cake and meal

1. 概述 花生仁饼粕是花生脱壳后,经机械压榨或溶剂浸提油后的副产物,以中国、印度、英国产量最高。2018年,我国花生饼粕产量为1 000万t,主产区为山东省,产量约占全国的四分之一,其次为河南、河北、江苏、四川等地,是当地畜禽的重要蛋白质来源。

花生脱壳取油的工艺可分为浸

negative effects on palatability and may be toxic. Cotton kernel cake and meal is a good source of tryptophan for pigs, but its methionine content is low, so it is generally not suitable for suckling pigs and piglets. Cottonseed cake and meal with free gossypol content less than 0.05% can be used up to 10%-20% in pig diets, and up to 3%-5% for sows. It should be careful to use cottonseed cake and meal with free gossypol content higher than 0.05%.

Cottonseed cake and meal has no poisoning problem if used for ruminants, and it can be used as a good source of protein for ruminants. Appropriate addition of cottonseed meal to dairy cow feed can increase the milk fat rate. If the addition amount exceeds 50% of the concentrate, the palatability will be affected, and the milk fat will become hard. Cottonseed cake and meal is a feeding material which induces constipation, and it must be used with soft stool feed materials like sesame meal. The general addition dosage is 20%-35% of the concentrated feed. When feeding young cattle, it is advisable to use less than 20% of the concentrated feed, and it needs to be matched with high-quality roughage containing high carotene. Beef cattle can use cottonseed cake and meal as the main protein feed, but it should be supplied with high-quality roughage, extra carotene and calcium to obtain good weight gain. Generally, it can be added up to 30%-40% in concentrated feed.

Free gossypol can cause obstacles to the germ cells of breeding animals, especially male animals. Therefore, the use of cottonseed cake and meal in male breeding animals should be prohibited, and it should be used as little as possible in female breeding animals.

1. Summary. Peanut cake and meal is the byproduct of oil extraction with dehulled peanuts (*Arachis hypogaea* L.) by pressing method or solvent method. The main countries which produce peanut cake and meal are China, India, and England. The total production of peanut cake and meal in China was 10 million tons in 2018, and the main producing areas were Shandong (accounting for 25% of the national output), Henan, Hebei, Jiangsu and Sichuan provinces.

There are four main processing methods for oil extrac-

提法、机械压榨法、预压浸提法和土法夯榨法。用机械压榨法和土法夯榨法榨油后的副产品为花生饼；用浸提法和预压浸提法榨油后的副产物为花生粕。

2. 营养特性 花生仁饼的粗蛋白质含量约为44%，花生仁粕的粗蛋白含量约为47%，但是63%为不溶于水的球蛋白，可溶于水的白蛋白仅占7%。氨基酸组成不平衡，赖氨酸、蛋氨酸含量偏低，精氨酸含量在所有植物性饲料中最高，赖氨酸与精氨酸之比在100∶380以上，饲喂家畜时适于和精氨酸含量低的菜籽饼粕、血粉等配合使用。在无鱼粉的玉米-豆粕型饲料中，产蛋鸡的第一、二、三、四位限制性氨基酸依次是蛋氨酸、亮氨酸（肉仔鸡为赖氨酸）、精氨酸、色氨酸。蛋氨酸、赖氨酸、色氨酸有合成品，可直接添加补充，精氨酸无合成品，可用花生仁饼粕补其不足。花生仁饼粕的有效能值在饼粕类饲料中最高。无氮浸出物中大多为淀粉、糖分和戊聚糖。残余脂肪熔点低，脂肪酸以油酸为主，不饱和脂肪酸占53%-78%。钙、磷含量低，B族维生素较丰富，尤其烟酸含量高，约174 mg/kg，核黄素含量低，胆碱含量为1 500-2 000 mg/kg。

花生仁饼粕中含有少量胰蛋白酶抑制因子且极易感染黄曲霉而产生黄曲霉毒素，容易引起动物中毒。《饲料卫生标准》规定花生饼粕黄曲霉毒素 B_1 含量不得大于0.05 mg/kg。

3. 原料标准 我国行业标准《饲料用花生（仁）饼粕》（NY/T 132—1989，NY/T 133—1989）规定：感官要求为花生饼为小瓦块状或圆扁块状，花生粕为黄褐色或浅

tion with peanut (kernel): extration method, mechanical press method, pre-pressure extraction method and native compaction method. Peanut cake is the byproduct from oil extraction with mechanical press and native compaction method. Peanut meal is the byproduct from oil extraction with extraction and pre-pressure extraction method.

2. Nutritional characteristics. Peanut cake has about 44% of crude protein content, and peanut meal has 47% of crude protein content. However, 63% of crude protein is water-insoluble globulin and water-soluble albumin only accounts for 7%. The composition of amino acids is unbalanced. The content of lysine and methionine is low. The content of arginine is the highest among all plant feeds. The ratio of lysine to arginine is above 100∶380. It is suitable for feeding livestock with rapeseed meal and blood meal with low acid content. In the corn-soybean meal feed without fish meal, the first, second, third and fourth restrictive amino acids of laying hens are methionine, leucine (lysine for broilers), arginine and tryptophan in order. Methionine, lysine, and tryptophan have synthetic products, which can be added directly. Arginine has no synthetic products, and peanut meal can be used to make up for its deficiency. The effective energy value of peanut meal is the highest in meal feed. Most of the nitrogen free extracts are starch, sugar and pentosan. The residual fat has a low melting point, the fatty acid is mainly oleic acid, and unsaturated fatty acids account for 53%-78%. The content of calcium and phosphorus is low, and the B vitamins are rich, especially the high content of niacin (about 174 mg/kg), and the content of riboflavin is low, and the content of choline is 1,500-2,000 mg/kg.

Peanut cake and meal contains a small amount of trypsin inhibitor, and it is very easy to be infected with Aspergillus flavus, which produces aflatoxin and causes animal poisoning. The Feed Hygienic Standard stipulates that the content of aflatoxin B_1 in peanut meal shall not exceed 0.05 mg/kg.

3. Material standards. According to the national industrial standards NY/T 132—1989 and NY/T 133—1989 Peanut Cake and Meal for Feed, peanut cake in good quality should be in small tiles or round flat blocks, and peanut meal in good quality should be in irregular fragments or

褐色不规则碎屑状,色泽新鲜一致;无发霉、变质、结块及异味异臭;水分含量不得超过12.0%。

4. 饲用价值　为避免黄曲霉素中毒,幼雏应避免使用花生仁饼粕,可用于成鸡。因其适口性较好,可提高鸡的食欲,育成期可用到6%,产蛋鸡可用到9%,若补充赖氨酸、蛋氨酸或鱼粉、豆粕、血粉配合使用,效果更好。在鸡饲粮中添加蛋氨酸、硒、胡萝卜素、维生素或提高饲粮蛋白质水平,都可以降低黄曲霉素的毒性。

花生仁饼粕是猪的优良蛋白饲料,适口性极好。因赖氨酸、蛋氨酸含量低,饲喂价值不及大豆饼粕。在满足育肥猪的赖氨酸、蛋氨酸需要的前提下,花生饼粕可代替全部大豆饼粕,但为了防止下痢和体脂变软,用量宜低于10%。为防止黄曲霉毒素中毒,最好不用于哺乳仔猪。花生仁饼粕对奶牛、肉牛的饲用价值与大豆饼粕相当。花生仁饼粕有通便作用,采食过多容易导致便软。经高温处理的花生仁饼粕,其蛋白质溶解度下降,可提高过瘤胃蛋白量,提高氮沉积量。

(五) 亚麻仁饼粕
Linseed cake and meal

1. 概述　亚麻仁饼是亚麻籽经脱油后的副产物。亚麻在我国西北、华北地区种植较多,主要产区在内蒙古、吉林、河北北部、宁夏、甘肃等沿长城一带,是当地食用油的主要来源。我国年产亚麻仁饼粕超过30万t,以甘肃最多。因亚麻籽中常混有芸芥籽及菜籽等,所以部分地区又将亚麻称为胡麻。

coarse power, with consistent colour (yellow-brown or light brown). It should be free of mildew, deterioration, agglomeration, and peculiar smell. The content of water should be lower than 12.0%.

4. Feeding values. In order to avoid aflatoxin poisoning, young chicks should not be fed with peanut (kernel) meal which can be used for adult chickens. Because of its good palatability, it can increase the appetite of chickens. It can be used up to 6% during the breeding period and 9% for laying hens. If supplemented with lysine, methionine or fish meal, soybean meal, blood meal, the effect will be better. Adding methionine, selenium, carotene, vitamins or increasing dietary protein levels in chicken diets can reduce the toxicity of aflatoxin.

Peanut cake and meal is an excellent protein feed for pigs with excellent palatability. Since the content of lysine and methionine is low, the feeding value is not as good as soybean cake and meal. On the premise of meeting the needs of lysine and methionine for fattening pigs, peanut cake and meal can replace all soybean cake and meal, but in order to prevent diarrhea and softening of body fat, the dosage should be less than 10%. In order to prevent aflatoxicosis, suckling piglets should not be fed with. The feeding value of peanut cake and meal for dairy and beef cattle is equivalent to that of soybean cake and meal. Peanut cake and meal has a laxative effect, and excessive intake can easily lead to soft stool. Peanut cake and meal treated at high temperature is reduced in protein solubility, which can increase the amount of rumen protein and the amount of nitrogen deposition.

1. Summary. Linseed cake and meal is the by-product of oil extraction with flaxseed (*Linum usitatissimum L.*). Flax is planted more in Northwest and North China. The main producing areas are Inner Mongolia, Jilin, north of Hebei, Ningxia, Gansu and other areas along the Great Wall. Linseed is the main source of local cooking oil. The annual production of linseed cake and meal in our country exceeds 300,000 tons, and Gansu produces the most. Because linseeds are often mixed with mustard seeds and

2. 营养特性 亚麻仁饼粕粗蛋白含量一般为32%-36%,氨基酸组成不平衡,赖氨酸、蛋氨酸含量低,富含色氨酸,精氨酸含量高,赖氨酸与精氨酸之比为100∶250。因此,饲料中使用亚麻仁饼粕时,应添加赖氨酸或搭配赖氨酸含量较高的饲料。粗纤维含量高,为8%-10%,热能值较低,代谢能仅为9.0 MJ/kg。脂肪中亚麻酸含量可达30%-58%。钙、磷含量较高,硒含量丰富,是优良的天然硒源之一。维生素中胡萝卜素、维生素D含量少,但B族维生素含量丰富。

亚麻仁饼粕中的抗营养因子包括生氰糖苷、亚麻籽胶、抗维生素B_6因子。生氰糖苷因在自身所含亚麻酶作用下生成氢氰酸而有毒。亚麻籽胶含量为3%-10%,它是一种可溶性糖,主要成分为乙醛糖酸,完全不能被单胃动物消化利用,故饲粮中亚麻仁饼粕用量过多时会影响畜禽食欲。

3. 原料标准 我国行业标准《饲料用亚麻仁饼粕》(NY/T 216—1992,NY/T 217—1992)规定:亚麻仁饼为褐色大圆饼,厚片或粗粉状,亚麻仁粕为浅褐色或深黄色不规则碎块状或粗粉状;具油香味,无发霉、变质、结块及异味;水分含量不得超过12.0%;不得掺入其他物质。

4. 饲用价值 鸡饲料中应尽量少用或不用亚麻仁饼粕。用量达5%时,即可造成食欲下降,生长受阻,用量达10%即有死亡现象。亚麻仁饼粕经水浸、高压蒸汽处理后添加可缓解其毒害。

用作猪饲料,其饲用价值高于

rapeseeds, therefore, linum is also called flax in some areas.

2. Nutritional characteristics: The crude protein content of linseed cake and meal is generally 32%-36%, the amino acid composition is not balanced, the content of lysine and methionine is low, the content of tryptophan and arginine is high, and the ratio of lysine to arginine is 100∶250. Therefore, when linseed cake and meal is used in feed, lysine should be added or matched with feed with higher lysine content. The crude fiber content is high (8%-10%), the thermal energy value is low, and the metabolizable energy is only 9.0 MJ/kg. The content of linolenic acid in fat can reach 30%-58%. The calcium and phosphorus content is high, and the selenium content is rich. It is one of the fine natural selenium sources. The content of carotene and vitamin D is low, but the content of B vitamins is rich.

The antinutritional factors in linseed cake and meal include cyanogenic glycosides, linseed gum, and anti-vitamin B_6 factors. Cyanogenic glycosides are toxic due to the formation of hydrocyanic acid under the action of the linsease contained in themselves. The content of linseed gum is 3%-10%. It is a kind of soluble sugar. Its main component is glyoxylic acid. It cannot be digested and utilized by monogastric animals. Therefore, excessive consumption of linseed cake and meal in the diet will affect the appetite of livestock and poultry.

3. Material standards. According to the national industrial standards NY/T 216—1992 and NY/T 217—1992 Linseed Cake and Meal for Feed, the linseed cake is a large brown round cake, thick slice or coarse powder. The good-quality linseed meal is light brown or dark yellow irregularly broken piece or coarse powder, with oil fragrance. It should be free of mildew, deterioration, agglomeration, and peculiar smell. The content of water should be lower than 12.0%. It shall not be mixed with other substances.

4. Feeding value. Linseed cake and meal should be used as little as possible or not used in chicken feed. When the amount reaches 5%, it can cause loss of appetite and hinder growth. The amount of 10% will cause death. Linseed cake and meal can be added after water immersion and high-pressure steam treatment to relieve its toxicity.

Used as pig feed, its feeding value is higher than that

芝麻饼粕和花生仁饼粕,但氨基酸不平衡,需同其他优质蛋白质饲料配合使用,补充其缺乏的氨基酸后,可获得良好的饲养效果。育肥猪饲料中可用至8%,不会影响生长和饲料效率,但过多使用则会造成腹脂熔点下降,引起软脂现象,并导致维生素B_6缺乏症。在母猪饲料中适当添加可预防便秘。

亚麻仁饼粕是反刍动物良好的蛋白质来源,适口性好,牛羊饲料中均可使用,可提高肉牛育肥效果,提高奶牛产奶量,且饲喂亚麻仁饼粕可使反刍动物被毛光泽改善。犊牛、羔羊、成年牛羊及种用牛羊均可使用,并可作为唯一蛋白质来源,配合其他蛋白质饲料使用可预防乳脂变软。

of sesame meal and peanut meal, but the amino acid is not balanced. It needs to be used in conjunction with other high-quality protein feeds. After supplementing amino acids it lacks, good feeding effects can be obtained. It can be used up to 8% in fattening pig feed, which will not affect growth and feed efficiency, but excessive use will cause the melting point of abdominal fat to drop, resulting in soft fat, and vitamin B_6 deficiency. Appropriate addition to sow feed can prevent constipation.

Linseed cake and meal is a good source of protein for ruminants and has good palatability. It can be used in cattle and sheep feed. It can improve the fattening effect of beef cattle and increase the milk production of dairy cows. Moreover, feeding linseed cake and meal can improve ruminant coats' shine. It can be used for calves, lambs, adult cattle and sheep and breeding cattle and sheep, and can be used as the only protein source. It can be used with other protein feeds to prevent milk fat from softening.

(六) 其他植物饼粕
Other plant cake and meal

1. 棕榈仁饼 棕榈仁饼为棕榈果实提油后的副产品。粗蛋白质含量低,仅14%–19%,属于粗饲料。赖氨酸、蛋氨酸及色氨酸均缺乏,脂肪酸属于饱和脂肪酸。肉鸡和仔猪不宜使用,生长育肥猪可用15%以下,奶牛使用可提高奶酪质量,但大量使用会影响适口性。

2. 椰子粕 椰子粕又称椰子干粕,是将椰子胚乳部分干燥为椰子干,再提油后所得的副产品。椰子粕呈淡褐色或褐色,纤维含量高而有效能值低;氨基酸组成欠佳,缺乏赖氨酸、蛋氨酸及组氨酸,但精氨酸含量高;所含脂肪属饱和脂肪酸,B族维生素含量高。椰子粕易滋生霉菌而产生毒素。椰子粕一般不用于肉鸡饲粮,因适口性不好,所以雏鸡和仔猪饲粮应尽量少用,在其他鸡料中用量宜在5%以下;在育肥猪饲粮中用量在10%以下。椰子

1. Palm kernel cake is a by-product of plam fruit oil extraction. The crude protein content is low, only 14%–19%, which belongs to roughage. Lysine, methionine and tryptophan are all deficient, and fatty acids are saturated fatty acids. It is not suitable for broilers and piglets. It can be used in growing and fattening pigs by less than 15% in feed. It can improve the quality of cheese in dairy cows, but heavy use will affect palatability.

2. Coconut meal, also known as dried coconut meal, is a by-product obtained by partially drying coconut endosperm into dried coconut and then extracting the oil. Coconut meal is light brown or brown, with high fiber content and low effective energy value; poor amino acid composition, lack of lysine, methionine and histidine, but high arginine content. Fats are saturated fatty acids, with high content of vitamin B. Coconut meal is prone to mold and produce toxins. Coconut meal is generally not used in broiler diets. Due to its poor palatability, it should be used as little as possible in chick and piglet diets. The dosage in other chicken diets should be less than 5%. In fattening pig diets, the dosage should be less than 10%. Coconut meal is a

粕为反刍动物良好的蛋白质来源，但为防止便秘，精饲料中使用20%以下为宜。

3. 苏籽饼 苏籽饼为紫苏种子榨油后的产品。其粗蛋白质含量为35%-38%，赖氨酸含量高；粗纤维含量高，有效能值低；含有抗营养因子：单宁和植酸。机榨法取油的苏籽饼含紫苏特有的臭味，适口性不好，对猪、鸡应注意限量饲喂。

good source of protein for ruminants. But to prevent constipation, it is advisable to use less than 20% in concentrated feed.

3. Perilla oil cake is the product of perilla seeds after oil extraction. Its crude protein content is 35%-38%, its lysine content is high, its crude fiber content is high, and its effective energy value is low. It contains antinutritional factors: tannin and phytic acid. The perilla oil cake obtained from the machine-extruded oil contains the peculiar smell of perilla, and the palatability is not good. For pigs and chickens, the amount should be limited if used in feed.

三、其他植物性蛋白质饲料
Other plant protein feeds

（一）玉米蛋白粉
Corn gluten meal

1. 概述 玉米蛋白粉是玉米加工的主要副产物之一，为玉米除去淀粉、胚芽、外皮后剩下的产品，我国年产玉米蛋白粉约为16.4万t。

2. 营养特性 玉米蛋白粉粗蛋白质含量为40%-60%，氨基酸组成不佳，蛋氨酸、精氨酸含量高，赖氨酸和色氨酸严重不足，赖氨酸、精氨酸之比达100∶200，与理想比值相差较远。粗纤维含量低，易消化，代谢能与玉米近似或高于玉米，为高能高蛋白质饲料。矿物质含量少，铁较多，钙、磷较低。维生素中胡萝卜素含量较高，B族维生素少。富含色素，主要是叶黄素和玉米黄质，前者是玉米含量的15-20倍，是较好的着色剂。

3. 原料标准 饲料用玉米蛋白粉国家行业标准（NY/T 685—2003）规定：玉米蛋白粉呈淡黄色至黄褐色，色泽均匀，粉状或颗粒状，无发霉、结块、虫蛀，具有本制品固有气味，无腐败变质；水分含量不

1. Summary. Corn gluten meal is one of the main by-products of corn processing. It is the product left after corn starch, germ and husk are removed. The annual output of corn gluten meal in China is about 164 thousand tons.

2. Nutritional characteristics. In corn gluten meal, the crude protein content is 40%-60%, the amino acid composition is not good, the content of methionine and arginine is high, the lysine and tryptophan are seriously insufficient, and the ratio of lysine to arginine is 100∶200. The ideal ratio is far from each other. The crude fiber content is low, easy to digest, and its metabolizable energy is similar to or higher than that of corn. It is high-energy and high-protein feed. It has less mineral content, more iron, lower calcium and phosphorus. The content of carotene in vitamins is higher, but the B vitamins are less. Corn gluten meal is rich in pigments, mainly lutein and zeaxanthin, with the former being 15-20 times of the content in corn and is a good coloring agent.

3. Material standard. The national industry standard of corn gluten meal for feed (NY/T 685—2003) stipulates, corn gluten meal is light yellow to yellowish-brown, uniform in color, powdery or granular; free of mold, agglomeration, or moth-eaten with the inherent odor of this product, and without corruption and deterioration; the moisture

得超过 12.0%；不含砂石等杂质，不得掺入非蛋白氮等物质。若加入抗氧化剂、防霉剂等添加剂,应在标签上做相应说明。

4. 饲用价值　玉米蛋白粉用于鸡饲料可节省蛋氨酸,着色效果明显。玉米蛋白粉太细,故配合饲料中用量不宜过大,否则影响采食量,以 5% 以下为宜,颗粒化后可用至 10% 左右。

玉米蛋白粉对猪适口性好,易消化吸收,与大豆饼粕配合使用可在一定程度上平衡氨基酸,用量在 15% 左右,大量使用时应添加合成氨基酸。

玉米蛋白粉可用作奶牛、肉牛的部分蛋白质饲料原料,因其密度大,故可配合密度小的原料使用。在精饲料中添加以 30% 为宜,过多会影响生产性能。

在使用玉米蛋白粉的过程中,应注意霉菌含量,尤其是黄曲霉毒素含量要在合理范围内。

content must not exceed 12.0%. Impurities such as sand and gravel shall not be mixed with non-protein nitrogen and other substances. If additives such as antioxidants and antifungal agents are added, corresponding instructions should be made on the label.

4. Feeding values. Corn gluten meal can save methionine when used in chicken feed, and the coloring effect is obvious. Since the corn gluten meal is too fine, the dosage in the compound feed should not be too large, otherwise it will affect the feed intake. It is better to use less than 5%, and it can be used to about 10% in dosage after granulation.

Corn gluten meal has good palatability for pigs and is easy to digest and absorb. When used with soybean meal, it can balance amino acids to a certain extent. The dosage is about 15%. When used in large quantities, synthetic amino acids should be added.

Corn gluten meal can be used as part of the protein feed material for dairy cows and beef cattle. Because of its high density, it can be used with low-density raw materials. It is appropriate to add 30% in concentrated feed, while too much addition will affect production performance.

In the process of using corn gluten meal, pay attention to the content of mold, especially the content of aflatoxin should be within the limit.

(二) 干酒糟及其可溶物(DDGS)
Distillers dried grains with solubles

1. 概述　脱水酒精糟(DDG)又名干酒精糟,为酵母发酵的谷物籽实生产酒精时,经粉碎、发酵、蒸馏,再过滤出酒精以后所获得的干酒糟。脱水后的酒精发酵副产品中的可溶性物质(DDS)由除去固形物部分的残液浓缩和干燥而得。DDGS 是 DDG 和 DDS 的混合物,为含可溶物的谷物干酒精糟,是目前市场上的主要产品。

早期的 DDGS 主要来源于饮料企业生产的副产品,营养价值低,变异大且来源有限。20 世纪 70 年代,巴西、美国等石油资源短缺而农业资源丰富的国家开始用谷物发酵

1. Summary. Distiller's dried grain (DDG) is also known as dry distiller's grain, which is obtained after crushing, fermenting, distilling, and filtering out the alcohol when producing alcohol from grain seeds fermented by yeast. The soluble substances (dried distiller's solubles, DDS) in the byproducts of alcoholic fermentation after dehydration are obtained by concentrating and drying the residual liquid after removing the solid part. DDGS (distillers dried grains with solubles) is a mixture of DDG and DDS, which is the main product on the market at present.

Early DDGS mainly comes from by-products produced by beverage companies, with low nutritive value, large variation, and limited sources. In the 1970s, countries such as Brazil and the United States that were short of oil resources and rich in agricultural resources began to use grain fermen-

生产燃料酒精,以代替部分纯汽油。燃料酒精性价比高且环保的优点在适应社会需要的前提下迅速发展。20世纪90年代以来,随着燃料酒精工业的迅速崛起和在世界范围内的进一步推广,以玉米为原料生产燃料酒精的规模不断扩大,以DDGS为主要形式的副产物产量逐渐增多。同时,随着加工工艺的改进,其营养价值得到提高,在动物生产中的应用更加广泛。玉米DDGS成为国际市场上最重要的酒精加工副产品。2008年,美国玉米DDGS产量达到1870万t,我国燃料酒精业在全球能源危机的前提下也迅速发展,DDGS产量达304万t,其中燃料酒精生产的DDGS为88万t。

目前国内外比较成熟的工艺是:将玉米酒精蒸馏废液先经过滤,然后滤渣干燥,滤清液同时浓缩,最后将干燥的滤渣和浓缩的滤清液混合干燥,即得DDGS。

2. 营养特性　在生产酒精的过程中,玉米中占籽实2/3的淀粉发酵成酒精和二氧化碳,剩余的产物中浓缩了除淀粉和糖以外的营养物质,与玉米有了很大区别,最大限度地保留了玉米中的蛋白质,其含量是玉米的3倍以上,故为蛋白质饲料。

DDGS颜色由浅黄色至深褐色,呈多样性,一般玉米中同时含有高粱、小麦等谷物生产的DDGS色泽都较深。DDGS呈粉状或碎屑状,高品质的DDGS有发酵香味,加热过度后因为发生美拉德反应而具有烟熏味及糊味。DDGS浓缩了玉米中除淀粉和糖以外的蛋白质、脂肪、维生素、纤维及发酵中产生的未知生长因子、糖化物、酵母等营养

tation to produce fuel alcohol to replace part of pure gasoline. The advantages of fuel alcohol, which is cost-effective and environmentally friendly, have developed rapidly under the premise of adapting to the needs of the society. Since the 1990s, with the rapid rise of the fuel alcohol industry and its further promotion worldwide, the production of fuel alcohol using corn as a raw material has continued to increase, and the output of by-products in the form of DDGS has gradually increased. At the same time, with the improvement of processing technology, its nutritive value has been improved, and its application in animal production has become more extensive. Corn DDGS has become the most important by-product of alcohol processing in the international market. In 2008, the output of corn DDGS in the United States reached 18.7 million tons. The fuel alcohol industry in my country also developed rapidly under the premise of the global energy crisis. The output of DDGS reached 3.04 million tons, of which DDGS produced by fuel alcohol was 880,000 tons.

At present, the mature process at home and abroad is filtering the waste liquor of corn alcohol distillation, then drying the filter residue, concentrating the filtrate at the same time, and finally mixing and drying the dried filter residue and the concentrated filtrate to obtain DDGS.

2. Nutritional characteristics. In the process of alcohol production, corn starch, which accounts for 2/3 of the seed kernel, is fermented into alcohol and carbon dioxide. The remaining products are concentrated with nutrients other than starch and sugar, which are very different from corn, and the corn is retained to the greatest extent. The content of protein is more than 3 times that of corn, so it is protein feed.

The color of DDGS varies from light yellow to dark brown. Generally, DDGS produced from grains such as sorghum and wheat are darker in color. DDGS is powdery or crumbly. High-quality DDGS has a fermented flavor. After overheating, it has a smoky and pasty flavor due to Maillard reaction. DDGS concentrates the protein, fat, vitamins, fiber in corn except starch and sugar, as well as the unknown growth factors, saccharides, yeast and other nutrients produced during fermentation. The high level of central detergent fiber and low lignin content in DDGS make it a source

物质。DDGS 中高水平的中性洗涤纤维和较低的木质素含量使其成为奶牛高消化率纤维的来源。有研究报道,不同酒精厂生产的 DDGS 之间营养水平差异很大。除 DDGS 干物质变异系数在 5% 以下,粗蛋白质、粗脂肪、粗纤维和一些氨基酸变异系数小于 10% 外,赖氨酸、蛋氨酸和磷变异系数分别高达 17.3%、13.6% 和 11.7%。但同一个酒精厂生产的 DDGS 变幅较小。

3. 饲用价值　DDGS 是一种很好的蛋白质资源,在家禽、猪、奶牛、肉牛、羊及水产饲料中都可应用,可弥补蛋白质资源的不足。DDGS 是不同阶段猪能量、蛋白质及其他营养成分的良好来源。DDGS 含有较高的粗蛋白质、丰富的 B 族维生素及未知生长因子,对于生长猪的生长发育很有利,是饲喂生长猪的一种质优价廉的蛋白质饲料。

DDGS 是必需脂肪酸——亚油酸和鸡的第一限制性氨基酸——蛋氨酸的重要来源,是种鸡和产蛋鸡的良好饲料原料之一。饲粮中添加适量的 DDGS 可以提高家禽的生产性能和蛋黄颜色。新鲜或干燥 DDGS 中的脂肪和有限纤维替代可溶性碳水化合物和淀粉,有助于维持瘤胃微生态的平衡和稳定瘤胃 pH 值。因此,新鲜或干燥 DDGS 能减少瘤胃酸中毒。DDGS 在过瘤胃蛋白质、适口性和有效纤维的安全性方面有其独特性,是奶牛和肉牛良好的饲料来源,可以替代奶牛饲粮中的部分玉米和豆粕。但用量在 30% 以上时采食量和产奶量下降。

4. 存在的问题　在 DDGS 使用中仍存在如下问题:① 营养成分不稳定。原料(品种、水分、杂质)、DDGS 中 DDG 和 DDS 的比例、酒

of high digestibility fiber for dairy cows. Studies have reported that the nutritional levels of DDGS produced by different alcohol plants vary greatly. Except for the coefficient of variation of DDGS dry matter which is below 5%, the coefficient of variation of crude protein, crude fat, crude fiber and some amino acids is less than 10%, the coefficient of variation of lysine, methionine and phosphorus is as high as 17.3%, 13.6% and 11.7% respectively. However, the DDGS produced by the same alcohol plant has a smaller variation.

3. Feeding value. DDGS is a good protein resource, which can be used in poultry, pigs, dairy cows, beef cattle, sheep and aquatic feeds, and can make up for the lack of protein resources. DDGS is a good source of energy, protein and other nutrients for pigs at different stages. DDGS contains high crude protein, rich B vitamins and unknown growth factors, which is very beneficial to the growth and development of growing pigs. It is a high-quality and cheap protein feed for growing pigs.

DDGS is an important source of essential fatty acids—linoleic acid and chicken's first limiting amino acid—methionine, and is one of the good feed materials for breeders and laying hens. Adding the right amount of DDGS to the diet can improve the performance and egg yolk color of poultry. Since the fat and limited fiber in fresh or dried DDGS can replace soluble carbohydrates and starches to help maintain the balance of rumen microbiota and stabilize rumen pH value, fresh or dried DDGS can reduce rumen acidosis. DDGS is unique in terms of rumen bypass protein, palatability (aroma produced during fermentation) and safety of effective fiber. DDGS is a good source of feed for dairy and beef cattle and can replace part of the corn and soybean cake and meal in dairy cattle diets. But the feed intake and milk production will decrease when the dosage is more than 30%.

4. Present issues. The following problems still exist in the use of DDGS: ① The nutrient composition is unstable. Raw materials (variety, moisture, impurities), the ratio of DDG and DDS in DDGS, alcohol fermentation process,

精发酵工艺等都影响其营养成分含量。② 色泽不一致。加工工艺中加热温度(高温和低温)和物料进入干燥器的流量(慢速和快速)是颜色变异的主要因素。③ 安全隐患。贮存不当易造成 DDGS 中霉菌毒素不同程度地增多。④ 缺乏快速评价 DDGS 质量的方法。

etc. all affect its nutrient content. ② The color is inconsistent. The heating temperature (high and low temperature) and the flow rate (slow and fast) of the material entering the dryer during processing are the main factors of color variation. ③ Safety hazards. Improper storage can easily cause mycotoxins in DDGS to increase to varying degrees. ④ Lack of methods to quickly evaluate the quality of DDGS.

第二节　动物性蛋白质饲料
Section 2　Animal protein feed

动物性蛋白质饲料主要指由水产类加工副产品、畜禽加工和乳品业副产物等调制成的蛋白质饲料。该类饲料的主要营养特点是：蛋白质含量高(40%－85%)，氨基酸组成比较平衡，并含有促进动物生长的动物性蛋白因子；碳水化合物含量低，不含粗纤维；粗灰分含量高，钙、磷含量高，比例适宜；维生素含量丰富(特别是维生素 B_2 和维生素 B_{12})；脂肪含量较高，虽然能值高，但脂肪易氧化酸败，不宜长时间贮藏。

Animal protein feed mainly refers to protein feed prepared from by-products of aquatic processing, livestock and poultry processing and dairy industry by-products. The main nutritional characteristics of this type of feed are：high protein content (40%－85%), relatively balanced amino acid composition. It contains animal protein factors (APF) that promotes animal growth. It has low carbohydrate content, no crude fiber, high crude ash content, high calcium and phosphorus content and appropriate ratio. Its vitamin content is rich (especially vitamin B_2 and vitamin B_{12}), fat content is high. Although its energy value is high, the fat is easy to oxidize and rancid, and it is not suitable for long-term storage.

一、水产加工副产物饲料
By-product feed of aquatic processing

(一) 鱼粉
Fish meal

1. 概述

鱼粉是以新鲜的全鱼或鱼品加工过程中所得的鱼杂碎为原料，经或不经脱脂加工制成的洁净、干燥和粉碎的产品。全世界的鱼粉生产国主要有秘鲁、智利、日本、丹麦、美国、挪威等，其中秘鲁与智利的出口量约占总贸易量的 70%。中国的

1. Summary

Fish meal is a clean, dried and crushed product made from fresh whole fish or fish offal from fish processing, with or without degreasing processing. The world's fish meal producing countries mainly include Peru, Chile, Japan, Denmark, the United States, Norway, etc., among which Peru and Chile's exports account for about 70% of the total trade volume. The production of fish meal in China

鱼粉产量不高,主要生产地在山东、浙江,其次为河北、天津、福建、广西等省(自治区、直辖市)。2009 年,我国进口 130.8 万 t 鱼粉,其中约 56%来自秘鲁,从智利进口量约占 26%。

(1) 鱼粉的分类方法主要有三种:① 根据来源将鱼粉分为国产鱼粉和进口鱼粉。显然,这种方法比较粗略,反映不出鱼粉的品质。② 按原料性质、色泽分类,将鱼粉分为普通鱼粉(橙白或褐色)、白鱼粉(浅黑褐或浓黑色)、鲸鱼粉(浅黑色)和鱼粕(鱼类加工残渣)等六种。③ 按原料部位与组成可分为全鱼粉(以全鱼为原料制成)、调整鱼粉(全鱼粉 + 鱼溶浆)、粗鱼粉(调整鱼粉 + 肉骨粉或羽毛粉)、鱼精粉(鱼溶浆 + 吸附剂)等六种。上述分类方法因国家不同而异,我国饲料行业目前还没有标准,多种方法都采用。

(2) 目前,国内外鱼粉的加工多根据鱼脂肪含量采用不同的方法,包括"高脂鱼"和"低脂鱼"两种加工工艺。

① 高脂鱼的加工工艺:是对脂肪含量较高的鱼先进行脱脂然后再干燥制粉的加工过程。首先,用蒸煮或干热风加热的方法,使鱼体组织蛋白质发生热变形而凝固,促使体脂分离溶出。然后,对固形物体进行螺旋压榨法压榨,将固体部分烘干制成鱼粉。干燥方法分为干热风法和蒸汽法两种。干热风的温度因热源形式不同,可从 100 - 400 ℃ 不等;蒸汽法为间接加热,干燥速度慢,但鱼粉质量好。整鱼经过去油、去浸汁、干燥、粉碎后的产品,蛋白质含量在 50% - 60% 不等。榨出

is not high. The main production areas are Shandong and Zhejiang, followed by Hebei, Tianjin, Fujian, Guangxi and other provinces (autonomous regions and municipalities). In 2009, China imported 1.308 million tons of fish meal, about 56% of which were from Peru and about 26% of which were from Chile.

(1) There are three main classification methods for fish meal: ① According to the source, fish meal can be divided into domestic fish meal and imported fish meal. Obviously, this method is relatively rough and does not reflect the quality of fish meal. ② According to the nature of raw materials and color classification, fish meal can be divided into six types: ordinary fish meal (orange white or brown), white fish meal (light black brown or dark black), whale meal (light black) and fish meal (fish processing residue), etc. ③ According to the part and composition of the raw materials, fish meal can be divided into six types: whole fish meal (made from whole fish), adjusted fish meal (whole fish meal plus fish lysate), coarse fish meal (adjusted fish meal plus meat and bone meal or feather meal) and mix fish soluble meal (fish soluble plus adsorbent), ect. The above classification methods vary from country to country. There is currently no standard for the feed industry in my country, and many methods are used.

(2) At present, the processing of fish meal at home and abroad mostly adopts different methods according to the fat content of the fish, and is divided into two processing technologies: "high-fat fish" and "low-fat fish".

① The processing technology of high-fat fish. It is the process of drying and flouring after defatting fish with higher fat content. First of all, using steaming or dry hot air heating method, the fish body tissue protein is thermally deformed and solidified, which promotes the separation and dissolution of body fat. Then, the solid object is squeezed by screw pressing, and the solid part is dried to make fish meal. Drying methods are divided into two types: dry hot air method and steam method. The temperature of the dry hot air can vary from 100 to 400 ℃ due to different heat sources. The steam method is indirect heating, the drying speed is slow, but the quality of fish meal is good. The protein content of the whole fish after de-oiling, de-soaking, drying, and crushing is 50% - 60%. The squeezed juice is

汁液经酸化、喷雾干燥或加热浓缩成鱼膏。鱼膏也可用鱼类内脏生产，原料经加酶水解、离心分离、去油水解液浓缩即制成鱼膏。制成的鱼膏可直接桶装出售，也可用淀粉或糠麸作为吸附剂，再经干燥、粉碎后出售，后者称为鱼汁吸附饲料或混合鱼溶粉，其营养价值因载体而异。

②低脂鱼的加工工艺：是对脂肪相对含量低的鱼及其他海产品的加工过程。根据原料的种类一般分为全鱼粉和杂鱼粉两类。全鱼粉是对脂肪含量少的鱼进行整体直接加热干燥，失去部分水分后再进行脱脂，固形物经第二次干燥至水分含量达18%时，粉碎制成鱼粉。通常每100 kg全鱼约可出全鱼粉22 kg，粗蛋白质含量在60%左右。杂鱼粉是将小杂鱼、虾、蟹以及鱼头、尾、鳍、内脏等直接干燥粉碎后的产品，又称鱼干粉，含粗蛋白45%-55%不等；或在产鱼旺季，先采用盐腌，再经脱盐，然后干燥粉碎制得。这种鱼粉往往脱盐不彻底（含盐10%以上），使用不当易造成畜禽食盐中毒。

2. 营养特性

鱼粉的主要营养特点是蛋白质含量高，一般脱脂全鱼粉的粗蛋白质含量高达60%以上。氨基酸组成齐全、平衡，尤其是主要氨基酸与猪、鸡体组织氨基酸组成基本一致。钙、磷含量高，比例适宜。微量元素中碘、硒含量高。富含维生素 B_{12}、脂溶性维生素A、维生素D、维生素E和未知生长因子。所以，鱼粉不仅是一种优质蛋白源，还是一种不易被其他蛋白质饲料完全取代的动物性蛋白质饲料。但鱼粉的营养成分因原料质量和加工工艺不同变异

acidified, spray-dried or heated and concentrated into fish soluble. Fish soluble can also be produced from fish innards. The raw materials are hydrolyzed with enzymes, centrifuged, and concentrated by degreasing hydrolysate to make fish soluble. The prepared fish soluble can be sold directly in barrels, and starch or bran can be used as adsorbents, and then dried and crushed for sale. The latter is called fish juice adsorption feed or compound fish soluble powder. The nutritive value varies with carriers.

② The processing technology of low-fat fish. It is the processing process of fish and other seafood with relatively low fat content. According to the types of raw materials, it is generally divided into two types: whole fish meal and mixed fish meal. Whole fish meal is obtained through directly heating and drying the whole fish with low fat content. After losing part of the water, it is defatted. When the solid matter is dried for the second time till the water content is 18%, it is crushed to make fish meal. Usually, every 100 kg of whole fish can produce about 22 kg of whole fish meal, and the crude protein content is about 60%. Mixed fish meal is a product made by directly drying and crushing small trash fish, shrimp, crab and fish heads, tails, fins, viscera, etc. It is also called dried fish meal, which contains 45%-55% of crude protein. In peak fish seasons, it is made by salting, desalting, then drying and crushing. This kind of fish meal is often incompletely desalinated (more than 10% of salt), and improper use can easily cause salt poisoning in livestock and poultry.

2. Nutritional characteristics

The main nutritional feature of fish meal is its high protein content. Generally, the crude protein content of defatted whole fish meal is as high as 60%. The amino acid composition is complete and balanced, especially the main amino acid is basically the same as the amino acid composition of pig and chicken body tissues. The calcium and phosphorus content is high and the ratio is appropriate. The content of iodine and selenium in trace elements is high. Fish meal is rich in vitamin B_{12}, fat-soluble vitamin A, vitamin D, vitamin E and unknown growth factors. Therefore, fish meal is not only a high-quality protein source, but also an animal protein feed that is not easily replaced by other protein feeds. However, the nutritional components of fish

较大。

通常真空干燥法或蒸汽干燥法制成的鱼粉,蛋白质利用率比用烘烤法制成的鱼粉约高10%。鱼粉中一般含有6%-12%的脂类,其中不饱和脂肪酸含量较高,极易被氧化产生异味。进口鱼粉因生产国的工艺及原料而异,质量较好的是秘鲁鱼粉及白鱼鱼粉,粗蛋白质含量可达60%以上。国产鱼粉由于原料品种、加工工艺不规范,产品质量参差不齐。

鱼浸膏中含水分约为50%,粗蛋白质30%,含硫氨基酸、色氨酸等含量均低于鱼粉。

3. 质量标准

(1) 鱼粉的品质鉴别:

① 色泽与气味:不同种类的鱼粉色泽存在差异,正常鲱鱼粉呈淡黄或淡褐色,沙丁鱼粉呈红褐色,鳕鱼粉等白鱼粉呈淡黄色或灰白色。蒸煮不透、压榨不完全、含脂较高的鱼粉颜色较深。各种鱼粉均具有鱼腥味;如果具有酸、臭及焦灼腐败味,则品质欠佳。

② 定量检测:鱼粉中水分含量一般为10%左右。水分过高不宜贮藏,过低则可能发生加热过度,会导致氨基酸利用率降低。粗蛋白质含量一般应在60%左右。正常鱼粉的胃蛋白酶消化率应在88%以上。粗脂肪含量一般不应超过12%,大于21%可能存在加工不良或原料不新鲜,这样的鱼粉贮藏时易发生酸败,出现异味,并影响其他营养物质的消化利用。

一般进口鱼粉含盐约2%,国产鱼粉含盐量应小于5%,但有些国产鱼粉含盐量很高,易造成畜禽食盐中毒,故检测鱼粉含盐量非常

meal vary greatly due to the quality of raw materials and processing techniques.

Generally, the protein utilization rate of fish meal made by vacuum drying method or steam drying method is about 10% higher than that of fish meal made by baking method. Fish meal generally contains 6%-12% of lipids, among which the content of unsaturated fatty acids is relatively high, which is easily oxidized to produce peculiar smell. Imported fish meal varies according to the process and raw materials of the producing country. The better quality is Peruvian fish meal and white fish meal, with a crude protein content of more than 60%. The quality of domestic fish meal is uneven due to the variety of raw materials and irregular processing technology.

The water content of fish extract is about 50%, the crude protein is 30%, and the content of sulfur-containing amino acids and tryptophan is lower than fish meal.

3. Material standard

(1) Classification of fish meal

① Color and smell. Different types of fish meal are different in color. Normal herring meal is light yellow or light brown, sardine meal is reddish brown, and white fish meal such as cod fish meal is light yellow or off-white. Fish meal that has not been cooked thoroughly, squeezed incompletely, and has a higher fat content is darker. All kinds of fish meal have a fishy smell. If it has the smell of acid, stink and burnt, the quality is not good.

② Quantitative detection. The moisture content in fish meal is generally about 10%. If the water content is too high, it is not suitable for storage. If the water content is too low, overheating may occur, which will reduce the utilization of amino acids. The crude protein content should generally be around 60%. The pepsin digestibility of normal fish meal should be above 88%. The crude fat content should generally not exceed 12%. If it is greater than 21%, the reason may be poor processing or stale raw materials. Such fish meal is prone to rancidity and peculiar smell when stored, and affects the digestion and utilization of other nutrients.

Generally, imported fish meal contains about 2% of salt, while domestic fish meal should contain than 5% of salt. However, some domestic fish meal contains high salt content, which may cause salt poisoning in livestock and

重要。全鱼粉粗灰分含量多在16%-20%，超过20%疑为非全鱼粉。

（2）鱼粉的标准：中国已正式颁布饲料用鱼粉质量的国家标准（GB/T 19164—2003），包括鱼粉的感官要求、理化指标和微生物指标。我国鱼粉专业标准适用于以鱼、虾、蟹等水产动物或鱼品加工过程中所得的鱼头、尾、内脏等为原料，进行干燥、脱脂、粉碎或先经蒸煮再压榨、干燥、粉碎而制成的作为饲料用的鱼粉。结构蓬松，无结块、霉变、酸败味，红鱼粉呈黄棕色、黄褐色；白鱼粉呈黄白色。

4. 饲用价值

因鱼粉中不饱和脂肪酸含量较高并具有鱼腥味，故在畜禽饲粮中使用量不可过多，否则会导致畜产品产生异味。家禽饲粮中使用鱼粉过多可导致禽肉、蛋产生鱼腥味，因此，当鱼粉中脂肪含量约为10%时，用量应控制在10%以下。火鸡宰前8周应停喂鱼粉。生长育肥猪饲粮中鱼粉用量应控制在8%以下，否则会使体脂变软、肉带鱼腥味。幼龄畜禽饲粮中鱼粉添加量应小于10%，成年畜禽小于5%。为降低成本，猪育肥后期饲粮可不添加鱼粉。

加工贮藏不当时，鱼粉中可产生肌胃糜烂素，并引起硫胺素缺乏症等。此外，计算配方时应考虑鱼粉的含盐量，以防食盐中毒。

（二）虾粉、虾壳粉、蟹粉
Shrimp meal, shrimp shell meal and crab meal

1. 概述

虾粉、虾壳粉是指利用新鲜小虾或虾头、虾壳，经干燥、粉碎而成

poultry, so it is very important to detect the salt content of fish meal. The crude ash content of whole fish meal is mostly between 16%-20%, and more than 20% is suspected to be non-whole fish meal.

(2) The national standard for fish meal (GB/T 19164—2003) summarizes sensory requirements, physical and chemical indicators, and microbiological indicators of fish meal. The professional standard of fish meal in China applies to the fish meal made as feed by drying, degreasing, crushing or drying, degreasing, crushing after steaming the fish, shrimp, crab and other aquatic animals or fish heads, tails, viscera, etc. obtained from fish processing. The structure of fish meal is fluffy, without agglomeration, mildew, and rancidity. The red fish meal is yellow-brown and the white fish meal is yellow-white.

4. Feeding value

Fish meal contains a high content of unsaturated fatty acids and has a fishy smell, therefore it should not be used too much in livestock and poultry diets, otherwise it will cause peculiar smell in livestock products. Excessive use of fish meal in poultry diets can cause fishy taste in poultry meat and eggs. Therefore, when the fat content of fish meal is about 10%, the amount in chicken diets should be controlled below 10%. Fish meal should be stopped 8 weeks before turkey slaughter. The amount of fish meal in the feed for growing and finishing pigs should be controlled below 8%, otherwise the body fat will become soft and the meat will smell fishy. The amount of fish meal added in feed for young livestock and poultry should be less than 10%, and less than 5% for adult livestock and poultry. In order to reduce cost, fish meal may not be added to the post-finishing pig feed.

Improper processing and storage can produce gizzarosin in fish meal and cause thiamine deficiency. In addition, the salt content of fish meal should be considered when calculating the formula to prevent salt poisoning.

1. Summary

Shrimp meal and shrimp shell meal refer to a kind of powdery products with fresh color and no spoilage and

的一类色泽新鲜、无腐败异臭的粉末状产品。蟹粉是指用蟹壳、蟹内脏及部分蟹肉加工生产的一种产品。这类产品的共同特点是含有一种被称为几丁质的物质,其化学组成类似纤维素,很难被动物消化。几丁质又名壳多糖、甲壳素,在昆虫、甲壳类(虾、蟹)等动物的骨骼中与碳酸钙相伴存在,可占甲壳有机物的50%-80%,在酵母、霉菌等微生物中也有发现。随着科学技术的发展,人们发现几丁质是由β-1、4键连接的氨基葡萄糖多聚体,分解产物为2-氨基葡萄糖,并证实几丁质对于虾、蟹壳的形成具有重要作用,还可用作蛋白质的凝聚剂和鱼生长促进剂。另外,还可用作医用缝合线、电影胶卷、纤维制品的原料。

2. 营养特性及饲用价值

这类产品中的成分随品种、处理方法、肉和壳的组成比例不同而异。一般虾蟹粉粗蛋白质含量约40%,虾壳粉、蟹壳粉粗蛋白质含量约为30%,其中1/2为几丁质氮。粗灰分含量为30%左右,并含有大量不饱和脂肪酸、胆碱、磷脂、固醇和具有着色效果的虾红素。

虾壳粉、蟹壳粉不仅可为畜禽提供蛋白质,还有一些其他特殊作用。鸡饲料中添加3%,有助于肉鸡脚趾和蛋黄着色。在猪饲料中添加3%-5%,可刺激肠道中双岐乳酸杆菌的生长,提高仔猪的抗病力,改善猪肉色泽。在虾料中添加10%-15%,也可取得良好的促生长效果。有报道指出,几丁质的水解产物N-乙酰氨基葡萄糖(壳多糖)可降低血中胆固醇含量,并具抗感染、促进消化、提高增重等功能。利用这类饲料时,应注意含盐量和新鲜度。

odor, which are made by drying and crushing fresh shrimps or shrimp heads and shells. Crab meal refers to a product produced by processing crab shells, crab offal and part of crab meat. The common feature of these products is that they contain a substance called chitin, whose chemical composition is similar to cellulose, which is difficult for animals to digest. Chitin, also known as tunicin, is accompanied by calcium carbonate in the bones of insects, crustaceans (shrimp, crab) and other animals, and can account for 50%-80% of the organic matter in crustaceans. It is also found in microorganisms such as yeasts and molds. With the development of science and technology, people have discovered that chitin is a glucosamine polymer linked by β-1,4 bonds, and the decomposition product is 2-glucosamine. It has been confirmed that chitin is important for the formation of shrimp and crab shells. It can also be used as a protein coagulant and fish growth promoter. In addition, it can also be used as a raw material for medical sutures, film rolls, and fiber products.

2. Nutritional characteristics and feeding value

The ingredients in these products vary with the variety, processing method, and composition ratio of meat and shell. Generally, the crude protein content of shrimp and crab meal is about 40%, and the crude protein content of shrimp and crab shell meal is about 30%, of which 1/2 is chitin nitrogen. The crude ash content is about 30%, and it contains a lot of unsaturated fatty acids, choline, phospholipids, sterols and astaxanthin with coloring effect.

Shrimp shell meal and crab shell meal not only provide protein for livestock and poultry, but also have some other special effects. Adding 3% to the chicken feed will help the toes and egg yolk color of the broiler. Adding 3%-5% to pig feed can stimulate the growth of Lactobacillusbifidus in the intestines, improve the disease resistance of piglets, and improve the color of pork. Adding 10%-15% to shrimp feed can also achieve good growth-promoting effects. It has been reported that N-acetylglucosamine (chitin), a hydrolysate of chitin, can reduce blood cholesterol levels, and has the functions of anti-infection, promoting digestion, and increasing weight gain. When using this type of feed, attention should be paid to the salt content and freshness.

二、畜禽副产物饲料
Livestock and poultry by-product feed

畜禽副产物饲料是指屠宰厂或肉联加工厂处理屠宰屠体后所得的副产物,经灭菌等加工处理而成。

Livestock and poultry by-product feed refers to the by-products obtained after the slaughter carcasses are processed by the slaughterhouse or meat processing plant, which are processed through sterilization and other processing.

(一) 肉骨粉与肉粉
Meat and bone meal, and meat meal

1. 概述

肉骨粉是由洁净、新鲜的动物组织和骨骼(不得含排泄物、胃肠内容物及其他外来物质)经高温高压蒸煮灭菌、干燥、粉碎制成的产品。肉粉是以纯肉屑或碎肉制成的饲料。骨粉是由洁净、新鲜的动物骨骼经高温高压蒸煮灭菌、脱脂(或)经脱胶、干燥、粉碎后的产品。2018年,我国肉骨粉的总产量为56万t。

2. 加工工艺

根据加工过程,肉骨粉和肉粉的加工方法主要有湿法生产和干法生产两种。

(1) 湿法生产:直接将蒸汽通入装有原料的加压蒸煮罐内,通过加热使油脂形成液状,经过滤与固体分离,再通过压榨法进一步分离出固体部分,经烘干、粉碎后即得成品。

(2) 干法生产:将原料初步捣碎,装入具有双壁层的蒸煮罐中,用蒸汽间接加热分离出油脂,然后将固体部分适当粉碎,用压榨法分离残留油脂,再将固体部分干燥后粉碎即得成品。

3. 营养特性

因原料组成和肉、骨的比例不同,肉骨粉的质量差异较大,其主要成分为:粗蛋白质20%-50%,赖氨

1. Summary

Meat and bone meal is a product made from clean, fresh animal tissues and bones (without excrement, gastrointestinal contents and other foreign substances) through high temperature and high pressure cooking, sterilization, drying, and crushing. Meat meal is a feed made of pure meat or minced meat. Bone meal is a product made from clean, fresh animal bones that are sterilized by high temperature and high pressure cooking, defatted (or) degummed, dried, and crushed. In 2018, the total output of meat and bone meal in China was about 560 thousand tons.

2. Processing technology

According to the processing process, the main processing methods of meat and bone meal and meat meal are wet production and dry production.

(1) Wet production. Directly pass steam into a pressure cooking tank containing raw materials, heat the oil to form a liquid state, filter and separate the solid, and then further separate the solid part by pressing, and the finished product is obtained after drying and crushing.

(2) Dry production. The raw materials are initially crushed, put into a double-walled cooking tank, and the fat is separated by indirect heating with steam, then the solid part is appropriately crushed, the residual fat is separated by the pressing method, and then the solid part is dried and crushed to obtain the finished product.

3. Nutritional characteristics

The quality of meat and bone meal varies greatly due to the composition of raw materials and the ratio of meat to bone. Its main components are: 20%-50% of crude

酸1%-3%,含硫氨基酸3%-6%,色氨酸低于0.5%;粗灰分为26%-40%,含钙7%-10%,含磷3.8%-5.0%,是动物良好的钙磷供源;含脂肪8%-18%;维生素B_{12}、烟酸、胆碱含量丰富,维生素A、维生素D含量较少。

4. 原料标准

我国于2006年颁布了国家标准《饲料用骨粉及肉骨粉》(GB/T 20193—2006)。其中饲料用肉骨粉质量指标为:总磷≥11%,粗脂肪≤3%,水分≤5%,酸价≤3 mg/kg。

5. 饲用价值

肉骨粉和肉粉可与谷类饲料搭配,补充谷类饲料蛋白质的不足。但由于肉骨粉主要由肉、骨、腱、韧带、内脏等组成,还包括毛、蹄、角、皮及血等废弃物,所以品质变异很大。若以腐败的原料制成产品,则品质更差,甚至可导致中毒。加工过程中热处理过度的产品其适口性和消化率下降。贮存不当时,脂肪易氧化酸败,影响适口性和动物产品品质。因此,其总体饲养效果较鱼粉较差。

肉骨粉的原料很易感染沙门氏菌,所以在加工处理畜禽副产品过程中要进行严格消毒。例如:英国曾经由于对动物副产物未进行正确的处理,用感染有传染性沙门氏菌的禽的副产物制成的肉粉饲喂家禽,导致禽蛋和肉仔鸡感染。另外,用患病家畜的副产物制成的肉粉应尽量不喂同类动物。目前,由于疯牛病的原因,许多国家已禁止用反刍动物副产物制成的肉粉饲喂同类动物。

protein, 1%-3% of lysine, 3%-6% of sulfur-containing amino acids, and less than 0.5% of tryptophan; 26%-40% of crude ash, 7%-10% of calcium, 3.8%-5.0% of phosphorus. It is a good source of calcium and phosphorus for animals. It has 8%-18% of fat and rich vitamin B_{12}, niacin and choline. The content of vitamin A and vitamin D is less.

4. Material standard

China promulgated the national standard Bone Meal, and Meat and Bone Meal for Feed (GB/T 20193—2006) in 2006. The quality indicators of meat and bone meal for feed are as follows: no less than 11% of total phosphorus, no more than 3% of crude fat, no more than 5% of moisture, no more than 3 mg/kg of acid value.

5. Feeding values

Meat and bone meal and meat meal, as a kind of protein feed ingredients, can be matched with cereal feeds to supplement the lack of protein in cereal feeds. However, because meat and bone meal is mainly composed of meat, bones, tendons, ligaments, internal organs, etc., as well as wastes such as hair, hooves, horns, skin, and blood, the quality varies greatly. If the product is made from corrupt raw materials, the quality will be worse, and it may even cause poisoning. The palatability and digestibility of products that are over-heated during the processing will decrease. If stored improperly, the fat will be easily oxidized and rancid, which will affect the palatability and the quality of animal products. Therefore, the overall feeding effect is worse than that of fish meal.

The raw materials of meat and bone meal are susceptible to Salmonella, so strict disinfection should be carried out during the processing of livestock and poultry by-products. For example, the United Kingdom used to feed poultry with meat meal made from the by-products of poultry infected with Salmonella due to improper handling of animal by-products, resulting in poultry eggs and broiler infections. In addition, meat meal made from by-products of diseased livestock should not be fed to similar animals as much as possible. At present, due to bovine spongiform encephalopathy, many countries have banned the use of meat meal made from ruminant by-products to feed similar animals.

（二）血粉
Blood meal

1. 概述

血粉是以畜、禽血液为原料,经干燥等加工处理制成的产品。动物血液一般占活体重的4%-9%,血液中的固形物约达20%。血粉在加工过程中有部分损失,以100 kg体重计算,牛的血粉产量为0.6-0.7 kg,猪为0.5-0.6 kg。动物血粉的资源量非常丰富,开发利用这一资源十分重要。

2. 加工工艺

利用全血生产血粉的方法主要有喷雾干燥法、蒸煮法、晾晒法、发酵法。

3. 营养特性

血粉粗蛋白质含量一般在80%以上。赖氨酸含量居天然饲料之首,达6%-9%,色氨酸、亮氨酸、缬氨酸含量也高于其他动物性蛋白质,但缺乏异亮氨酸、蛋氨酸,总的氨基酸组成非常不平衡。血粉的蛋白质、氨基酸利用率与加工方法、干燥温度和时间的长短有很大关系。通常持续高温会使氨基酸的利用率降低,低温喷雾法生产的血粉优于蒸煮法生产的血粉。血粉含钙、磷少,含铁多,约2 800 mg/kg。

4. 原料标准

我国商业行业标准《饲料用血粉》(SB/T 10212—1994)的技术要求中规定:感官要求血粉为干燥粉粒状物;具有本制品固有气味,无腐败变质气味;暗红色或褐色;能通过2-3 mm筛;不含砂石等杂质。理化指标以粗蛋白质、粗纤维、水分和粗灰分为质量控制指标,各项质量指标均以90%干物质为基础。供饲料和工业用的血粉,分蒸煮血粉

1. Summary

Blood meal is a product made from livestock and poultry blood after drying and other processing treatments. Animal blood generally accounts for 4%-9% of living body weight, and the solid content in blood is about 20%. There is a partial loss of blood meal during processing. Calculated by 100 kg body weight, the blood meal output of cattle is 0.6-0.7 kg, and that of pigs is 0.5-0.6 kg. The resource of animal blood meal is very rich, and it is important to develop and utilize it.

2. Processing technology

The main methods of producing blood meal from whole blood are spray drying method, cooking method, drying method and fermentation method.

3. Nutritional characteristics

The crude protein content in blood meal is generally above 80%. The content of lysine ranks first in natural feed, reaching 6%-9%. The content of tryptophan, leucine and valine is also higher than other animal proteins, but it lacks isoleucine and methionine, and the total amino acid composition is very unbalanced. The protein and amino acid utilization rate of blood meal has a great relationship with processing method, drying temperature and length of time. Generally, continuous high temperature will reduce the utilization of amino acids, and blood meal produced by low-temperature spray method is better than blood meal produced by cooking method. Blood meal contains less calcium and phosphorus, and more iron (about 2,800 mg/kg).

4. Material standards

The technical requirements of China's commercial industry standard Blood Meal for Feed (SB/T 10212—1994) stipulates the blood meal to be dry powder and granular. It has the inherent odor of this product with no spoilage smell. It should be dark red or brown in color, can pass 2-3 mm sieve, does not contain impurities such as sand and stone. The physical and chemical indicators are controlled by crude protein, crude fiber, moisture and crude ash, and all quality indicators are based on 90% of dry matter in content. Blood meal for feed and industrial use is divided into two types:

与喷雾血粉两类（不包括"发酵血粉"）。

5. 饲用价值

（1）血粉适口性差，氨基酸组成不平衡，并具黏性，过量添加易引起腹泻，因此饲粮中血粉的添加量不宜过高。一般仔鸡、仔猪饲粮中用量应小于2%，成年猪、鸡饲粮中用量不应超过4%。血粉对幼龄反刍动物适口性差，育成牛和成牛饲料中可少量使用，以6%－8%为宜。

（2）不同种类动物的血源及新鲜度是影响血粉品质的一个重要因素。使用血粉要考虑新鲜度，防止微生物污染。

（3）由于血粉自身的氨基酸利用率不高，氨基酸组成也不理想，故应根据血粉的营养特性科学利用，在设计饲料配方时尽可能与异亮氨酸含量高和缬氨酸含量低的饲料配伍。

（三）血浆蛋白粉与血球蛋白粉

Plasma protein powder and hemoglobulin powder

由于加工的血粉适口性差、膻味重、氨基酸组成不平衡、消化利用率低，故其在动物饲粮中的应用受到限制。20世纪90年代后，各国均改进了血粉加工工艺，以提高其利用效果。血浆蛋白粉、血球蛋白粉就是动物血液深加工的两种制品。

（四）水解羽毛粉

Hydrolysed feather meal

1. 概述

水解羽毛粉是将家禽羽毛经过清洗、水解、干燥、粉碎或膨化制成的产品。一般每羽成年鸡可得风干

steamed blood meal and sprayed blood meal (excluding "fermented blood meal").

5. Feeding value

(1) Blood meal has poor palatability, unbalanced amino acid composition, and stickiness. Excessive addition can easily cause diarrhea. Therefore, the amount of blood meal in the diet should not be too high. Generally, the amount used in the feed for broilers and piglets should be less than 2%, and the amount used in the feed for adult pigs and chickens should not exceed 4%. Blood meal has poor palatability to juvenile ruminants. It can be used in a small amount in feed for growing and growing cattle, and 6%－8% is appropriate.

(2) The blood source and freshness of different kinds of animals is an important factor affecting the quality of blood meal. Freshness should be considered when using blood meal to prevent microbial contamination.

(3) Since the amino acid utilization rate of blood meal itself is not high, and the amino acid composition is not ideal, it should be used scientifically according to the nutritional characteristics of blood meal. When designing the feed formula, it should be compatible with feeds with high isoleucine content and low valine content as much as possible.

The processed blood meal has poor palatability, heavy taint, unbalanced amino acid composition, and low digestibility and utilization rate, therefore its application in animal diets is limited. After the 1990s, various countries improved the blood meal processing technology to improve its utilization effect. Plasma protein powder and hemoglobulin powder are two products of deep processing of animal blood.

1. Summary

Hydrolysed feather meal is a product made by cleaning, hydrolyzing, drying, crushing or puffing poultry feathers. Generally, each adult chicken can get 80－150 g of air-

羽毛80-150 g,为体重的4%-5%,是一种潜力很大的蛋白质饲料资源。禽类羽毛是皮肤的衍生物。羽毛蛋白质中85%-90%为角蛋白质,属于硬蛋白质,具有很大的稳定性,不经加工处理很难被动物利用。通过水解可破坏其中的双硫键,使不溶性角蛋白质变为可溶性蛋白质,有利于动物消化利用。

2. 加工工艺

(1) 高压水解法:又称蒸煮法,是羽毛粉加工的常用方法。一般水解条件为:温度115-200 ℃,压力207-690 kPa,时间为0.5-1 h。水解能使羽毛的二硫键发生裂解。在加工过程中若加入2% HCl可使分解加速,但水解后需将水解物用清水洗至中性。

(2) 酶解法:是利用蛋白酶类水解羽毛蛋白。选用高活性的蛋白水解酶,在适宜的反应条件下,使角蛋白质裂解成易被动物消化吸收的短分子肽,然后脱水制粉。这种水解羽毛粉蛋白质的生物学效价相对较高。蛋白酶水解条件依水解酶的种类而异。

(3) 膨化法:在温度240-260 ℃、压力$1.0-1.5\times10^3$ kPa下进行膨化加工。成品外形呈棒状,质地疏松、易碎,但氨基酸利用率无明显提高。

3. 营养特性

羽毛粉中含粗蛋白质80%-85%,胱氨酸含量为2.93%,居所有天然饲料之首。据分析,其缬氨酸、亮氨酸、异亮氨酸的含量分别为6.05%、6.78%、4.21%,高于其他动物性蛋白质,但赖氨酸、蛋氨酸和色氨酸的含量相对缺乏。由于胱氨酸在代谢中可代替50%的蛋氨酸,所以在饲粮配方中添加适量水解羽毛粉可补充蛋氨酸不足。同时,水

dried feathers, which is 4%-5% of body weight, so feather meal is a protein feed resource with great potential. Avian feathers are derivatives of skin. 85%-90% of feather protein is keratin, which belongs to hard protein, has great stability and is difficult to be used by animals without processing. Through hydrolysis, the disulfide bonds can be destroyed, and the insoluble keratin protein can be turned into soluble protein, which is beneficial to animal digestion and utilization.

2. Processing technology

(1) High-pressure hydrolysis method. Also known as cooking method, it is a common method for feather meal processing. The general hydrolysis conditions are: temperature at 115-200 ℃, pressure of 207-690 kPa, and 0.5-1 h. Hydrolysis can cleave the disulfide bonds of feathers. If 2% of HCl is added during processing, the decomposition can be accelerated, but after hydrolysis, the hydrolysate needs to be washed with water to neutrality.

(2) Enzymatic hydrolysis. It uses proteases to hydrolyze feather protein. Using high-activity proteolytic enzymes, under suitable reaction conditions, the keratin is cleaved into short molecular peptides that are easily digested and absorbed by animals, and then dehydrated to make powder. The biological potency of this hydrolyzed feather meal protein is relatively high. The conditions of protease hydrolysis vary with the type of hydrolase.

(3) Puffing method. Puffing process is carried out at a temperature of 240-260 ℃ and a pressure $1.0\times10^3-1.5\times10^3$ kPa. The finished product is rod-shaped, loose and fragile, but the amino acid utilization rate is not significantly improved.

3. Nutritional characteristics

Feather meal contains 80%-85% of crude protein and 2.93% of cysteine, ranking first among all natural feeds. According to analysis, the content of valine, leucine, and isoleucine is 6.05%, 6.78%, and 4.21% respectively, which are higher than other animal proteins, but the content of lysine, methionine and tryptophan is relatively insufficient. Since cysteine can replace 50% of methionine in metabolism, adding an appropriate amount of hydrolyzed feather meal to the diet formula can supplement the lack of methionine. At the same time, hydrolyzed feather meal also

解羽毛粉还具有平衡其他氨基酸的功能,故应充分利用这一资源。此外,水解羽毛粉的过瘤胃蛋白质含量约为70%,是反刍动物良好的过瘤胃蛋白质源,营养价值与棉籽饼相当。

4. 原料标准

目前还没有制定饲料用羽毛粉的国家标准。北京市地方标准《饲料用羽毛粉》(DB/1100 B4610—89)中规定:粗蛋白质≥80%,粗灰分<4%,胃蛋白酶消化率≥90%。

5. 饲用价值

养殖生产中,水解羽毛粉常因蛋白质生物学效价低、适口性差、氨基酸组成不平衡而被限量利用。水解羽毛粉的赖氨酸、蛋氨酸、色氨酸、组氨酸含量明显低于鲱鱼粉;可溶性血液蛋白质中精氨酸、异亮氨酸、甘氨酸、酪氨酸和蛋氨酸的含量低于鲱鱼粉。

水解羽毛粉因氨基酸组成不平衡,适口性差,一般在单胃动物饲料中的添加量不应过高,控制在5%-7%比较合适。研究表明,羽毛粉在蛋鸡、肉鸡饲粮中的添加量以4%为宜;在生长猪饲粮中,以3%-5%为宜;在鱼、鹿饲粮中用量以3-10%为宜;在火鸡饲粮中用量一般以2.5%-5.0%为宜;奶牛饲粮中应控制在5%以下。

has the function of balancing other amino acids, so this resource should be fully utilized. In addition, the rumen by-pass protein content of hydrolyzed feather meal is about 70%, which is a good ruminant protein source for ruminants, and its nutritive value is equivalent to cottonseed cake.

4. Material standard

At present, our country has not formulated a national standard for feather meal for feed. Beijing local standard Feather Meal for Feed (DB/1100 B4610—89) stipulates it should contain 80% of crude protein, 4% of crude ash, and its pepsin digestibility should be over 90%.

5. Feeding value

In aquaculture production, hydrolyzed feather meal is often used in limited quantities due to low protein biological potency, poor palatability, and unbalanced amino acid composition. Lysine, methionine, tryptophan, and histidine of hydrolyzed feather meal are significantly lower than herring meal; the content of arginine, isoleucine, glycine, tyrosine and methionine in the protein of soluble blood meal is lower than that of herring meal.

Hydrolyzed feather meal has poor palatability due to the unbalanced amino acid composition. Generally, the addition amount in monogastric animal feed should not be too high, and it is appropriate to control it within 5%-7%. Studies have shown that 4% of feather meal is appropriate in the diets of laying hens and broilers and 3%-5% is appropriate in the diets of growing pigs. For diets of fish and deer, and in the turkeys diet, 3%-10% is appropriate. For the dairy cow's diet, the addition amount should be controlled below 5%.

第三节 非蛋白氮饲料

Section 3 Non-protein nitrogen feed

凡含氮的非蛋白质可饲物质均可称为非蛋白氮饲料,包括饲料用的尿素、双缩脲、氨、铵盐及其他合成的简单含氮化合物。作为简单的纯化合物质,NPN不能为动物提供

Any nitrogen-containing non-protein feeble material can be called non-protein nitrogen (NPN) feed. NPN includes feed used urea, biuret, ammonia, ammonium salt and other synthetic simple nitrogen compounds. As a simple pure chemical substance, NPN cannot provide energy for ani-

能量，其作用只是供给瘤胃微生物合成蛋白质所需的氮源，以节省饲料蛋白质。世界各国广泛用其作为反刍动物的蛋白质补充料。

mals. Its function is only to supply the nitrogen source required by rumen microbes to synthesize protein to save feed protein. It is widely used as a protein supplement for ruminants all over the world.

一、尿素
Urea

1. 概述

尿素为白色、无臭、结晶状物质。味微咸苦，易溶于水，吸湿性强。纯尿素含氮量为46%，一般商品尿素的含氮量为45%。每千克尿素相当于2.8 kg粗蛋白质，或相当于7 kg豆饼的粗蛋白质含量。瘤胃细菌能产生活性很强的脲酶，当尿素进入瘤胃后，很快被脲酶水解为氨和二氧化碳。尿素水解后产生的氨和蛋白质降解产生的氨，均可用于合成菌体蛋白。菌体蛋白在真胃和小肠内经酶的作用，转化为游离氨基酸，在小肠内被吸收利用。

2. 影响尿素利用的因素

（1）饲粮中易被消化吸收的碳水化合物的数量是影响微生物尿素利用效率的最主要因素。谷物饲料中的碳水化合物在瘤胃中很易被发酵成淀粉和糖，同样，瘤胃中的尿素也可以迅速转化为氨，为微生物利用。

（2）供给反刍动物适量的天然饲料蛋白质，其水平应占饲粮的7%－12%，以促进菌体蛋白的合成。粗饲料中粗纤维含量高，不利于利用尿素的微生物繁殖，也达不到充分使用尿素氮的目的。

（3）供给适量的硫、钴、锌、铜、锰等微量元素，可为微生物合成含硫氨基酸和吸收利用氮素提供有利

1. Summary

Urea is a white, odorless, crystalline substance which is slightly salty and bitter in taste, soluble in water, and highly hygroscopic. The nitrogen content of pure urea is 46%, and the nitrogen content of general commercial urea is 45%. Each kilogram of urea is equivalent to 2.8 kg of crude protein, or equivalent to the crude protein content of 7 kg of soybean cake. Rumen bacteria can produce highly active urease. When urea enters the rumen, it is quickly hydrolyzed by urease into ammonia and carbon dioxide. Both the ammonia produced by the hydrolysis of urea and the ammonia produced by the degradation of feed protein can be used to synthesize bacterial protein. Bacterial protein is converted into free amino acids by enzymes in the real stomach and small intestine, and is absorbed and utilized in the small intestine.

2. Factors affecting urea utilization

(1) The number of carbohydrates that are easily digested and absorbed in the diet is the most important factor affecting the efficiency of urea utilization. Carbohydrates in grain feed are easily fermented into starch and sugar. Similarly, urea in the rumen can be quickly converted into ammonia as well. These nutrients can be used by microorganisms.

(2) Ruminants should be provided with an appropriate amount of natural feed protein, which should account for 7%－12% of the diet to promote the synthesis of bacterial protein. The high content of crude fiber in roughage is not conducive to the reproduction of microorganisms using urea, and it also fails to achieve the purpose of fully using urea nitrogen.

(3) Supplying appropriate amounts of trace elements such as sulfur, cobalt, zinc, copper, manganese, etc., can provide favorable conditions for microorganisms to synthe-

条件。

（4）供给适量的维生素，特别是维生素A、维生素D，以保证微生物的正常活性。

（5）控制尿素在瘤胃中的分解速度。瘤胃微生物合成蛋白质的速度比 NPN 的分解速度慢，因此，必须抑制脲酶，使尿素分解缓慢或加速微生物合成。

（6）尿素的饲喂对象为6个月以上的反刍动物，用量不能超过饲粮总氮的 1/3，或占精料补充料干物质的 1%。产奶量高于 27 kg/d 的奶牛饲粮中不应添加尿素。

3. 尿素的毒性

尿素本身并不具有毒性，但用量过多可引起氨中毒。当饲料中尿素水平过高时，反刍动物吸收的氨量会超过肝脏降解氨的量，氨就会参与动物体的循环。大脑组织对氨很敏感，当血氨水平高于正常量时，会导致神经症状的发生，氨中毒主要表现为：气喘，走路不稳，运动失调，甚至死亡。氨中毒可通过加酸而得到缓解，将醋酸溶入冷水饲喂反刍动物可减少氨的吸收，并稀释瘤胃中的氨。尿素不宜单独饲喂，应与精饲料合理搭配。生豆粕、生大豆、南瓜等饲料含有大量脲酶，切不可与尿素一起饲喂，以免引起中毒。

size sulfur-containing amino acids and absorb and utilize nitrogen.

(4) Supplying proper amounts of vitamins, especially vitamin A and vitamin D can ensure the normal activity of microorganisms.

(5) To control the decomposition rate of urea in the rumen is necessary. The rate of protein synthesis by rumen microorganisms is slower than that of NPN decomposition. Therefore, urease must be inhibited to slow down the decomposition of urea or accelerate the synthesis of microorganisms synchronously.

(6) Urea is fed to ruminants of over 6 months old, and the dosage cannot exceed 1/3 of the total nitrogen in the diet, or 1% of the dry matter of the concentrate supplement. It should not be added to the diet of dairy cows with milk yield higher than 27 kg/d.

3. Toxicity of urea

Urea itself is not toxic, but too much dosage can cause ammonia poisoning. When the level of urea in the feed is too high, the amount of ammonia absorbed by ruminants will exceed the amount of ammonia degraded by the liver, and ammonia will participate in the animal body the loop. The brain tissue is very sensitive to ammonia. When the blood ammonia level is higher than the normal amount, it will cause neurological symptoms. Ammonia poisoning mainly manifests as asthma, unstable walking, and movement disorders, even death. Ammonia poisoning can be alleviated by adding acid. Feeding ruminants with acetic acid in cold water can reduce the absorption of ammonia and dilute the ammonia in the rumen. Urea should not be fed alone and should be properly matched with concentrated feed. Raw soybean meal, raw soybeans, pumpkins and other feeds contain a large amount of urease, which must not be fed together with urea to avoid poisoning.

二、胺盐类
Amine salts

为降低尿素在瘤胃中的水解速度和延缓氨的生成速度，目前比较有效的方法和产品有以下几种。

1. 脂肪酸尿素　脂肪酸尿素

In order to reduce the rate of hydrolysis of urea in the rumen and delay the rate of ammonia production, currently more effective methods and products are as follows.

1. Fatty acid urea (urea-fatty-acid), also known as

又称脂肪脲,是脂肪膜包被的尿素,目的是提高能量、改善适口性和降低尿素分解速度。含氮量一般大于30%,呈浅黄色颗粒。

2. 腐脲 腐脲是尿素和腐殖酸按4∶1的比例在100-150 ℃温度下生产的一种黑褐色粉末,含氮24%-27%。

3. 羧甲基纤维素尿素 用1份羧甲基纤维素钠盐包被9份尿素,再以20%的水拌成糊状,制粒(直径12.5 mm),经24 ℃干燥2 h即成。用量可占牛饲粮的2%-5%。

4. 氨基浓缩物 氨基浓缩物是用20%尿素、75%谷实和5%膨润土混匀,在高温、高湿和高压下制成。

5. 磷酸脲(尿素磷酸盐) 尿酸脲为20世纪70年代国外开发的一种含磷非蛋白氮饲料添加物。含氮10%-30%,含磷8%-19%。毒性低于尿素,对牛、羊的增重效果明显。

6. 铵盐 包括无机铵盐(如碳酸氢铵、硫酸铵、多磷酸铵、氯化铵)和有机铵盐(如醋酸铵、丙酸铵、乳酸铵、丁酸铵)两类。

7. 液氨和氨水 液氨又称无水氨,一般由气态氨液化而成,含氮82%。氨水是氨的水溶液,含氮15%-17%,具刺鼻气味,可用来处理青贮饲料及糟渣等饲料。

fatty urea, is urea coated with a fatty membrane. The purpose is to increase energy, improve palatability and reduce the rate of urea decomposition. The nitrogen content is generally greater than 30%, and it is light yellow particles.

2. Humus is a dark brown powder produced by urea and humid acid at a temperature of 100-150 ℃ at a ratio of 4∶1, with a nitrogen content of 24%-27%.

3. Coat nine parts of urea with one part of carboxymethyl cellulose sodium salt, then mix with 20% of water to form a paste, granulate (12.5 mm in diameter), and dry at 24 ℃ for 2 hours. The dosage can account for 2%-5% of cattle feed.

4. Amino-concentrate is made by mixing 20% of urea, 75% of grains and 5% of bentonite under high temperature, high humidity and high pressure.

5. Urea phosphate is a phosphorus-containing non-protein nitrogen feed additive developed abroad in the 1970s. It contains 10%-30% of nitrogen and 8%-19% of phosphorus. The toxicity is lower than that of urea, and the weight gain effect on cattle and sheep is obvious.

6. Ammonium salts include inorganic ammonium salts (such as ammonium bicarbonate, ammonium sulfate, ammonium polyphosphate, ammonium chloride) and organic ammonium salts (such as ammonium acetate, ammonium propionate, ammonium lactate, ammonium butyrate).

7. Liquid ammonia (ammonia liquor), also known as anhydrous ammonia, is generally liquefied from gaseous ammonia and contains 82% of nitrogen. Ammonia solution is an aqueous solution of ammonia with 15%-17% of nitrogen and a pungent odour. It can be used to treat feed such as silage and slag.

第八章 矿物质饲料
Chapter 8　Mineral feed

本章介绍了常量矿物质饲料，包括钙源性饲料、磷源性饲料、钠源性饲料以及含硫饲料和含镁饲料等，概要介绍了沸石、麦饭石、稀土、膨润土、海泡石、凹凸棒石和泥浆等天然矿物质饲料。

矿物质是构成动物体组织、维持正常生命活动和生产畜产品不可缺少的重要物质，具有调节体内渗透压、保持体液酸碱平衡、维持神经和肌肉正常兴奋性等作用。

各种动植物饲料中均含有一定量的矿物质元素，由于动物采食饲料的多样性，可在某种程度上满足其对矿物质的需要。现代化、集约化的动物养殖，多是在舍饲条件下饲养高产动物，动物对矿物质的需要量提高，常规的基础饲料不能满足动物对矿物质的需要，必须在其饲粮中另行添加所需的矿物质。

根据我国《饲料工业术语》（GB/T 10647—2008）的定义，矿物质饲料是指可供饲用的天然的、化学合成的或经特殊加工的无机饲料原料或矿物质元素的有机络合物原料。

This chapter introduces the macroelement feeds including calcium source feed, phosphorus source feed, sodium source feed, sulfur-containing feed and magnesium-containing feed, etc. The natural mineral feeds such as zeolite, maifan stone, rare earth, bentonite, sepiolite, attapulgite and mud are briefly introduced in this chapter.

Minerals are important substances that constitute body tissues, maintain normal life activities and produce animal products for animals. They have the functions of regulating the osmotic pressure in the body, maintaining acid-base balance of body fluids, and maintaining normal excitability of nerves and muscles, etc. for animals.

Various animals and plants feed all contain a certain amount of mineral elements. Due to the diversity of animal feeds, they can meet their mineral needs to a certain extent. Modern and intensive animal breeding is mainly to raise high-yielding animals under the conditions of feedlot, and the demand for minerals of animals is increased. The routine basic feed cannot meet the needs of animals for minerals, so minerals required must be added to their diets.

According to the definition of China's Feed Industry Terms (GB/T 10647—2008), mineral feed refers to the natural, chemically synthesized or specially processed inorganic feed raw materials or organic complex materials of mineral elements.

第一节　常量矿物质饲料
Section 1　Macroelement feed

根据含量一般将矿物质元素分为常量元素和微量元素。动物需要

According to the content, mineral elements are generally divided into macroelements and microelements. The mac-

的常量元素包括钙、磷、钠、钾、氯、镁、硫等。常用的基础饲料中,一般钾含量较高,能满足需要,不需要额外补充,而其他元素含量一般不能满足动物需要,必须利用矿物质饲料进行补充。因此,常量矿物质饲料包括钙源性饲料、磷源性饲料、钠源性饲料以及含硫饲料和含镁饲料等。

roelements needed by animals include calcium, phosphorus, sodium, potassium, chlorine, magnesium, sulfur, etc. In the common basic feed, the content of potassium is generally high, which can meet the needs without additional supplement, while the content of other elements can not meet the needs of animals, so mineral feed must be used for supplement. Therefore, macroelement feed usually includes calcium source feed, phosphorus source feed, sodium source feed, sulfur-containing feed and magnesium-containing feed.

一、钙源性饲料
Calcium source feed

钙是骨骼、牙齿及畜产品(鸡蛋、牛奶)的主要成分。动物缺钙会出现佝偻病、软骨病等缺乏症,并且影响生产性能,如蛋鸡产蛋减少、蛋壳变薄等。通常,天然植物性饲料中的含钙量与各种动物的需要量相比均感不足,特别是产蛋家禽、泌乳牛和生长幼畜表现更为明显。因此,动物饲粮中应注意钙的补充。

钙源性饲料是能给动物提供钙元素饲料的总称。常用的钙源性饲料有石灰石粉、贝壳粉、蛋壳粉、石膏等。

Calcium is the main component of bones, teeth and animal products (eggs, milk). Calcium deficiency in animals will lead to rickets and osteomalacia and other deficiencies, and affect the production performance, such as egg yield reduction of laying hens, eggshell thinning and so on. Generally, the calcium content in natural plant feed is insufficient compared with the requirements of various animals, especially for laying poultry, lactating dairy cows and growing young animals. Therefore, attention should be paid to calcium supplementation in animal diets.

Calcium source feed refers to the general term that can provide calcium element feed for animals. The common calcium source feed includes limestone meal, shell meal, eggshell meal, gypsum, etc.

(一) 石灰石粉
Limestone meal

石灰石粉又称石粉,为天然的碳酸钙($CaCO_3$),多呈白色或灰白色粉末,一般含钙35%以上,是补充钙最廉价、最方便的矿物质原料。按干物质计,石灰石粉的成分与含量如下:灰分96.9%,钙35.89%,氯0.03%,铁0.35%,锰0.027%,镁0.06%。

天然石灰石,只要铅、汞、砷、氟的含量不超过安全系数,均可用作

Limestone meal, also known as stone meal, is a kind of natural calcium carbonate ($CaCO_3$), which is white or gray white powder. Generally, it contains more than 35% of calcium. It is the cheapest and most convenient mineral material to supplement calcium. In terms of dry matter, the composition and content of limestone meal are as follows: 96.9% of ash, 35.89% of calcium, 0.03% of chlorine, 0.35% of iron, 0.027% of manganese, 0.06% of magnesium.

Natural limestone, as long as the content of plumbum, mercury, arsenic and fluorine does not exceed the safety

饲料。一般认为,饲料级石灰石粉中镁的含量不宜超过0.5%,重金属如砷等的含量更有严格限制。

石灰石粉的用量依畜禽种类及生长阶段而定,一般畜禽配合饲料中用量为0.5%-2%,蛋鸡和种鸡料可达7%-8%。石灰石粉过量使用会降低饲粮有机养分的消化率,还对青年鸡的肾脏有害,使泌尿系统尿酸盐过多沉积而发生炎症,甚至形成结石。若蛋鸡饲粮中石灰石粉过量,蛋壳上会附着一层薄薄的细粒,影响蛋的合格率。石灰石粉最好与贝壳粉按1:1的比例配合使用。

石灰石粉作为钙的来源,粒度以中等为好,一般猪为0.5-0.7 mm,禽为0.6-0.7 mm。对蛋鸡来讲,石灰石粉不可过细,粒度可达1.5-2.0 mm。较粗的粒度有助于保持血液中钙的浓度,满足形成蛋壳的需要,从而增加蛋壳的强度,减少蛋的破损率。如果石灰石粉过细,则会造成蛋壳颜色变浅,质脆、起沙,无光泽,软、破蛋比例增加。但要注意粗粒会影响饲料的混合均匀度。

将石灰石锻烧成氧化钙(CaO,生石灰),加水调制成石灰乳,再与二氧化碳作用生成沉淀碳酸钙,此产品细而轻,又称轻质碳酸钙。其含碳酸钙在95%以上,钙含量介于34%-38%。除用作钙源外,石灰石粉还广泛用作微量元素预混合饲料的稀释剂或载体。

(二) 贝壳粉
Shell meal

软体动物的外套膜具有一种特

factor, can be used as feed. It is generally considered that the content of magnesium in feed-grade limestone meal should not exceed 0.5%, and the content of heavy metals such as arsenic is more strictly limited.

The dosage of limestone meal depends on the type and growth stage of livestock and poultry. The dosage of limestone meal in general livestock formula feed is 0.5%-2%, and it can reach 7%-8% in the formula of laying hens and breeders. Excessive use of limestone meal will reduce the digestibility of organic nutrients in the diet, and it is also harmful to the kidneys of young chickens, causing inflammation and even stone formation due to excessive deposition of urate in the urinary system. If the limestone meal in the diet of laying hens is excessive, a thin layer of fine particles will be attached to the eggshell, which will affect the qualified rate of eggs. Limestone meal is best used with shell meal in a ratio of 1:1.

As the source of calcium, limestone meal has a medium particle size, generally 0.5-0.7 mm for pigs and 0.6-0.7 mm for poultry. For laying hens, limestone meal can not be too fine, and the particle size can be within 1.5-2.0 mm. Coarse particle size helps to maintain the concentration of calcium in the blood to meet the needs of eggshell formation, thereby increasing eggshell strength and reducing egg breakage. If the limestone meal is too fine, the eggshell will become lighter in color, brittle, sandy, dull, and the proportion of soft and broken eggs will increase. But it should be noted that coarse grains will affect the mixing uniformity of feed.

Limestone is burned into calcium oxide (CaO, limestone), which is added into water to produce lime milk, which then reacts with carbon dioxide to form precipitated calcium carbonate. This product is fine and light, also known as light calcium carbonate. The calcium carbonate content in it is above 95%, and the calcium content in it is between 34% and 38%. In addition to being used as calcium source, limestone meal is also widely used as diluent or carrier for microelement premix feed.

The mantle of mollusks has a special kind of gland cells

殊的腺细胞，其分泌物可形成保护身体柔软部分的钙化物，称为贝壳。贝壳粉是各种贝类外壳（蚌壳、牡蛎壳、蛤蜊壳、螺蛳壳等）经加工粉碎而成的粉状或粒状产品，多呈灰白色、灰色、灰褐色。主要成分为碳酸钙，其含量为90%-95%，含钙量应不低于33%。贝壳粉中还含有动物体内所必需的微量矿物质元素，如铜、镁、钾、硝、磷、锰、铁、锌等。此外，在贝壳中还含有少量氨基酸成分。

贝壳粉能促进畜禽骨骼生长，增强消化功能，增加蛋、奶的产量和改善其品质，提高抗病能力。研究表明，在蛋鸡饲粮中添加贝壳粉可以明显提高产蛋率，改善蛋壳质量，减少破蛋、软蛋。午后选择性地添加贝壳粉，能高年老蛋鸡（48周龄以后）的产蛋率、周产蛋量。

不同畜禽对贝壳粉的粒度要求为：猪，25%通过0.36 mm筛；蛋鸡，70%通过1.80 mm筛；肉鸡，60%通过0.30 mm筛。

贝壳粉内常掺杂砂石和泥土等杂质；若贝肉未除尽，加之贮存不当，堆积日久易出现发霉、腐臭等情况。所以，选购及应用时要特别注意。贝壳粉还具有吸附性、分散性、矿化性、无毒性等特点，可作为饲料添加剂的优良载体。

whose secretion can form calcifications that protect the soft parts of the body, which is called shell. Shell meal is a powdered or granular product of various shellfish shells (clam shell, oyster shell, clam shell, snail shell, etc.), which is mostly off-white, gray or gray-white. The main component is calcium carbonate, whose content is 90%-95% and calcium content should not be less than 33%. Shell meal also contains mineral microelements necessary for animals, such as copper, magnesium, potassium, nitrate, phosphorus, manganese, iron and zinc, etc. In addition, the shell also contains a small amount of amino acids.

Shell meal can promote bone growth of livestock and poultry, enhance their digestive function, increase egg and milk yield, improve egg's quality and livestock and poultry's disease resistance ability. Studies have shown that the addition of shell meal in the diet of laying hens can significantly improve the egg production rate, improve eggshell quality, and reduce broken and soft eggs. Adding shell meal selectively in the afternoon could improve the laying rate and weekly egg production of old laying hens of 48 weeks of age.

The particle size requirements of different livestock and poultry for shell meal are: for pigs, 25% through a 0.36 mm sieve; for laying hens, 70% through a 1.80 mm sieve and for broilers, 60% through a 0.30 mm sieve.

The shell meal is often mixed with impurities such as sand and soil. If the shell meat is not completely removed, coupled with improper storage, moldy and rotten odors are prone to occur on a long accumulation time. Therefore, special attention should be paid to the purchase and application. Shell meal also has the characteristics of adsorption, dispersion, mineralization and non-toxicity, which can be used as an excellent carrier of feed additives.

（三）蛋壳粉
Eggshell meal

鸡蛋的可食部分为88%，以蛋壳为主的副产物，包括残留蛋清、蛋壳内膜，占12%左右。因此，蛋壳中除含有34%-36%的钙外，尚含有7%左右的粗蛋白质和0.1%左

The edible part of eggs is 88%, and the main by-products eggshells, including residual egg white and eggshell inner membrane, account for about 12%. Therefore, in addition to 34%-36% of calcium, eggshells still contain about 7% of crude protein and about 0.1% of phosphorus. Rea-

右的磷。对蛋壳的合理利用既可以减少环境污染,又可以节省资源。

禽蛋加工厂或孵化厂废弃的蛋壳,经干燥灭菌、粉碎后即得蛋壳粉。蛋壳粉是理想的钙源饲料,其利用率高,用于产蛋鸡和产蛋鸭饲料可增加蛋壳硬度。值得注意的是,蛋壳干燥的温度应超过82 ℃,以消除传染病原。

(四) 石膏
Gypsum

石膏的主要成分为硫酸钙($CaSO_4 \cdot XH_2O$),通常是二水硫酸钙($CaSO_4 \cdot 2H_2O$),为灰色或白色的结晶粉末。石膏原料分为天然石膏和化学石膏,其中90%为天然石膏。化学石膏是指磷肥生产的副产品、磷酸工业的副产品等,最主要的是氟石膏和磷石膏,但因其含有高量的氟、砷、铝等而品质较差,使用时应加以处理。

石膏含钙量为20%-23%,含硫16%-18%,既可提供钙,又是硫的良好来源。石膏粉是防治鸡病的良药。在饲料中添加1%-2%的石膏粉,有预防鸡啄羽、啄肛的作用。在饲粮中添加0.5%-1.0%的石膏粉,并配以维生素AD粉补钙,可用于治疗产蛋期笼养疲劳症,比常用的增加骨粉、贝壳粉效果更好。石膏有生肌敛疮功效,在使用抗生素或其他药物治疗大肠杆菌病的同时,在饲粮中添加适量石膏粉可促进肠道和器官炎症的消除。石膏粉以2%的比例拌料,可用于治疗传染性法氏囊病的后遗症。

石膏粉内有掺杂滑石粉的问题,要注意识别。

此外,大理石、白云石、白垩石、方解石、熟石灰、石灰水等均可作为

sonable utilization of eggshell can not only reduce environmental pollution, but also save resources.

Eggshell meal is obtained from egg processing plant or hatchery waste eggshell after drying, sterilization and crushing. Eggshell meal is an ideal calcium source feed with high utilization rate, which can increase eggshell hardness for laying hens and laying ducks. It is worth noting that the drying temperature of eggshell should exceed 82 ℃ to eliminate infectious diseases.

The main component of gypsum is calcium sulfate ($CaSO_4 \cdot XH_2O$), usually calcium sulfate dihydrate ($CaSO_4 \cdot 2H_2O$), which is gray or white crystalline powder. Gypsum raw materials are divided into natural gypsum and chemical gypsum, of which 90% are natural gypsum. Chemical gypsum refers to the by-products of phosphate fertilizer production and phosphoric acid industry. Fluorogypsum and phosphogypsum are the most important ones, but their quality is poor due to the high content of fluorine, arsenic and aluminum, which should be treated before use.

Gypsum contains 20%-23% of calcium and 16%-18% of sulfur, which can provide calcium as well as a good source of sulfur. Gypsum powder is a good medicine to prevent chicken disease. Adding 1%-2% of gypsum powder in feed can prevent chicken feather pecking and anus pecking. Adding 0.5%-1.0% of gypsum powder to the diet and supplementing calcium with vitamin AD powder is good for the the treatment of cage-feeding fatigue during egg production, which is better than the commonly used bone meal and shell meal. Gypsum has the effect of promoting myogenicity and astringing sores. While using antibiotics or other drugs to treat E. coli diseases, adding an appropriate amount of gypsum powder in the diet can promote the elimination of intestinal and organ inflammation. Gypsum powder mixed with 2% proportion can be used to treat sequelae of infectious bursal disease.

The problem of doped talcum powder in calcium sulfate dihydrate should be recognized.

In addition, marble, dolomite, chalk stone, calcite, lime, lime water can be used as calcium supplement feed.

补钙饲料。至于利用率很高的葡萄糖酸钙、乳酸钙等有机酸钙，因其价格较高，多用于水产饲料，畜禽饲料中应用较少。其他还有甜菜制糖的副产品滤泥，也属于碳酸钙产品，是由石灰乳清除甜菜糖汁中的杂质经二氧化碳中和沉淀而成，除含碳酸钙外，还有少量的有机酸钙盐和其他微量元素。滤泥钙源性饲料尚未被很好地开发利用，如果以加工甜菜量的4%计，全国每年可生产40万～50万t此类钙源性饲料。

钙源性饲料很便宜，但不能用量过多，否则会影响钙磷平衡，使钙和磷的消化、吸收和代谢都受到影响。微量元素预混料常常使用石粉或贝壳粉作为稀释剂或载体，使用量占配比较大时，配料时应注意把其含钙量计算在内。

As for organic acid calcium such as calcium gluconate and calcium lactate with high utilization rate, because of its high price, they are mostly used for aquatic feed, and less used in livestock and poultry feed. The filter mud, a by-product of sugar beet production, is also a calcium carbonate product. It is formed by the neutralization and precipitation of carbon dioxide through the removal of impurities in sugar beet juice by lime milk. In addition to calcium carbonate, there are also a small amount of organic acid calcium salts and other microelements. The filter mud calcium-derived feed has not been well developed and utilized. If 4% of sugar beet is processed, 400,000 to 500,000 tons of such calcium-derived feed can be produced annually in China.

Calcium source feed is very cheap, but can not be used too much, otherwise it will affect the balance of calcium and phosphorus, and affect calcium and phosphorus digestion, absorption and metabolism. Stone meal or shell meal is often used as diluent or carrier in microelement premix. When the proportion of usage is large, the calcium content should be calculated in the batching.

二、磷源性饲料
Phosphorus source feed

富含磷的矿物质饲料主要是饲料级磷酸盐，如磷酸钙类、磷酸钠类、骨粉及磷矿石等。在畜禽饲粮中使用这类饲料，可促进体内新陈代谢，提高生产性能，增强机体抗病能力。但要注意，除了不同磷源有着不同的利用率外，还要注意这类原料中有害物质如氟、铝、砷等是否超标。

Mineral feeds rich in phosphorus are mainly feed-grade phosphates, such as calcium phosphate, sodium phosphate, bone meal and phosphate rock. The use of such feeds in livestock and poultry diets can promote metabolism in the body, improve production performance, and enhance the body's ability of disease resistance. However, it should be noted that in addition to the different utilization rates of different phosphorus sources, it is also necessary to notice that the harmful substances such as fluorine, aluminum, arsenic in such raw materials should not exceed the standard.

（一）磷酸钙类
Calcium phosphates

在饲料级磷酸盐中钙盐占95%以上，主要有磷酸氢钙（磷酸二钙）、磷酸二氢钙（磷酸一钙）、脱氟磷酸钙（磷酸三钙）和磷酸一二钙。

Calcium salts account for more than 95% of feed-grade phosphates, mainly including calcium hydrogen phosphate (dicalcium phosphate), calcium dihydrogen phosphate (monocalcium phosphate), defluorinated calcium phosphate (tricalcium phosphate) and monocalcium phosphate.

1. 磷酸二氢钙　磷酸二氢钙，又称磷酸一钙或过磷酸钙，纯品为白色结晶粉末，多为一水盐。市售品是以湿式法磷酸液（脱氟精制处理后再使用）或干式法磷酸液作用于磷酸二钙或磷酸三钙制成的，常含有少量未反应的碳酸钙及游离磷酸，呈酸性，吸湿性强。磷酸二氢钙含磷22%左右，钙15%左右，主要用于水产动物饲料。本品磷高钙低，在配制饲粮时可用于调整钙磷平衡。使用磷酸二氢钙时应注意脱氟处理，含氟量不得超过标准（HG 2861—1997）。

2. 磷酸氢钙　也称磷酸二钙，为白色或灰白色粉末或粒状产品，是一种枸溶性磷酸盐，又分为无水盐（$CaHPO_4$）和二水盐（$CaHPO_4 \cdot 2H_2O$）两种，一般生产中使用的多是后者，其稍溶于水，易溶于酸，不溶于乙醇，能溶于柠檬酸铵溶液。由于 $CaHPO_4 \cdot 2H_2O$ 属于一种热敏性物质，其结晶水的键合力很脆弱，在反应或干燥的过程中很容易因受热（75 ℃）而失去结晶水，变成一水或无水物磷酸氢钙（$CaHPO_4 \cdot H_2O$ 或 $CaHPO_4$）。磷酸氢钙一般是在干式法磷酸液或精制湿式法磷酸液中加入石灰乳或磷酸钙而制成，市售品中除含有无水磷酸二钙外，还含有少量的磷酸一钙及未反应的磷酸钙。

磷酸氢钙是世界上产量最大、使用最普遍的饲料磷酸盐品种，在我国产量占饲料磷酸盐总量的90%以上。磷酸氢钙同磷酸一钙、磷酸三钙相比，具有质优价廉的优势，其外观、流动性比磷酸一钙、

1. Calcium dihydrogen phosphate. Calcium dihydrogen phosphate, also known as monocalcium phosphate or calcium superphosphate, is pure white crystalline powder, mostly monohydrate salt. The commercial products are made of wet-process phosphoric acid solution (used after defluorination refining) or dry-process phosphoric acid solution acting on dicalcium phosphate or tricalcium phosphate. They often contain a small amount of unreacted calcium carbonate and free phosphoric acid, which are acidic and have strong hygroscopicity. Calcium dihydrogen phosphate contains about 22% of phosphorus and about 15% of calcium, which is mainly used for aquatic animal feed. The product has high phosphorus and low calcium, which can be used to adjust calcium and phosphorus balance in formulated diets. Attention should be paid to defluorination when using calcium dihydrogen phosphate, and the fluorine content should not exceed the standard (HG 2861—1997).

2. Calcium hydrogen phosphate. Calcium hydrogen phosphate, also known as dicalcium phosphate, is a white or off-white powder or granular product. It is a citrate-soluble phosphate, which is also divided into anhydrous salt ($CaHPO_4$) and dihydrate salt ($CaHPO_4 \cdot 2H_2O$). The latter is generally used in production, which is slightly soluble in water, easily soluble in acid, insoluble in ethanol, and soluble in ammonium citrate solution. Since $CaHPO_4 \cdot 2H_2O$ is a kind of thermally sensitive material, the bond force of its crystal water is very fragile. During the reaction or drying process, it tends to lose the crystal water due to heating (75 ℃) and become monohydrate or anhydrous calcium hydrogen phosphate ($CaHPO_4 \cdot H_2O$ or $CaHPO_4$). Calcium hydrogen phosphate is generally made by adding lime milk or calcium phosphate to dry phosphoric acid solution or refined wet phosphoric acid solution. The commercial products contain a small amount of monocalcium phosphate and unreacted calcium phosphate in addition to anhydrous dicalcium phosphate.

Calcium hydrogen phosphate is the largest and most widely used feed phosphate variety in the world, which accounts for more than 90% of the total feed phosphate in China. Compared with monocalcium phosphate and tricalcium phosphate, calcium hydrogen phosphate has the advantage of high quality and low price, and its appearance and

酸三钙好。其磷钙为1∶1.29，与动物骨骼中的磷钙比最为接近，同时又能全部溶解于动物胃酸中，其生物效价约比磷酸三钙高10%，钙、磷吸收率较骨粉高10%。在饲料中的添加量为：猪、禽2%，反刍动物（牛）2%～3%。饲料级磷酸氢钙应注意脱氟处理，含氟量不得超过其质量标准（HG 2636—2000）。

常见的磷酸氢钙掺假物有石粉、骨粉、滑石粉、磷矿粉、磷酸三钙、农用过磷酸钙、磷酸的混合物以及石粉加磷酸的混合物等。这些掺假产品，有的含钙磷较高，但畜禽利用率低，还含有大量的氟，严重危害畜禽健康，使用时应注意检测。

3. 脱氟磷酸钙　又称磷酸三钙，纯品为白色无臭粉末。饲料用磷酸三钙常由磷酸废液制造，灰色或褐色，并有臭味，分为一水盐 $[Ca_3(PO_4)_2 \cdot H_2O]$ 和无水盐 $[Ca_3(PO_4)_2]$ 两种，以后者居多。磷酸三钙经脱氟处理后，称脱氟磷酸钙，为灰白色或茶褐色粉末，含钙29%以上，含15%～18%的磷，含氟量在0.12%以下。

该产品在国外主要用于家禽（肉鸡和火鸡）饲料。其作为饲料磷源对肉鸡的生长、成骨作用与磷酸氢钙相似，可以替代其他磷源应用于肉鸡饲料，从而降低饲料成本。磷酸三钙由于相对生物学效价较低，国内使用量很少。

脱氟磷酸钙与磷酸氢钙相比，还具有生产工艺简单、原料易得、投资少、能耗低和成本低的优点。

fluidity are better than monocalcium phosphate and tricalcium phosphate. Its phosphorus-calcium ratio is 1∶1.29, which is the closest to that in animal bones, and can be completely dissolved in animal gastric acid. Its biological potency is about 10% higher than that of tricalcium phosphate, and the absorption rate of calcium and phosphorus is 10% higher than that of bone meal. The addition amount in feed is: 2% for pigs and poultry, and 2%–3% for ruminants (cattle). Feed-grade calcium hydrogen phosphate should be used after defluorination treatment and fluoride content should not exceed its quality standard (HG 2636—2000).

Common calcium hydrogen phosphate dopants include stone meal, bone meal, talc meal, phosphate rock meal, tricalcium phosphate, agricultural superphosphate, mixture of phosphoric acid and mixture of stone meal and phosphoric acid. Some of these adulterated products contain high calcium and phosphorus, but the utilization rate for livestock and poultry is low, and they also contain a large amount of fluorine, which seriously endangers the health of livestock and poultry. They should be tested before using.

3. Defluorinated calcium phosphate　Defluorinated calcium phosphate, also known as tricalcium phosphate, is white odorless meal. Tricalcium phosphate for feed is usually made of waste phosphoric acid, gray or brown, and has a bad smell. It can be divided into monohydrate $[Ca_3(PO_4)_2 \cdot H_2O]$ and anhydrous $[Ca_3(PO_4)_2]$, the latter being the majority. After defluorination treatment, tricalcium phosphate is called defluorinated calcium phosphate, which is gray white or tea brown powder, containing more than 29% of calcium, 15%–18% of phosphorus and less than 0.12% of fluorine.

The product is mainly used for poultry (broiler and turkey) feed abroad. As a feed phosphorus source, it has similar effects on growth and osteogenesis of broilers as calcium hydrogen phosphate, and can replace other phosphorus sources in broiler feed, so as to reduce feed cost. Tricalcium phosphate is rarely used in China due to its low biological potency.

Compared with calcium hydrogen phosphate, defluorinated calcium phosphate also has the advantages of simple production process, easy availability of raw materials, low

4. 磷酸一二钙 磷酸一二钙 [$CaHPO_4 \cdot 2H_2O, Ca(H_2PO_4)_2 \cdot H_2O$] 是磷酸二氢钙与磷酸氢钙的共晶结合物,是一种水溶性磷酸盐与枸溶性磷酸盐相结合的矿物质饲料,其中磷酸二氢钙是水溶性磷酸盐,约占磷酸一二钙分子量的60%。磷酸一二钙是20世纪90年代末期,由欧洲科学家研制的一种新磷源,用于取代传统的饲料磷酸盐。随着人类对环境保护的重视程度越来越高,欧美国家在很多厂家大力推广使用粒状产品。粒状产品具有以下优点:① 产品总磷较高,与传统磷酸氢钙相比可减少添加量,增加饲料配方空间,便于提高饲料品质,降低饲料生产成本。② 产品水溶性磷含量高,且颗粒在动物肠胃中停留时间较长,有利于吸收利用。③ 生物学效价较高(>75%),动物粪便中残留的磷较少,在提高磷资源利用率的同时有利于环保。④ 粒状产品在使用中不易起尘,能减少物料在运输和加工过程中的损失,有利于改善加工环境。⑤ 产品密度为 $0.8-0.9$ g/cm³,为多棱形晶体,在预混料时有较好的亲和力,不会产生沉淀或浮顶等不均现象。⑥ 产品呈微酸性,可改变动物口感,提高动物的采食量。

国家发展和改革委员会于2005年7月发布了饲料级磷酸一二钙产品的化工行业标准(HG/T 3776—2005)。

(二)磷酸钾类
Potassium phosphate

1. 磷酸二氢钾 磷酸二氢钾又称磷酸一钾,分子式为 KH_2PO_4,

4. Monocalcium phosphate. Monocalcium phosphate [$CaHPO_4 \cdot 2H_2O$, $Ca(H_2PO_4)_2 \cdot H_2O$] is the eutectic combination of calcium dihydrogen phosphate and calcium hydrogen phosphate. It is a kind of mineral feed combined with water-soluble phosphate and citrate-soluble phosphate. Calcium dihydrogen phosphate is water-soluble, accounting for about 60% of the molecular weight of monocalcium phosphate. Monocalcium phosphate is a new phosphorus source developed by European scientists in the late 1990s to replace traditional feed phosphate. With the increasing emphasis on environmental protection, many European and American countries have vigorously promoted the use of granular products in many manufacturers. The granular products have the following advantages: ① The total phosphorus of the product is higher. Compared with the traditional calcium hydrogen phosphate, it can reduce the addition amount, increase the feed formula space, facilitate the improvement of feed quality, and reduce feed production cost. ② The product has high water-soluble phosphorus content and long residence time of particles in animal gastrointestinal tract, which is conducive to absorption and utilization. ③ The biological potency is higher (>75%), and the residual phosphorus in animal feces is less, which is beneficial to environmental protection while improving the utilization rate of phosphorus resources. ④ Granular products are not prone to dust, which can reduce the loss of materials in transportation and processing, and improve the processing environment. ⑤ The density of the product is $0.8-0.9$ g/cm³, which is a polygonal crystal. It has good affinity when premixed and will not cause unevenness such as precipitation or floating roof. ⑥ The product is slightly acidic, which can change the animal's taste and increase the animal's intake.

In July 2005, National Development and Reform Commission issued the chemical industry standard of feed-grade monocalcium phosphate (HG/T 3776—2005).

1. Potassium dihydrogen phosphate, also known as monopotassium phosphate, is a colorless tetragonal or white

含磷22%以上,钾28%以上,为无色四方结晶或白色结晶性粉末,水溶性好,易被动物吸收利用,可同时提供磷和钾。适当使用有利于动物体内的电解质平衡,促进动物生长发育和生产性能的提高。饲料级磷酸二氢钾应符合HG 2860—1997的要求。磷酸二氢钾有潮解性,宜保存于干燥处。

2. 磷酸氢二钾 磷酸氢二钾也称磷酸二钾,分子式为$K_2HPO_4 \cdot 3H_2O$,呈白色结晶或无定型粉末。磷酸氢二钾一般含磷13%以上,钾34%以上,应用同磷酸一钾。

crystalline meal with molecular formula of KH_2PO_4, containing more than 22% of phosphorus and 28% of potassium. It has good water solubility and is easily absorbed and used by animals, and can provide phosphorus and potassium at the same time. Appropriate use is beneficial to the electrolyte balance in animals and promotes the growth and production performance of animals. Feed-grade potassium dihydrogen phosphate should meet the requirements of HG 2860—1997. Potassium dihydrogen phosphate is deliquescent and should be kept in a dry place.

2. Dipotassium hydrogen phosphate, also known as dipotassium phosphate, has a molecular formula of $K_2HPO_4 \cdot 3H_2O$. It is white crystalline or amorphous powder. Dipotassium hydrogen phosphate generally contains more than 13% of phosphorus and 34% of potassium. The application is the same as monopotassium phosphate.

(三) 磷酸钠类
Sodium phosphate

1. 磷酸二氢钠 磷酸二氢钠又称磷酸一钠,有无水物(NaH_2PO_4)及二水物($NaH_2PO_4 \cdot 2H_2O$)两种,均为白色结晶性粉末。无水物含磷约25%,含钠约19%。因其不含钙,故在钙要求低的饲料中可充当磷源,在调整高钙、低磷配方时,不会改变钙的比例。磷酸二氢钠有潮解性,宜保存于干燥处。

2. 磷酸氢二钠 磷酸氢二钠又称磷酸二钠,分子式为$Na_2HPO_4 \cdot XH_2O$,呈白色无味的细粒状,无水物一般含磷18%-22%,钠27%-32.5%,应用同磷酸一钠。

1. Sodium dihydrogen phosphate, also known as monosodium phosphate, is white crystalline powder with anhydrous (NaH_2PO_4) and dihydrate ($NaH_2PO_4 \cdot 2H_2O$). Anhydrous contains about 25% of phosphorus and 19% of sodium. Because it does not contain calcium, it can be used as a phosphorus source in feeds with low calcium requirements, and the proportion of calcium will not be changed when adjusting the formula of high calcium and low phosphorus. Sodium dihydrogen phosphate is deliquescent and should be kept in a dry place.

2. Disodium hydrogen phosphate is also called disodium phosphate. Its molecular formula is $Na_2HPO_4 \cdot XH_2O$, which is white, odorless and fine-grained. Anhydrous generally contains 18%-22% of phosphorus and 27%-32.5% of sodium. The application is the same as monosodium phosphate.

(四) 其他磷酸盐
Other phosphates

1. 磷酸铵 本品为饲料级磷酸或湿式处理的脱氟磷酸中和后的产品,含氮9%以上,含磷23%以上,含氟量不可超过含磷量的1%,

1. Ammonium phosphate. The product is neutralized by feed-grade phosphoric acid or wet defluorinated phosphoric acid. It contains more than 9% of nitrogen and 23% of phosphorus. The fluorine content cannot exceed 1% of

含砷量不可超过 25 mg/kg,含铅等重金属应在 30 mg/kg 以下。对于反刍动物,本品可用来补充磷和氮,但磷酸铵中所含的氮量换算成粗蛋白质后,其量在饲粮中不可超过 2%。对于非反刍动物,本品仅能当磷源使用,且要求其所提供的氮换算成粗蛋白质后,其量在饲粮中不可超过 1.25%。

2. 磷酸液为磷酸的水溶液,一般以 H_3PO_4 表示,应保证最低含磷量,含氟量不可超过含磷量的 1%。本品具有强酸性,使用不方便,可在青贮时喷加,也可以与尿素、糖蜜及微量元素混合制成牛用液体饲料。

3. 磷酸脲 分子式为 $H_3PO_4 \cdot CO(NH_2)_2$,由尿素与磷酸作用生成,呈白色结晶性粉末,易溶于水,其水溶液呈酸性。本品利用率较高,是反刍动物良好的饲料添加剂。

4. 磷矿石粉为磷矿石粉碎后的产品,常含有超过允许量的氟和其他如砷、铅、汞等杂质。用作饲料时,必须进行脱氟处理,使其符合允许量标准。

5. 此外,磷酸盐类还有磷酸氢二铵、磷酸氢镁、三聚磷酸钠、次磷酸盐、焦磷酸盐等。除了饲料级磷酸钙盐外,其他磷酸盐由于价格偏高,主要用于特种水产和反刍动物饲料中。

(五) 骨粉
Bone meal

骨占动物体重的 20%~30%,骨粉是以骨为原料加工而成的一种矿物质饲料。其中含有丰富的矿物质,主要是羟磷灰石晶体

the phosphorus content. The arsenic content cannot exceed 25 mg per kg. The heavy metals such as lead should be less than 30 mg per kg. For ruminant animals, the product can be used to supplement phosphorus and nitrogen, but the amount of nitrogen contained in ammonium phosphate can not exceed 2% in the diet when converted to crude protein. For non-ruminant animals, this product can only be used as a phosphorus source, and the nitrogen provided by it is required to be converted into crude protein, and its amount in the diet should not exceed 1.25%.

2. Phosphoric acid solution is an aqueous solution of phosphoric acid is generally expressed as H_3PO_4, and the minimum phosphorus content should be guaranteed, and the fluorine content should not exceed 1% of the phosphorus content. The product has strong acidity and is inconvenient to use. It can be sprayed during silage and mixed with urea, molasses and microelements to make liquid feed for cattle.

3. Urea phosphate. The molecular formula is $H_3PO_4 \cdot CO(NH_2)_2$, which is generated by the interaction of urea and phosphoric acid. It is white crystalline meal, easily soluble in water, and its aqueous solution is acidic. This product has high utilization rate and is a good feed additive for ruminants.

4. Phosphate rock powder. Phosphate rock powder is a product of smashed phosphate rock, which often contains more than the allowable amount of fluorine and other impurities such as arsenic, lead and mercury. When used as feed, it must be defluoridated to meet the allowable standard.

5. In addition, phosphates include diammonium hydrogen phosphate, magnesium hydrogen phosphate, sodium tripolyphosphate, hypophosphate, pyrophosphate, etc. In addition to feed-grade calcium phosphate, other phosphates are mainly used in special aquatic products and ruminant feed due to their high prices.

Bone accounts for 20%–30% of animal weight. Bone meal is a mineral feed made from bones. It is rich in minerals, mainly hydroxyapatite crystal $[Ca_{10}(PO_4)_6(OH)_2]$ and anhydrous calcium hydrogen phosphate ($CaHPO_4$).

[Ca$_{10}$(PO$_4$)$_6$(OH)$_2$]和无水型磷酸氢钙(CaHPO$_4$),在其表面吸附了 Ca^{2+}、Mg^{2+}、Na$^+$、Cl$^-$、HCO$_3^-$、F$^-$及柠檬酸根等离子。骨粉除了含有钙和磷外,还含有少量粗蛋白质和动物必需的微量元素,如 Co、Cu、Fe、Mn、Si、Zn 等。

骨粉中钙磷比例为 2∶1,是动物体内吸收钙磷的最佳比例。另外,骨粉中的钙以羟磷灰石晶体的形式存在,该结晶与胶原纤维结合在一起,当胶原纤维被酶解后,羟磷灰石晶体部分被解离,钙转化为极易被动物吸收的氨基酸钙。因此,骨粉是补充家畜钙、磷需要的良好来源,被称为钙磷平衡调节剂。

骨粉按加工方法可分为煮骨粉、蒸制骨粉、脱胶骨粉和焙烧骨粉等。

1. 煮骨粉　原料骨经开放式锅炉煮沸,直至附着组织脱落,再经粉碎而制成。这种方法制得的骨粉色泽发黄,骨胶溶出少,蛋白质和脂肪含量较高,但易吸湿腐败,适口性差,不易久存。

2. 蒸制骨粉　原料骨在高压(2.03 kPa)蒸汽条件下加热,除去大部分蛋白质及脂肪,使骨骼变脆,加以压榨、干燥、粉碎而制成。一般含钙 24%、磷 10% 左右,含粗蛋白质 10%。这种骨粉含蛋白质较少,但其色泽洁白,易于消化,无特殊气味。

3. 脱胶骨粉　也称特级蒸制骨粉,制法与蒸制骨粉基本相同。用 40.5 kPa 压力蒸制处理或利用抽出骨胶的骨骼经蒸制处理而得到,由于骨髓和脂肪几乎全部除去,故无异臭,色泽洁白,可长期贮存,

Ca^{2+}, Mg^{2+}, Na$^+$, Cl$^-$, HCO$_3^-$, F$^-$ and citrate ions are adsorbed on its surface. In addition to calcium and phosphorus, bone meal also contains a small amount of crude protein and microelements necessary for animals, such as Co, Cu, Fe, Mn, Si and Zn.

The ratio of calcium and phosphorus in bone meal is 2∶1, which is the best proportion for the absorption of calcium and phosphorus in animals. In addition, calcium in bone meal exists in the form of hydroxyapatite crystals, which are combined with collagen fibers. When collagen fibers are enzymatically hydrolyzed, part of hydroxyapatite crystals are dissociated, and calcium is easily converted into amino acid calcium that is easily absorbed by animals. Therefore, bone meal is a good source of calcium and phosphorus for livestock, which is called calcium and phosphorus balance regulator.

Bone meal can be divided into boiled bone meal, steamed bone meal, degummed bone meal and roasted bone meal according to the processing method.

1. Boiled bone meal　The raw material bone is boiled in an open boiler until the attached tissue falls off, and then crushed. The bone meal prepared by this method is yellow in color with less bone glue dissolution, and higher in protein and fat content. However, it tends to absorb moisture and has poor palatability, so it is not easy to store for a long time.

2. Steamed bone meal　Raw bones are heated under high pressure (2.03 kPa) steam, through which most of the protein and fat is removed and the bones become brittle, which are then pressed, dried and crushed. Generally it contains 24% of calcium, 10% of phosphorus and 10% of crude protein. This kind of bone meal contains less protein, but its color is white, easy to digest, and has no special smell.

3. Degummed bone meal　It is also called special-grade steamed bone meal. The preparation method is basically the same as that of steamed bone meal. It is obtained by steaming treatment with 40.5 kPa pressure or by steaming the bones from which bone glue is extracted. Since the bone marrow and fat are almost completely removed, there

是一种质量稳定可靠、卫生安全的优质钙磷饲料。

4. 焙烧骨粉（骨灰） 将骨骼堆放在金属容器中经烧制而成，这是利用被细菌污染的废弃骨骼的可靠方法，充分烧透既可灭菌又易粉碎。

骨粉是我国配合饲料中常用的磷源饲料。骨粉的含氟量较低，只要杀菌消毒彻底，便可安全使用。一般在猪、鸡饲料中添加量为1%－3%。值得注意的是，用简易方法生产的骨粉，即不经脱脂、脱胶和热压灭菌而直接粉碎制成的生骨粉，因含有较多的脂肪和蛋白质，易腐败变质。尤其是品质低劣、有异臭、呈灰泥色的骨粉，常携带大量病菌，用于饲料易引发疾病传播。有的兽骨收购场地，为避免蝇蛆繁殖，喷洒敌敌畏等药剂而使骨粉带毒，这种骨粉绝对不能用作饲料。

欧洲一些国家因饲料安全问题已明令禁止在饲料中添加骨粉。我国禁止在反刍动物饲料中添加使用肉骨粉等动物源性原料。随着矿物质磷酸盐的逐步成熟，动物骨粉由于质量不稳定、易传染疾病等原因，在饲料工业中的用量将呈现逐渐减少的趋势。

is no odor, and the color is white, which can be stored for a long time. It is a high-quality calcium and phosphorus feed with stable quality, reliable hygiene and safety.

4. Roasted bone meal (bone ash) It is prepared by stacking bones in metal containers and firing them. This is a reliable method to use the abandoned bones contaminated by bacteria. It can be sterilized and easily crushed after being fully burned.

Bone meal is a commonly used phosphorus source feed in formulated feeds in China. The fluorine content of bone meal is low, and it can be used safely as long as it is sterilized thoroughly. The percentage added in pig and chicken feed is generally 1%－3%. It is worth noting that the bone meal produced by a simple method, that is, the raw bone meal is directly crushed without degreasing, degumming and hot pressing sterilization, is prone to spoilage due to its high content of fat and protein. In particular, bone meal with poor quality, peculiar smell, and plaster color, often carries a large number of bacteria, which can easily cause disease transmission when used in feed. In some animal bone purchase sites, in order to avoid maggots reproduction, dichlorvos and other agents are sprayed to make bone poisonous. Such bone powder can not be used as feed.

Some European countries have explicitly banned the addition of bone meal in feed due to feed safety issues. It is forbidden to add animal-derived raw materials such as meat and bone meal in ruminant feed in China. With the gradual maturity of mineral phosphates, the amount of animal bone meal in the feed industry will gradually decrease due to its unstable quality and infectious diseases.

三、钠源性饲料
Sodium source feed

畜禽常用的植物性饲料中钠的含量较低，一般不能满足需要。钠是动物不可缺少的重要矿物质元素。缺乏时，会使家畜的食欲降低、生长缓慢、被毛粗糙无光泽。产奶量及繁殖性能降低，饲料利用率下降，有时会出现异食癖。因此，进行

The content of sodium in the plant feed commonly used by livestock and poultry is low and generally cannot meet the needs. Sodium is an indispensable important mineral element for animals. Lack of sodium will cause decrease of the appetite, slow growth, and rough and dull coat, lower milk yield and reproductive performance, lower feed utilization rate, and sometimes pica occurs of livestock. Therefore,

钠源性饲料补充是关键所在。

supplement of sodium from sodium source feed is necessary.

（一）氯化钠
Sodium chloride

氯化钠一般称为食盐,包括海盐、井盐和岩盐3种。精制食盐含氯化钠99%以上,粗盐含氯化钠95%。纯净的食盐含氯60.3%、钠39.7%,此外尚含有少量的钙、镁、硫等杂质。食用盐为白色细粒,工业用盐为粗粒结晶。

大部分植物性饲料的钠和氯的含量较少,相反含钾丰富。因此,以植物性饲料为主的畜禽,应增加食盐的饲喂。食盐除了具有维持体液渗透压和酸碱平衡的作用外,还是一种良好的调味剂,可刺激唾液分泌,提高饲料适口性,增强动物食欲。

食盐的供给量要根据家畜的种类、体重、生产能力、季节和饲粮组成等综合考虑。一般食盐在风干饲粮中的用量为:牛、羊、马等草食家畜约为1%,猪和家禽以0.25%-0.5%为宜。饲料中使用食盐时应注意:① 最好用细盐均匀拌入饲料中。② 充分考虑增加摄取食盐量的因素,除饲料中日常添加的食盐外,还应考虑鱼粉、鱼干、地下水中的含盐量。③ 添加食盐要保证充足的饮水。④ 夏季可适当提高食盐用量。⑤ 添加食盐要适量,并要准确计量,谨防食盐中毒。

草食家畜需要钠和氯较多,对食盐的耐受量较大,很少发生食盐中毒。但是猪和家禽,尤其是家禽,饲粮中食盐过量或混合不匀易引起中毒。雏鸡饲料中添加0.7%以上的食盐,则会导致生长受阻,甚至有死亡现象。产蛋鸡饲料中含盐超过

Sodium chloride is generally called salt, including sea salt, well salt and rock salt. Refined salt contains more than 99% of sodium chloride, and crude salt contains 95% of sodium chloride. The pure salt contains 60.3% of chlorine and 39.7% of sodium, and also contains a small amount of impurities such as calcium, magnesium and sulfur. Edible salt is white fine-grained, and industrial salt is coarse-grained crystal.

Most of the plant feeds contain less sodium and chlorine. On the contrary, they are rich in potassium. Therefore, livestock and poultry, which are mainly plant-based, should be supplemented with salt in the feed. In addition to maintaining body fluid osmotic pressure and acid-base balance, salt is also a good condiment, which can stimulate saliva secretion, improve feed palatability and enhance animal appetite.

The supply of salt should be considered comprehensively according to the type, weight, production capacity, season and feed composition of livestock. Generally, the amount of salt used in air-dried diets is about 1% for cattle, sheep, horses and other herbivorous livestock, and 0.25%-0.5% for pigs and poultry. Attention should be paid to the use of salt in feed: ① It is best to mix fine salt into feed evenly. ② Fully consider the factors of increasing the intake of salt. In addition to the daily salt added in feed, the salt content in fish meal, dried fish and groundwater should also be considered. ③ Sufficient drinking water should be ensured when adding salt. ④ The salt consumption can be appropriately increased in summer. ⑤ Salt should be added in accurately measured amount to prevent salt poisoning.

Herbivorous livestock need more sodium and chlorine, and have a greater tolerance to salt, and rarely suffer from salt poisoning. However, pigs and poultry, especially poultry, are prone to poisoning due to excessive salt or uneven mixing in diets. Adding more than 0.7% of salt to chicken feed can cause growth retardation and even death. When the salt content in the feed of laying hens exceeds 1%, it can

1%时，可引起饮水增多，粪便变稀，产蛋率下降。猪一次采食超过每千克体重1g左右的食盐，同时饮水不足，就会发生食盐中毒。

在缺碘地区，可采用碘化食盐，如无出售，可以自配，在食盐中混入碘化钾，使碘的含量达到0.007%。配合时，要注意使碘分布均匀，如配合不匀，可引起碘中毒。再者碘易挥发，应注意密封保存。补饲食盐时，除了直接拌在饲料中外，也可以以食盐为载体，制成微量元素添加预混料。在缺硒、铜、锌地区，也可以分别制成含亚硒酸钠、硫酸铜、硫酸锌或氧化锌的食盐砖或食盐块供放牧家畜舔食。由于食盐吸湿性强，在相对湿度75%以上时即开始潮解，所以作为载体的食盐必须保持含水量在0.5%以下，并妥善保管。

lead to drinking more water, diluted feces and decreased egg production rate. Pigs fed with more than 1g of salt per kilogram of body weight at a time without enough water may suffer from salt poisoning.

In iodine-deficient areas, iodized salt can be used. If it is not sold, it can be self-prepared. Potassium iodide is mixed in the salt to increase the iodine content to 0.007%. In coordination, attention should be paid to making the iodine evenly distributed. If the coordination is not uniform, it can cause iodine poisoning. Moreover, iodine is volatile and should be sealed and stored. When supplementing salt, salt can also be used as a carrier to make a premix for adding trace elements in addition to direct addition. In areas with selenium deficiency, copper deficiency and zinc deficiency, salt bricks or salt blocks containing sodium selenite, copper sulfate, zinc sulfate or zinc oxide can also be made for licking by grazing livestock. Due to the strong hygroscopicity of salt, it begins to decompose when the relative humidity is above 75%, the salt used as a carrier must maintain the water content below 0.5% and be properly preserved.

（二）碳酸氢钠
Sodium bicarbonate

碳酸氢钠又名小苏打，分子式为$NaHCO_3$，为无色结晶粉末，无味，略具潮解性，其水溶液因水解而呈微碱性，受热易分解放出二氧化碳。碳酸氢钠含钠27%以上，生物利用率高，是优质的钠源性矿物质饲料之一。

碳酸氢钠不仅可以补充钠，更重要的是具有缓冲作用，其营养生理作用实质为维持体内电解质平衡和酸碱平衡，促进畜禽对饲料的消化吸收和利用，增强机体免疫力和抗应激的能力。研究证实，奶牛和肉牛饲粮中添加碳酸氢钠可以调节瘤胃pH值，给微生物提供一个良好的生长和繁殖环境，防止精料型饲粮引起的代谢性疾病，提高增重、

Sodium bicarbonate, also known as baking soda, is a colorless crystalline powder with a molecular formula of $NaHCO_3$. It is tasteless and slightly deliquescent. Its aqueous solution is slightly alkaline due to hydrolysis, and it tends to decompose and release carbon dioxide when heated. Sodium bicarbonate contains more than 27% of sodium and has high bioavailability. It is one of the high-quality sodium source mineral feeds.

Sodium bicarbonate can not only supplement sodium, but more importantly, it has a buffering effect. Its nutritional and physiological function is essentially to maintain the electrolyte balance and acid-base balance in the body, promote the digestion, absorption and utilization of feed by livestock and poultry, and enhance the immunity and anti-stress ability of the body. Studies have confirmed that the addition of sodium bicarbonate in the diets of cows and beef cattle can regulate rumen pH value, provide a good growth and reproduction environment for microorganisms, prevent

产奶量和乳脂率。在奶牛和肉牛饲粮中，碳酸氢钠的一般添加量为 0.5%-2.0%，与氧化镁配合使用效果更佳。蛋鸡饲粮中添加碳酸氢钠（0.1%-1.0%），具有提高产蛋量、改善蛋壳质量的作用；种鸡饲粮中添加 0.4% 碳酸氢钠，可提高受精率 4%-5%；2 周龄后的肉鸡饲粮中添加 0.7% 的碳酸氢钠，日增重提高 5% 左右；夏季在肉鸡和蛋鸡饲粮中添加 0.5% 左右的碳酸氢钠，可减缓热应激，防止生产性能的下降。在猪饲料中添加 0.5%-0.8% 的碳酸氢钠，可使肥育猪增重加快，仔猪成活率提高，特别是在饲料中赖氨酸不足时，碳酸氢钠既可弥补赖氨酸的不足，又可使猪加快生长，使饲料利用率提高 8%-10%。

添加碳酸氢钠应注意：为保持适宜的电解质平衡，应适当减少食盐添加量；碳酸氢钠不很稳定，不宜久置，最好随拌随喂。

（三）硫酸钠
Sodium sulfate

硫酸钠又名芒硝，分子式为 Na_2SO_4，为白色粉末，含钠 32% 以上，含硫 22% 以上，生物利用率高，既可补钠又可补硫，特别是补钠时不会增加气含量，是优良的钠、硫源之一。据报道，在蛋鸡饲粮中添加 0.3% 的硫酸钠，可提高产蛋率 2.4%，蛋重提高 3.7%；添加 1% 的硫酸钠，可有效控制啄羽、啄肛等恶癖。一般建议，高温季节的产蛋鸡和肉仔鸡饲粮中硫酸钠最大添加量为 0.5%，其他时间鸡群可按 0.3%-0.4% 添加。在猪饲粮中添加 0.4%-0.6%，可使其食欲旺盛，被

metabolic diseases caused by concentrate diets, and improve weight gain, milk yield and milk fat rate. In dairy and beef cattle diets, the general addition of sodium bicarbonate is 0.5%-2.0%, and the effect is better when used in conjunction with magnesium oxide. Adding sodium bicarbonate (0.1%-1.0%) to the diet of laying hens can improve egg production and eggshell quality. The fertilization rate can be increased by 4%-5% by adding 0.4% of sodium bicarbonate to the diet of breeder chickens. Adding 0.7% of sodium bicarbonate to the diet of broilers after 2 weeks of age increases the daily gain by about 5%. The addition of about 0.5% of sodium bicarbonate in broiler and laying hens diets in summer can alleviate the heat stress and prevent the decline of production performance. Adding 0.5%-0.8% of sodium bicarbonate into pig feed can accelerate the weight gain of finishing pigs and improve the survival rate of piglets. Especially when lysine is insufficient in feed, sodium bicarbonate can not only make up for the deficiency of lysine, but also accelerate the growth of pigs and increase the feed utilization rate by 8%-10%.

In order to maintain the appropriate electrolyte balance, the amount of salt added should be appropriately reduced when adding sodium bicarbonate. Sodium bicarbonate is not very stable and should not be kept for a long time, and it is best to feed it with mixing.

Sodium sulfate, also known as mirabilite, is a white powder with molecular formula Na_2SO_4. The content of sodium is more than 32%, and the content of sulfur is more than 22%. The biological utilization rate is high. It can supplement sodium and sulfur, especially when supplementing sodium, it will not increase the gas content. It is one of the excellent sources of sodium and sulfur. It is reported that adding 0.3% of sodium sulfate in the diet of laying hens can increase egg production rate by 2.4% and egg weight by 3.7%. Adding 1% of sodium sulfate can effectively control feather pecking and anus pecking. It is generally recommended that the maximum addition amount of sodium sulfate in the diets of laying hens and broilers in high temperature season is 0.5%, and the addition amount of sodi-

毛光亮。家兔饲粮中添加0.2%的硫酸钠，具有提高生产性能和毛皮质量的作用。在反刍动物饲粮中添加硫酸钠，对改善动物饲料中的氮素及其他营养物质如粗纤维的消化吸收利用，促进体内蛋白质尤其是含硫氨基酸的生物合成和氧化还原过程作用明显。一般按饲粮干物质量的0.5%-0.8%添加含水结晶硫酸钠，添加过量易引起硫中毒。

硫酸钠作为添加剂使用时应注意以下几点：① 添加硫酸钠时，应根据饲料和畜禽种类等选择适宜量。② 应在饲粮粗蛋白质稍低（低1%-2%）而缺乏含硫氨基酸的情况下添加。③ 不能单独添加硫酸钠，应与蛋氨酸同时添加才能起协同作用，并要注意二者的比例。④ 添加硫酸钠时，需注意饲料中钠和氯的含量。⑤ 蛋鸡饲粮中添加硫酸钠，必须注意含钙量要符合标准，否则会影响蛋壳品质。⑥ 必须混合均匀。⑦ 仔猪饲粮中不宜添加硫酸钠。

um sulfate can be 0.3%-0.4% in other periods. Adding 0.4%-0.6% to the diet of pigs can increase their appetite and coat quality. Adding 0.2% of sodium sulfate in rabbit diet can improve production performance and fur quality. Adding sodium sulfate in ruminant diets can improve the digestibility and utilization of nitrogen and other nutrients such as crude fiber in animal feed, and promote the biosynthesis and redox process of proteins, especially sulfur-containing amino acids. Generally water crystalline sodium sulfate is added according to 0.5%-0.8% of the dry weight of the diet and excessive amount easily leads to sulfur poisoning.

The following points should be paid attention to when sodium sulfate is used as an additive: ① When sodium sulfate is added, the appropriate amount should be selected according to feed, livestock and poultry, etc. ② Sodium sulfate should be added when the dietary crude protein is slightly low (1%-2% lower) and lack of sulfur-containing amino acids. ③ Sodium sulfate cannot be added alone, and methionine should be added at the same time to play a synergistic effect, and the proportion of the two should be noted. ④ When adding sodium sulfate, attention should be paid to the content of sodium and chlorine in feed. ⑤ When adding sodium sulfate to the diet of laying hens, it must be noted that the calcium content should meet the standard, otherwise it will affect the eggshell quality. ⑥ It must be mixed evenly. ⑦ Sodium sulfate should not be added in piglet diets.

四、其他常量矿物质饲料
Other macroelement feeds

（一）含硫饲料
Sulfur-containing feed

硫是动物所必需的矿物质元素之一，而又不能在体内合成，必须由饲料来供给。一般认为动物所需的硫是有机硫，如蛋白质中的含硫氨基酸等。因此，蛋白质饲料是动物的主要硫源，但其成本较高。有研究认为，无机硫如硫酸钠、硫酸钾、

Sulfur is one of the essential mineral elements for animals, which cannot be synthesized in vivo and must be supplied by feed. It is generally believed that the sulfur required by animals is organic sulfur, such as sulfur-containing amino acids in protein. Therefore, protein feed is the main source of sulfur for animals, but its cost is high. Some studies believe that inorganic sulfur such as sodium sulfate,

硫酸钙、硫酸镁等对动物也具有一定的营养意义。同位素试验表明，反刍动物瘤胃中的微生物能有效地利用无机硫化合物合成含硫氨基酸和维生素。研究表明，适当增加饲粮中的无机硫含量可减少雏鸡对含硫氨基酸的需要，并有利于合成生命活动所必需的牛磺酸，从而促进生长。

用于补充硫的来源主要有含硫氨基酸、硫酸钠、硫酸钾、硫酸镁等。不同的含硫饲料硫的利用率不同，反刍动物对蛋氨酸的硫利用率为100%，硫酸钠为54%，硫为31%，且硫的补充量不宜超过饲粮干物质的0.05%。对幼雏而言，硫酸钠、硫酸钾、硫酸镁均可充分利用，而硫酸钙利用率较差。经国内外试验证实，在硫酸盐中，硫酸钠的生物利用率最高，其硫元素能被单胃动物肠道迅速吸收，并且大部分被利用，其不仅能提高鸡的生产性能，还能节省一部分含硫氨基酸。许多国家都广泛使用硫酸钠作为饲料蛋白质营养强化剂，用于畜禽生产，取得了较好效果。

potassium sulfate, calcium sulfate, and magnesium sulfate also have certain nutritional significance for animals. Isotope experiments show that microorganisms in ruminant's rumen can effectively utilize inorganic sulfur compounds to synthesize sulfur-containing amino acids and vitamins. Studies have shown that an appropriate increase in the content of inorganic sulfur in the diet can reduce the chick's need for sulfur-containing amino acids, and is conducive to the synthesis of taurine necessary for life activities, thereby promoting growth.

The main sources of sulfur supplementation include sulfur-containing amino acids, sodium sulfate, potassium sulfate, magnesium sulfate, etc. The utilization rates of sulfur in different sulfur-containing feeds are different. The sulfur utilization rate of methionine, sodium sulfate and sulfur for ruminants is 100%, 54% and 31% respectively. Moreover, the supplement amount of sulfur should not exceed 0.05% of the dry matter of feed. For young chicks, sodium sulfate, potassium sulfate and magnesium sulfate can be fully utilized, while the utilization rate of calcium sulfate is poor. It is confirmed by tests at home and abroad that in sulfate, sodium sulfate has the highest bioavailability, and its sulfur can be quickly absorbed by the intestines of monogastric animals. Most of it is used, which can not only improve the production performance of chickens, but also save a part of sulfur-containing amino acids. Sodium sulfate is widely used in many countries as a dietary protein nutrition enhancer for livestock and poultry production, and has achieved good results.

（二）含镁饲料
Magnesium-containing feed

饲料中含镁丰富，一般都在0.1%以上，因此不必另外添加。但早春牧草中镁的利用率很低，有时会导致放牧家畜因缺镁而出现"草痉挛"，故对放牧的牛羊以及以玉米作为主要饲料并补加非蛋白氮喂的牛，常需要补加镁。在热应激和饲粮中添加不饱和脂肪酸的情况下，泌乳奶牛的饲粮镁浓度应提高。许多研究也证明，为了提高饲粮中

Feed is rich in magnesium, generally above 0.1%, so no additional magnesium is required. However, the utilization rate of magnesium in early spring forage is very low, which sometimes leads to "grass spasm" of grazing livestock due to lack of magnesium. Therefore, it is often necessary to supplement magnesium for grazing cattle and sheep and cattle fed with corn as the main feed and supplemented with non-protein nitrogen. The dietary magnesium concentration of lactating dietary cows should be increased in the case of heat stress and the addition of unsaturated fatty acids

养分的消化率和动物的生产性能，额外添加镁离子是必要的。另外，为了防止饲喂奶牛高能量饲料时造成酸中毒，可在其饲粮中添加占干物质 0.8% 的碳酸氢钠和 0.4% 的氧化镁作为缓冲剂。用于补充镁的矿物质主要有氧化镁、硫酸镁、碳酸镁、磷酸镁和氯化镁。不同的矿物质镁源的吸收率变化很大，在正常瘤胃 pH 值下，矿物质来源的镁离子如氧化镁的溶解度很低，而硫酸镁和氯化镁中的镁离子溶解度高，在瘤胃中的吸收率也高。饲料工业中使用的氧化镁一般为菱镁矿在 800-1 000 ℃ 煅烧的产物，其化学组成为：MgO 85.0%，CaO 7.0%，SiO_2 3.6%，Fe_2O_3 2.5%，Al_2O_3 0.4%，烧失量 1.5%，吸收率为 50%。

in diets. Many studies have also proved that in order to improve the digestibility of nutrients in diets and animal production performance, additional magnesium ions are necessary. In addition, in order to prevent acidosis caused by feeding high-energy feed for dairy cows, 0.8% of sodium bicarbonate and 0.4% of magnesium oxide can be added as buffers in their diets. The minerals used to supplement magnesium mainly include magnesium oxide, magnesium sulfate, magnesium carbonate, magnesium phosphate and magnesium chloride. The absorption rate of magnesium from different minerals varies greatly. Under normal rumen pH value, the solubility of magnesium ions from minerals such as magnesium oxide is very low, and the solubility of magnesium ions in magnesium sulfate and magnesium chloride is high, and the absorption rate in the rumen is also high. Magnesium oxide used in feed industry is usually the product of magnesite calcined at 800-1,000 ℃, and its chemical composition is: 85.0% of MgO, 7.0% of CaO, 3.6% of SiO_2, 2.5% of Fe_2O_3, 0.4% of Al_2O_3, 1.5% of loss on ignition, and its absorption rate is 50%.

第二节　天然矿物质饲料
Section 2　Natural mineral feed

随着饲料工业的发展，又有许多天然矿物质被用作饲料，其中使用较多的有沸石、麦饭石、稀土、膨润土、海泡石、凹凸棒石和泥炭等，这些天然矿物质饲料多属非金属矿物。

With the development of feed industry, many natural minerals are used as feed, among which zeolite, maifanite, rare earth, bentonite, sepiolite, attapulgite and peat are used more frequently. These natural minerals are mostly non-metallic minerals.

一、沸石
Zeolite

沸石是沸石族矿物的总称，已知的天然沸石有 40 余种。天然沸石是含碱金属和碱土金属的含水铝硅酸盐类。沸石大多呈三维硅氧四面体及三维铝氧四面晶体格架结构，晶体内部具有许多孔径均匀一

Zeolite is the general term of zeolite minerals, and there are more than 40 known natural zeolites. Natural zeolites are bydrous aluminosilicates containing alkali metals and alkaline earth metals. Zeolites are mostly three-dimensional silicon-oxygen tetrahedrons and three-dimensional aluminum-oxygen tetrahedral crystal lattice structures. There

致的孔道和内表面积很大的孔穴（500－1 000 m²/g），孔道和孔穴两者的体积占沸石总体积的50%以上。通常情况下，晶体孔道和孔穴中含有金属阳离子和水分子，且与格架结构结合得比较弱。

沸石的特殊结构使之具有良好的选择性吸附、阳离子交换、催化激活、耐酸和热稳定等性能，这些性能使沸石在畜牧业中广泛被应用于微量元素添加剂的载体或稀释剂、畜禽粪便的除臭剂、畜舍环境的净化剂。大量研究表明，沸石能吸附动物体内的有害微生物、毒素和氨，延长饲料在消化道内的滞留时间，因而能减少动物发病率，提高饲料的转化率、生产性能和经济效益。

沸石中除了一些矿物质微量元素外不含任何营养成分，添加过多会降低饲料的营养水平，影响动物生长发育和生产性能。在实际生产中应根据动物的种类、生理阶段、饲粮营养水平和健康状况等来确定其添加量，一般在各类畜禽饲粮中的添加量为：猪饲料中5%－7%，5%时效果最佳；鸡饲料中3%－6%，以3%为佳；母牛饲料中8%左右；羔羊饲料中以5%为宜；鱼饲料中3%－5%。

天然沸石种类很多，品位不一，不同来源和品位的沸石饲用效果也不尽相同。在畜牧养殖业和饲料工业中尤以斜发沸石和丝光沸石使用价值最大。天然沸石本身是无毒无害的。饲料级沸石粉中砷含量小于1 mg/kg，铅含量小于20 mg/kg，氟含量小于300 mg/kg，均低于国家规定的食品卫生标准。

are many pores with uniform pore size and large internal surface area ($500-1,000 \text{ m}^2/\text{g}$) in the crystal. The volume of pores and holes accounts for more than 50% of the total volume of zeolite. Usually, the crystal pores and cavities contain metal cations and water molecules, and are weakly bound to the lattice structure.

The special structure of zeolite enables its good selective adsorption, cation exchange, catalytic activation, acid resistance and thermal stability. These properties make zeolite widely used in animal husbandry as a carrier or diluent for trace element additives, deodorizer of livestock manure and purifying agent of livestock house environment. A large number of studies have shown that zeolite can adsorb harmful microorganisms, toxins and ammonia in animals and prolong the retention time of feed in the digestive tract, thus reducing the incidence of animals and improving feed conversion rate, production performance and economic benefits.

Zeolite does not contain any nutrients except some mineral microelements, so excessive addition will reduce the nutritional level of feed and affect animal growth and production performance. In the actual production, the addition amount should be determined according to the animal type, physiological stage, dietary nutrition level and health status. Generally, the addition amount in various livestock and poultry diets is: 5%-7% for pigs, with 5% the best; 3%-6% for chickens, preferably 3%; about 8% for cows; 5% for lambs; and 3%-5% in fish feed.

There are many kinds of natural zeolites with different grades, and the feeding effects of zeolites from different sources and grades are also different. In animal husbandry and feed industry, clinoptilolite and mordenite are the most valuable. Natural zeolite itself is nontoxic. The content of arsenic, lead and fluorine in feed grade zeolite powder is less than 1 mg/kg, 20 mg/kg and 300 mg/kg, respectively, which are all lower than the national food hygiene standards.

二、麦饭石
Maifan stone

麦饭石因其外观似麦饭团而得名,是一种经过蚀变、风化或半风化,具有斑状或似斑状结构的硅酸盐岩石,其主要化学成分为二氧化硅(SiO_2)和三氧化二铝(Al_2O_3),二者占70%–80%。麦饭石含有K、Na、Ca、Mg、Cu、Zn、Fe、Se等对动物有益的常量元素和微量元素,且由于其具有多孔性海绵状结构,使这些元素的溶出性好。因此,麦饭石具有多种作用:①调节畜禽新陈代谢,促进生长发育。②调节胃肠功能,有利于营养物质的吸收,提高饲料转化率。③增加肉、蛋、奶中微量元素及氨基酸含量,提高畜产品品质。④清除动物体内有害气体和各种金属离子及有害菌,提高动物抗病能力。⑤提高肝脏功能,增加血清中的抗体,增强机体免疫能力等。

在畜牧生产中,麦饭石一般以其水浸液和细粉的方式用作饲料添加剂,添加量以0.5%–2.0%为宜。麦饭石还可用作微量元素及其他添加剂的载体或稀释剂,可减少活性成分损失。麦饭石还有降低饲料中棉籽饼毒素的作用。在水产养殖上,麦饭石可用来改良鱼塘水质,使水的化学耗氧量和生物耗氧量下降,溶解氧提高,提高鱼虾的成活率和生长速度。麦饭石资源丰富,来源广泛,安全可靠,无毒无害,价格便宜,对动物组织器官、生理生化功能以及动物产品品质均无不良影响。

Maifanite is named after its appearance of wheat rice balls. It is a silicate rock with porphyritic or porphyritic structure after alteration, weathering or semi-weathering. Its main chemical components are SiO_2 and Al_2O_3, accounting for 70%–80%. Maifanite contains K, Na, Ca, Mg, Cu, Zn, Fe, Se and other macroelements and microelements which are beneficial to animals, and because of its porous sponge structure, these elements have good dissolution properties. Therefore, maifanite has a variety of functions: ① regulate the metabolism of livestock and poultry and promote growth and development; ② regulate gastrointestinal function, facilitate the absorption of nutrients, and improve feed conversion rate; ③ increase the content of microelements and amino acids in meat, egg and milk, and improve the quality of livestock products; ④ remove harmful gases, various metal ions and harmful bacteria in animals to improve the disease resistance of animals; ⑤ improve liver function, increase antibodies in the serum, and enhance the boy's immunity, etc.

In livestock production, maifanite is generally used as feed additives in the form of waterextract and fine powder, and the appropriate addition percentage is 0.5%–2.0%. Maifanite can also be used as a carrier or diluent for microelements and other additives, which can reduce the loss of active components. Maifanite also has the effect of reducing toxins in cottonseed cake in feed. In aquaculture, maifanite can be used to improve the water quality of fish pond, reduce the chemical ammonia consumption and biological oxygen consumption of water, increase the dissolved oxygen, and improve the survival rate and growth rate of fish and shrimp. Maifanite is rich in resources, widely sourced, safe and reliable, non-toxic and harmless, cheap, and it has no adverse effects on animal tissues and organs, physiological and biochemical functions and animal product quality.

三、膨润土
Bentonite

膨润土由酸性火山凝灰岩变化而成,俗称白黏土,又名斑脱岩,是蒙脱石类黏土岩组成的一种含水的层状结构铝硅酸盐矿物。膨润土的主要化学成分为 SiO_2、Al_2O_3、H_2O 以及少量的 Fe_2O_3、FeO、MgO、CaO、Na_2O 和 TiO_2 等,其中含有动物生长发育所必需的多种常量和微量元素。

膨润土的结构决定其具有高分散性、悬浮性、膨润性、黏结性、吸附性、阳离子交换性等许多优良特性。另外,由于膨润土无毒,对人、畜、植物等无害,其用于饲料添加剂,具有提高饲料混合均匀度,改进饲料的松散性,增进食欲,延缓饲料通过动物消化道的时间,吸附消化道内的重金属、有害气体和细菌等多种作用。

研究表明,在畜禽饲粮中添加1%~3%的膨润土具有提高生产性能等方面的效果。在反刍动物饲粮中使用非蛋白氮时,膨润土作为凝固剂和稀释剂,可提高其使用安全性和利用率。饲喂奶牛高精料和青贮饲料时添加膨润土,可提高产奶量和钙、铁、铜、锰、锌的利用率。在肉鸡和蛋鸡饲粮中添加膨润土,可提高平均增重、成活率、产蛋率、蛋重,改善蛋壳品质,并使蛋中的铁、铜、钴、锰含量提高。在猪饲粮中添加膨润土能提高生长速度,降低料肉比。在长毛兔饲粮中添加少许膨润土,可提高产毛量。饲料中添加膨润土,还具有吸附黄曲霉毒素、降低游离棉酚毒性、保护饲料热处理时植酸酶活性的作用。

Bentonite is a kind of water-bearing layered aluminosilicate mineral composed of montmorillonite clay rocks, which is changed from acidic volcanic tuff, commonly known as white clay and also known as bentonite. The main chemical components of bentonite are SiO_2, Al_2O_3, H_2O and a small amount of Fe_2O_3, FeO, MgO, CaO, Na_2O and TiO_2, etc., which contain a variety of macroelements and microelements necessary for animal growth and development.

The structure of bentonite determines that it has many excellent properties such as high dispersion, suspension, swelling, adhesion, adsorption, and cation exchange. In addition, because bentonite is non-toxic and harmless to human, livestock, plants and so on, it can be used as feed additives to improve the uniformity of feed mixing and the looseness of feed, increase appetite, delay the time of feed passing through the animal digestive tract, and adsorb heavy metals, harmful gases and bacteria in the digestive tract.

Studies have shown that adding 1%-3% of bentonite in livestock and poultry diets has the effect of improving production performance. When non-protein nitrogen is used in ruminant diets, bentonite can be used as a coagulant and diluent to improve its safety and utilization rate. The milk yield and the utilization rate of calcium, iron, copper, manganese and zinc can be increased by adding bentonite when feeding dairy cattle high concentrate and silage. The addition of bentonite into the diets of broilers and laying hens can increase the average weight gain, survival rate, egg production rate, egg weight, improve eggshell quality, and increase the content of iron, copper, cobalt and manganese in eggs. Adding bentonite in pig diets can increase growth rate and reduce feed conversion ratio. Adding a little bentonite to the wool rabbits' diets can increase the wool yield. The addition of bentonite in feed also has the effects of adsorbing aflatoxin, reducing the toxicity of free gossypol, and protecting phytase activity during heat treatment of feed.

四、凹凸棒石
Attapulgite

凹凸棒石是一种镁铝硅酸盐,呈三维立体全链结构及特殊的纤维状晶体型,具有离子交换、胶体、吸附、催化等化学特性。凹凸棒石的主要成分除二氧化硅(约60%左右)外,尚含有多种畜禽必需的微量元素:铜21 mg/kg,铁1 310 mg/kg,锌21 mg/kg,锰1 382 mg/kg,钴11 mg/kg,钼0.9 mg/kg,硒2 mg/kg,氟361 mg/kg,铬13 mg/kg。

凹凸棒石可用作微量元素载体、稀释剂和畜禽舍净化剂等。研究表明,在畜禽饲料中应用凹凸棒石,可提高饲料利用率,改善动物生产性能和健康状况,降低生产成本。凹凸棒石还具有一定的提高机体抗氧化能力的作用,能显著提高血浆中SOD酶、GSH-PX酶活性和肝脏中SOD酶活性,显著降低血浆和肝脏中丙二醛的含量。

Attapulgite is a kind of magnesium-aluminate silicate with a three-dimensional full-chain structure and a special fibrous crystal type. It has chemical properties such as ion exchange, colloid, adsorption and catalysis. In addition to silica (about 60%), the main components of attapulgite also contain a variety of essential microelements for livestock and poultry: 21 mg/kg of copper, 1,310 mg/kg of iron, 21 mg/kg of zinc, 1,382 mg/kg of manganese, 11 mg/kg of cobalt, 0.9 mg/kg of molybdenum, 2 mg/kg of selenium, 361 mg/kg of fluorine, and 13 mg/kg of chromium.

Attapulgite can be used as a microelement carrier, a diluent and a purifying agent for livestock and poultry houses. Studies have shown that the application of attapulgite in livestock and poultry feed can increase feed utilization, improve animal production performance and health status, and reduce production costs. Attapulgite also has a certain role in improving the body's antioxidant capacity. It can significantly increase the activity of SOD enzyme, GSH-PX enzyme in plasma and the activity of SOD enzyme in liver, and significantly reduce the content of malondialdehyde in plasma and liver.

五、其他天然矿物质饲料
Other natural mineral feeds

(一) 稀土
Rare earth

稀土元素是15种镧系元素及与其化学性质相似的钪、钇等17种元素的总称。化学组成一般为48%的铈、25%的镧、16%的钕、2%的钐、5%的镨;此外,剩余的钷、铕、钆、铽、镝、钬、铒、铥、镱、镥、钪、钇等12种元素约占稀土元素的4%。

Rare earth elements are the general term for 15 kinds of lanthanide elements and 17 kinds of elements such as scandium and yttrium with similar chemical properties. The chemical composition is generally 48% of cerium, 25% of lanthanum, 16% of neodymium, 2% of samarium, and 5% of praseodymium. Other 12 elements which account for 4% of the total are mega, europium, gadolinium, terbium, dysprosium, holmium, erbium, thulium, ytterbium, lutetium, scandium, and yttrium.

稀土是一类有益的辅助性营养元素：一方面参与动物体内物质代谢；另一方面对酶有不同程度的激活作用。稀土和微量元素有互作关系，可提高血中 Cu、Mn、Zn、Fe、Se 等微量元素的含量，间接影响动物机体的代谢，达到提高动物生产性能的目的。

研究表明，稀土作为饲料添加剂，在肉鸡饲粮中添加 20 – 100 mg/kg 可不同程度提高其增重和成活率，并降低料肉比。在蛋鸡饲粮中添加 16 mg/kg 和 32 mg/kg 可提高其全期成活率 5% – 10%，并使产蛋高峰延长 4 周以上，饲料消耗下降 8.8% – 12.6%。稀土还具有提高鸡饲粮能量利用效率、干物质消化率和蛋白质消化率，抑制病菌（白痢杆菌、大肠杆菌、葡萄球菌、痢疾杆菌），减少疾病的作用。

目前使用的稀土饲料添加剂有无机稀土和有机稀土两种类型。无机稀土主要有硝酸稀土、碳酸稀土、氯化稀土和硫酸稀土，目前常用的是硝酸稀土；有机稀土主要包括有机酸稀土（如柠檬稀土添加剂）、维生素 C 稀土、稀土酵母、稀土壳聚糖和氨基酸稀土螯合剂等。此外，根据添加剂中所含稀土元素的种类，还可以分为单一稀土饲料添加剂和复合稀土饲料添加剂。稀土化合物无"三致"作用，放射性远低于国家标准，经消化道摄入的毒性很低，在饲料中添加适量的稀土是安全可行的。使用微量稀土作为饲料添加剂，肉、蛋、鱼、虾等产品中的稀土含量均未见增加。

Rare earth is a kind of beneficial auxiliary nutrients. On the one hand, it participates in material metabolism in animals; on the other hand, it has different degrees of activation on enzymes. Rare earth and microelements have interaction relationship, which can increase the content of Cu, Mn, Zn, Fe, Se and other microelements in blood, indirectly affecting the metabolism of animal body, and achieving the purpose of improving animal production performance.

Studies have shown that adding 20 – 100 mg/kg of rare earth as a feed additive in broiler diets can increase the weight gain and survival rate to varying degrees, and reduce the feed conversion ratio. Adding 16 mg/kg and 32 mg/kg in the diet of laying hens can increase the survival rate of the whole period by 5% – 10%, prolong the egg production peak by more than four weeks, and reduce the feed consumption by 8.8% – 12.6%. Rare earth also has the effect of improving the energy utilization efficiency, dry matter digestibility and protein digestibility of chicken diets, inhibiting pathogens (Shigella, Escherichia coli, Staphylococcus, Shigella), and reducing the occurence of diseases.

At present, there are two types of rare earth feed additives: inorganic rare earth and organic rare earth. Inorganic rare earth mainly includes rare earth nitrate, rare earth carbonate, rare earth chloride and rare earth sulfate. At present, rare earth nitrate is commonly used. Organic rare earth mainly includes organic acid rare earth (such as lemon rare earth additive), vitamin C rare earth, rare earth yeast, rare earth chitosan and amino acid rare earth chelating agent. In addition, according to the types of rare earth elements contained in the additives, it can also be divided into single rare earth feed additives and composite rare earth feed additives. Rare earth compounds have no "triadic" effect, and the radioactivity is much lower than that of the national standard. The toxicity through the digestive tract is very low. It is safe and feasible to add an appropriate amount of rare earth in the feed. Using trace rare earth as a feed additive, the rare earth content in products such as meat, eggs, fish and shrimp has not increased.

（二）海泡石
Sepiolite

海泡石是一种纤维状、富镁黏土矿物，属特种稀有矿石，含有丰富的畜禽必需矿物质元素。化学成分因质量不同而差异很大，一般以二氧化硅为主，占30%-60%，另外还有三氧化二铝、氧化钙、三氧化二铁、氧化镁、氧化钾、氧化钠等。海泡石可吸附自身重200%-250%的水分。海泡石呈灰白色，有滑感、无毒、无臭，具有特殊的层链状晶体结构，因此具有独特的吸附性、自由流动性、抗胶凝性、化学惰性和无毒性，被广泛用作微量元素载体或稀释剂、饲料抗结块剂、饲料黏合剂、吸附剂、环境除臭剂、动物生长促进剂等。

研究表明，在颗粒饲料加工中，添加2%-4%的海泡石可以增加各种成分间的黏合力，促进其凝聚成团。当加压时海泡石显示出较强的吸附性能和胶凝作用，有助于提高颗粒的硬度及耐久性。饲料中的脂类物质含量较高时，用海泡石作黏合剂最适宜。海泡石还能显著减少非淀粉多糖的负面影响，改变空肠黏性，提高有机物质消化率以及阻碍胃肠道对黄曲霉毒素的吸收。在畜禽饲料中添加1%-1.5%的海泡石，可加快畜禽生长和肥育，提高肉、蛋、奶的含量和饲料的生物效价，也可以防止氨引起的中毒和慢性疾病。

Sepiolite is a fibrous magnesium-rich clay mineral, which is a special rare ore and contains abundant essential mineral elements for livestock and poultry. The chemical components vary greatly due to their different quality. Generally, silicon dioxide is the main component, accounting for 30%-60%. In addition, aluminum oxide, calcium oxide, iron oxide, magnesium oxide, potassium oxide and sodium oxide are also included. Sepiolite can adsorb 200%-250% of its own weight of water. Sepiolite is grayish-white, slippery, non-toxic, odorless, and has a special layer-chain crystal structure. Therefore, sepiolite has unique adsorption, free mobility, anti-gelatinization, chemical inertness, and non-toxicity, and is widely used as microelement carrier or diluent, feed anti-caking agent, feed binder, adsorbent, environmental deodorant, and animal growth promoter, etc.

Studies have shown that adding 2%-4% of sepiolite in pellet feed processing can increase the adhesion between various components and promote their agglomeration. The sepiolite shows strong adsorption and gelation when pressed, which helps to improve the hardness and durability of the particles. When the lipid content in feed is high, sepiolite is the most suitable binder. Sepiolite can also significantly reduce the negative effects of non-starch polysaccharides, change the viscosity of jejunum, improve the digestibility of organic matter and hinder the absorption of aflatoxin in the gastrointestinal tract. Adding 1%-1.5% of sepiolite in livestock and poultry feed can accelerate the growth and fattening of livestock and poultry, increase the content of meat, eggs and milk and the biological potency of feed, and also prevent ammonia-induced poisoning and chronic diseases.

（三）泥炭
Peat

泥炭又称草炭或草煤，是在沼泽的形成过程中由未被完全分解的植物残体在腐水和缺氧环境下腐解堆积保存而形成的天然有机沉积物，含有丰富的营养成分，其中木质

Peat, also known as peat or grass coal, is a natural organic sediment formed during the formation of swamps by the decomposition and accumulation of incompletely decomposed plant residues in water and anoxic environments. It contains rich nutrients, including 30%-40% of lignin, 30%

素30%-40%，多糖类30%-33%，粗蛋白质4%-5%，腐殖酸10%-40%。另外，还含有钙、镁、硅、铁、锰、锌、硼等多种矿物质元素。

泥炭一般不直接用作饲料，需先进行分离与转化，才能成为畜禽可食的饲料。另外，可对泥炭加工处理后用泥炭腐殖酸作为饲料添加剂，或利用泥炭中的水解物质作为培养基制取饲料酵母和生产泥炭发酵饲料、泥炭糖化饲料等。用泥炭或加工成复合腐殖酸饲料添加剂饲喂畜禽，具有调理胃肠机能和促进新陈代谢、提高生产性能和饲料利用率、减少疾病等作用。泥炭无毒无害，方便、价廉，具有独特的理化性质，对于发展生态农业、改善生态环境、生产无公害产品具有重要利用价值。

-33% of polysaccharides, 4%-5% of crude protein, 10%-40% of humic acid. In addition, it also contains calcium, magnesium, silicon, iron, manganese, zinc, boron and other mineral elements.

Peat is generally not used as feed directly. It needs to be separated and transformed before it can be used as feed for livestock and poultry. In addition, peat humic acid can be used as a feed additive after peat processing, or the hydrolysis substances in peat can be used as a medium to produce feed yeast and peat fermentation feed and peat saccharification feed. Feeding livestock and poultry with peat or compound humic acid feed additive can regulate gastrointestinal function, promote metabolism, improve production performance and feed utilization, and reduce diseases. Peat is non-toxic, harmless, convenient to use, inexpensive and has unique physical and chemical properties. It has important utilization value for the development of ecological agriculture, the improvement of ecological environment and the production of pollution-free products.

第九章 维生素饲料
Chapter 9 Vitamin feed

维生素是动物维持生理机能所必需的一类低分子有机化合物。对单胃动物来说，大多数维生素不能或不完全能由体内合成而满足需要，必须从食物或饲料中补充。反刍动物瘤胃微生物可合成 B 族维生素和维生素 K_2。

各种青绿饲料中含有丰富的维生素。在粗放饲养条件下，因饲喂大量青绿饲料，动物不会缺乏维生素。随着畜禽生产水平的大幅度提高，饲养方式的工厂化、集约化，一方面动物对维生素的需要量增加；另一方面，由于动物脱离了阳光、土壤和青绿饲料等自然条件，仅仅依靠饲料中的天然来源不能满足动物对维生素的需要，必须另外补充。

维生素添加剂是指工业合成或由天然原料提纯精制（或高度浓缩）的各种单一维生素制剂和由其生产的复合维生素制剂。目前已用于饲料的维生素至少有 16 种，即：维生素 A（包括胡萝卜素）、D（包括 D_2、D_3）、E（包括 α-生育酚、β-生育酚和 γ-生育酚）、K、B_1、B_2、B_6、B_{12}、烟酸和烟酰胺、泛酸、胆碱、叶酸、生物素、维生素 C、肌醇和肉毒碱。

大多数维生素的衡量单位常以毫克(mg)（如 B 族维生素）或微克(ug)表示。维生素 A、D、E 采用统一的国际单位(IU)表示。

由于大多数维生素都有不稳定、易氧化或易被其他物质破坏失

Vitamin is a kind of low molecular organic compound which is necessary for animals to maintain physiological function. For monogastric animals, most vitamins can not be completely synthesized in the body to meet the needs and must be supplemented from food or feed. Microorganisms in the rumen of ruminants can synthesize B vitamins and vitamin K_2.

Various green feeds are rich in vitamins. Under the condition of extensive feeding, the animals will not be lack of vitamins because of a lot of green forage in feed. With the dramatic improvement of livestock and poultry production levels and the industrialization and intensification of feeding methods, on the one hand, animals' demand for vitamins has increased; on the other hand, since animals are separated from natural conditions such as sunlight, soil and green forage, natural sources in feed alone can not meet the animals' needs for vitamins and must be supplemented.

Vitamin additives refer to various single vitamin preparations and multivitamin preparations produced by industrial synthesis or purified (or highly concentrated) from natural raw materials. There are at least 16 kinds of vitamins used in feed, namely: vitamin A (including carotene), vitamin D (D_2 and D_3), vitamin E (including α-tocopherol, β-tocopherol and γ-tocopherol), vitamin K, vitamin B_1, vitamin B_2, vitamin B_6, vitamin B_{12}, niacin and niacinamide, pantothenic acid, choline, folic acid, biotin, vitamin C, alcohol and carnitine.

Most vitamins are measured in milligrams (mg) (such as B vitamins) or in micrograms (ug). Vitamins A, D, and E are expressed in international units (IU) of the series.

Since most vitamins are unstable, easily oxidized or destroyed by other substances and in accordance with the

效的特点和饲料生产工艺上的要求,几乎所有的维生素制剂都经过了特殊加工处理或包装。例如:制成稳定的化合物或利用稳定物质包被等。为了满足不同使用的要求,在剂型上还有粉剂、油剂、水溶性制剂等。此外,商品维生素饲料添加剂还有各种不同规格含量的产品,可归纳为四类:

(1) 纯制剂。稳定性较好的维生素单一浓度制剂多为含其化合物在95%以上的纯品制剂,如维生素B_1、B_2、B_6、叶酸、烟酸、泛酸钙、维生素K_3等的纯品制剂化合物纯度为95%-99%。其特点:浓度高、稳定性较差、流动性差。

(2) 经包被处理的制剂。此类制剂又称稳定型制剂。脂溶性维生素及维生素C极不稳定,其饲用制剂除加抗氧化剂外,常用明矾、鱼胶、阿拉伯胶、糊精、蔗糖、葡萄糖、乳糖、淀粉等稳定物质进行包被以提高其稳定性,主要有微粒胶囊和微粒粉剂两种产品。

微粒胶囊为颗粒状,流动性好。微粒粉剂(或称喷雾干燥粉末)比微粒胶囊抗氧化性能更好,硬度高,能抵抗机械损伤,粒度适中,单位饲料中颗粒较多,微粒表面粗糙而不规则,易吸附混匀。

(3) 稀释制剂。稀释制剂是利用脱脂米糠等载体或稀释剂制成的各种浓度的维生素预混合饲料。这类制剂中几乎都加有抗氧化剂。

(4) 复合维生素制剂。是根据动物需要,将多种维生素配制成的维生素预混料。

requirements of the feed production process, almost all vitamin preparations are specially processed or packaged. For example: making stable compounds or using stable substances to coat, etc. In order to meet the requirements of different applications, there are powders, oils, water-soluble preparations on the dosage form. In addition, commercial vitamin feed additives also have a variety of products with different specifications and content, which can be classified into four categories.

(1) Pure preparations. Vitamin single-concentration preparations with better stability are mostly pure preparations containing more than 95% of their compounds, such as vitamins B_1, B_2, B_6, folic acid, niacin, calcium pantothenate, vitamin K_3, etc. The purity of pure preparation is 95%-99%. It has the characteristics of high concentration, poor stability and poor fluidity.

(2) Coated preparations. Such preparations are also known as stable preparations. Fat-soluble vitamins and vitamin C are extremely unstable. In addition to adding antioxidants to their feeding preparations, alum, fish glue, Arabic gum, dextrin, sucrose, glucose, lactose, starch and other stable substances are commonly used to coat them to improve their stability. There are mainly two kinds of microparticle capsules and microparticle powders.

The microparticle capsule is granular and has good fluidity. The microparticle powder (or spray drying powder) has better anti-oxidation performance than the microparticle capsule and high hardness. It can resist mechanical damage. It has moderate particle size and more particles in unit feed. Its surface of particles is rough and irregular and easy to absorb and mix.

(3) Dilution preparations. The dilution preparation is the vitamin premixed feed with various concentrations made of carrier or diluent such as defatted rice bran. Almost all of these preparations contain antioxidants.

(4) Multivitamin preparations. It is a vitamin premix prepared with multiple vitamin according to the needs of animals.

第一节 脂溶性维生素饲料
Section 1 Fat-soluble vitamin feed

包括维生素 A、D、E、K 四种。脂溶性维生素的特点是稳定性都很差,其制剂基本都是经过稳定性处理的制剂。饲用维生素 A、D、E 制剂一般为经包被处理的稳定型制剂,维生素 K 制剂多为化学稳定剂处理的制剂。

It includes vitamins A, D, E and K. Fat-soluble vitamins are characterized by poor stability, and their preparations are basically treated with stability. Dietary vitamin A, D and E preparations are generally coated stable preparations, while vitamin K preparations are mostly prepared with chemical stabilizers.

一、维生素 A 和 β-胡萝卜素
Vitamin A and β-carotene

维生素 A 又名视黄醇。缺氧时对热稳定,有氧时对热不稳定,易被紫外线破坏。胡萝卜素是植物中存在的维生素 A 原。作为饲料中维生素 A 的补充物,主要有维生素 A 和 β-胡萝卜素制剂,生产上应用的多为维生素 A 制剂。

1. 饲用维生素 A 制剂。维生素 A 制剂有天然物和化工合成两类。天然物主要是鱼肝油及其制品,化工合成的主要有维生素 A 醇、维生素 A 乙酸酯和维生素 A 棕榈酸酯和维生素 A 丙酸酯制剂。

用于饲料的目前主要为化工合成的维生素 A 乙酸酯和维生素 A 棕榈酸酯的制剂。分子式和相对分子质量分别为 $C_{22}H_{32}O_2$、328.5 和 $C_{36}H_{60}O_2$、524.9。二者均为黄色,维生素 A 乙酸酯为粉状结晶(熔点 57-60 ℃),易吸湿,遇热或酸性物质、见光或吸潮后易分解。产品规格有 30 万 IU/g、40 万 IU/g 和 50 万 IU/g。维生素 A 棕榈酸酯室温下呈油脂状团块(熔点 28-29 ℃)。酯化后维生素 A 添加剂的制作可

Vitamin A is also known as retinol. It is stable to heat when hypoxic, and unstable to heat in the presence of oxygen. It is easy to be destroyed by ultraviolet rays. Carotene is the provitamin present in plants. As a supplement of vitamin A in feed, there are mainly vitamin A and β-carotene preparations, and most vitamin A preparations is used in production.

1. Dietary vitamin A preparations. There are two types of vitamin A preparations: natural products and chemical synthesis. The natural products are mainly cod liver oil and its products, and the chemical synthesis mainly includes vitamin A alcohol, vitamin A acetate, vitamin A palmitate and vitamin A propionate preparations.

The vitamin preparations used for feed are mainly chemically synthesized preparations of vitamin A acetate and vitamin A palmitate. The molecular formula and relative molecular weights are $C_{22}H_{32}O_2$, 328.5 and $C_{36}H_{60}O_2$, 524.9. Both of them are yellow. Vitamin A acetate is a powdery crystal (melting point 57-60 ℃). It tends to absorb moisture and decompose when exposed to heat or acidic substances, or exposed to light or moisture. The product specifications are 300,000 IU/g, 400,000 IU/g and 500,000 IU/g. Vitamin A palmitate is a greasy mass (melting point 28-29 ℃) at room temperature. The preparation of vitamin A additive after esterification can be made by

采用微型胶囊技术,也可使用吸附方法。微型胶囊技术的步骤是先在乳化器内加入阿拉伯胶,并加入油液状的维生素 A 酯,进行乳化,形成微粒。再移至反应罐中,加入明胶水溶液,利用电荷关系,使乳化微粒和明胶水溶液之间发生交联作用,形成被明胶包被的微粒。随后,再加糖衣、疏水剂,再用淀粉包被,即制成微型胶囊。在制作工艺中还可加入抗氧化剂,避免维生素 A 氧化。吸附方法是先对油液状维生素 A 酯乳化,并用抗氧化剂稳定,再以干燥的小麦麸和硅酸盐进行吸附。

维生素 A 添加剂,因工艺条件不同,其粒度的大小也有差别。国外虽规定一般在 0.1－1.0 mm,实际上多在 0.177－0.590 mm(80 至 30 目)。可在水中弥散的维生素 A 添加剂,粒度更小,最大不得超过 0.35 mm。维生素 A 添加剂应有较好的牢固度,耐磨损,表面应粗糙,以便容易混合均匀。维生素 A 添加剂的制作工艺质量,可用显微化学方法检视,也可将维生素 A 添加剂与氯化胆碱混合在一起,贮存一定时间后,用仪器测定维生素 A 的损失量。也可用生物学方法进行实用性检验。

紫外线和氧都可促使维生素 A 醋酸酯和维生素 A 棕榈酸酯分解。湿度和温度较高时,稀有金属盐可使分解速度加快。含有 7 个水的硫酸亚铁($FeSO_4 \cdot 7H_2O$)可使维生素 A 醋酸酯的活性损失严重。与氯化胆碱接触时,活性将受到严重损失。在 pH 值 4 以下环境中和在强碱环境中,维生素 A 很快分解。维生素 A 酯经包被后,可使损失减少。维生素 A 制成微型胶囊或颗粒后,活性的稳定性有了很大提高,但是,它仍然是最易受到损害的添

microcapsule technology or by adsorption method. The procedure of microcapsule technology is to add gum arabic into the emulsifier first. The oil-like vitamin A ester is added to emulsify and form particles which are then moved to the reaction tank together with the gelatin aqueous solution, and use the charge relation to cause the crosslinking effect between the emulsion and the gelatin water to form the particles covered by gelatin. Then, sugar coating, hydrophobic agent, and starch coating are added to make microcapsules. Antioxidants can also be added in the production process to avoid oxidation of vitamin A. The adsorption method is to emulsify the oil-liquid vitamin A ester first, stabilize it with antioxidant, and then adsorb it with dried wheat bran and silicate.

Vitamin A additives are different in particle size due to the different process conditions. Although foreign regulations of their granularity are generally between 0.1 and 1.0 mm, in fact, they are usually within 0.177－0.590 mm (80 to 30 mesh). The vitamin A additive that can be dispersed in water has a smaller particle size and the maximum size should not exceed 0.35 mm. Vitamin A additives should have good firmness, abrasion resistance, and the surface should be rough for easy mixing and uniformity. The quality of the preparation process of vitamin A additives can be inspected by microchemical methods, or the vitamin A additives can be mixed with choline chloride, and after a certain period of storage, the loss of vitamin A can be measured by instruments. Biological methods can also be used for practical tests.

Both ultraviolet rays and oxygen can promote the decomposition of vitamin A acetate and vitamin A palmitate. When humidity and temperature are high, the decomposition rate of rare metal salts can be accelerated. $FeSO_4 \cdot 7H_2O$ containing 7 mg/l water can cause severe loss of activity of vitamin A acetate. When in contact with choline chloride, its activity will be severely impaired. Vitamin A is quickly broken down under pH value 4 or under strong alkalinity. After vitamin A ester is coated, the loss can be reduced. The stability of the activity has been greatly improved after vitamin A is made into micro-capsules or particles. However, it is still one of the most vulnerable additives. Special attention should be paid in use and storage.

加剂之一,在使用和贮存时,应特别注意。

常见的维生素 A 制剂有:① 维生素 A 油,大多是由海鱼的鱼肝中提取,加入抗氧化剂后制成微囊作添加剂,也称鱼肝油。其中含维生素 A 850 IU/g 和维生素 D 65 IU/g。② 包被型稳定维生素制剂,维生素 A 乙酸酯和维生素 A 棕榈酸酯的微粒胶囊、微粒粉剂是最常用的饲用维生素 A 制剂。包被型维生素 A 制剂含有效成分一般在 10 万~50 万 IU/g,由制药厂生产的高浓度制剂一般为 50 万 IU/g。

维生素 A 稳定性很差,遇光、氧、酸等可迅速被破坏,特别是同时湿热或某些微量元素存在时会加速破坏。包被型制剂较稳定,但在湿热和微量元素存在时仍会被破坏。

饲用维生素 A 除上述高浓度单项制剂外,还有以脱脂米糠、黄豆细粉等作载体的单项预混料。此外,稳定型维生素 AD 或维生素 ADE 粉剂应用广泛,可以避免其中两种或三种物质分离。

2. β-胡萝卜素制剂。β-胡萝卜素分子式为 $C_{40}H_{56}$,相对分子质量为 536.9;其纯品为红棕色至深紫色的结晶性粉末;对光、氧和酸十分敏感;不溶于水,微溶于脂肪和油,溶于丙酮、石油醚等有机溶剂。因稳定性差,商品 β-胡萝卜素制剂同维生素 A 制剂一样,多为各种包被材料处理的稳定制剂。

以添加 β-胡萝卜素补充维生素 A 很不经济,但对处于不良情况下的某些繁殖母畜具有维持正常繁殖性能的作用。通常在动物发情不明显、妊娠率低、妊娠后交配、分娩困难和产弱子等情况下添加。

The common vitamin A preparations are: ① Vitamin A oil, mostly extracted from the fish liver of marine fish, is made into microcapsules as additives after adding antioxidants, also known as cod liver oil, which contains vitamin A 850 IU/g and vitamin D 65 IU/g. ② Coated stable vitamin preparations, microparticle capsules and microparticle powders of vitamin A acetate and vitamin A palmitate are the most commonly used dietary vitamin A preparations. Coated vitamin A preparations generally contain effective ingredients between 100,000 and 500,000 IU/g, and the high-concentration preparations produced by pharmaceutical factories are generally 500,000 IU/g.

The stability of vitamin A is very poor, and can be quickly destroyed when exposed to light, oxygen, acid, especially when it is exposed to heat or moisture or certain trace elements it will deteriorate more quickly. The coated preparation is more stable, but will still be damaged in the presence of moisture, heat and trace elements.

Apart from high-concentration single preparations of vitamin A for feed, there are also single premixes with defatted rice bran and soybean powder as carriers. In addition, stable vitamin AD or vitamin ADE powders are widely used, which can avoid the separation of two or three substances.

2. β-carotene preparations. The molecular formula of β-carotene is $C_{40}H_{56}$. The relative molecular weight is 536.9. Its pure products are reddish brown or dark purple crystalline powder. They are very sensitive to light, oxygen and acids, insoluble in water, slightly soluble in fat and oil, soluble in acetone, petroleum ether and other organic solvents. As a result of poor stability, commercial β-carotene preparations, as well as vitamin A preparations, are mostly stable preparations coated with a variety of materials.

Supplementing vitamin A with β-carotene is uneconomical, but has the effect of maintaining normal reproductive performance in certain breeding females under adverse conditions. It is usually added when the animal is not in estrus, has a low pregnancy rate, copulates after pregnancy, has difficulty giving birth and gives birth to weak offspring.

二、维生素 D
Vitamin D

维生素 D 又名钙(或骨)化醇,抗佝偻病维生素等。维生素 D 的两种主要形式是维生素 D_2(麦角钙化醇)和维生素 D_3(胆钙化醇)。维生素 D_2 可由紫外线照射处理饲用酵母而得。维生素 D_3 对禽类的活性远高于维生素 D_2,对于其他动物维生素 D_3 效果也很好,而且维生素 D_3 较维生素 D_2 稳定性好。因此,饲料中添加的多为维生素 D_3。

在配合饲料中,维生素 D_3 的稳定性虽比维生素 D_2 好,但它与热、潮湿和某些无机元素、氧化剂等直接接触时,也很容易被破坏失效。因此,也需要进行特殊的防氧化和包被处理,处理方式与维生素 A 基本相同,包括维生素 D 微粒胶囊、微粒粉剂、β-环糊精包被物和维生素 D 油等制剂。稳定的维生素 AD 制剂为常用的商品性维生素 D 添加剂形式。

维生素 D 添加剂有如下几种:

1. 维生素 D_2 和维生素 D_3 的干燥粉剂,外观呈奶油色粉末,含量为 50 万 IU/g 或 20 万 IU/g。

2. 维生素 D_3 微粒是饲料工业中使用的主要维生素 D_3 添加剂,其原料为胆固醇。这种胆固醇可从羊毛脂中分离制得,然后经酯化、溴化、再脱溴和水解即得 7-脱氢胆固醇。

经紫外线光照射得维生素 D_3。维生素 D_3 添加剂是以含量为 130 万 IU/g 以上的维生素 D_3 为原料,酯化后,配以一定量的 BHT 及乙氧喹啉抗氧化剂,采用明胶和淀粉等

Vitamin D is also known as calciferol, anti-rickets vitamin, etc. The two main forms of vitamin D are D_2 (ergocalciferol) and D_3 (cholecalciferol). Vitamin D_2 can be obtained from feeding yeast treated with ultraviolet radiation. Vitamin D_3 is much more active than vitamin D_2 in poultry, and it is also very effective for other animals. Besides, vitamin D_3 is more stable than vitamin D_2. Therefore, vitamin D_3 is mostly added in feed.

In the compound feed, the stability of vitamin D_3 is better than vitamin D_2, but it is also easy to be destroyed when it is in direct contact with heat, moisture and some inorganic elements, oxidants and so on. Therefore, special anti-oxidation and encapsulation treatments are also required. The treatment methods are basically the same as those of vitamin A, including preparations such as vitamin D microcapsules, micropowders, β-cyclodextrin coatings and vitamin D oils. Stable vitamin AD preparations are commonly used in the form of commercial vitamin D additives.

Vitamin D additives are available in the following categories:

1. Dry powder of vitamin D_2 and vitamin D_3, which is cream-colored powder with a content of 500,000 IU/g or 200,000 IU/g.

2. Vitamin D_3 particles are the main vitamin D_3 additives used in the feed industry, and their raw materials are cholesterol. This cholesterol can be isolated from lanolin and then esterified, brominated, debrominated and hydrolyzed to obtain 7-dehydrocholesterol.

Vitamin D_3 is obtained by ultraviolet light. Vitamin D_3 additive is made of vitamin D_3 with a content of more than 1.3 million IU/g of vitamin D_3 as raw material. After esterification, it is mixed with a certain amount of BHT and ethoxyquinoline antioxidants, using gelatin and starch and

辅料,经喷雾法制成的微粒。产品规格有 50 万 IU/g、40 万 IU/g 和 30 万 IU/g。

3. 维生素 A/D 微粒是以维生素 A 乙酸酯原油与含量为 130 万 IU/g 以上的维生素 D_3 为原料,配以一定量的 BHT 及乙氧喹啉抗氧化剂,采用明胶和淀粉等辅料,经喷雾法制成的微粒。每单位质量中维生素 A 乙酸酯与维生素 D_3 之比为 5:1。

维生素 D_3 酯化后,又经明胶、糖和淀粉包被,稳定性好,在常温(20-25 ℃)条件下,在含有其他维生素添加剂的预混剂中,贮存一年,甚至 24 个月,也没有什么损失。但是,如果温度为 35 ℃,在预混剂中贮存 24 个月,活性将损失 35%。如添加剂制作工艺较差,储存期不能过长。

other excipients, and the particles are made by spraying. The product specifications are 500,000 IU/g, 400,000 IU/g and 300,000 IU/g.

3. Vitamin A/D particles are made by spray method with vitamin A acetate crude oil and vitamin D_3 with a content of more than 1.3 million IU/g as the raw materials, combined with a certain amount of BHT and ethoxyquinoline antioxidants together with gelatin, starch and other excipients. The ratio of vitamin A acetate to vitamin D_3 per unit mass is 5:1.

After the esterification of vitamin D_3, it is coated with gelatin, sugar and starch to maintain good stability. Under the condition of normal temperature (20 - 25 ℃), it is stored for one year or even 24 months in premix containing other vitamin additives without any loss. However, if the temperature is 35 ℃, its activity will be lost by 35%, after it is stored in premix for 24 months. If the additive manufacturing process is poor, the storage period should not be too long.

三、维生素 E
Vitamin E

维生素 E 属乙酸酯,较游离维生素 E 稳定。作为非抗氧化剂的饲用维生素 E 为 α-生育酚乙酸酯,其中自然界存在的 D-α-生育酚乙酸酯效价最高,化工合成的维生素 E 是 dl-α-生育酚形式的产品。

维生素 E 在动物性饲料中含量极少,仅人和牛的初乳及蛋类中有一定含量。通常主要存在于植物性饲料中,植物油尤其小麦胚油是维生素 E 的丰富来源。另外,大多数青绿饲料、籽实胚芽、调制良好的青干草、谷物籽实饲料、酵母、糠麸等均是维生素 E 的良好来源。动物体内不能合成维生素 E,只能通过外源的供给来满足其生长、生产需要。

维生素 E 在饲料中极易被氧

Vitamin E is acetate, which is more stable than free vitamin E. As a non-antioxidant, dietary vitamin E is α-tocopherol acetate, of which d-α-tocopherol acetate has the highest titer in nature, and vitamin E synthesized by chemical industry is a product in the form of dl-α-tocopherol.

Vitamin E content in animal feed is very low. Only human and bovine colostrum and eggs have a certain content. Usually found in plant feeds, vegetable oil, especially wheat germ oil are rich sources of vitamin E. In addition, most of the green forage, seed germ, well-prepared green hay, grain seed feed, yeast, bran and so on are good sources of vitamin E. Animal body can not synthesize vitamin E. It can only meet its growth, production needs through exogenous supply.

Vitamin E is easily destroyed by oxidation in feed. Al-

化破坏,虽对其他维生素可起到保护作用,但自身却失去生理活性。因此,一般非抗氧化用维生素E制剂中也应添加抗氧化剂和进行其他稳定性处理。维生素E制剂也有油剂、粉剂(微粒胶囊、微粒粉剂、β-环糊精包被物)和可溶性粉剂。

though it can play a protective role on other vitamins, it loses its physiological activity. Therefore, antioxidants and other stability treatments should also be added to general non-antioxidant vitamin E preparations. Vitamin E preparations are also available as oils, powders (microcapsules, micropowders, β-cyclodextrin coatings), and soluble powders.

四、维生素 K_3
Vitamin K_3

由于人工合成的维生素 K_3 制剂效价高,又是水溶性结晶,性质较稳定,故饲用维生素 K 多是 K_3 制剂。

目前饲用维生素 K_3 制剂有亚硫酸氢钠甲萘醌(MSB)、亚硫酸氢钠甲萘醌复合物(MSBC)、亚硫酸氢二甲基嘧啶甲萘醌(MPB)和亚硫酸烟酰胺甲萘醌(MNB),其活性成分为甲萘醌。

1. 亚硫酸氢钠甲萘醌。MSB 多含 3 个结晶水,其分子式为 $C_{11}H_8O_2 \cdot NaHSO_3 \cdot 3H_2O$,含活性成分约 52%。为白色或灰色结晶性粉末,无臭或微有特异臭味,有吸湿性,遇光易分解。MSB 对皮肤和呼吸道黏膜有刺激性。

MSB 商品制剂是含 MSB 94%-96% 的高浓度产品,其稳定性差,但价格便宜。用明胶包被处理的 MSB 微粒胶囊制剂一般含 MSB 50%,稳定性好,且无刺激性。

2. 亚硫酸氢钠甲萘醌复合物。MSBC 的化合物成分与 MSB 相同,为 $C_{11}H_8O_2 \cdot NaHSO_3 \cdot 3H_2O$,二者的区别在于形成亚硫酸氢钠结合物时,MSBC 添加了过量的亚硫酸氢钠提高甲萘醌的稳定性。此制剂常含有较多的游离亚硫酸氢钠,因而活性成分甲萘醌含量较低,一般在 30%-40%。我国饲料添加剂标

Due to the high potency, water-soluble crystallization and stable properties of the synthetic vitamin K_3 preparations, dietary vitamin K is mostly K_3 preparations.

Currently, dietary vitamin K_3 preparations include menadione sodium bisulfite (MSB), menadione sodium bisulfite complex (MSBC), menadione dimethylpyrimidinol bisulfite (MPB) and menadione nicotinamide bisulfite (MNB), whose active ingredient is menaphone.

1. Menadione sodium bisulfite. MSB contains more than 3 crystalline waters, and its molecular formula is $C_{11}H_8O_2 \cdot NaHSO_3 \cdot 3H_2O$, containing about 52 percent of active ingredients. It is white or gray crystalline powder, odorless or slightly odorous, hygroscopic, easy to decompose when exposed to light. MSB is irritating to skin and respiratory mucosa.

MSB commercial preparations are high-concentration products containing 94%-96% MSB, which have poor stability but are cheap. MSB microcapsule preparations coated with gelatin generally contain 50 percent of MSB, and have good stability and no irritation.

2. Menadione sodium bisulfite complex. The compound composition of MSBC is the same as that of MSB, which is $C_{11}H_8O_2 \cdot NaHSO_3 \cdot 3H_2O$. The difference between them is that when forming sodium bisulfite binding compound, MSBC adds excessive sodium bisulfite to improve the stability of menadione. This preparation often contains more free sodium bisulfite, so the content of active ingredient menaphone is low, generally between 30 percent and 40 percent. China's feed additive standard (GB7294—

准（GB7294—2017）要求含 $C_{11}H_8O_2 \cdot NaHSO_3 \cdot 3H_2O$ 60%-75%，即活性成分31.1%-39.1%。MSBC稳定性较好，是目前应用最广泛的维生素K_3制剂。

3. 亚硫酸氢二甲基嘧啶甲萘醌(MPB)。MPB为稳定性最好的K_3制剂，含活性成分45.5%。在饲料制粒过程中能保持较高的活性，但具有一定毒性，且价格较贵。因此，目前应用不及MSBC广泛。因具有毒性，应限制使用。美国食品与药物管理局规定，以MPB作为营养性添加剂使用时，鸡与火鸡用量不得超过全价饲料的2 mg/kg，生长肥育猪不得超过10 mg/kg。

4. 亚硫酸烟酰胺甲萘醌。MNB分子式为$C_{17}H_{16}O_6N_2S$，相对分子质量376.23。为白色结晶粉末，微溶于水(1.3 g/100 ml)，易溶于乙烷和氯仿。MNB是在MSB基础上，结合烟酰胺分子团，取代MSB中的钠离子及三个结晶水，已全面消除了引起产品不稳定的因素——水分子的存在；大分子团遮蔽了异构体转化的路径，抑制异构体的生成，同时辅助基团的存在有利于生物体内维生素K_3有更高的稳定性和生物活性，同时还可补充烟酰胺。目前，含MNB 96%和50%的产品，分别含甲萘醌43.9%和22.9%，含烟酰胺30%和16%。

2017) requires 60% to 75% of contain $C_{11}H_8O_2 \cdot NaHSO_3 \cdot 3H_2O$, namely between 31.1% and 39.1% of active ingredient. MSBC is the most widely used vitamin K_3 preparation with good stability.

3. Menadione dimethylpyrimidinol bisulfite. MPB is the most stable K_3 preparation, containing 45.5% of active ingredient. It can maintain high activity in the process of feed granulation, but it has certain toxicity and is more expensive. Therefore, the current application is not as extensive as MSBC. Use should be restricted due to its toxicity. The US Food and Drug Administration (FDA) stipulates that when using MPB as a nutritional additive, dosage may not exceed 2 mg/kg of full feed for chickens and turkeys, and may not exceed 10 mg/kg for growing and finishing pigs.

4. Menadione nicotinamide bisulfite. The molecular formula of MNB is $C_{17}H_{16}O_6N_2S$, and its relative molecular weight is 376.23. It is a white crystalline powder, slightly soluble in water (1.3 g/100 ml), easily soluble in ethane and chloroform. MNB, based on MSB, combined with niacinamide molecule group to replace sodium ion and three crystal water in MSB, has completely eliminated the unstable factor of the product — the presence of water molecule. The macromolecules block the path of isomer transformation and inhibit the formation of isomers. At the same time, the presence of auxiliary groups is conducive to the higher stability and bioactivity of vitamin K_3 in organisms, and can also supplement nicotinamide. Currently, products of 96 and 50 percentage of content contain 43.9% and 22.9% of menaphone, and 30% and 16% of nicotinamide respectively.

第二节　水溶性维生素饲料

Section 2　Water-soluble vitamin feed

目前饲料中添加的有维生素B_1、B_2、B_6、B_{12}、烟酸和烟酰胺、泛酸、胆碱、叶酸、生物素、维生素C、肌醇、肉毒碱。

At present, vitamins B_1, B_2, B_6, B_{12}, niacin and nicotinamide, pantothenic acid, choline, folic acid, biotin, vitamin C, inositol, carnitine are added in the feed.

一、硫胺素（维生素 B_1）
Thiamine（Vitamin B_1）

维生素 B_1 又称硫胺素、抗神经炎素。大多数常用饲料中，维生素 B_1 均较丰富，特别是各类谷物、磨粉工业的副产品（如糠麸、米糠、麦麸、榨油后的油饼或饼粕）、苜蓿等中含量更多，青绿饲料和优质干草、豆类、大蒜、胚芽饼等也是维生素 B_1 的重要来源。根茎类饲料中含量较少，木薯淀粉、精米和面粉中含量很低。蛋黄、肝、肾、瘦猪肉、蚕蛹、酒精、酵母等动物性食物和饲料也是维生素 B_1 的较好来源。肉骨粉中维生素 B_1 含量很低。结晶型的或饲料中的维生素 B_1 极易被热、碱破坏，但在酸性溶液中加热到 120 ℃ 并不分解。因此，谷物干燥或加工会使可利用的维生素 B_1 浓度降低。

硫胺素为嘧啶衍生物，具有阳离子特性，能同许多阴离子形成盐或复杂的有机化合物。用于饲料的主要是由化学合成法制得的硫胺素盐酸盐（盐酸硫胺素）和硝酸盐（单硝酸硫胺素）。

1. 盐酸硫胺素。盐酸硫胺素分子式为 $C_{12}H_{17}ClN_4OS \cdot HCl$，含有效成分 78.7%，为白色结晶或结晶性粉末，略有特异性臭味。易溶于水，具有吸湿性。在 pH 值 3.5 以下时稳定性较好，但在中性或碱性条件下不稳定，对热、氧化剂、还原剂、金属盐类敏感，特别是在有水分存在的条件下稳定性更差。

2. 单硝酸硫胺素。单硝酸硫胺素分子式为 $C_{12}H_{17}N_5O_4S$，含有效成分 81.1%，为白色或微黄色结晶性粉末，无臭或略有特异性臭味。微溶于水，吸湿性小。在中性和碱

Vitamin B_1 is also known as thiamine and aneurin. In most commonly used feeds, especially in all kinds of grains, by-products of milling industry (such as bran bran, rice bran, wheat bran, oil cake or cake after oil pressing) and alfalfa, etc., vitamin B_1 is rich. Green forage and high-quality hay, beans, garlic, germ cake are also important sources of vitamin B_1. The content of rhizomes is low, while the content of tapioca starch, white rice and flour is low. Egg yolk, liver, kidney, lean pork, silkworm pupa, alcohol, yeast and other animal food and feed are also good sources of vitamin B_1. The content of vitamin B_1 in meat and bone meal is very low. Vitamin B_1 in crystalline or feed is easily damaged by heat and alkali, but does not decompose when heated to 120 ℃ in acidic solution. Therefore, grain drying or processing can reduce the concentration of vitamin B_1 available.

Thiamine is a pyrimidine derivative with cationic properties and can form salts or complex organic compounds with many anions. The feed is mainly made by chemical synthesis of thiamine hydrochloride (thiamine hydrochloride) and nitrate (thiamine mononitrate).

1. Thiamine hydrochloride. The molecular formula of thiamine hydrochloride is $C_{12}H_{17}ClN_4OS \cdot HCl$, which contains 78.7% of active ingredients. It is white crystal or crystalline powder with a little specific odor. It is soluble in water and hygroscopic. It has good stability under pH 3.5, but is unstable under neutral or alkaline conditions. It is sensitive to heat, oxidants, reducing agents and metal salts, especially in the presence of water.

2. Thiamine mononitrate. The molecular formula of thiamine mononitrate is $C_{12}H_{17}N_5O_4S$, which contains 81.1% of active ingredients. It is white or yellowish crystalline powder with no odor or slightly specific odor. It is slightly soluble in water and has little hygroscopicity. Under

性条件下不稳定,但对热、氧化剂、还原剂较盐酸硫胺素敏感性差。在饲料中的配伍性较好,在预混料和配合饲料的加工和贮存过程中较稳定,特别是在加有吸湿性强的氯化胆碱的维生素与微量元素的复合预混料中以及饲料的制粒、膨化和宠物罐头饲料的加工过程中的损失率远低于盐酸硫胺素。

neutral and alkaline conditions it is not stable, but less sensitive to heat, oxidants, reducing agents than thiamine hydrochloride. It has good compatibility in feed, and is more stable during the processing and storage of premix and compound feed, especially in the compound premix of vitamins and trace elements with strong hygroscopic choline chloride, as well as pelleting, puffing and canned pet feed. The loss rate during processing is far lower than that of thiamine hydrochloride.

二、核黄素(维生素 B_2)
Riboflavin (Vitamin B_2)

商品维生素 B_2 为核黄素及其酯类,用作饲料的主要是由微生物发酵或化学合成的核黄素。干燥的结晶状核黄素对氧化剂、酸、热极稳定,但遇碱、光则迅速分解,特别是在碱性溶液中或紫外线下分解更快。因此,必须密封、避光保存,在室温(25 ℃)下至少可贮存一年。

在预混料中,应尽量避免与碱性物质配伍,特别是同时含有较多游离水的条件下,核黄素损失量增加。在避光的干粉料中,核黄素稳定性较好。饲料的制粒和膨化加工对核黄素有一定破坏。

维生素 B_2 广泛存在于生物体中,一般动物性食品或饲料中含量较高,如肝脏、肾、心、乳品(干乳酪、干乳清、干脱脂乳)、鱼粉、肉粉、蛋、干酵母、真菌、鸡粪(细菌合成)、蚕蛹粉、血粉等。植物性饲料中含维生素 B_2 的也很多,如青绿饲料、优质草粉(尤其是苜蓿粉)、麦类、糠麸、胚芽、黄玉米等。但相对而言,谷物块根、块茎、油饼类饲料中核黄素贫乏。

饲用维生素 B_2 制剂除纯品外,还有以大豆皮粉或玉米芯粉等作为载体或稀释剂制成的多种不同浓度的产品。纯品维生素 B_2 含量在

Commercial vitamin B_2 is riboflavin and its esters. Riboflavin used as feed is mainly fermented or chemically synthesized by microorganisms. Dry crystalline riboflavin is extremely stable to oxidants, acids and heat, but rapidly breaks down when exposed to alkali and light, especially in alkaline solutions or under ultraviolet light. Therefore, it must be sealed, stored away from light and stored at room temperature (25 ℃) for at least one year.

In premix, the combination with alkaline substance should be avoided as far as possible, especially when more free water is also contained, the amount of riboflavin loss will increase. The stability of riboflavin is better in dry powders that are protected from light. The granulation and expansion of feed will damage riboflavin to some extent.

Vitamin B_2 is widely found in organisms, and the content of general animal food or feed is relatively high, such as liver, kidney, heart, dairy (dry cheese, dry whey, dry skimmed milk), fish meal, meat meal, egg, dry yeast, fungus, chicken feces (bacterial synthesis), silkworm pupa powder, blood meal, etc. There are also many plant feeds containing vitamin B_2, such as green forage, high-quality grass meal (especially alfalfa meal), wheat, bran, germ, yellow corn and so on. But comparatively speaking, the riboflavin is poor in grain roots, tubers and oil cakes.

In addition to the pure dietary vitamin B_2 preparations, there are also a variety of products with different concentrations made with soybean husk powder or corn cob powder as carrier or diluent. The pure vitamin B_2 whose content is

96%以上,有静电作用,易吸附于加工设备上,在配制饲料时需预处理。经稀释处理的产品无静电作用,流动性好。

above 96% with the electrostatic effect is easy to be adsorbed on the processing equipment. It needs to be pretreated in the preparation of feed. The diluted product has no electrostatic effect and has good fluidity.

三、泛酸
Pantothenic acid

游离泛酸极不稳定,极易吸湿,在自然界很少存在。因此,用于饲料者多选用稳定性好的泛酸钙。此外,在液体饲料中,泛酸和泛醇也有应用。

饲用泛酸钙产品有右旋泛酸钙和外消旋泛酸钙两种。由于仅D型泛酸及其盐类具有生物活性,dl-泛酸钙效价为D-泛酸钙的50%。D-泛酸钙的生物活性为泛酸的92%。

D-泛酸钙为白色吸湿性粉末,无臭,味微苦,易溶于水。在阴冷、干燥条件下较稳定,吸湿后或在水溶液中会水解,效价降低。在酸、中性条件下更易被破坏,对酸特别敏感,对热中等敏感,但对氧化、还原作用和光稳定。因此,在预混料和配合饲料中应避免与吸湿性强、呈酸性反应的硫酸盐、氯化物等组分共存。D-泛酸钙在干粉配合饲料中损失不大,但混后再粉碎,损失会增加。

右旋泛醇(D-泛醇)为无色黏稠液体,长期贮存可形成结晶,能同水混溶。在酸性(pH值3-7)液体中稳定性较好,因而在此条件下可选用。其效价与泛酸相当。

泛酸的钙盐具有较强的吸湿性,包装的容器必须具有较好的防潮性。在稀释产品中常添加防结块剂,以增加流动性,防止结块。

Free pantothenic acid is extremely unstable and highly hygroscopic, and it rarely exists in nature. Therefore, stable calcium pantothenate is selected for feed. In addition, pantothenic acid and panthenol are also used in liquid feed.

Dietary calcium pantothenate products include calcium dextrorotate and calcium racemic pantothenate. Since only D-type pantothenic acid and its salts are biologically active, the titer of dl-calcium pantothenate is 50% of that of D-calcium pantothenate. The Biological activity of D-calcium pantothenate is 92% of that of pantothenic acid.

D-calcium pantothenate is a white hygroscopic powder. It is odorless with slightly bitter taste, and easily soluble in water. It is more stable in cold and dry conditions, hydrolyzed after hygroscopic or in aqueous solution, and its titer decreases. It is more vulnerable to damage under acid and neutral conditions, especially sensitive to acid and moderately sensitive to heat, but stable to oxidation, reduction and light. Therefore, the premix and compound feed should avoid the coexistence of sulfate, chloride and other components with strong hygroscopicity and acidic reaction. The loss of D-calcium pantothenate in the dry powder mixed feed is small, but the loss is increased after mixing and grinding.

Dextranol (D-panthenol) is a colorless, viscous liquid that forms crystals in long-term storage and is miscible with water. It has good stability in acidic (pH 3-7) liquid, so it can be used under these conditions. Its potency is comparable to that of pantothenic acid.

The calcium salt of pantothenic acid has stronger hygroscopic property, therefore the packing container must be moisture-proof. Anti-caking agents are often added to diluted products to increase fluidity and prevent agglomeration.

四、维生素 B_5（烟酸和烟酰胺）
Vitamin B_5 (niacin and nicotinamide)

烟酸在自然界分布甚广。在鱼粉、饲用酵母、乳、肾、肝脏、肉骨粉、肉类等动物性食品和饲料中含量丰富。几乎所有植物性饲料中都含有不同量的烟酸，如麸皮、青绿饲料、麦类、高粱、玉米、大豆、糠麸、稻米、饼粕类、豆科牧草等中都含量丰富。但在玉米、高粱、小麦等谷物中烟酸处于结合状态，不易被畜禽在消化道中吸收利用。

用于饲料的维生素 B_5 有烟酸和烟酰胺两种形式的产品。二者均为白色或微黄色粉末，无臭。烟酸味微酸，溶于水、乙醇，易溶于碱性溶液，无吸湿性，流动性好。烟酰胺味苦，易溶于水、乙醇，溶于甘油，吸湿性强，流动性差。

烟酸和烟酰胺在干燥和水溶液中都很稳定，几乎不受热、光、氧化、还原影响。酸、碱对二者有轻微影响。在与微量元素配合时，烟酸适宜同呈酸性反应的硫酸盐、氯化物和硝酸盐配合；而烟酰胺适宜同呈中性或碱性反应的氧化物配合。

由于烟酰胺具有较强的吸湿性，主要用于配制液体饲料和水溶性制剂，其他饲料中则选用烟酸。烟酸的溶解度可满足配制犊牛、乳猪、羔羊的代乳料要求，因此，无需选用烟酰胺。

烟酸在各种饲料中的稳定性都很好，在配合饲料的加工（如制粒、灭菌及与微量元素混合）和贮存过程中损失均很少，但膨化处理对烟酸的破坏较大，一般流失为 10%–20%。

Niacin is widely distributed in nature. It is abundant in fish meal, feeding yeast, milk, kidney, liver, meat and bone meal, meat and other animal food and feed. Almost all plant feeds contain different amounts of niacin, such as bran, green forage, wheat, sorghum, corn, soybean, bran, rice, cake, legumes, etc. which are rich in content. However, niacin in corn, sorghum, wheat and other grains is in a state of binding, and is not easy to be absorbed and utilized by livestock and poultry in the digestive tract.

Vitamin B_5 used in feed are in forms of niacin and niacinamide. Both are white or yellowish powder, odorless. Niacin is slightly acidic, soluble in water and ethanol, easily soluble in alkaline solution, non-hygroscopic and has good fluidity. Niacinamide tastes bitter. It is easily soluble in water, ethanol, and glycerol with strong hygroscopicity and poor fluidity.

Niacin and niacinamide are stable in dry and aqueous solutions and are almost unaffected by heat, light, oxidation, reduction. Acid and alkali have a slight effect on both. In combination with trace elements, niacin is suitable for the combination of sulphates, chlorides and nitrates that react with acidity, while niacinamide is suitable for the combination with oxides that react with neutrality or alkalinity.

Nicotinamide is mainly used to prepare liquid feed and water-soluble preparation because of its strong hygroscopicity, and niacin is selected in other feeds. The solubility of niacin can meet the requirements of milk substitutes for calves, suckling pigs and lambs, and nicotinamide is not needed.

The stability of niacin in all kinds of feeds is very good, and the loss during the processing (such as granulation, sterilization and mixing with trace elements) and storage of compound feeds is very small, but the expansion treatment has a great damage to niacin, which generally loses between 10% and 20%.

五、维生素 B_6
Vitamin B_6

尽管吡哆醛、吡哆胺与吡哆醇对动物有相同的生物学效价，但前二者的稳定性差。特别是在光、加工和贮存温度、酸、碱度和水分的影响下，稳定性更差。因此，通常作为补充维生素 B_6 的均为吡哆醇的盐酸盐。盐酸吡哆醇的生物活性相当于吡哆醇的82.3%。

商品盐酸吡哆醇多为含量在58%以上的产品，是一种白色至微黄色、无臭的结晶性粉末，味酸苦，易溶于水，但无吸湿性。此产品在干燥避光条件下稳定性好，但对光、碱敏感，特别是在水溶液中或吸湿条件下，遇光或碱迅速分解。

盐酸吡哆醇在应用干燥、惰性载体和各种维生素预混料中稳定性很好，在与氯化胆碱和微量元素矿物质共存，特别是呈碱性反应的微量元素氧化物和碳酸盐共存时，盐酸吡哆醇迅速而大量地被破坏。

Although pyridoxal, pyridoxamine and pyridoxine have the same biological potency to animals, the stability of the former two is poor. Especially under the influence of light, processing and storage temperature, acid, alkalinity and water, the stability is worse. Thus, pyridoxine hydrochloride is commonly used as a vitamin B_6 supplement. The bioactivity of pyridoxine hydrochloride is equivalent to 82.3% of that of pyridoxine.

The commercial products containing more than 58% of pyridoxine hydrochloride are white or slightly yellow. It is odorless crystalline powder with sour and bitter taste, easily soluble in water, but non-hygroscopic. This product has good stability under dry and dark conditions, but it is sensitive to light and alkali, especially in aqueous solution or hygroscopic conditions. It will decompose rapidly when exposed to light or alkali.

Pyridoxine hydrochloride is very stable in the application of dry, inert carriers and various vitamin premixes. When it coexists with choline chloride and trace element minerals, especially oxides and carbonates that react with alkalinity, pyridoxine hydrochloride, it is rapidly and massively destroyed.

六、生物素
Biotin

生物素又称维生素H。生物素的补充物为右旋生物素（D-生物素）制剂，纯品一般含D-生物素98%以上，是一种近白色结晶性粉末，在冷水中溶解度低，随水温升高其溶解度增加，但高温时稳定性受到影响。

生物素是稳定性较好的一种维生素，对氧化、还原、微量元素都很稳定，强酸、强碱、紫外线对生物素稍有影响，生物素对热敏感。

鱼、奶、肝脏、肾、蛋黄（或全蛋）、肉等动物性食物或饲料中生

Biotin is also called vitamin H. The supplement of biotin is dextrin (D-biotin) preparation. The pure product generally contains more than 98% of D-biotin. It is a nearly white crystalline powder with low solubility in cold water and high solubility with the increase of water temperature, but its stability is affected at high temperature.

Biotin is a kind of vitamin with good stability, which is very stable to oxidation, reduction and trace elements. Strong acid, alkali and ultraviolet light have a little influence on biotin which is sensitive to heat.

Biotin is abundant in fish, milk, liver, kidney, egg yolk (or whole egg), meat and other animal food or feed.

物素的含量丰富。畜禽对动物性来源生物素的利用率也较植物性来源者为高。同时，生物素也广泛分布于植物性饲料中，例如花生、废糖蜜、玉米、小麦及其他谷类、米糠、马铃薯、豆粕、青草等中含量较多。所有动物肠道内的细菌都能合成生物素，反刍动物瘤胃微生物的合成数量较多，足够反刍家畜利用。而猪、鸡、鱼等动物常需要在日粮中供给适量生物素，尤其是雏鸡和仔猪及鱼类更需要补加生物素。

生物素的饲用商品制剂一般为含D生物素1%或2%的预混料。其产品有两种形式，即载体吸附型生物素和与一定载体（如糊精）混合后经喷雾干燥制得的喷雾干燥型生物素制剂。喷雾干燥型粒度较前者小，其水溶性和吸湿性因载体不同而不同。两种产品在干燥密闭条件下都较稳定。

The utilization rate of animal-derived biotin by livestock and poultry is also higher than that of plant-derived ones. At the same time, biotin is also widely distributed in plant feeds, such as peanuts, waste molasses, corn, wheat and other grains, rice bran, potatoes, soybean meal, grass, etc. The bacteria in the intestinal tract of all animals can synthesize biotin, and ruminant microorganisms have a large amount of synthesis, which is sufficient for the use of ruminant livestock. But pigs, chickens, fish and other animals, especially chicks, piglets and fish that need to be supplemented with biotin in daily diets.

Dietary commercial preparations of biotin are generally premixed materials containing 1 or 2 percent of biotin. It has two forms, namely, the carrier-adsorbed biotin and the spray-dried biotin prepared by spray drying after being mixed with a certain carrier (such as dextrin). The spray-dried has a smaller particle size than the former, and its water-solubility and hygroscopicity are different with different carriers. Both products are stable under dry and airtight conditions.

七、维生素 B_{12}
Vitamin B_{12}

维生素 B_{12} 又称氰钴胺素，是一种暗红色针状结晶细粉，无臭无味，溶于水和乙醇。维生素 B_{12} 在弱酸和中性条件下稳定性好；应避光贮存，不宜与有还原作用的维生素C等物配伍。

维生素 B_{12} 在包括含有微量元素的预混料、配合料中都比较稳定，月损失率约为1%-2%。制粒、膨化对维生素 B_{12} 的损失有增加，制粒约为2%-4%，膨化约为2%-6%。

动物性饲料是维生素 B_{12} 的主要来源，如肝脏、动物粪便（如牛、羊和鸡的粪便）的发酵产物、鸡舍垫草、鱼粉、鱼副产品、抗生素药渣、肉粉、蛋、奶（脱脂奶）、贝类、反刍动物的器官、牛羊肠内容物、肉骨

Vitamin B_{12} is also called cyanocobalamin. It is a dark red needle-like crystalline powder, odorless, tasteless, soluble in water and ethanol. Vitamin B_{12} has good stability under weak acid and neutral conditions. It should be stored away from light and should not be compatible with reducing vitamin C and other substances.

Vitamin B_{12} is relatively stable in premix and blends containing trace elements, and the monthly loss rate is about 1% to 2%. The loss of Vitamin B_{12} is increased by granulation (about 2% to 4%) and swelling (about 2% to 6%).

Animal feed is the main source of vitamin B_{12}, such as liver, animal manure (such as cattle, sheep and chicken manure) fermentation products, bedding, fish meal, fish by-products, antibiotics residues, meat powder, egg, milk (skimmed milk), shellfish, ruminant organs, intestinal contents of cattle and sheep, meat and bone meal, crude ca-

粉、粗酪蛋白、血粉等都是维生素B_{12}的良好来源。一般植物性饲料中不含维生素B_{12}，但链丝菌培养物及菌体中以及污泥、土壤等中都含有一定量的维生素B_{12}。

维生素B_{12}添加剂的主要商品形式有氰钴胺、羟基钴胺等，主要通过发酵法生产。另外，在生产链霉素时，从灰色链丝菌的发酵液废液中也可提取得到维生素B_{12}，外观为红褐色细粉。作为饲料添加剂有1%、2%和0.1%等剂型。维生素B_{12}容易受到盐酸硫胺素和抗坏血酸的损害。

商品维生素B_{12}纯品含维生素B_{12}95%以上。由于饲料中添加量极少，饲用维生素B_{12}商品制剂多加有载体或稀释剂。含维生素B_{12}0.1%或1%-2%的预混料粉剂产品，其颜色、吸湿性以及其他特性随维生素B_{12}的含量、载体的特性而不同。如：以玉米淀粉作为稀释剂的产品吸湿性较以碳酸钙为稀释剂的产品强。

sein, blood meal, etc. Generally, plant feed does not contain vitamin B_{12}, but there is a certain amount of vitamin B_{12} in streptomyces cultures and bacteria as well as in sludge and soil.

The main commercial forms of vitamin B_{12} additives include cyanocobalamin, hydroxycobalamin, etc., which are produced by fermentation. In addition, in the production of streptomycin, vitamin B_{12} can also be extracted from the waste liquor of the fermentation broth of streptomycin gray, with a reddish brown fine powder appearance. As feed additives, there are 1%, 2% and 0.1% formulations. Vitamin B_{12} is susceptible to damage from thiamine hydrochloride and ascorbic acid.

Commercial vitamin B_{12} pure product contains more than 95% of vitamin B_{12}. Because the amount added in feed is very few, commercial vitamin B_{12} preparations for feed are often added with carriers or diluents. The color, hygroscopicity and other characteristics premix powder products containing 0.1% or 1% to 2% of vitamin B_{12} are different with the content of vitamin B_{12}, the characteristics of carrier. For example, the hygroscopicity of products with corn starch as diluent is stronger than that with calcium carbonate as diluent.

八、叶酸
Folic acid

叶酸又称蝶酰谷氨酸。为黄色或橙黄色结晶粉末，无臭、无味，几乎不溶于冷水，随着水温的升高以及在酸性或碱性溶液中，其溶解度增加；但迅速下降，特别是在酸性溶液中，损失更快。叶酸能被紫外线分解，在干燥、避光条件下稳定性较好，密封包装贮存于阴凉、干燥处至少可保存一年。

商品制剂主要有两种剂型。高浓度制剂含叶酸量以干物质计算，不少于95%，含水量一般低于8.5%。此产品为极细粉末，易凝集成团，流动性差，应用时需要预混合处理。另一类为加有一定载体或包

Folic acid is also called pteroylglutamic acid. It is yellow or orange yellow crystalline powder, odorless, tasteless, almost insoluble in cold water. With the rise of the water temperature or put in acidic or alkaline solutions, its solubility increases; but it declines rapidly, especially in acidic solutions with rapid loss. Folic acid can be decomposed by ultraviolet ray. It has good stability under dry and dark conditions. It can be stored in a cool and dry place in sealed package for at least one year.

There are two main types of commercial preparations. The folic acid content of a high concentration preparation is calculated on the basis of dry matter of not less than 95% of content, and the water content is generally less than 8.5%. This product is a very fine powder, easy to agglutinate into a group, and has poor fluidity. Pre-mixing is required for

被材料加工制成的、含叶酸80%左右的喷雾干燥型制剂或微囊制剂。以糊精作为载体的喷雾干燥型制剂为微颗粒状粉末，流动性好，在预混料或配合饲料中，易扩散混匀。以明胶或异丙醇和乙基纤维素作为包被材料制成的微囊制剂稳定性好，特别是乙基纤维素包被制剂稳定性优于明胶包被制剂。

叶酸在预混料和配合饲料中的稳定性较差，主要受光照和含水量的影响，吸湿性强的微量矿物质硫酸盐、氯化物、氯化胆碱等对叶酸的效价影响大，因此要尽量避免与这些物质配伍。粉碎、制粒、膨化处理对叶酸的破坏更大：经包被处理的叶酸产品在饲料的加工和贮存过程中稳定性虽有提高，但在饲料的膨化、高压灭菌处理时损失仍很大。

application. The other type is spray-dried preparation or microcapsule preparation containing about 80% of folic acid, which is made with a certain carrier or coating materials. The spray-dried preparation with dextrin as a carrier is a microgranular powder with good fluidity and is easy to diffuse and mix in premix or compound feed. Microcapsule preparations made of gelatin or isopropyl alcohol and ethyl cellulose as the coating materials have good stability, especially the stability of ethyl cellulose coated preparation is better than that of gelatin coated preparations.

The stability of folic acid in premix and compound feed is poor, which is mainly affected by light and water content. The highly hygroscopic trace minerals such as sulfate, chloride, choline chloride have a great impact on the potency of folic acid, so we should try to avoid compatibility with these substances. Crushing, granulating and expanding treatments have greater damage to folic acid: although the stability of the coated folic acid products has been improved in the process and storage of feed, the loss is still great in the process of expanding and autoclastic treatment of feed.

九、胆碱
Choline

胆碱的饲料添加物主要是氯化胆碱，含胆碱86.8%，其商品制剂有液体和干粉剂两类产品。

液体氯化胆碱制剂一般为含氯化胆碱70%以上的水溶液，为无包透明的黏性液体，有轻微异臭，粉剂为以70%氯化胆碱液体制剂加入一定的载体（如玉米芯粉、脱脂米糠粉、稻壳粉、二氧化硅、无水硅酸盐等）和抗结块剂制成含氯化胆碱50%的产品。依载体不同，为白色或黄褐色粉末或颗粒，有特异臭味，流动性依载体不同而不同，一般有机载体产品流动性较差，而二氧化硅、硅酸盐产品流动性较好。

氯化胆碱是最稳定的维生素，在饲料的加工和贮存期间损失很少，但氯化胆碱制剂都具有很强的

Choline feed additives are mainly choline chloride containing 86.8% of choline. Its commercial preparations includes liquid and dry powder products.

Liquid choline chloride preparations are commonly aqueous solutions containing more than 70% of choline chloride. They are transparent sticky liquids without a package and have a slight odor. The powder is a 70% of choline chloride liquid preparation with a certain carrier (such as corn core powder, defatted rice powder, rice husk powder, silicon dioxide, anhydrous silicate, etc.) and anti-caking agents are made into products containing 5% of choline chloride. Depending on the carrier, it is white or yellowish-brown powder or particles with specific odor, and the fluidity varies according to the carrier. Generally, organic carrier products have poor fluidity, while silica and silicate products have good fluidity.

Choline chloride is the most stable vitamin and has little loss during feed processing and storage. However, choline chloride preparations are highly hygroscopic and have a seri-

吸湿性，对许多活性成分，特别是对许多维生素的有效性有严重影响，应尽量避免与其他活性成分接触。此外，氯化胆碱在饲料中的添加量大，因此，一般不加入维生素预混料中，多直接加入配合饲料。

甜菜碱作为甲基供体可替代部分胆碱用于畜禽、鱼和观赏动物饲料中。甜菜碱为黄色结晶，商品制剂含甜菜碱97%以上，作为甲基供体的效果为50%氯化胆碱的2.3倍，对维生素的稳定性无影响，但甜菜碱不能防止鸡胚骨短粗症的发生。

ous effect on the effectiveness of many active ingredients, especially many vitamins, so contacting with other active ingredients should be avoided. In addition, the amount of choline chloride added in the feed is large, so it is generally not added to vitamin premix but directly added to compound feed.

Betaine as methyl donor can substitute some choline for livestock, poultry, fish and ornamental animals. Betaine is yellow crystal, and commercial preparations contain more than 97% of betaine. The effect of betaine as methyl donor is 2.3 times that of 50% of choline chloride, which has no effect on the stability of vitamins. However, betaine cannot prevent the occurrence of chicken tibia phrenia.

十、维生素 C
Vitamin C

维生素 C 又称抗坏血酸。目前，常用的添加物有 L-抗坏血酸、L-抗坏血酸钠、L-抗血酸钙、L-抗坏血酸-2-多磷酸盐，是一种有效、稳定性好的补充物。

L-抗坏血酸为白色或类白色结晶性粉末，无臭，味酸，易溶于水，在干燥、密闭条件下相当稳定；但在湿热条件下，极易被氧化剂、碱、微量元素等破坏。L-抗坏血酸在成分复杂的预混料和配合饲料中，特别是与氯化胆碱等吸湿性极强的成分共存时保存率低，更不耐粉碎、制粒、膨化、灭菌等加工处理。包被处理的 L-抗坏血酸的稳定性有一定提高，但仍易被粉碎、制粒、膨化、灭菌等工序破坏。

L-抗坏血酸钙、L-抗坏血酸钠均为白色粉末，易溶于水，稳定性较抗坏血酸好，因此作为饲料添加剂较 L-抗坏血酸普遍。L-抗坏血酸钙的活性相当于 81.6% L-抗坏血酸，L-抗坏血酸钠相当于 90% L-抗坏血酸。

Vitamin C is also called ascorbic acid. At present, the commonly used additives are L-ascorbic acid, L-sodium ascorbate, L-calcium ascorbate, and L-ascorbic acid-2-polyphosphate. It is an effective and stable supplement.

L-ascorbic acid is white or almost white crystalline powder, odorless, sour in taste, and easily soluble in water. It is quite stable under dry and airtight conditions; but in humid and hot conditions, it is easily destroyed by oxidant, alkali, trace elements and so on. L-ascorbic acid has a low retention rate in premix and compound feeds with complex components, especially when co-existing with highly hygroscopic components such as choline chloride, etc., and is less resistant to grinding, granulation, expansion, sterilization and other processing. The stability of the coated L-ascorbic acid is improved to some extent, but it still tends to be destroyed by grinding, granulation, expansion, sterilization and other processes.

Both L-calcium ascorbate and L-sodium ascorbate are white powders, easily soluble in water, and more stable than ascorbic acid, so they are more common as feed additives than L-ascorbic acid. The activity of L-calcium ascorbate is equivalent to 81.6% of that of L-ascorbic acid, and L-sodium ascorbate is equivalent to 90% of that of L-ascorbic acid.

十一、肌醇
Inositol

肌醇即环己六醇,其分子式为 $C_6H_{12}O_6$,相对分子质量 180.16。水产饲料中常需添加肌醇。用作饲料添加剂者为化学合成肌醇,其产品为含肌醇 97% 以上的白色结晶或结晶性粉末,无臭,具有甜味,易溶于水。肌醇很稳定,在饲料中不易被破坏。

Inositol, or cyclohexanol, has a molecular formula of $C_6H_{12}O_6$ and a relative molecular weight of 180.16. Inositol is often added to aquatic feeds. Inositol is chemically synthesized as feed additive. The product is a white crystal or crystalline powder containing over 97% of inositol, odorless, sweet and easily soluble in water. Inositol is very stable and is not easily damaged in feed.

第三节 维生素饲料的合理应用
Section 3 Reasonable application of vitamin feed

一、饲粮中维生素添加量的确定
Determination of dietary vitamin supplementation

确定方法主要有以下几种:

1. 日粮组成及各种养分的含量和相互关系。
2. 饲料中维生素拮抗因子。
3. 饲料中固有维生素的利用率。
4. 动物的饲养方式。放牧动物可从草、虫及其他天然饲料中获得大部分维生素,而舍饲动物主要由饲料中获得。平养鸡垫料中的微生物可合成维生素 B,粪便中也含有许多维生素,而笼养鸡则无法由垫料、粪便中获得这些维生素,因而它们的需要量增加。
5. 环境条件(温度等)。
6. 动物的健康状况及应激。增加维生素 A、维生素 E、维生素 C 和某些 B 族维生素等,能增加动物的抗病和抗应激能力,以此目的添加的维生素需增加一倍或更高的添加量。

The following methods are usually taken:

1. Dietary composition and the content and correlation of various nutrients.
2. Vitamin antagonistic factors in feed.
3. Utilization rate of inherent vitamins in feed.
4. The way animals are raised. Grazing animals get most of their vitamins from grass, insects, and other natural feeds, while feedlot animals get most of their vitamins from feeds. Microorganisms in the bedding of flat-raised chickens can synthesize vitamin B, and many vitamins are also found in feces. However, caged-chickens cannot obtain these vitamins from bedding and feces, so their needs increase.
5. Environmental conditions (temperature, etc.).
6. Health status and stress of animals. Increasing vitamin A, vitamin E, vitamin C and some B vitamins can increase animals' resistance to disease and stress, so the amount of vitamins added for this purpose should be doubled or more.

7. 维生素在各种饲料加工过程中的损失,包括原料、预混合料和配合饲料的加工处理、贮存条件及时间、饲料中各种化学物质与微生物等的影响。

7. Loss of vitamins in various feed processing processes, including raw materials, pre-mixture and mixed feed processing, storage conditions and time, and the effects of various chemicals and microorganisms in the feed, etc.

二、维生素饲料的选择
Selection of vitamin feed

维生素饲料的选择,应根据其使用目的和生产工艺,综合考虑制剂的稳定性、加工特点、质量规格和价格等因素。一般用于生产预混合料时,生产条件、技术好,可选择纯品或药用级制剂;生产条件差,无预处理工艺、设备的情况下,应尽量选择稳定性好、流动性适中、含量低的经保护性处理、预处理的产品;若用于生产液体饲料或宠物罐头饲料,必须选择水溶性制剂。

The choice of vitamin feed should be based on its use purpose and production process, and the factors such as stability, processing characteristics, quality specifications and price of the preparation should be considered comprehensively. Generally in the production of pre-mixed materials, if production conditions and technologies are good, pure products or pharmaceutical grade preparations can be selected. In the case of poor production conditions with no pretreatment technology and equipment, the products with good stability, moderate fluidity and low content after protective treatment and pretreatment should be selected as far as possible. Water-soluble preparations must be selected if used in the production of liquid feed or canned pet feed.

三、维生素饲料的配伍
Compatibility of vitamin feed

在生产预混合料时,应注意原料(包括载体)的搭配,尤其是生产高浓度预混合饲料时,应根据维生素的稳定性和其他成分的特性,合理搭配,注意配伍禁忌,以减少维生素在加工贮存过程中的损失。总的说来,大部分维生素添加剂对微量矿物元素不稳定。在潮湿成含水量较高的条件下,维生素对各种因素的稳定性均下降。因此,要避免维生素与矿物质共存,特别要避免同时与吸湿性强的氯化胆碱共存。

在选用商品"多维"时,要注意其含维生素的种类,若某种或某几种维生素不含在内而又需要者,必

In the production of pre-mixed materials, attention should be paid to the mix of raw materials (including carriers). Especially in the production of high-concentration pre-mixed feed, reasonable combinations and compatibility restrictions should be followed according to the stability of vitamins and the characteristics of other components to reduce the loss of vitamins in the process of processing and storage. Generally speaking, most vitamin additives are unstable to trace mineral elements. Under humid conditions with high moisture content, the stability of vitamins to various factors is reduced. Therefore, it is necessary to avoid the coexistence of vitamins and minerals, especially to avoid coexisting with choline chloride, which is high hygroscopic.

In the selection of commodities "multivitamin", it is necessary to pay attention to the types of vitamins they contain. If a certain or a few vitamins are not included but are

须另外添加。"多维"中往往不含氯化胆碱和维生素C,有的产品中缺生物素和泛酸等。此外,在饲料中添加抗维生素 B_1 的抗球虫药(如氨丙啉)时,维生素 B_1 的用量不宜过多,若每千克日粮中维生素 B_1 含量达10 mg时,抗球虫剂效果会降低。

needed, they must be added separately. "Multivitamin" often does not contain choline chloride and vitamin C. Some products are lack of biotin and pantothenic acid, etc. In addition, when anti-vitamin B_1 anticoccidioides (such as aminopropyl) are added to the feed, the dosage of vitamin B_1 should not be too much. If the vitamin B_1 content in every kilogram of diet reaches 10 mg, the effect of anti-coccidioides will be reduced.

四、维生素饲料的添加方法
Method of adding vitamin feed

不同维生素饲料产品的特性不同,添加方法也不同。一般干粉饲料或预混合料,可选用粉剂直接加入混合机混合。当维生素制剂浓度高,在饲料中的添加量小或原料流动性差时,则应先进行稀释或预处理,再加入主混合机混合。液态维生素制剂的添加必须由液体添加设备喷入混合机或先进行处理变为干粉剂。对某些稳定性差的维生素,在生产颗粒饲料或膨化饲料时,选择制粒、膨化冷却后再喷涂在颗粒表面的添加方法,能减少维生素的损失。

Different vitamin feed products have different characteristics and are suitable for different ways of addition. Generally, dry powder feed or pre-mixture can be mixed directly by adding powder to the mixer. When the concentration of vitamin preparation is high, the amount of addition in the feed is small or the raw material liquidity is poor, and it should be diluted or pretreated first, and then added to the main mixer for mixing. The liquid vitamin preparation must be sprayed into the mixer by the liquid adding equipment or treated first before it becomes dry powder. For some vitamins with poor stability, in the production of pelleted feed or expanded feed, the addition method of granulation, expanding cooling and then spraying on the surface of particles can reduce the loss of vitamins.

五、维生素饲料产品的包装贮存
Packaging and storage of vitamin feed products

维生素饲料产品应密封、隔水包装,真空包装更佳。维生素饲料产品需贮藏在干燥、避光、低温条件下。高浓度单项维生素制剂一般可贮存1~2年,不含氯化胆碱和维生素C的维生素预混合料贮存期不超过6个月,含维生素的复合预混合料,贮存期最好不宜超过3个月。所有维生素饲料产品,开封后需尽快用完。

Vitamin feed products should be packed in sealed, water-proof packaging. Vacuum packaging is more suitable. Vitamin feed products should be stored in dry, dark, low temperature conditions. High-concentration single vitamin preparations can generally be stored for 1 to 2 years. Vitamin pre-mixture without choline chloride and vitamin C should not be stored more than 6 months. Compound premixes containing vitamins should not be stored more than 3 months. All vitamin feed products must be used up as soon as possible after opening.

第十章 饲料添加剂
Chapter 10　Feed additives

第一节　非营养性饲料添加剂
Section 1　Non-nutritive feed additives

非营养性添加剂是指加入饲料中用于改善饲料利用效率、保持饲料质量和品质、有利于动物健康或代谢的一些非营养性物质。主要包括饲料药物添加剂、益生素、酶制剂、酸化剂、中草药及植物提取成分、防霉剂、饲料调制和调质添加剂等。

Non-nutritive additives refer to some non-nutritive substances added in feed to improve feed utilization efficiency, maintain feed quality, and benefit animal health or metabolism. It mainly includes medicated additives, probiotics, enzyme preparations, acidulants, Chinese herbal medicines and plant extracts, antimold agents, feed preparation and conditioning additives, etc.

一、药物添加剂
Medicated additives

药物添加剂是指为防治动物疾病并改善动物产品质量、提高产量而掺入载体或稀释剂的一种或多种兽药的预混物。

Medicated additives refer to the premixes of one or more veterinary drugs mixed with carriers or diluents to prevent and control animal diseases, improve the quality of animal products, and increase yield.

（一）药物添加剂的分类
Classification of medicated additives

目前，世界上生产的抗生素已达 200 多种，作为饲料添加剂的有 60 多种。世界各国的分类方法不尽相同，一般可按它的抗菌谱和作用对象、来源及化学结构来分。按其化学结构可分为以下几种。

At present, there are more than 200 kinds of antibiotics produced in the world, and more than 60 kinds of antibiotics are used as feed additives. The classification methods in different countries in the world are not identical. Generally, it can be classified according to its antibacterial spectrum and target, source of origin and chemical structure. According to its chemical structure, it can be divided into the following categories.

(1) 多肽类:杆菌肽锌、硫酸黏杆菌素、恩拉霉素。

(2) 四环素类:土霉素、金霉素。大多数国家已禁止使用。

(3) 大环内酯类:泰乐菌素(渔业禁用)。

(4) 含磷多糖类:黄霉素。

(5) 聚醚类抗生素:莫能菌素、盐霉素、马杜霉素。

(6) 氨糖基苷类:新霉素、越霉素A、潮霉素B。

(7) 化学合成类:磺胺类、喹乙醇。

(1) Polypeptides: zinc bacitracin, colistin sulfate, enramycin.

(2) Tetracyclines: oxytetracycline and chlortetracycline. It has been banned in most countries.

(3) Macrolides: tylosin (prohibited in fisheries).

(4) Phosphorus-containing polysaccharides: flavomycin.

(5) Polyether antibiotics: monensin, salinomycin, maduramycin.

(6) Amino glycosides: neomycin, destomyin A, hygromycin B.

(7) Chemical synthesis: sulfonamides, olaquindox.

(二) 常用药物添加剂
Commonly used medicated additives

1. 抗生素

抗生素是微生物(细菌、真菌、放线菌等)的代谢产物,对特异性的微生物具有抑制或杀灭作用。其来源有微生物、植物提取、化学方法合成或半合成等。动物生产中所饲用的抗生素包括促生长类的抗生素和用于加药饲料的抗生素。前者是指那些以亚治疗剂量应用于健康动物饲料中,以改善动物营养状况,促进动物生长,提高饲料效率的抗生素。后者主要用于治疗,即动物在疾病状态下使用的饲料,可以在有兽医处方的情况下加入某些抗生素。

(1) 作用及机制:抗生素是通过影响细菌菌体的代谢过程,改变细菌细胞的菌体形态而发挥其抑菌和杀菌作用的,其作用机制主要包括以下四种类型:

抑制核酸的合成:某些抗生素(如源于放线菌)能抑制 DNA 解旋酶的活性,使 DNA 失去模板功能,从而抑制它的复制和转录。

抑制蛋白质的合成:四环素类、

1. Antibiotics

Antibiotics are metablites of microorganisms (bacteria, fungi, actinomycetes, etc.), which can inhibit or kill specific microorganisms. Their sources are microorganisms, plant extraction, chemical synthesis or semi-synthesis, etc. Antibiotics used in animal production include growth-promoting antibiotics and antibiotics for medicated feed. The former refers to those antibiotics used in healthy animal feed at sub-therapeutic doses to improve animal nutrition, promote animal growth, and improve feed efficiency. The latter is mainly used for therapeutic purpose, that is, the feed used by the animal in a disease state can be added with certain antibiotics on a veterinary prescription.

(1) Action and mechanism: Antibiotics exert their antibacterial and bactericidal effects by affecting the metabolic process of bacterial cells and changing the morphology of bacterial cells. The mechanism of action mainly includes the following four types:

Inhibiting nucleic acid synthesis. Certain antibiotics (such as derived from actinomycetes) can inhibit the activity of DNA helicase and make DNA lose its template fuction, thereby inhibiting its replication and transcription.

Inhibiting protein synthesis. Antibiotics such as tetracy-

氨基糖苷类和大环内酯类等抗生素均可影响病原微生物体内蛋白质的合成,但它们的作用位点和作用环节各不相同。如:大多数氨基糖苷类抗生素都能阻碍蛋白质合成的起始延伸和终止阶段;四环素类抗生素则阻断肽链的延长;大环内酯类抗生素抑制肽链的移位反应。

改变细胞膜的通透性:多肽类抗生素如多黏菌素 E、短杆菌肽 S 等具有表面活性剂的作用,能降低细菌细胞膜的表面张力,改变细胞膜的通透性,甚至破坏膜的结构,使氨基酸、单糖、无机盐离子等外漏,影响细胞正常代谢,致使细菌死亡。多烯类抗生素,如制霉菌素、两性霉素等能选择性作用于细胞膜含固醇的微生物,与膜结合后形成膜-多烯化合物,引起细胞质膜的通透性增加,导致胞内代谢物外泄而使细胞死亡。

干扰细胞壁的合成:细菌细胞壁中含有肽聚糖(黏肽),如革兰氏阳性菌细胞壁的组成中肽聚糖占细胞壁干重的 50%-80%。革兰氏阴性菌细胞壁的组成中肽聚糖占细胞壁干重的 1%-10%。青霉素、杆菌肽、万古霉素等抗生素作用于细胞壁肽聚糖合成的不同阶段,使细胞壁缺损,抗渗透压能力降低,引起菌体变形、破裂而死亡。

(2)耐药机制:细菌对抗生素的耐药性引起人们极大的关注。研究表明,其机制有以下两方面:第一,细菌本身遗传特征的传播后所产生的抗性;第二,细菌通过生化功能产生抗性,其主要有三种类型:产生导致抗生素灭活的酶,降低细胞透过抗生素的能力,耐药菌降低细胞透过抗生素的能力。

clines, aminoglycosides and macrolides can affect the protein synthesis in pathogenic microorganism, but their sites of action and links are different. For example, most aminoglycoside antibiotics can hinder the initiation and extension, termination phase of protein synthesis; tetracycline antibiotics can block the elongation of the peptide chain; macrolide antibiotics can inhibit the translocation reaction of the peptide chain.

Changing the permeability of the cell membrane. Polypeptide antibiotics such as polymyxin E, gramicidin S have surfactants effects, which can reduce the surface tension of the bacterial cell membrane, change the permeability of the cell membrane, and even destroy the membrane structure, lead to the leakage of amino acids, monosaccharides, inorganic salt ions, etc., affect the normal metabolism of cells and cause the death of bacteria. Polyene antibiotics, such as nystatin, amphotericin, etc. can selectively act on microorganisms containing sterols in the cell membrane, and form a membrane-polyene compound after binding to the membrane, causing increase of the permeability of cytoplasmic membrane, leading to excretion of intracellular metabolism substance and cell death.

Interfering with cell wall synthesis. Bacterial cell walls contain peptidoglycans (mucopeptides). For example, in the composition of gram-positive bacterial cell walls, peptidoglycans account for 50%-80% of the dry weight of the cell wall. In the composition of the cell wall of gram-negative bacteria, peptidoglycan accounts for 1%-10% of the dry weight of the cell wall. Antibiotics such as penicillin, bacitracin, vanguolein act on the different stages of cell wall peptidoglycan synthesis, causing cell wall defects, reducing osmotic resistance, and causing bacterial cells degeneration, rupture and death.

(2) Resistance mechanism. The resistance of bacteria to antibiotics has caused great concern. The studies have shown that its mechanism has the following two aspects: firstly, the resistance produced by the spread of the genetic characteristics of the bacteria; secondly, the bacteria develop resistance through biochemical functions, which mainly have three types: producing enzymes that cause the inactivation of antibiotics, reducing the ability of cells to penetrate antibiotics, and drug-resistant bacteria that reduce the ability

2. 抗球虫药

主要是通过作用于球虫生活史的不同阶段而达到抑制与杀灭球虫的目的。抗球虫药的作用机制主要包括以下三方面。

（1）影响虫体的正常功能：如氯苯胍对大鼠肝细胞线粒体内氧化磷酸化作用和ATP酶的活性有抑制作用；莫能菌素能与碱金属离子相互作用，特别是抑制钾离子向肝细胞线粒体内转移，导致球虫的正常生理功能发生紊乱，使某些物质代谢和三磷酸腺苷水解受阻，使虫体因体内供能不足而附着无力，从而被排出体外。

（2）竞争性对抗虫体代谢：如磺胺类药物的结构和原虫体的对氨基苯甲酸相似，可竞争性地抑制对氨基苯甲酸的作用，干扰叶酸的合成代谢过程，最终影响核酸的代谢——一碳集团的代谢而发挥抗球虫作用。

（3）抑制核酸合成：如二甲氧甲基嘧啶、乙胺嘧啶等药物是二氢叶酸还原酶的抑制剂，该酶被抑制后二氢叶酸不能被还原为四氢叶酸，从而阻碍了合成嘌呤、嘧啶核苷酸，最终使核酸合成减少，虫体的繁殖受到抑制。这些药物与磺胺类药物分别作用于叶酸合成的不同环节，二者合用有增效作用。

3. 驱蠕虫药

按药物的驱虫谱可分为抗线虫药、抗吸虫药和抗绦虫药。目前世界各国批准使用的驱蠕虫类药物仅有两种，均为氨基苷类抗生素，即越霉素A和潮霉素B。目前，药物添加剂虽在动物生产中发挥了很大作用，但其在畜禽产品中的残留及不

of cells to penetrate antibiotics.

2. Anticoccidial drugs

Anticoccidial drugs are mainly used to inhibit and kill coccidia by acting on different stages of the life history of coccidia. The mechanism of action of anticoccidial drugs mainly includes the following three aspects.

（1）Affecting the normal function of the worm body. For example, robenidine has an inhibition on the oxidative phosphorylation and ATPase activity in the mitochondria of rat liver cells; monensin can interact with alkali metal ions, especially inhibit potassium ions to transfer into the mitochondria of the liver cells, cause the normal physiological functions of the coccidia to be disordered, and hinder the metabolism of certain substances and the hydrolysis of adenosine triphosphate, resulting in the worms' weak attachment due to insufficient supply of energy in the body, and thus excreted from the body.

（2）Competitive resistance to worm metabolism. For example, the structure of sulfonamides is similar to the para-aminobenzoic acid of protozoa, which can competitively inhibit the effect of para-aminobenzoic acid, interfere with the anabolic process of folic acid, and ultimately affect the metabolism of nucleic acids—the metabolism of the one-carbon group to play an anticoccidial effect.

（3）Inhibiting nucleic acid synthesis. For example, drugs such as dimethoxymethylpyrimidine and pyrimethamine are inhibitors of dihydrofolate reductase. After the enzyme is inhibited, dihydrofolate cannot be reduced to tetrahydrofolate, which hinders the synthesis of purine and pyrimidine nucleotides, and ultimately reduces nucleic acid synthesis and inhibits the reproduction of the worms. These drugs and sulfonamides act on different links of folic acid synthesis, and the combination of the two has a synergistic effect.

3. Anthelmintic drugs

The anthelmintic spectrum of the drugs can be divided into anti-matode drugs, anti-trematode drugs and anti-tapeworm drugs. At present, there are only two anthelmintic drugs approved for use in various countries in the world, both of which are aminoglycoside antibiotics, namely destomycin A and hygromycin B. At present, although medicated additives have played a great role in animal production,

良效应也困扰着人们。寻找无(低)药残,无(低)污染,能替代不良药物添加剂(特别是抗生素)的物质,已成为当今动物营养学研究的重点之一。

their residues and adverse effects in livestock and poultry products also trouble people. Looking for substances that have no (low) drug residues, no (low) pollution, and can replace undesirable medicated additives (especially antibiotics) has become one of the focuses of animal nutrition research today.

二、益生素
Probiotics

益生素(probiotics)一词源于希腊语,即pro(有益于)+bio(生命)+ics(制剂),意思是有利于生命。狭义的益生素是指益生菌,即可以直接饲喂动物,并通过调节动物肠道微生物平衡达到预防疾病、促进动物生长和提高饲料利用率的活性微生物或其培养物,我国又称其为微生态制剂或饲用微生物添加剂。通常将能选择性地促进大肠有益菌增殖、抑制有害菌增殖的非消化食物成分称为益生元或化学益生素。广义的益生素是对益生菌和益生元的统称。

The word probiotics is derived from the Greek word pro (good for) and bio (life) and ics (preparation), which means benifical to life. In a narrow sense, probiotics refer to proboitics bacteria, which are active microorganisms or their cultures that can be directly fed to animals and can prevent diseases, promote animal growth and improve feed utilization by adjusting the animal's intestinal microbial balance. They are also called microecologics or microbial feed additives in China. Generally, non-digestible food components that can selectively promote the proliferation of beneficial bacteriae and inhibit the proliferation of harmful bacteria in the large intestine are called prebiotics or chemical probiotics. In a broad sense, probiotics are the collective term for probiotics bacteria and prebiotics.

(一) 益生素的种类及特征
Types and characteristics of probiotics

益生素菌种很多,各国都在筛选自己的菌种。我国农业部第105号公告公布的容许使用的饲料添加剂品种目录中,饲料级微生物添加剂有12种。目前,我国常用的益生素菌种有6种:芽孢杆菌、乳酸杆菌、粪链球菌、酵母菌、黑曲霉、米曲霉。益生素的分类因依据不同有多种。根据制剂的用途及作用机制可分为微生物生长促进剂和微生态治疗剂;依活菌制剂的组成可分为单一制剂和复合制剂。目前较多使用的分类方法是依据微生物的菌种类型分为无孢子杆菌、芽孢杆菌、酵母菌及霉菌类以及光合细菌。

There are many species of probiotics, and each country is screening its own syrains. There are 12 kinds of feed-grade microbial additives in the list of allowable feed additives published in the Annouement No. 105 of China Ministry of Agriculture. At present, there are 6 kinds of probiotics commonly used in China, including Bacillus, Lactobacillus, Streptococcus faecalis, yeast, Aspergillus niger, Aspergillus oryzae. Proboitics can be classified into a variety of kinds depending on the different base. The preparations can be divided into microbial growth promoters and microecological therapeutics according to their use and mechanism of action. They can be divided into single preparations and compound preparations according to the composition of the active bacteria preparation. At present, the most commonly used classification method is based on the types of mi-

1. 无孢子杆菌

属厌氧或兼性厌氧，耐酸但不耐热，均为无孢子生成菌，主要有乳酸杆菌、双歧杆菌和乳酸球菌。

2. 芽孢杆菌

属需氧或兼性厌氧，以内生孢子形式存在，稳定性好，具有较强的蛋白酶和淀粉酶活性，同时还具有平衡和稳定乳酸杆菌的作用。目前使用的主要是枯草芽孢杆菌、地衣芽孢杆菌和东洋芽胞杆菌。

3. 酵母菌及霉菌类

在动物消化道中零星存在，一般不繁殖，不定居，但可参与消化道内的物质代谢，具有广泛的酶活性，含有丰富的维生素、蛋白质、未知因子。酵母及其培养物在近十年来才作为益生素进行开发。霉菌主要与细菌类制成复合益生素。

4. 光合细菌

光合细菌能在厌氧光照条件下同化二氧化碳，有些菌还有固氮作用。光合细菌不仅为生物体宿主提供丰富的蛋白质、维生素、矿物质等营养物质，而且可以产生辅酶Q等生物活性物质，提高宿主的免疫力。光合细菌在改善水体环境、增加水产动物体重、色泽等方面有很好的作用。

（二）化学益生素

Chemical probiotics

20世纪80年代的学者们开始对动物肠道固有的有益菌产生了研究兴趣，认为在饲料中添加一些不能被机体消化，而且只能被肠道有益菌利用并能促使其增殖的物质，就可克服益生素活性难以保证的缺陷。这类物质多属短链带分支的糖

crobes, which can be devided into Ansporobacterium, Bacillus, yeasts and molds, and photosynthetic bacteria.

1. Asporobacterium

They are anaerobic or facultative anaerobic, acid-tolerant but heat-labile, and are non-sporogenic bacteria, mainly including Lactobacillus, Bifidobacterium and Lactococcus.

2. Bacillus

They are aerobic or facultative anaerobic existing in the form of endospores. They have good stability and strong activity of protease and amylase, and also have the effect of balancing and stabilizing Lactobacilli. The main strains currently used are Bacillus subtilis, Bacillus licheniformis and Bacillus Oriental.

3. Yeasts and molds

They are found sporadically in the digestive tract of animals. They generally do not reproduce or settle down, however, they can be involved in the metabolism substances in the digestive tract. They have a wide range of enzyme activities, and are rich in vitamins, protein, and unknown factors. Yeast and its culture have only been developed as a probiotic in the past ten years. Molds are mainly combined with bacteria to make compound probiotics.

4. Photosynthetic bacteria

Photosynthetic bacteria can assimilate carbon dioxide under anaerobic light conditions, and some bacteria have the effect of fixing nitrogen. Photosynthetic bacteria not only provide rich protein, vitamins, minerals and other nutrients for the host organism, but also produce biologically active substances such as coenzyme Q to improve the host's immunity. Photosynthetic bacteria play a very good role in improving the water environment, increasing the weight and color of aquatic animals.

In the 1980s, scholars began to develop interest in studying the beneficial bacteria inherent in the intestinal tract of animals. They believed that adding some substances that could not be digested by the body and could only be used by the beneficial bacteria in the intestine tract to promote their proliferation could overcome the defect of unguaranteed activity of probiotics substances. These substances are mostly

类物质,被称为化学益生素,本质上为低聚糖。它作为一种新型添加剂,在饲料中的应用研究取得了很大进展。

short-chain and branched carbohydrate substances, which are called chemical probiotics, and are essentially oligosaccharides. As a new type of additive, its application research in feed has made great progress.

(三) 益生素的作用机制
The mechanism of action of probiotics

益生素的作用机制在理论上的进展还很小,现阶段的研究主要是基于一些假说。

The mechanism of action of prebiotics has made little progress in theory, and the current research is mainly based on some hypotheses.

(1) 优势菌群学说:正常微生物群对整个肠道菌群起决定作用,当肠道内菌群比例失调,此时腐败菌和致病菌大量繁殖,动物体就容易产生疾病。

(1) The dominant flora theory

The normal microbiota plays a decisive role in the entire intestinal flora. When the proportion of the intestinal flora is out of oroportion, spoilage bacteria and pathogenic bacteria will multiply, and the animal is prone to disease.

(2) 微生物夺氧学说:需氧或兼性厌氧芽孢杆菌消耗氧气造成厌氧环境,可以扶植和促进正常菌群的生长繁殖,同时抑制需氧病原菌和兼性厌氧病原菌的生长。

(2) The microbial oxygen deprivation theory

Aerobic or facultative anaerobic Bacillus consumes oxygen to create an anaerobic environment, which can support the shoots and promote the growth and reproduction of normal flora, while inhibiting the growth of aerobic pathogens and facultative anaerobic pathogens.

(3) 菌群屏障学说:多数有益菌能分泌一种凝集素,该物质能专一性结合到肠黏膜上皮细胞产生的糖蛋白上,使有益菌在肠黏膜表面形成一个生物学屏障,竞争性抑制了致病菌、条件性致病菌的定植、入侵。

(3) The microbial shield theory

Most beneficial bacteria can secrete a kind of lectin which specifically binds to the glycoproteins produced by intestinal mucosal epithelial cells, so that beneficial bacteria form a biological barrier on the surface of the intestinal mucosa, and competitively inhibit the colonization and invasion of pathogenic bacteria and conditional pathogens.

(4) "三流运转"理论:益生素能促进肠道相关淋巴组织的活动,抑制腐败微生物的过度生长和毒性物质的产生,促进肠蠕动,维持黏膜结构完整,从而保证微生物系统中基因流、能量流和物质流的正常运转。

(4) The "three-stream operation" theory

Probiotics can promote the activities of intestinal-related lymphoid tissues, inhibit the overgrowth of spoilage microorganisms and the production of toxic substances, promote intestinal peristalsis, and maintain the integrity of the mucosal structure, thereby ensuring the normal operation of gene flow, energy flow and material flow in the microbial system.

(四) 影响益生素作用效果的因素
Factors affecting the effect of probiotics

1. 影响益生素作用效果的因素很多,包括动物种类、动物年龄与生理状态,环境卫生状况;益生素的

1. There are many factors that affect the effect of probiotics, including animal species, animal age and physiological state, and environmental sanitation; types and dosage of

种类,使用剂量;饲料加工储藏条件及饲料中其他饲料添加剂(如抗生素、矿物元素)的使用情况等。但这些因素与益生素使用效果之间的定量关系目前还不清楚。益生素的一般添加量为每克饲料中 $10^6 - 10^7$ 个活菌。

2. 益生素是活菌制剂,为保持益生素中活菌的数量和活力,使之在动物体内能够充分发挥作用,应注意以下几点:

(1) 益生素应保存于阴凉、通风干燥处,以防温度、湿度和紫外线对活菌的破坏作用。

(2) 对饲料混合时产生的瞬时高温敏感的益生素(如乳酸杆菌、双歧杆菌),应先制成预混料后再混入饲料中;饲料中的水分对活菌有较大影响,故无微胶囊保护的活菌制剂须在与饲料混合后当天用完,以免降低其使用效果。

(3) 益生素与其他饲料添加剂的混合使用应先进行试验以防其他添加剂降低益生素的效果。

(4) 对于容易失活的乳酸菌等菌株,应采用微胶囊包埋技术对活菌体进行包埋,从而使更多的菌体到达肠道,真正起到益生素的作用。

(五) 益生素的安全性
The safety of probiotics

益生素的首要问题是安全性,其次才是有效性问题。直接饲用益生素应具备以下几个条件:

(1) 益生素应是非致病性活菌或由微生物发酵而产生的无毒副作用的有机物质。

(2) 益生素应是能对宿主及机体内有害菌群在提高生长率或抗菌方面产生有利影响的。

(3) 益生素应是活的微生物,

probiotics; processing and storage conditions of feed and the use of other feed additives (such as antibiotics, mineral elements) in feed. However, the quantitative relationship between these factors and the effects of probiotics is currently unclear. The general dosage of probiotics is 10^6 to 10^7 live bacteria per gram of feed.

2. Probiotics are live bacteria preparations. In order to maintain the number and vitality of live bacteria in probiotics and make them fully play their role in animals, the following stipulations should be abided by:

(1) Probiotics should be stored in a cool, ventilated and dry place to prevent the destruction of live bacteria by temperature, humidity and ultraviolet rays.

(2) Probiotics (such as Lactobacillus and Bifidobacterium) that are sensitive to transient high temperature during feed mixing should be premixed first and then mixed into the feed. The water in the feed has a great influence on the live bacteria, so the live bacteria preparations without microcapsule protection should be used up on the same day after mixing with the feed, so as not to reduce its use effect.

(3) The mixed use of probiotics with other feed additives should be tested first to prevent other additives from reducing the effect of probiotics.

(4) For lactic acid bacteria and other strains that are easy to be inactivated, microencapsulation embedding technology should be used to embed the live bacteria, so that more bacteria can reach the intestinal tract and really play the role of probiotics.

The first issue of probiotics is safety, and the second issue is effectiveness. Probiotics for direct feeding should meet the following conditions.

(1) Probiotics should be non-pathogenic live bacteria or organic substances that are fermented by microorganisms without toxic side effects.

(2) Probiotics should have a beneficial effect on the growth rate or antimicrobial activity of the host and harmful bacteria in the body.

(3) Probiotics should be live microorganisms, and

且要求与正常有益菌群能共存，并且自身具有抗逆能力。

（4）益生素应能在肠道环境中只对有益菌群有利，而且其代谢产物不对宿主产生不利影响。

（5）益生素应有较好的包被技术可以顺利躲过胃液的水解，并且在生产条件下，可以长期储存，并保持良好的稳定性。

they must coexist with normal beneficial bacteria and have the ability to resist stress.

(4) Probiotics should only be beneficial to beneficial bacteria in the intestinal environment, and their metabolic products should not adversely affect the host.

(5) Probiotics should have a better encapsulated technology that can smoothly escape the hydrolysis of gastric juice, and can be stored for a long time and maintain good stability under the condition of the production.

三、酶制剂
Enzyme preparations

饲用酶制剂是将一种或多种利用生物技术生产的酶与载体和稀释剂采用一定的生产工艺制成的一种饲料添加剂，可提高饲料的消化利用率，改善畜禽的生产性能，减少粪便中氮、磷、硫等给环境造成的污染，转化和消除饲料中的抗营养因子，并充分利用新的饲料资源。

Feed enzyme preparation is a feed additive made of one or more enzymes, carriers and diluents produced by biotechnology through a certain production process, which can improve the digestibility and utilization of feed, improve the production performance of livestock and poultry, reduce environmental pollution caused by nitrogen, phosphorus, and sulfur in manure, transform and eliminate anti-nutritional factors in feed, and make full use of new feed resources.

（一）种类
Type

饲料工业上使用的酶制剂主要是消化糖类和植酸磷的酶，也有些产品包含有蛋白酶和脂酶。

The enzyme preparations used in the feed industry are mainly enzymes that digest carbohydrates and phytate phosphorus, and some products also contain proteases and lipases.

1. 消化糖类的酶　这类酶包括淀粉酶和非淀粉多糖（NSP）酶。非淀粉多糖酶又包括半纤维素酶、纤维素酶和果胶酶。半纤维素酶包括木聚糖酶、甘露聚糖酶、阿拉伯木聚糖酶和聚半乳糖酶；纤维素酶包括 C1 酶、Cx 酶和 β-葡聚糖酶。

2. 蛋白酶　该酶使蛋白质水解为小分子物质。根据最适 pH 的不同，可分为酸性、中性和碱性蛋白酶。由于动物胃液呈酸性，小肠液多为中性，所以饲料中多添加酸性蛋白酶和中性蛋白酶，其主要作用是将饲料蛋白质水解为氨基酸。

1. Enzymes for digesting carbohydrates. These enzymes include amylase and non-starch polysaccharide (NSP) enzymes. Non-starch polysaccharide enzymes include hemicellulase, cellulase and pectinase. Hemicellulase includes xylanase, mannanase, arabinoxylanase and polygalactase. Cellulase includes C_1, Cx and β-glucanase.

2. Protease. This enzyme hydrolyzes protein into small molecules. According to the optimum pH, it can be divided into acidic, neutral and alkaline proteases. Since animal gastric juice is acidic and small intestinal juice is mostly neutral, acidic protease and neutral protease are added to feed, whose main function is to hydrolyze feed protein into amino acids.

3. 脂肪酶 该酶是水解脂肪分子中甘油酯键的一类酶的总称。微生物产生的脂肪酶通常在 pH 值为3.5—7.5时水解力最好,最适温度为38 ℃。因此,微生物脂肪酶非常适用于饲料。脂肪酶一般从动物消化液中提取。外源性脂肪酶的作用与动物的年龄有关,生长动物体内的脂肪酶足以满足自身需要,但幼畜日粮中添加脂肪酶可能有益。

4. 植酸酶 植酸磷难以被单胃动物消化利用,还通过螯合作用降低动物对锌、锰、铁、钙等矿物元素和蛋白质的利用率,是一种天然抗营养因子。植酸酶可显著地提高磷的利用率,促进动物生长和提高饲料营养物质转化率。

3. Lipase. Lipase is the general term for a class of enzymes that hydrolyze the glyceride bonds in fat molecules. The lipase produced by microorganisms usually has the best hydrolysis power when the pH value ranges from 3.5 to 7.5, and the optimum temperature is 38 ℃. Therefore, microbial lipase is very suitable for feed. Lipase is generally extracted from animal digestive juices. The role of exogenous lipase is related to the age of the animal. The lipase in growing animals is sufficient to meet their own needs, but it may be beneficial to add lipase to the diet of young animals.

4. Phytase. Phytate phosphorus is difficult to be digested and utilized by monogastric animals. It also reduces the utilization rate of zinc, manganese, iron, calcium and other mineral elements and protein by chelaing. It is a natural antinutritional factor. Phytase can significantly increase the utilization rate of phosphorus, promote animal growth and increase the conversion rate of feed nutrients.

(二) 酶制剂的作用
The effect of enzyme preparations

1. 补充内源酶的不足,激活内源酶的分泌。
2. 破碎植物细胞壁,提高养分消化率。
3. 消除抗营养因子。
4. 降低消化道食糜黏度。

1. Replenish the deficiency of endogenous enzymes and activate the secretion of endogenous enzymes.
2. Crush plant cell walls and improve nutrient digestibility.
3. Eliminate antinutritional factors.
4. Reduce the viscosity of digestive tract chyme.

(三) 饲用酶的选用原则
The selection principle of feed enzymes

饲用酶制剂的应用效果主要取决于酶的组分、活性与动物种类和日粮的匹配性。酶的活性高、品种适宜、与动物日粮匹配性好,应用的效果必然好。但由于酶的种类多,动物日粮的组成也比较复杂,具体选用时问题还比较多,针对具体情况可从以下几个方面考虑:

(1) 动物日粮组成的特异性。
(2) 畜禽的特征:
选用酶要适应胃和小肠的生理特点。通常情况下猪和牛的消化道

The application effect of feed enzyme preparations mainly depends on the compatibility of enzyme composition and activity with animal species and diet. With high enzyme activity, suitable variety, and good compatibility with animal diets, the application effect must be good. However, due to the variety of enzymes, the complex composition of animal diets, there are still many specific problems when selecting them. The following aspects can be considered for specific situations:

(1) The specificity of animal diet composition.
(2) The characteristics of livestock and poultry.
The selection of enzymes should be adapted to the physiological characteristics of the stomach and small intes-

要比禽长，禽类对饲料的消化利用率较低，酶的添加比例需要高些；水产动物饲料中的添加比例则应更高些。另外，幼龄动物消化系统发育不完善，各种消化酶的分泌量不足，需要添加饲用酶，且量要高。成年畜禽消化酶分泌充足，一般不需要添加酶制剂特别是高剂量，但如果日粮的营养水平较低，其内的抗营养因子含量高，可选用以消除抗营养因子为主的复合酶。

（3）酶的特性：不但要考虑酶的合理搭配和酶的稳定性，还要考虑储藏期问题。

（4）饲用酶的应用和影响因素：目前酶制剂在动物养殖中的应用越来越广泛。在水产动物饲料中添加酶制剂可有效地减轻水质污染，降低动物排泄物中氮、磷的浓度，改善饲养环境。在肉鸡日粮中添加复合酶制剂可使体增重提高0.75%－5.33%（平均2.98%），饲料转化率提高1.92%－8.3%（平均4.69%）；对蛋鸡的采食量、产蛋性能及死亡率作用也很明显。仔猪由于其消化系统发育不完善，又缺少分解饲料中非淀粉多糖的消化酶，导致断奶仔猪营养不良，发生腹泻等，添加饲用酶制剂后可明显减少甚至逆转这类不良现象。在牛、羊饲料中添加酶制剂尤其是纤维素酶可明显提高牛羊对饲料中粗纤维的利用率。在奶牛饲料中添加真菌纤维素酶，可使泌乳量和饲料利用率提高。在育肥牛的饲料中添加胰淀粉酶和蛋白酶，可使日增重提高。日粮种类、动物年龄和种类、动物消化道内环境、饲用酶制剂的性质、加工过程及与其他添加剂的相互作用

tine. Under normal circumstances, the digestive tract of pigs and cattle is longer than that of poultry. Poultry has a lower digestibility and utilization rate of feed, and the ratio of enzymes needs to be higher. The addition ratio in aquatic animal feed should be much higher. In addition, the digestive system of young animals is not well developed, and the secretion of various digestive enzymes is insufficient, and feed enzymes need to be added, and the amount should be high. Adult livestock and poultry have sufficient secretion of digestive enzymes. Generally, there is no need to add enzyme preparations, especially high-dose. However, if the nutritional level of the diet is low, and the content of antinutritional factors is high, the compound enzyme mainly used to eliminate antinutritional factors can be selected.

（3）The characteristics of enzymes. Not only the reasonable combination and the stability of the enzymes but also the storage period should be considered.

（4）Application and influencing factors of feed enzymes. Enzyme preparations are now being used widely in animal production. Adding enzymes to aquatic animal feeds can effectively reduce water pollution, lower the concentration of nitrogen and phosphorus in animal excreta, and improve the feeding environment. Adding enzyme preparation into broiler diets can increase body weight gain by 0.75%-5.33% (average 2.98%), feed conversion ratio by 1.92%-8.3% (average 4.69%). The effect on feed intake, laying performance and mortality of laying hens are also obvious. Piglets have a poor digestive system and lack of digestive enzymes to break down non-starch polysaccharides in feed, leads to malnutrition and diarrhea in weaned piglets. These undesirable phenomena can be significantly reduced even reversed after feeding enzyme preparation. Addition enzyme preparation, especially cellulase into the feed of cattle and sheep can significantly improve the utilization rate of crude fibre. Adding fungal cellulose in the feed of dairy cows can improve the milk yield and feed utilization rate. The daily weight gain can be increased by adding amylase and protease into the feed of fatting cattle. The type of diet, the age and type of animal, the environment in the digestive tract of animals, the nature of the enzyme preparation, the process and the interaction with other additives all influence the effect of the enzyme preparation. The effective

等，都会影响酶制剂的作用效果。酶制剂对幼龄动物的应用效果要好于成年动物，同一种酶制剂作用于不同的动物产生的效果往往不同。饲料加工过程中，过高的温度可能破坏其中添加酶的活性。

of enzyme preparations are better on young animals than that in adult animals, and the effects of the same enzyme preparation on different animals are often different. High temperature may destroy the activity of added enzymes in the processing of feeds.

四、酸化剂
Acidulants

能够提高饲料酸度的一类物质被称为酸化剂。酸化剂已广泛应用到饲料生产中，在养殖业中取得了良好的经济效益。

The substances that can increase the acidity of feed is called acidulant. Acidulants have been widely used in feed production, and have achieved great economic benefit in aquaculture industry.

（一）常用酸化剂的种类及特性
Types and characteristics of commonly used acidulants

目前，国内外使用的酸化剂有单一酸化剂（有机酸化剂和无机酸化剂）和复合酸化剂两种。

1. 有机酸化剂　主要有柠檬酸、延胡索酸、乳酸、苹果酸、甲酸、乙酸、丙酸。现在广泛使用较好的是柠檬酸和延胡索酸。

（1）柠檬酸：最初从柠檬中提取。它可使动物胃肠道 pH 值下降，延缓排空速度；减少腹泻发生率。柠檬酸可与 Ca、P、Cu 等必需矿物元素结合，易被吸收利用，生物效价高。柠檬酸还可以促进胃液和胃消化酶的分泌，促进营养物质的吸收。同时，柠檬酸可作为防霉剂和抗氧化剂的增效剂，对饲料中的金属离子有封闭作用，从而使饲料不能被氧化。

（2）延胡索酸：又称富马酸，结构式 HOOC═CH—CH═COOH。延胡索酸可起到增重效果，能够提高有机物质的吸收率，减少机体的能量消耗，提高产品的沉积能。它还对细菌具有抑制或杀灭作用。

2. 无机酸化剂　包括盐酸和

At present, there are two kinds of acidulants at home and aboard: single acidulant (organic acidulant and inorganic acidulant) and compound acidulant.

1. Organic acidulant. Main products are citric acid, fumaric acid, lactic acid, malic acid, formic acid, acetic acid, and propionic acid. Citric acid and fumaric acid are widely used now.

(1) Citric acid. It is originally extracted from lemon. It can lower the pH value of the animal's gastrointestinal tract and delay the emptying speed, reducing the incidence of diarrhea. Citric acid can be combined with essential mineral elements such as Ca, P, Cu, etc., which is easily absorbed and utilized, and has high biological potency. Citric acid can also promote the secretion of gastric juice and digestive enzymes, and promote the absorption of nutrients. At the same time, citric acid can be used as a synergist of antifungal agents and antioxidants, and has a blocking effect on metal ions in the feed, so that the feed cannot be oxidized.

(2) Fumaric acid. It is also called fumaric acid, and structural formula is HOOC═CH—CH═COOH. Fumaric acid can increase the weight of animals and the absorption rate of organic substances, reduce the body's energy consumption, and increase the deposition energy of the products. It also has an inhibitory or killing effect on bacteria.

2. Inorganic acidulant. It includes hydrochloric acid

磷酸。其中磷酸具有双重作用，可作为酸化剂或磷的来源。

3. 复合酸化剂　是用几种特定的有机酸和无机酸复合而成，能迅速降低 pH 值，保持良好的缓冲能力，且生物性能和添加成本最佳。优化的复合体系是饲料酸化剂发展的一种趋势。

and phosphoric acid. Among them, phosphoric acid has a dual role and can be used as an acidifier or a source of phosphorus.

3. Compound acidulant. It is compounded with several specific organic and inorganic acids, which can quickly reduce pH value, maintain good buffering capacity, and have the best biological performance and lower cost. The optimized compound system is a trend in the development of feed acidulant.

（二）酸化剂的作用及机制
Roles and mechanisms of acidulants

1. 酸化剂能降低日粮、胃肠道的 pH 值，提高酶活性。动物饲料中添加酸化剂可使胃内 pH 值下降，从而激活胃蛋白酶，促进蛋白质分解。分解的产物可刺激十二指肠分泌胃蛋白酶，从而促进蛋白质分解吸收。如：添加柠檬酸和磷酸可以提高小肠内胰蛋白酶和淀粉酶活性。

2. 改善胃肠道微生物区系。酸化剂可降低胃肠道 pH 值来抑制有害微生物的繁殖，促进有益菌（如乳酸杆菌）繁殖。

3. 直接参与体内代谢，提高营养物质的消化率，抗应激，增强免疫机能。有些有机酸是能量转换过程中的中间产物，可参与代谢，如乳酸是糖酵解的终产物之一，并通过糖原异生和脂肪分解造成组织消耗。延胡索酸也可供动物应激时所需的能量。延胡索酸本身具有镇静作用，会抑制神经中枢，使机体活动减少。

4. 其他作用机制　如增强食欲，助消化，可能的原因是日粮中加酸能直接刺激口腔内的味蕾细胞，使唾液分泌增多而增加食欲。

1. Acidulants can lower the pH value of diets and gastrointestinal tract and increase the enzyme activity. The addition of acidulants in animal feed can lower the pH value in the stomach, thereby activating pepsin and promoting protein decomposition. The decomposition products can stimulate the duodenum to secrete pepsin, thereby promoting protein decomposition and absorption. For example, citric acid and phosphoric acid can increase the activity of trypsin and amylase in the small intestine.

2. Acidulants can improve the microflora of the gastrointestinal tract. Acidulants can lower the pH value of the gastrointestinal tract in order to inhibit the reproduction of harmful microorganisms and promote the reproduction of beneficial bacteria (such as Lactobacilli).

3. Acidulants can directly participate in the body metabolism, improve the digestibility of nutrients, resist stress, and enhance immune function. Some organic acids are intermediate products in the energy conversion process and can participate in metabolism. For example, lactic acid is one of the end products of glycolysis, and causes tissue consumption by gluconeogenesis and lipolysis. Fumaric acid can also provide energy for animals under stress. Fumaric acid itself has a sedative effect, which can inhibit the nerve center and reduce the body's activity.

4. Other mechanisms of action. Acidulants can enhance animals' appetite and help digestion, the possible reason of which is that the addition of acid in the diet can directly stimulate the taste bud cells in the oral cavity, increasing the secretion of saliva and increasing appetite.

（三）影响酸化剂使用效果的因素
Factors affecting the effectiveness of acidulants

1. 酸化剂的种类和用量

由于各种酸化剂的分子质量、溶解度、解离常数、能量值等不同，在使用效果上也有所差异。酸化剂的作用效果还与其用量有关。用量不足，起不到应有的酸化效果；用量过多，则可引起动物生产性能下降。酸化剂用量过多造成生产性能下降的可能原因有：第一，影响适口性，降低采食量；第二，可能会改变体内的酸碱平衡；第三，胃内过低的pH值会降低胃酸和胃蛋白酶分泌，对小肠酶活性也有不利影响。

2. 日粮的种类和组成

日粮类型不同，其酸化效果不同。在玉米—豆粕型简单日粮中加入有机酸，仔猪可明显提高日增重，而在加入乳制品的复杂日粮中的酸化效果不明显，实质上是其中的大豆蛋白和酪蛋白的差异，同时也与乳糖存在与否及其用量有关。日粮组成不同，其酸结合力不同。矿物质和高蛋白饲料的酸结合力强，与谷类饲料相比，消化时需要较低的pH值。日粮酸结合力高，可降低酸化的效果。高$CaCO_3$水平（每千克干物质59g钙）几乎完全阻止了胃中食糜的酸化，因此，在仔猪料中应尽量使用酸结合力低的钙源饲料。高CP组（>20%）添加酸化剂没有效果，而低CP组（16%）添加酸则提高了仔猪日增重，这也可能是因为高蛋白质饲料具有较强酸结合力的缘故。

3. 年龄或体重的影响

仔猪饲料酸化的重要理论依据是仔猪胃酸分泌不足。随着仔猪年龄和体重的增长，消化道机能逐步完善，胃酸分泌逐步增强，因此，加

1. Types and dosages of acidulants

The effectiveness of acidulant varies according to their molecular mass, solubility, dissociation constants and energy values. The effect of acidulant is also related to its dosage. Insufficient dosage will not get the desired acidifying effect, while excessive dosage will cause a decrease in animal performance. The possible reasons for the decrease in production performance caused by excessive dosage of acidulant are: firstly, it affects palatability and reduces feed intake; secondly, it may change the acid-base balance in the body; thirdly, low pH value in the stomach reduces gastric acid and pepsin secretion, and also has a negative effect on the activity of small intestinal enzymes.

2. Types and composition of diets

The effect of acidification varies according to the types of diets. The addition of organic acids to a simple corn-soybean diet significantly increases the daily weight gain of piglets, but the acidication effect in the complex diet with dairy products is not obvious. It is the essential difference between soybean and casein. It is also related to the presence or absence of lactose and its dosage. Different dietary compositon has different acid binding capacity. The acid binding capacity of mineral and high protein feed is high, and requires a lower pH value for digestion than cereal diets. The diet has a high acid binding capacity, which can reduce the acidification effect. High $CaCO_3$ levels (59 g calcium per kg dry matter) almost completely prevent the acidification of the gastric chyme, therefore, feeds with a low acid binding capacity should be used in piglet diets as much as possible. The addition of acidulant to the high CP group (>20%) has no effect, while the addition of acidulant to the low CP group (16%) of the piglet increases its daily gain. This may also be due to the high acid-binding power of the high protein feed.

3. Effects of age or weight

An important theoretical basis for piglet feed acidification is the insufficiency of gastric acid secretion. With the increase of ages and body weight of piglets, the digestive tract fuction gradually improves and gastric acid secretion

酸效果将降低。研究表明,在仔猪早期断奶后的头1-2周内酸化效果明显,3周以后效果逐步降低,4周以后基本没有效果。

4. 酸化剂与抗生素、高铜和NaHCO₃的相互作用

在以延胡索酸酸化的日粮中添加 NaHCO₃,进一步提高了日增重。这表明延胡索酸与 NaHCO₃ 之间存在互作,可能与 NaHCO₃ 调节体内酸碱平衡的作用有关。有机酸与抗菌剂合并使用效果往往优于单独使用。另外,有机酸、抗生素和高铜联合使用,效果最好。三者各有不同的功能,可能具有互补或加性效应。

gradually increases. Therefore, the effect of adding acid will be reduced. The studies have shown that the acidification effect of piglets is obvious in the first 1 to 2 weeks after early weaning, then gradually decreases after 3 weeks and is basically ineffective after 4 weeks.

4. Interaction of acidulant with antibiotics, high copper and NaHCO₃

The addition of NaHCO₃ to diets acidified with fumaric acid further increased daily gain. This indicates that there is an interaction between fumaric acid and NaHCO₃, which is possibly related to the role of NaHCO₃ in regulating acid-base balance in the body. The combination of organic acids and antimicrobial agents is often more effective than that of using it alone. In addition, organic acids, antibiotics and high copper are most effective when used in combination. They have different functions and may have complementary or additive effects.

五、中草药添加剂
Chinese herbal medicine additives

抗生素的长期使用,会使一些菌株产生耐药性,给人类某些疾病的预防和治疗带来困难。中草药添加剂有几个明显的优点:第一,非特异性的抗菌作用,它不但能调节机体的免疫机能,也能直接杀菌;第二,几乎无残留或残留量低;第三,病原微生物不易产生耐药性;第四,资源丰富,可就地取材。

The long-term use of antibiotics can cause some strains of bacteria to develop resistance, making the prevention and treatment of certain human diseases difficult. Chinese herbal medicine additives have several obvious advantages: firstly, non-specific antibacterial effect, which can not only regulate the body's immune function, but also directly sterilise; secondly, almost no residue or low residue; thirdly, pathogenic microorganisms are less likely to develop drug resistance; fourthly, the resources are abundant, and materials can be obtained locally.

(一) 中草药添加剂的种类
Types of Chinese herbal medicine additives

1. 增食促生长剂

主要有健胃消化的中草药,以改善饲料的适口性,提高采食量,促进消化液的分泌。属这方面的中草药有神曲、麦芽、山楂、陈皮、青皮、枳实、枳壳、苍术、厚朴、豆蔻等。

1. Feeding and growth promoting agents

There are mainly Chinese herbal medicines for stomach digestion to improve the palatability of feed, increase feed intake, and promote the secretion of digestive juice. Chinese herbal medicines belonging to this type include medicated leaven, malt, hawthorn, tangerine, green peel, Citrus aurantium, orange fruit, Atractylodes, Magnolia officinalis, Cardamom, etc.

2. 增产剂

它可提高动物产品的质量,如促进多产蛋的增蛋剂,包括淫羊藿、水牛角、沙苑子(扁茎黄芪)等;催肥剂如远志、柏子仁、酸枣仁等,还有五味子、山药、山楂、松针粉、鸡冠花、钩吻等;以及促进乳腺发育和乳汁合成分泌、增加产奶量的催乳剂,有王不留行、四叶参、通草、马鞭草、鸡血藤等。

3. 免疫增强剂

天然中草药添加剂以调整整体、平衡阴阳为特点,具有多功能、双向调节的作用,是较理想的免疫增强剂。这类免疫增强剂有:黄芪、人参、党参、白术、茯苓、当归、白芍等。

4. 抗寄生虫剂

具有增强机体抗寄生虫侵害能力和驱除体内寄生虫的作用,如槟榔、贯众、使君子、南瓜子、乌梅等对绦虫、蛔虫、姜片吸虫等寄生虫有驱逐作用。

5. 抗应激剂

是指具有缓解、防治由应激原引起的应激综合征的中草药添加剂。如抗热的中草药:柴胡、黄连、茵陈、栀子、黄芩等;抗惊厥的天然中草药:天麻、白芍、酸枣仁、地龙;镇静催眠的中草药:茯神、刺五加、臭梧桐、冰片等。

(二) 中草药的作用机制
Mechanisms of action of Chinese herbal medicine

天然中草药的作用有:增强免疫功能,提高抗病力;抑菌抗病毒作用;增加抗应激功能;增强食欲,提高养分物质的利用和促生长作用;

2. Yield increasing agents

It can improve the quality of animal products, such as egg enhancer that promotes egg production, including epiphyllum, buffalo horn, and Astragalus complanus (Astragalus astragali), etc.; fattening agents such as Polygala tenuifolia, cypress kernels, and jujube kernels, Schisandra, yam, hawthorn, pine needle powder, cockscomb, Gelsemium, etc. There are also prolactins that promote breast development and milk synthesis and secretion, and increase milk production, such as cowherb, four leaf ginseng, Tetrapanax papyriferus, verbena, Caulis spatholobi, etc.

3. Immune-boosting agents

Natural Chinese herbal medicine additives are characterised by the adjustment of the whole body and the balance of yin and yang, and have a multi-functional, bidirectional regulatory effect. They are ideal immune-boosting agents. These immune enhancers include: Astragalus, Ginseng, Radix Codonopsis, Atractylodes, Poria, Radix Angelicae Sinensis, Radix Paeoniae Alba, etc.

4. Anti-parasitic agents

They can enhance the body's ability to resist parasitic attacks and expel parasites from the body. For example, betel nut, cyrtornium fortunei, the fruit of rangoom creeper, pumpkin seeds, umeboshi, etc. have an expelling effect on parasites such as tapeworm, roundworm and fassiolopsis.

5. Anti-stress agents

They are Chinese herbal medicine additives with the ability to relieve and prevent stress syndromes caused by stressors. For example, anti-fever herbs include radix bupleuri, Coptis chinecis, capillary artemisia, Fructus Gardeniae, Scutellaria baicalensis, etc.; anticonvulsant herbs include gastrodia, peony root, Jujube kernel, pheretima; sedative and hypnotic herbs include Poria cocos (Schw.) Wolf, acanthopanax, harlequin glorbower leaf, borneol, etc.

Natural Chinese herbal medicine has several functions: enhancing immune function and disease resistance; antibacterial and antiviral effects; increasing stress resistance; enhancing appetite, improving nutrient utilization and promo-

提高畜产品品质。

1. 抗菌作用

许多植物提取物具有抗菌作用,如金银花、连翘、蒲公英、黄连、大蒜、鱼腥草等对金黄色葡萄球菌、溶血链球菌、肺炎链球菌等革兰氏阳性菌具有抑制作用。牛至属植物具有抗菌作用。其主要抗微生物成分是挥发油中所含的酚类化合物——香芹酚和百里酚。牛至提取物抗菌的作用机制主要是通过对细菌细胞壁结构蛋白的变性和凝固来实现的,从而使细胞产生渗漏,水分失去平衡,最终导致细菌死亡。香芹酚和百里酚也能加速肠绒毛表面腔上皮细胞的更新率,减少病原体对上皮细胞的感染并提高养分的吸收利用。

2. 抗病毒作用

许多植物里的有效成分主要是激活非特异性免疫系统,抑制或杀灭病毒。体外实验和临床实践证明,金银花、连翘、鱼腥草、贯众、黄芩、大青叶、板蓝根、黄柏、丹皮等对流感病毒亚甲型有抑制作用。多糖在体内外都具有抗病毒作用,表现为多糖抑制体内外病毒繁殖,保护细胞免于感染,提高感染动物的存活率。黄芪多糖能促进淋巴细胞的转化,提高免疫球蛋白的含量,抑制病毒的繁殖。

3. 免疫增强作用

很多中草药含有多糖、有机酸、生物碱、苷类和挥发油等成分,这些物质在动物体内刺激免疫系统,产生抗病能力,从而起到防病治病的作用。如:多糖能促进胸腺应激反应,刺激网状内皮系统,提高宿主对癌细胞的特异性抗原免疫反应能力。有机酸类则能增强吞噬细胞的特异吞噬功能。

ting growth; improving the quality of livestock products.

1. Antibacterial effects

Many plant extracts have antibacterial effects, such as honeysuicle, fruntus forsythiae, dandelion, coptis, garlic and Houttuynia cordata, which have inhibitory effects on Gram-positive bacteria such as Staphylococcus aureus, Streptococcus haemolyticus and Streptococcus pneumoniae. Oregano plants have antibacterial effects. Its main anti-microbial components are the phenolic compounds contained in the volatile oil—carvacrol and thymol. The antibacterial mechanism of action of oregano extracts is mainly through denaturation and coagulation of the structural proteins of the bacterial cell wall, resulting in leakage, loss of water balance and ultimately bacterial death. Carvacrol and thymol can also accelerate the rate of renewal of luminal epithelial cells on the surface of the intestinal villi, reduce the infection of epithelial cells by pathogens and improve the absorption and utilization of nutrient.

2. Antiviral effects

The active ingredients in many plants are mainly responsible for activating the non-specific immune system and inhibiting or killing viruses. In vitro experiments and clinical practice have proved that honeysuckle, fruntus forsythiae, houttuynia cordata, kanzhong, scutellaria, dacryophyllum, panax notoginseng, cypress and tannin have an inhibitory effect on influenza virus subtype A. Polysaccharides have antiviral effects both in vivo and in vitro. They can inhibit virus reproduction in vivo and in vitro, protect cells from infection, and increase survival rate of infected animals. Astragalus polysaccharides can promote the transformation of lymphocytes, increase the content of immunoglobulins and inhibit the reproduction of viruses.

3. Immune-enhancing effects

Many Chinese herbal medicines contain polysaccharides, organic acids, alkaloids, glycosides and volatile oils, which stimulate the immune system in animals and improve their resistance to disease, thus playing a role in disease prevention and treatment. For example, polysaccharides can stimulate the thymic stress response, stimulate the reticuloendothelial system and increase the host's specific antigenic immune response to cancer cells. Organic acids can enhance the specific phagocytic function of phagocytes.

4. 抗应激作用

人参、刺五加、黄芪等可提高动物对低压缺氧和常压缺氧以及中毒性缺氧的耐受力；中草药中黄芪成分可以通过提高肾上腺皮质的分泌功能，增强机体抗疲劳功能。

4. Anti-stress effect

Ginseng, Astragalus and Astragalus can improve the tolerance of animals to hypobaric hypoxia, atmospheric hypoxia, and toxic hypoxia. Astragalus in herbal medicine can enhance the body's anti-fatigue function by increasing the secretion function of the adrenal cortex.

（三）中草药提取工艺及技术
Extraction process and technology of Chinese herbal medicine

中草药成分复杂，既有有效成分，也有无效成分和有毒成分。为了更好地利用植物，需要提取它的有用成分。常用的提取方法有煎煮法、回流法、浸渍法、渗漉法等。但是这些提取方法存在损失大、周期长、工序多、效率低的缺点。近几年来，出现了几种新型的提取方法，在一定程度上克服了上面的缺点。

1. 临超界流体萃取技术

其原理是利用流体（溶剂）在临界点附近某区域（超临界区）内与待分离混合物中的溶质具有异常相平衡行为和传递性能，且溶质的溶解能力随压力和温度的改变而在相当宽的范围内变动，利用这种作溶剂，可以从多种液态或固态混合物中萃取出待分离组分。常用的临超界流体为CO_2。它的最大的优点是可以在近常温的条件下提取分离，几乎保留产品中的全部有效成分，无有机溶剂残留，产品纯度高，操作简单节能。

2. 半仿生提取法

它是将整体药物研究与分子药物研究法相结合，从生物药剂学的角度，模拟口服给药及药物经胃肠道转运的原理，为经消化道中给药的中药制剂设计的一种新的提取方法。其优点是能够提取和保留更多的有效成分，能缩短生产周期，降低成本。

Chinese herbal medicines are complex in composition, with active, ineffective and toxic components. In order to make better use of plants, it is necessary to extract its effective components. Common extraction methods include decoction, reflux, maceration and percolation. However, these extraction methods have the disadvantages of large losses, long cycle time, multiple processes and low efficiency. In recent years, several new extraction methods have emerged, which overcome the above shortcomings to a certain extent.

1. Supercritical fluid extraction

The principle is to use the fluid (solvent) in a region near the critical point (supercritical zone) and the solute in the saparation mixture with abnormal phase equilibrium behaviour and transfer properties. Moreover, the solubility of the solute varies with the change of pressure and temperature in a wide range. Using this as a solvent, the components can be separated from a variety of liquid or solid mixture of components. A commonly used critical superfluid is CO_2. The most important advantage of this fluid is that it can be extracted and separated at near room temperature, and almost all the active ingredients in the product are retained with no organic solvent residues, high product purity, simple operation, and energy saving.

2. Semi-bionic extraction method

It is a new extraction method designed for Chinese medicinal preparations that are administered through the gastrointestinal tract by combining holistic drug research with molecular drug research methods, and simulating the principles of enclosure administration and drug transport through the gastrointestinal tract from the perspective of biopharmacology. The advantage of this method is that more active ingredients can be extracted and retained, and the production cycle can be shortened, and costs can be reduced.

3. 超声提取技术

其基本原理主要是利用超声波的空化作用加速植物有效成分的浸出提取，它具有提取时间短、产率高、无须加热等优点。

4. 微波萃取技术

其原理是在微波场中，吸收微波能量的差异使得物质的某些区域或萃取体系中的某些组分被选择性加热，使得被萃取物质从基体或体系中分离。微波萃取具有设备简单、萃取效率高、使用范围广、污染小等特点。

3. Ultrasonic extraction technology

The basic principle is to use the ultrasound cavitation effect to accelerate the extraction of active ingredients from plants. It has the advantages of short extraction time, high yield and no need of heating, etc.

4. Microwave extraction technology

The principle is that in the microwave field, the difference in the absorption of microwave energy makes some regions of the material or some components of the extraction system be selectively heated, so that the extracted material is separated from the matrix or system. Microwave extraction has the advantages of simple equipment, high extraction efficiency, wide application range, and less pollution.

六、饲料品质调节剂
Feed quality regulators

（一）抗氧化剂
Antioxidants

1. 抗氧化剂的作用机制

不同的抗氧化剂具有不同的作用机制：一些是借助于还原反应，降低饲料内部及其周围的氧含量；一些可以放出氢离子将自动氧化过程中所产生的过氧化物破坏分解；还有一些可能与所产生的过氧化物（游离基）相结合，使自动氧化过程中的连锁反应中断，从而阻止氧化过程的进行。

2. 饲料中常用的抗氧化剂

（1）乙氧基喹啉（乙氧喹）：是一种黏滞、黄褐至褐色的液体。乙氧基喹啉在动物肠道吸收后，绝大部分在体内进行脱乙基反应，经肾脏迅速排出体外，不在体内蓄积，广泛用于油脂、鱼粉、维生素及预混料、配合饲料等。乙氧基喹啉在饲料中的添加量不超过 150 mg/kg。

（2）丁羟甲氧苯（丁羟基茴香醚）：为白色或微黄褐色结晶或结

1. Mechanisms of action of antioxidants

Different antioxidants have different mechanisms of action: some can reduce the oxygen content in and around the feed by means of reduction reactions, some can emit hydrogen ions to break down the peroxides produced during auto-oxidation, and others may combine with the peroxides produced (free radicals) to interrupt the chain reaction in the auto-oxidation process, thus preventing the process from proceeding.

2. Antioxidants commonly used in feed

(1) Ethoxyquin (ethoxy quinoline). It is a viscous, yellowish brown or brown liquid. After the ethoxyquin is absorbed in the intestinal tract of animals, most of it undergoes a deethylation reaction in the body, and is rapidly excreted from the body through the kidneys, and does not accumulate in the body. It is widely used in fats, fish meal, vitamins and premixes, compound feeds, etc. The addition amount of ethoxyquin in the feed does not exceed 150 mg/kg.

(2) Butylated hydroxyanisole (butyl anisole). It is white or slightly yellowish brown crystal or crystalline

晶性粉末。丁羟甲氧苯被动物代谢分解后,迅速从体内排出,通常不在畜禽体内蓄积,是目前广泛使用的油脂抗氧化剂。每千克油脂中的添加量为 100－200 mg。

（3）其他抗氧化物：还有丁羟甲苯、生育酚、抗坏血酸等。另外,特丁基对苯二酚是一种新型高效抗氧化剂,它的抗氧化效果和安全性是同类饲料抗氧化添加剂中较好的,在饲料行业中具有广泛的应用前景。

powder. After being metabolized and decomposed by animals, butylated hydroxybenzoate is rapidly excreted from the body and usually does not accumulate in livestock and poultry. It is currently widely used lipid antioxidant. The added amount per kilogram of oil is 100－200 mg.

(3) Other antioxidants. There are butylated hydroxytoluene, tocopherol, ascorbic acid and so on. In addition, tert-butyl hydroquinone (TBHQ) is a new type of high-efficiency antioxidant. Its antioxidant effect and safety are better than other similar feed antioxidant additives, and it has a wide application prospects in the feed industry.

（二）防霉剂

Antifungal agents

1. 防霉剂的作用及机制

其主要作用为两个方面：一是破坏霉菌的细胞壁和细胞膜,二是破坏或抑制细胞内酶的作用,降低酶的活性。

2. 常用于饲料的防霉剂

（1）丙酸及其盐类：丙酸为具有强烈刺激性气味的无色透明液。丙酸及其盐类是饲料中应用最为普遍的防霉剂,属酸性防霉剂。其效果为：丙酸＞丙酸铵＞丙酸钠＞丙酸钙。常见用法是：丙酸无毒,很易挥发,应用时多是让丙酸吸附在多孔硅胶或蛭石上,再混入饲料,不仅使丙酸慢慢释放,保持防霉效果的时间也会加长。湿度大和温度高时添加量多,可为每吨饲料 0.3－0.8 kg。

（2）苯甲酸（安息香酸）及苯甲酸钠：苯甲酸为白色片状或针状结晶或结晶性粉末,苯甲酸钠为白色颗粒或结晶性粉末。二者在体内参与代谢,不蓄积,毒性低,是安全的防霉剂。苯甲酸及苯甲酸钠的主要作用是能抑制微生物细胞内呼吸

1. The role and mechanism of antifungel agents

Its main funtion is in two aspects: one is to destroy the cell wall and cell membrane of moulds, the other is to destroy or inhibit the action of intracellular enzymes and reduce the activity of enzymes.

2. Antifungel agents commonly used in feed

(1) Propionic acid and its salts. Propionic acid is a knid of colourless, transparent liquid with a strong pungent ordor. Propionic acid and its salts are the most widely antifungel agents used in feed, belonging to acidic antifungel agents. So far as the effect is concerned, propionic acid is the best, and the calcium acid is the worst, and the ammonium propionic acid and sodium propionic acid ranks the second and the third respectively. Propionic acid is non-toxic and volatile. It is mostly used to adsorb propionic acid on porous silica gel or vermiculite, and then mixed with the feed, which not only releases propionic acid slowly, but also keeps the mould inhibiting effect for a long time. It can be added by 0.3－0.8 kg per ton of feed in high humidity and high temperature.

(2) Benzoic acid (benzoic acid) and sodium benzoate. Benzoic acid is white-flake-shaped or needle-shaped crystal or crystalline powder, while sodium benzoate is white granular or crystalline powder. Both are involved in metabolism in the body. They do not accumulate in the body, have low toxicity, and are safe antifungel agents. The main effects of benzoic acid and sodium benzoate are to

酶的活性以及阻碍乙酰辅酶 A 的缩合反应，使三羧酸循环受阻，代谢受到影响；还阻碍细胞膜的通透性。美国食品及药物管理局规定，最大使用量按苯甲酸计应小于 0.1%。另外，近年来研究较多的是双乙酸钠，具有成本低、性质稳定、使用条件宽、防霉防腐作用显著等优点。

inhibit the activity of respiratory enzymes in microbial cells and to impede the condensation of acetyl coenzyme A, then the tricarboxylic acid cycle is blocked and metabolism is affected, and the permeability of cell membranes is also impeded. The US Food and Drug Administration (FDA) states that the maximum use level should be less than 0.1% based on benzoic acid. In addition, in recent years more research is about sodium diacetate, which has the advantages of low cost, stable properties, wide use conditions, and remarkable anti-mildew and anti-corrosion effects.

（三）饲料青贮添加剂
Feed silage additives

已研究和应用的青贮添加剂主要有无机酸、有机酸及其盐类、醛类及其他防霉剂、接种物、酶、非蛋白氮、糖及含糖物以及矿物微量元素等。青贮料中常添加的无机酸有盐酸、硫酸和磷酸；青贮料中添加的有机酸主要有甲酸、乙酸、丙酸、苯甲酸、山梨酸等及其盐类，其中以甲酸及其盐类在青贮饲料中应用最为广泛。

The silage additives that have been researched and applied mainly include inorganic acids, organic acids and their salts, aldehydes and other antifungal agents, inoculants, enzymes, non-protein nitrogen, sugars and sugar-substance containing, and mineral trace elements. The inorganic acids often added in silage include hydrochloric acid, sulfuric acid and phosphoric acid. The organic acids added to silage mainly include formic acid, acetic acid, propionic acid, benzoic acid, sorbic acid, etc. and their salts, among which, formic acid and its salts are most widely used in silage.

（四）调味剂
Flavoring agents

调味剂是根据不同动物在不同生长阶段的生理特性和采食习性，为改善饲料的诱食性、适口性及饲料转化率，提高饲料质量而添加到饲料中的一种添加剂。

Flavoring agent is an additive added to feed in order to improve the attractivity, palatability and feed conversion rate of feed according to the physiological characteristics and feeding habits of different animals at different growth stages.

1. 调味剂的作用机制

巴甫洛夫指出"食欲即消化液"，没有食欲就不可能有消化液的分泌，而使饲料消化受阻。调味剂是通过香气与味觉作用，以动物采食行为和采食心理为基础，使动物受到刺激，产生食欲，分泌更多的消化液，提高采食量，促进消化酶的分泌，并可提高酶活性，进而提高饲料的利用率。

1. The mechanism of action of flavoring agents

Pavlov pointed out that "appetite is digestive juices". Without appetite, digestive juice can not be secreted, so that the feed digestion is hindered. Flavoring agents work through the effects of aroma and taste, and are based on animal eating behavior and eating psychology, which stimulate animals to produce appetite, secrete more digestive juices, increase feed intake, promote the secretion of digestive enzymes, and increase the enzyme activity, thereby improving the feed utilization rate.

2. 常用饲用调味剂及其在饲

2. Common feeding flavouring agents and their appli-

料中的应用

(1) 香料及引诱剂:香料是目前应用最为广泛的饲用调味剂之一。主要有以下几种:鸡用香料,多用于产蛋鸡和肉鸡。牛用香料,主要应用于产奶牛饲料和犊牛人工乳或代乳品中。奶牛喜欢柠檬、甘草、茴香、甜味等。猪用香料,主要用于人工乳、代乳料、补乳料和仔猪开食饲料,促进采食,防止断奶期间生产性能下降。添加的香料主要为乳香型、水果香型等。鱼用香料,可以增加采食量,在提高饵料利用率方面有很好的效果。

(2) 鲜味剂:谷氨酸钠为常用鲜味剂,按 0.1% 的添加量添加在猪用饲料或人工代乳料中能提高食欲。

(3) 甜味剂:常用的甜味剂有糖精及糖精钠,无色至白色结晶或白色风化粉末,无臭。稀溶液味甜,大于 0.026% 时则味苦。易溶于水 (66%,20 ℃),难溶于乙醇。其甜度约为蔗糖的 500 倍。能改善饲料的适口性,但在配合饲料中不可超过 150 mg/kg。

(4) 酸味剂:水溶性的有机酸如乳酸、甲酸、柠檬酸等均可用作酸味剂添加到饲料中以提高饲料的适口性,并可以调整幼畜胃肠道的 pH 值。

(五) 着色剂
Colorants

着色剂的作用:使饲料增色,提高商品质量;使动物产品增色,如蛋黄等;有利于促进动物采食。着色剂主要是化学合成的类胡萝卜素及其衍生物。

1. 叶黄素

由微生物生产的叶黄素为黄色至橙色,主要用于产蛋鸡和肉鸡饲

cation in feed

(1) Spices and attractants: Spices are one of the most widely used feed flavorings. The commonly used flavorings are chicken spice, bovine spice, spices for pigs, and fish feed. Chicken spice is mostly used for laying hens and broilers. Bovine spice is mainly used in dairy cow feed and calf artificial milk or milk replacer. Cows like lemon, licorice, fennel, sweet, etc. Spices for pigs are mainly used for artificial milk, milk replacer, milk supplement and piglet starter feed to promote feed intake and prevent the decline of production performance during weaning. The added flavors are mainly frankincense and fruit flavors. Fish feed can increase feed intake and has a good effect in improving the utilization rate of bait.

(2) Umami agents. Sodium glutamate is a commonly used umami agent. Adding 0.1% to pig feed or artificial milk replacer can increase appetite.

(3) Sweeteners. Commonly used sweeteners are saccharin and sodium saccharin, which are colourless or white crystal or white weathered powder, and odourless. Dilute solution tastes sweet, while bitter if content is more than 0.026%. It is easily soluble in water (66%, 20 ℃), but hardly soluble in ethanol. Its sweetness is about 500 times that of sucrose. It can improve the palatability of feed, but should not exceed 150 mg/kg in compound feed.

(4) Acidifiers. Water-soluble organic acids such as lactic acid, formic acid and citric acid, can be used as acidifiers to improve the palatability of feed and adjust the pH value of the gastrointestinal tract of young animals.

The role of colorants is to enhance the colour of feeds and improve the quality of products, enhance the colour of animal products, such as egg yolk, etc. and to promote animal feeding. Colorants are mainly chemically synthesised carotenoids and their derivatives.

1. Lutein

Lutein produced by microorganisms is yellow or orange and is mainly used in the feed for laying hens and broilers to

料中,以增加蛋黄及皮肤、喙、脚胫色泽。

2. 玉米黄

主要成分为玉米黄素及隐黄素。玉米黄素分子式 $C_{40}H_{56}O_2$,相对分子质量为 568.85,为血红色油状黏液,10 ℃以下时为橘黄色半凝固状物,溶于乙醚、石油醚、丙酮及酯类,不溶于水。它比维生素 B_2 更接近天然黄色。动物吸收后可部分转化为维生素 A,有一定的营养价值。

3. β-阿朴-8′-胡萝卜素酸乙酯

为橙黄色着色剂,是应用最为广泛的人工合成着色剂之一。主要用于蛋黄及肉鸡皮肤、喙、脚胫着色。利用率高,色素沉着好,为着色最有效的类胡萝卜素。

4. 其他着色剂还有橘黄色素、柠檬黄质、虾红质等。

(六) 黏结剂
Binding agents

在动物颗粒饲料和鱼虾饵料的生产过程中,添加一定的黏结剂,增加饲料的黏结性,有助于颗粒的成形和保证一定的颗粒硬度和耐久性,可提高水产饵料在水中的稳定性,减少加工过程中的粉尘。常见的黏结剂有:α-淀粉、褐藻酸钠(目前鱼、虾饵料中应用较普遍的黏结剂)。膨润土和膨润土钠(多用于畜、禽颗粒饲料)用量要求不超过配合饲料的2%。其他黏结剂还有琼脂、阿拉伯胶、瓜尔胶、蚕豆胶、陶土等。

increase the colour of egg yolk, skin, beak and shin.

2. Zeaxanthin

Its main components are zeaxanthin and cryptoxanthin. Zeaxanthin has a molecular formula of $C_{40}H_{56}O_2$ and a relative molecular mass of 568.85. It is blood-red oily mucilage or orange-yellow semi-consolidated substance below 10 ℃. It is soluble in ether, petroleum ether, acetone and esters, and insoluble in water. Zeaxantin is closer to the natural yellow than vitamin B_2. It can be partially converted into vitamin A after being absorbed by animals, which has some nutritive value.

3. Ethyl β-apo-8′-carotate

It is orange-yellow colarant, and is one of the most widely used synthetic colouring agents. It is mainly used for colouring egg yolk and broiler skin, beak and shin with high utilisation rate and good pigmentation. It is the most effective carotenoid for coloring.

4. Other colorants include orange pigment, lemon yellow pigment, shrimp red pigment, etc.

In the production of animal pellet feeds, fish and shrimp baits, the addition of certain binding agents can increase the adhesion of the feed, help to form the pellets and ensure a certain degree of hardness and durability of the pellets, improve the stability of aquatic baits in water and reduce the dust during processing. Common binding agents include: alpha-starch, sodium alginate (a common binding agent used in fish and shrimp baits). The content of bentonite and sodium bentonite (mostly used in pellet feed of livestock and poultry) should not exceed 2% of the feed. Other binding agents include AGAR, gum Arabic, guar gum, broad bean gum, clay, etc.

第十一章 配合饲料产品
Chapter 11　Formulated feed products

第一节　饲料配合概述
Section 1　Overview of feed composition

单一的饲料原料各有其特点，提供养分相对较为单一，有的以供应能量为主，有的以供应蛋白质、氨基酸或矿物质及维生素为主，还有些饲料原料是以特殊目的而添加到饲料中；有的原料粗纤维含量高，有的水分含量高，有些单一饲料原料还存在适口性差、含有抗营养因子等问题，不能直接饲喂动物，需要进一步加工后才能应用。因此，单一饲料原料普遍存在营养不平衡、饲喂效果差、不能满足动物生长的营养需要等问题。为了合理利用各种饲料原料，提高饲料的利用效率和营养价值，提升饲料产品的综合性能，有必要将各种饲料原料进行合理搭配，采用适当的加工工艺，充分发挥各种单一饲料的优点、避开其缺点，生产满足不同饲养目的且营养均衡的饲料产品，使饲料资源利用达到最大化。

Single feed raw materials have their own characteristics, whose nutrients are relatively single: some mainly supply energy, some mainly supply protein, amino acids or minerals and vitamins, and some feed materials are added to feed for special purposes. Some materials contain high crude fiber content, and some contain high water content, and some single feed materials also have poor palatability, anti-nutritional factors and other problems, which can't be fed directly to animals and need further processing before they can be used. Therefore, single feed raw materials generally have weak points such as nutrition imbalance, poor feeding effects and cannot meet the nutritional needs of animal growth. In order to make rational use of various feed raw materials, improve the comprehensive performance of feed products, it is necessary to make reasonable collocation of various feed raw materials and adopt appropriate processing technology, so as to give full play to the advantages of various single feed and avoid its disadvantages. In order to maximize the utilization of feed resources, we need to produce feed products with balanced nutrition to meet different feeding purposes.

一、饲料配合的概念
The concept of feed composition

饲料配合是指按一定的配方把不同来源的饲料依一定比例均匀地

Feed composition refers to the process in which feed from different sources is uniformly mixed in a certain pro-

混合在一起,并按规定的工艺流程生产饲料产品的过程。满足一头动物一昼夜所需各种营养物质而采食的各种饲料总量称为日粮。动物饲养中,将按百分比配合用于自由采食的配合饲料称为饲粮。饲料配合的依据包括使用目的、动物类型、生长阶段、生理要求、生产用途、营养需要标准、饲料营养价值、饲料法规和饲料管理条例等,生产的饲料产品在保证产品质量前提下,应有利于人类和动物的健康,有利于环境保护和维护生态平衡。此外,饲料配合还应考虑方便饲喂和用户使用,方便运输和保存,减轻用户劳力。

portion according to the scientific formula and the feed product is produced according to the specified technological process. The total amount of all kinds of feed that an animal needs for one day and night is called a ration. In animal feeding, the compound feed which is used for free eating according to percentage is called a diet. The basis of feed composition includes the use purpose, animal type, growth stage, physiological requirements, production use, nutrition requirements standard, feed nutritive value, feed regulations and feed management regulations, etc. Under the premise of ensuring product quality, the produced feed products should be beneficial to the health of humans and animals, environmental protection and the maintenance of ecological balance. In addition, the feed composition should also facilitate feeding and customer's use, facilitate transportation and storage, and reduce user's labor.

二、配合饲料的特点
Characteristics of formulated feed

配合饲料是在科学试验基础上并经过实践验证而设计和生产的,集中了动物营养和饲料科学的研究成果,并能把各种不同的组分(原料)均匀混合在一起,从而保证有效成分的稳定一致,提高了饲料的营养价值和经济效益。配合饲料主要具有以下特点:

1. 营养平衡:配合饲料是根据畜禽的营养需要,应用最新科研成果,按营养平衡原理调配的饲料,可以大大提高饲料的利用效率。

2. 饲用安全:配合饲料在生产过程中采用强力搅拌机,从而保证了微量成分的均匀混合,防止畜禽营养不足或过量引起的中毒。

3. 充分利用饲料资源:配合饲料是由多种饲料根据畜禽的生理特点和营养需要配制而成,因而除具有营养性外,也合理经济地利用了饲料资源,避免因使用单一饲料而

Formulated feed is designed and produced on the basis of scientific experiments and practical verification. It concentrates the research results of animal nutrition and feed science, and can mix various components (raw materials) evenly, so as to ensure the stability and consistency of virtual components, and improve the nutritive value and economic benefits of feed. Formulated feed has the following characteristics:

1. Nutritional balance. Based on the nutritional needs of livestock and poultry, formulated feed is a kind of feed which can greatly improve the utilization efficiency of feed by applying the latest scientific research results according to the principle of nutrient balance.

2. Feed safety. Formulated feed adopts a strong mixer in the production process to ensure the uniform mixing of trace ingredients and prevent poisoning caused by inadequate or excessive nutrition of livestock and poultry.

3. Making the best of feed resources. Formulated feed is made up of a variety of feeds according to the physiological characteristics and nutritional needs of livestock and poultry. Therefore, in addition to being nutritious, it can also use feed resources reasonably and economically and avoid

造成的浪费。

4. 提高劳动生产率：配合饲料的生产为养殖专业户和集约化经营者提供了良好条件，大大节省了饲料加工设备的投入，同时也降低了饲喂成本。

配合饲料虽然具有上述优点，但在生产过程中也受其他因素制约，如原料的价格、适口性、不同原料之间的拮抗、货架期等均会影响产品的质量，生产时应充分考虑这些因素。

waste caused by using a single feed.

4. Increasing labor productivity. The production of formulated feed provides good conditions for the professional farmers and intensive operators, and greatly saves the input of feed processing equipment, and also reduces the feeding costs.

Although formulated feed has the above advantages, but in the production process, it is also restricted by other factors, such as the price of raw materials, palatability, antagonism between different raw materials, shelf life, etc., which will affect the quality of formulated feed products. These factors should be fully considered in the production.

三、饲料配合产品的种类和组成
Types and composition of feed compound products

饲料配合产品的分类方法很多，按饲料生产过程或生产工艺可分为粉料、颗粒料、破碎料、膨化饲料、扁状饲料、液体饲料、漂浮饲料和块状饲料等；按饲喂对象可分为猪用饲料、家禽用饲料、反刍和草食动物饲料、水产动物饲料、实验动物饲料、特种经济动物饲料、伴侣和观赏动物饲料等；按营养成分和用途可分为配合饲料、浓缩饲料、精料补充料、预混合饲料等。

1. 配合饲料

配合饲料是指根据饲养动物的营养需要，将多种饲料原料和饲料添加剂按饲料配方经工业化加工的饲料。这类饲料产品亦称为全价配合饲料、完全配合饲料、全日粮配合饲料。通常可根据动物种类、年龄、生产用途等划分为各种类型。此种饲料可以全面满足饲喂对象的营养需要，用户不必另外添加任何营养性饲用物质而直接饲喂动物，但必须注意选择与饲喂对象相符合的配合饲料。在实际生产中，由于科学技术水平和生产条件的限制，许多

There are many classification methods for feed compound products. According to the production process or manufacturing of feed, it can be divided into powder feed, granule feed, crushing feed, expanded feed, flat feed, liquid feed, floating feed and block feed, etc. According to the feeding objects, it can be divided into pig feed, poultry feed, ruminant and herbivorous animal feed, aquatic animal feed, experimental animal feed, special economic animal feed, companion and ornamental animal feed, etc. According to nutrient composition and purpose, it can be divided into formula feed, concentrate feed, concentrate supplement, premix feed and so on.

1. Formula feed

According to the nutritional needs of the raised animals, formula feed is an industrially processed feed consisting of a variety of feed raw materials and feed additives according to the feed formula. This kind of feed product is also called complete formula feed. It is usually classified into various types according to animal species, age, production use, etc. This kind of feed can fully meet the nutritional needs of the feeding objects. The customers do not need to add any other nutrient feeding substances and can feed the animals directly, but must pay attention to selecting the formula feed which is consistent with the feeding objects. In actual production, many "complete feeds" are difficult to reach the "full price" in nutrition on account of the limita-

"完全饲料"难以达到营养上的"全价",故可根据饲料配合的水平区分为"全价配合饲料"与"混合饲料"。混合饲料由某些饲料原料经过简单加工混合而成,为初级配合饲料,主要考虑能量、蛋白质、钙、磷等营养指标,在许多农村地区常见。混合饲料可用于直接饲喂动物,但饲养效果不理想。

2. 浓缩饲料

浓缩饲料指主要由蛋白质饲料、矿物质饲料和饲料添加剂按一定比例配制的均匀混合物,又称平衡用配合料。浓缩饲料通常为全价饲料中除去能量饲料的剩余部分,故与能量饲料按规定比例混合即可制成配合饲料。

浓缩饲料一般占配合饲料的20%-40%。市场上将使用量在10%-20%的产品称为超级浓缩料或料精,其基本组成为一部分蛋白质饲料、添加剂预混料及具有特殊功能的物质。使用时还需要搭配能量饲料和部分蛋白质饲料。

3. 精料补充料

精料补充料指为补充以饲喂粗饲料、青饲料、青贮饲料等为主的草食动物的营养,而用多种饲料原料和饲料添加剂按一定比例配制的均匀混合物,由能量饲料、蛋白质饲料、矿物质饲料及饲料添加剂组成。精料补充料是为草食动物配制生产的,不单独构成饲粮,主要是用以补充采食饲草不足的那一部分营养。除草食动物采食的青、粗饲草及青贮饲料外,再给予适量的精料补充料,可全面满足饲喂对象的各种营养需要。当基础饲草改变时,应根据动物生产反应及时调整精料补充料的供给量。

4. 预混合饲料

预混合饲料指由矿物质饲料、氨基酸、微量元素、维生素、非营养

tions of scientific and technological level and production conditions, so they can be divided into "full price compound feeds" and "mixed feeds" according to the level of feed composition. Mixed feed is a kind of primary mixed feed made of some feed raw materials after simple processing. It mainly considers nutrition indicators such as energy, protein, calcium, phosphorus, which is common in many rural areas. The mixed feed can be used to feed animals directly, but the feeding effect is not ideal.

2. Concentrate feed

Concentrate feed refers to a homogeneous mixture mainly made up of protein feed, mineral feed and feed additives in a certain proportion, which is also called a balance formula feed. Concentrate feed is usually the remaining part of the energy feed which is removed in the full price feed, so it can be mixed with the energy feed according to the specified ratio to make the compound feed.

Concentrate feed generally accounts for 20% to 40% of the compound feed. In the market, the products with the usage amount between 10% and 20% are called super concentrate or concentrate. The basic composition is a part of protein feed, additive premix and substances with special functions. It also needs to be combined with energy feed and part of protein feed when used.

3. Concentrate supplement

Concentrate supplement is a homogeneous mixture of various feedstuff and feed additives in a certain proportion to supplement the nutrients of herbivores mainly feeding green fodder, silage, roughage, etc. It consists of energy feed, protein feed, mineral feed and feed additives. Concentrate supplement is prepared and produced for herbivores. It does not constitute a separate diet, but is mainly used to supplement the part of nutrients that are insufficient in forage grass. That is, in addition to the green, crude forage grass and silage feed, herbivore animals can be given an appropriate amount of concentrate supplement to fully meet the various nutritional needs of the feeding objects. The feed quantity of concentrate supplement should be adjusted in time according to the animal production response when changing the base forage.

4. Premix feed

Premix feed refers to the kind of or several kinds of mineral feed, amino acids, trace elements, vitamins and

性饲料添加剂等之中的一种（类）或多种（类）与载体或稀释剂按一定比例配制的均匀混合物，简称预混料。预混料不能直接饲喂动物，其作用是有利于微量的原料均匀分散于大量的配合饲料中。

预混合饲料包含单一预混合饲料、复合预混合饲料、添加剂预混合饲料、微量元素预混合饲料和维生素预混合饲料等产品。

复合预混合饲料或添加剂预混合饲料可视为配合饲料的核心，因为配合预混饲料的微量活性成分往往是决定配合饲料饲喂效果的因素。

non-nutritional feed additives as well as the homogeneous mixture made up of carriers or diluents in a certain proportion. Premix cannot be fed directly to animals, and its function is to disperse trace raw materials evenly in a large number of compound feeds.

Premix feed includes single premix, compound premix, additive premix, trace element premix, vitamin premix and other products.

Compound premix feed or additive premix feed can be regarded as the core of compound feed, because the trace active component of compound premix feed is often the determinant of the feeding effect of compound feed.

第二节　配合饲料的生产工艺
Section 2　Production technology of formulated feed

一、概述
Summary

配合饲料是在配方设计的基础上，按照一定的生产工艺流程生产出来的。配合饲料加工工艺由于产品形状、设备类型、生产规模以及使用方式等不同，加工能力和工艺组成有所差别。但不管何种类型，配合饲料基本加工工艺一般由原料的接收和清理、粉碎、配料混合、后处理（调质、膨化、制粒、干燥、过筛）、包装、运输和贮存等主要工段和工序以及液体添加系统、蒸汽系统、通风除尘等系统组成。

配合饲料产品有粉状和粒状两种形式。粉状配合饲料生产工艺相对简单，主要包括原料的接收清理、粉碎、计量配料、混合、称重打包、运输等工序。粒状配合饲料工艺前段

Compound feed is produced on the basis of formula design and in accordance with a certain production process. Due to the difference of product shape, equipment type, production scale and usage, the processing capacity and process composition of compound feed are different. However, regardless of the type, the basic processing technology of compound feed generally consists of the main sections and processes of receiving and cleaning, crushing, ingredient mixing, post-treatment (conditioning, expansion, granulation, drying, screening) of the raw materials, packaging, transportation and storage, as well as liquid adding system, steam system, ventilation and dust removal system.

Compound feed products have two forms of powder and granular. The production process of powder compound feed is relatively simple, mainly including receiving and cleaning, crushing of the raw materials, metering and batching, mixing, weighing and packaging, transportation and

与粉状配合饲料生产工艺相同,混合后需要通入蒸汽或加水进行调质,并用制粒机压制成所需要的各种形状的颗粒,并经冷却、筛选等处理设备再进行称重打包、运输等工序。粒状配合饲料虽然存在加工成本提高,同时还易引起热敏性原料如维生素、酶制剂等的损失,脂肪部分变质等缺点,但经过制粒后配合饲料具有明显的优势:

(1)高温调质可以有效杀菌,保证饲料安全卫生,延长饲料贮存时间。

(2)制粒过程的加热、挤压等可促进淀粉糊化,破坏饲料原料中的抗营养因子,提高饲料适口性。

(3)颗粒饲料产品按照营养需求配方生产,一般营养均衡,可满足动物生长需要,提高饲料报酬。

(4)与粉料相比,颗粒饲料可减少自然损失,改善饲养环境卫生,减少污染。

other processes. The former process of granular compound feed is the same as that of powder compound feed. After mixing, steam or water should be added for conditioning, then the granules of various shapes should be pressed by granulator, and weighed, packed and transported by cooling, screening and other processing equipment. Although granular compound feed has some disadvantages, such as higher processing cost, loss of heat sensitive raw materials like vitamins and enzymes, and partial deterioration of fat, compound feed has obvious advantages after pelletizing.

(1) High temperature conditioning can sterilize effectively, ensure the safety and sanitation of feed and prolong the storage time of feed.

(2) Heating and extrusion during granulation can promote starch gelatinization, destroy the antinutritional factors in feed raw materials, and improve the palatability of feed.

(3) Pellet feed products are produced in accordance with nutritional requirements and are generally balanced in nutrition, which can meet the needs of animal growth and improve feed conversion.

(4) Compared with powder feed, pellet feed can reduce natural loss, improve feeding environment and reduce pollution.

二、原料的接收、清理与储存
Reception, cleaning and storage of raw materials

1. 原料的接收

原料接收是配合饲料生产的第一道工序,各种原料以不同的方式运送到饲料厂内,通过检验、计量、清理等工序后入库或直接投入使用。原料的接收设备主要是机械输送设备及其附属设备和设施(如台秤、自动秤等称量设备、料仓及卸货台、卸料坑等设施)。由于原料的包装方式、运输方式不同,原料的物理形态也差异很大,接收工艺也不尽相同。散装的固体原料经台秤称重后自动卸入卸料坑,袋装原料用输送设备输送到房式仓,液体原料经检验合格后灌装储存。原料接收

1. Reception of raw materials

The reception of raw materials is the first process in the production of compound feed. All kinds of raw materials are transported to the feed factory in different ways and put into storage or directly put into use after inspection, measurement, cleaning and other processes. The reception of raw materials mainly includes mechanical transmission equipment and its accessories and facilities (such as weighbridge, automatic scale and other weighing equipment, silo and discharging platform, discharging pit and other facilities). Due to the different packaging and transportation methods of raw materials, the physical form of raw materials is also very different, the receiving process of raw materials is therefore not the same. The bulk receiving of solid raw materials can be automatically discharged into the discharging pit after be-

的基本任务包括:准确计量进厂原料的数量、品种和日期;正确取样并对样品进行初步的快速检验;对照合同检查数量和品质的符合情况,及时而准确地将符合规格的原料入库贮存。对于数量和质量不符合规定的原料,应与供货商协商解决,或提出索赔;对不符合安全贮存条件的原料进行必要的贮前处理。

2. 原料的清理

原料的清理是利用原料与杂质在物理性质上的差异进行分选除杂的过程,主要包括筛选、风选、磁选、色选等方法。配合饲料厂需要清理的饲料原料主要是植物性饲料,如饲料谷物、农产品加工副产品等。所用谷物及饼粕类饲料常含泥土、金属等杂质,需要清理出来,一方面保证成品的含杂量尽量在规定的最低限度,另一方面保证加工设备的安全运作,减少设备磨损及改善工作环境。液体饲料原料只需要过滤即可。饲料厂主要采用筛选和磁选方法清理杂质,并且在筛分以及其他加工过程中常辅以通风除尘,改善车间环境卫生。

清理工艺基本上是先筛选再磁选。粉状原料清理主要使用粉料清理筛和永磁筒,无须粉碎的粉料原料经清理后自己进入配料、混合工段,需要粉碎的则要求清理设备的产量应与粉碎机的要求匹配。粒状原料的清理主要使用圆筒初清筛和永磁筒,根据饲料厂生产工艺的不

ing weighed by platform scale. The bagged raw materials are transported to the warehouse by conveying equipment, and the liquid raw materials are filled and stored after passing the inspection. The basic assignments of receiving raw materials include: accurately measuring the quantity, variety and date of incoming raw materials; correctly sampling and rapidly testing the samples; checking the quantity and quality against the contract; putting the raw materials that meet the specifications in storage timely and accurately. If the quantity and quality of raw materials do not meet the requirements, it shall be settled by negotiating with the supplier or claiming for compensation. It is necessary to process the raw materials which do not meet the safe storage conditions before storage.

2. Cleaning of raw materials

Cleaning of raw materials is a process of separation and impurity removal by using the difference of physical properties between raw materials and impurities, which mainly includes screening, air separation, magnetic separation, color separation and other methods. The raw materials that need to be cleaned up in the feed mill are mainly plant feed, such as feed grain and by-products of agricultural products processing. It is necessary to clean out the dirt, metal and other impurities contained in the grain and cake feed. On the one hand, this method can limit the impurities in the finished products within the prescribed minimum. On the other hand, it can also ensure the safe operation of processing equipment, reduce equipment wear and improve the working environment. The liquid feed ingredients only need to be filtered. Screening and magnetic separation methods are mainly used to clean impurities in feed factories, and ventilation and dust removal are often supplemented in screening and other processes to improve workshop environmental sanitation.

The cleaning process is basically magnetic separation after screening. The cleaning of powdery raw materials mainly uses powder cleaning screen and permanent magnet cylinder. The powdery raw materials that do not need to be crushed enter the batching and mixing section after cleaning. If crushing is required, the output of the cleaning equipment should match the requirements of the crusher. According to the different production processes of the feed fac-

同,可分为入仓前清理和入仓后清理。

(1)仓前清理工艺:原料经接收装置后自己进入清理工段,使仓内贮存的原料为干净原料,可长时间储存,减少杂质对原料的污染,避免水分较高杂质霉变影响原料贮存。原料接收设备产量高时应配置产量较大的清理设备,以提高清理设备的使用效率。

(2)仓后清理工艺:由卸料坑进料,经提升机到刮板输送机进入立筒仓。生产时,原料经出仓螺旋输送机到提升机进入振动筛、磁选机进行清理。这种工艺由于清理工序在进仓后,清理设备的规格可以和主车间结合起来,其产量应与后续工段的设备(如粉碎机)相匹配。因此,相对仓前清理工艺,仓后清理工艺中的清理设备产量小,设备的利用率较高。但原料未经清理贮存可能会受到局部霉变杂质的影响,不利于长时间贮存。

3. 原料的储存

饲料厂的原料存储直接影响生产的正常进行和工厂的经济效益,各种原料应根据不同的特性采取不同的保管和贮藏措施。所有原料的贮存均应注意采取通风、干燥、防潮、防鼠害虫害、防发热霉变等措施,保证饲料原料贮存期间质量不变。

(1)原粮类饲料原料如玉米、小麦、大豆等因使用量大,通常采用料仓储存,料仓的形状根据仓体的截面形状主要有圆形和矩形(含正方形),建筑材料有钢筋混凝土、钢材、木材等多种,当前饲料厂原料仓

tory, it can be divided into cleaning before warehousing and cleaning after warehousing.

(1) Cleaning process before warehouse. Raw materials enter the cleaning section after receiving device, so that the raw materials stored in the warehouse are clean raw materials, which can be stored for a long time to reduce the pollution of impurities and avoid the influence of high moisture and impurities mildew on raw material storage. When the output of raw material receiving equipment is high, cleaning equipment with larger output should be equipped to improve the use efficiency of cleaning equipment.

(2) Cleaning process after warehouse. The materials are fed from the discharging pit, and then enter the vertical silo through the elevator to the scraper conveyor. In the process, the raw materials will be cleaned by the screw conveyor to the elevator and into the vibrating screen and magnetic separator. Since the cleaning process is after the warehouse entry, the specifications of the cleaning equipment can be combined with the main workshop. The output should match with the equipment of subsequent sections (such as grinder). Therefore, compared with the cleaning process before the warehouse, the output of cleaning equipment in the cleaning process after the warehouse is small, and the utilization rate of equipment is high. However, raw materials stored without cleaning may be affected by local mildew impurities, which is not conducive to long-term storage.

3. Storage of raw materials

The storage of raw materials directly affects the normal production and the industrial economic benefits. Different storage measures should be taken for different raw materials according to their different characteristics. The measures of ventilation, drying, moisture-proof, rodent and pest prevention, heat and mildew prevention should be taken to ensure the quality of raw materials during storage.

(1) Due to the large usage of raw grain feed materials such as corn, wheat, soybean, etc., they are usually stored in silos. According to the cross-section shape of bins, the shapes of silos are mainly round and rectangular (including square). There are many kinds of building materials such as reinforced concrete, steel, wood, etc. Most raw material

储以钢筋混凝土结构和钢结构的立筒仓居多，同时需要配备通风条件或倒仓设施，有条件的可配备烘干与制冷设备。原粮类饲料原料储存前应检验原料的水分、杂质和发芽等情况，实行按不同等级质量的分级存放。

（2）油料饼粕有片状、粉状和团状等，大小不均，散落性差，饲料厂通常采用袋装码垛储存或库内薄糠覆盖堆垛法贮藏。袋装码垛注意包堆要低些，留有一定的走道，便于检查，发现问题及时处理。饼状饲料一般采用库内薄糠覆盖堆垛，堆垛时饼间留有空间，油饼之间铺一层干糠，垛高因库房高度而异，四周与饼间隙用干糠塞满后密封贮存。在含油高、水分低的情况下，饼粕可能局部自热，严重的可发生自燃，故应加强管理。

（3）动物性蛋白质原料如鱼粉、肉骨粉、肠膜蛋白粉、蚕蛹粉等，此类饲料用量相对不大，可以采用袋装仓储。由于饲料中蛋白质含量高并含有盐分，容易吸湿，在储存过程中应尽量防止吸湿及腐败变质。如果是散装存放在仓内，则应尽快用完，以免堵塞出口不能卸料。

（4）石粉、磷酸氢钙和盐等矿物质原料一般都是包装存放，保持干燥，防止潮解。维生素的稳定性受贮存条件、微量元素和载体的影响较大，容易失效，故应单独存放，尽量低贮存。

storage of feed factories are vertical silos with reinforced concrete structure and steel structure. At the same time, ventilation conditions or storage facilities should be provided, and drying and refrigeration equipment can be provided if conditions permit. The moisture, impurities and germination of raw grain feed materials should be inspected before storage, and the raw materials should be stored according to different levels of quality.

（2）The oil cake has different shapes like flake, powder and ball, but it is uneven in size, poor in scattering, the feed factories usually adopt the method of bagged stacking storage or thin bran covering stacking in the warehouse. When palletizing bags, the height of the piles should be limited and a certain aisle should be left for frequent inspection and timely handling if problems occur. Cake feed are generally stacked with thin bran in the warehouse. When stacking, there is space between the cakes and a layer of dry bran should be laid between the oil cakes. The height of embankment varies with the height of warehouse. The gap between the cake and the surrounding area should be filled with dry bran and sealed for storage. In the case of high oil content and low moisture content, the cake may self-heat locally, and spontaneous combustion may occur in severe cases. It is necessary to strengthen the management.

（3）Animal protein raw materials such as fish meal, meat and bone meal, intestinal membrane protein powder, silkworm pupa powder, etc., can be stored in bags on account of the small consumption. Due to the high content of protein and salt in feed, it often absorbs moisture. Therefore, moisture absorption and deterioration should be prevented as far as possible during storage. If it is stored in bulk in the warehouse, it should be used up as soon as possible, so as to avoid blocking the outlet and not discharging.

（4）Mineral materials such as stone powder, calcium hydrogen phosphate and salt are generally packed and stored to keep them dry to prevent deliquescence. The stability of vitamins is greatly affected by storage conditions, trace elements and carriers, so it is easy to lose efficacy. It should be stored separately at low temperature as possible.

三、原料的粉碎
Crushing of raw materials

粉碎是饲料加工中必不可少的工序是利用机械作用将物料由大块破碎成小块的过程,主要包括切削、碾压、撞击、碾磨等。粉碎机是配合饲料生产的重要设备,电耗一般占生产线总能耗的1/3以上。目前,用于饲料原料的粉碎机主要有锤片式粉碎机、辊式磨粉机、爪式粉碎机、无筛锤式粉碎机、饼类粉碎机等,其中锤片式粉碎机应用最为广泛,有卧式和立式两种。粉碎的质量不仅影响产品的感官质量,还影响饲料的内在品质及饲喂效果。粉碎过的谷物的果种皮被撕裂,内部营养成分暴露,使其具有较好的适口性,畜禽更喜食。粉碎后的饲料具有较大的表面积,从而便于与消化酶接触,促进消化吸收。此外,饲料粉碎对于配合饲料的混合、制粒和膨化等都是必要的工序。粉碎的饲料易于混合成为均匀的粉体,粒度相近满足制粒、膨化的要求。

饲料原料粉碎的工艺流程应根据对粒度的要求和饲料的品种等条件而定。按原料粉碎次数,可分为一次粉碎工艺和二次粉碎工艺。按与配料工序的组合形式,可分为先配料后粉碎工艺与先粉碎后配料工艺。

1. 一次粉碎工艺

一次粉碎工艺是利用粉碎机将饲料原料经过一次粉碎即可满足粒度要求的粉碎工艺。原料进入待粉碎仓中,由喂料器喂入粉碎机中粉碎即可。该工艺的特点是工艺简单,粉碎机容易控制,缺点是粒度不

Crushing is an essential process in feed processing. It is a process of crushing materials from large pieces into small pieces by mechanical effect. It mainly includes cutting, rolling, impact, grinding, etc. Grinder is an important equipment in the production of compound feed, and the power consumption generally accounts for more than 1/3 of the total energy consumption of the production line. At present, the crushers used for feed materials mainly include hammer crusher, roller crusher, claw crusher, non-sieve hammer crusher, cake crusher and so on. Among them, hammer crusher are the most widely used, including horizontal and vertical types. The quality of crushing not only affects the sensory quality of the product, but also affects the feed internal quality and feeding effect. The crushed grain that seed coat is torn and internal nutrients is exposed has a better palatability for livestock and poultry. The crushed feed has a large surface area, which facilitates contact with digestive enzymes and promotes digestion and absorption. In addition, feed crushing is a necessary process for the mixing, granulation and expansion of compound feed. The crushed feed is easy to be mixed into uniform powder with similar granularity, which meets the requirements of granulation and expansion.

The technological process of crushing feed raw material should be determined according to the requirements for particle size and feed variety. According to the crushing times of raw materials, it can be divided into a primary crushing process and a secondary crushing process. According to the combination form of batching and batching process, it can be divided into batching before crushing and batching after crushing.

1. Primary crushing process

Primary crushing process is a crushing process through which the feed raw materials can meet the requirements of granularity through first crushing by using a grinder. The raw materials enter the powder bin and is fed into the grinder by the feeder for crushing. The process is simple and the grinder is easy to control. The disadvantages of the process

均匀,电耗较高。

2. 二次粉碎工艺

二次粉碎工艺是利用粉碎机,将饲料原料经过二次粉碎即可满足粒度要求的粉碎工艺。根据粉碎机的类型和数量可分为单机循环二次粉碎工艺、双机二次粉碎工艺和微粉碎工艺。

（1）单机循环二次粉碎工艺:用一台粉碎机将物料粉碎后进行筛分,筛上物再回流到原来的粉碎机参与第二次粉碎。

（2）双机二次粉碎工艺:该工艺的基本设置是采用两台筛片不同的粉碎机使物料得到二次粉碎,其中第二台粉碎机控制粉碎成品粒度。每台粉碎机上各设一道分级筛,将物料先经第一道筛筛理,符合粒度要求的筛下物直接进入混合机,筛上物进入第一台粉碎机,粉碎的物料再进入分级筛进行筛理。符合粒度要求的物料进入混合机,其余的筛上物进入第二台粉碎机粉碎,粉碎后进入混合机。

（3）微粉碎工艺:用于颗粒饲料生产时对原料的粉碎。进入待粉碎仓的物料通过第一次粗粉,经过微粉碎机粉碎的物料进入微细分级机分级和风网组成的分级处理系统,没有达到粒度要求的物料二次进入微粉碎机,使这部分物料进行循环粉碎。

are inhomogeneous granularity and high power-consumption.

2. Secondary crushing process

The secondary crushing process is a crushing process which can meet the requirements of granularity through secondary crushing of feed raw materials by using a grinder. According to the type and quantity of grinder, it can be divided into single cycle secondary crushing process, double machine secondary crushing process and micro crushing process.

（1）Single cycle secondary crushing process. It means that the material is crushed by a crusher for sieving, then the sieve contents are then returned to the original crusher for the second crushing.

（2）Double machine secondary crushing process. The basic setting of the process is to use two grinders with different sieves to make the material to be crushed twice, and the second grinder controls the granularity of the crushed product. The materials firstly pass through the first sieve, then the sieved materials that meet the requirements of particle size directly enter the mixer, and the materials on the sieve enter the first crusher, and the crushed materials enter the grading screen for screening. The materials that meet the requirements of particle size are put into the mixer, and the remaining sieve are put into the second grinder for crushing, and then into the mixer.

（3）Micro crushing process. It is used to crush the raw materials in the production of pellet feed. The materials entering the powder bin pass through the first coarse powder. The materials crushed by the micro grinder enter the hierarchical processing system composed of the classification of the micro classifier and the aspiration network. The materials that do not meet the granularity requirements enter the micro grinder for the second time, so that these materials can be crushed circularly.

四、配料及混合
Batching and mixing

1. 配料

配料是指根据饲料配方的要求,对多种饲料原料进行准确计量

1. Batching

Batching is the process of weighing and accurately measuring the amount of various feed raw materials accord-

的过程。配料是配合饲料生产的关键环节,是实现配方目标的重要工序。配料装置的核心是配料秤,其正确性、灵敏度、稳定性和不变性直接影响到配料的正确性和质量。

常用的配料工艺包括人工添加配料、容积式配料、一仓一秤配料、多仓一秤配料和多仓数秤配料等。

(1)人工添加配料 人工控制添加配料用于小型饲料加工厂和饲料加工车间。这种配料工艺是将参加配料的各种组分由人工称量,然后由人工将称量过的物料倒入混合机中。因为全部采用人工计量和人工配料,工艺极为简单,设备投资少,产品成本低,计量灵活。

(2)容积式配料 每只配料仓下面配置一台容积式配料器。

(3)一仓一秤配料 每个配料仓各自配置一台配料秤,配料秤的规格视原料特性、用量要求和生产规模而定,给料、称量和卸料各自单独完成。此工艺有利于缩短配料周期,减少配料误差。

(4)多仓一秤配料 所有饲料原料共用一台配料秤,由自动控制系统协调进料、称量、换料和卸料等过程。此工艺配料周期长,配料过程稳定性较差。

(5)多仓数秤配料 将所计量的物料按照其物理特性或称量范围分组,每组配上相应的配料秤。此工艺适用于大型饲料厂和预混料生产,配料周期缩短且配料精度和稳定性增加。

2. 混合

混合是配合饲料生产中将按照配料工序配好的各种物料混合均匀

ing to the requirements of feed formula. Batching is a key link in the production of compound feed and an important process to achieve the goal of formula. The core of batching device is batching scale and its validity, sensitivity, stability and invariability of batching scale directly affect the accuracy and quality of batching.

The batching processes we often use include manual adding batching, volumetric batching, one-stored-one-scale batching, multiple-stored-one-scale batching and multiple-stored-multiple-scale batching, etc.

(1) Manual adding batching. Manual control adding of batching is used in small feed processing plants and feed processing workshops. This batching process is to manually weigh the various components involved in the batching, and then manually pour the weighed materials into the mixer. The process is extremely simple, the equipment investment is small, the product cost is low, and the measurement is flexible because of using manual measurement and batching.

(2) Volumetric batching. Each batching bin is equipped with a volumetric batcher.

(3) One-stored-one-scale batching. Each batching bin is equipped with a batching scale. The specification of the batching scale depends on the characteristics of raw materials, consumption requirements and production scale, and the feeding, weighing and unloading are completed separately. This process can shorten the batching cycle and reduce the batching error.

(4) Multiple-stored-one-scale batching. There is one mixing scale for all feed raw materials. The automatic control system coordinates the process of feeding, weighing, refueling and unloading. The batching period of this process is long and the batching process stability is poor.

(5) Multiple-stored-multiple-scale batching. The metered materials are grouped according to their physical characteristics or weighing range, and each group is equipped with corresponding batching scales. This process is suitable for large-scale feed factory and premix production. The mixing cycle is shortened and the batching accuracy and stability are increased.

2. Mixing

Mixing is a process in which all kinds of materials are mixed evenly according to the batching process in the pro-

的过程,是保证按配方要求的营养物质均匀分布的重要环节。混合机也是配合饲料生产的关键设备。根据产品的要求,有预混合和最后阶段混合两道混合工序。预混合是根据配方将各种微量元素、氨基酸、维生素和非营养性添加剂等与载体和稀释剂预先混合在一起的过程。最后阶段混合是根据配方要求将原料组分配比混合,制成营养全面的配合饲料。混合完毕后,粉状配合饲料根据产品要求进入后道工序。可以直接进入包装工序,或为了改善配合饲料质量和外观特性,对粉状配合饲料进行制粒或膨化处理。

duction of compound feed and is an important link to ensure the uniform distribution of nutrient substance according to the formula requirements. The mixer is also the key equipment in the production of compound feed. According to the requirements of the products, there are two mixing processes: premixing and final stage mixing. Premixing is the process of various trace elements, amino acids, vitamins and non-nutritive additives with carriers and diluents according to the formula. Final stage mixing is to mix the raw materials according to the requirements of the formula to make nutritious compound feed. After mixing, the powdered compound feed is put into the later process according to the product requirements. It can be directly put into the packaging process, or pelletized or extruded to improve the quality and appearance of the compound feed.

五、制粒或膨化
Granulation or expansion

根据饲养实践和质量要求,某些种类的配合饲料需要进一步将混合好的粉状配合饲料进行制粒或膨化。颗粒或膨化饲料具有营养均匀全面、易消化吸收、动物不挑食、适口性好、不黏嘴、不分级、便于贮存和运输等优点。

1. 制粒工序

制粒是利用机械将粉状配合饲料经(或不经)调质后挤出压模模孔制成颗粒状饲料的过程。制粒工序分调质、制粒、冷却、破碎和筛分等环节。

(1) 调质

饲料的调质是通过湿热和压力处理而改变饲料理化性质的过程。这一过程使粉状饲料吸水软化,发生淀粉糊化、蛋白变性等变化,使饲料变得更具可塑性,易于成型。通过调质可以提高饲料的应用价值,提高制粒性能,改善颗粒产品质量,破坏和杀灭饲料中的有害因子,同时还可以添加液体原料组分。调质

According to the feeding practice and quality requirements, some kinds of compound feed need to be pelletized or expanded with mixed powdered compound feeds. Pellet or expanded feed has the advantages of uniform and comprehensive nutrition, easy digestion and absorption, good palatability, no sticking to the mouth, no grading, and easy storage and transportation.

1. Granulation process

Pelletizing is a process of extruding powdered compound feed through (or without) conditioning and extruding the die hole into granular feed. Granulation process is divided into conditioning, granulation, cooling, crushing and screening.

(1) Conditioning

The conditioning of feed is a process of changing the physicochemical property of feed through heat and humidity and pressure treatment. This process makes the powdered feed absorb water and soften, causing changes such as starch dextrinization and protein denaturation, which makes the feed more plastic and easy to shape. Through conditioning, the application value of feed can be improved, the pelleting performance can be improved, the quality of pellet products can be improved, the injurious factors in feed can

的好坏直接决定着颗粒饲料产品的质量。

（2）制粒

制粒加工是将调质好的粉状配合饲料制成颗粒饲料的过程，可以采用环模或平模进行挤压制粒。

① 环模制粒：调质均匀的物料连续均匀地分布在压辊和压模之间，物料由供料区、压紧区进入挤压区，被压辊钳入模孔连续挤压，形成柱状的饲料，随着压模回转，被固定在压模外面的切刀切成颗粒状饲料。

② 平模制粒：混合后的物料进入制粒系统后，位于压粒系统上部的旋转分料器均匀地把物料撒布于压模表面，然后由旋转的压辊将物料压入模孔并从底部压出，经模孔出来的柱状饲料由切刀切成需求的长度。

（3）冷却

颗粒饲料刚从制粒机出来时，含水量和温度分别高达16%－18%和75－90 ℃，使颗粒饲料容易变形破碎，贮藏时也会产生黏结和霉变现象。冷却器的作用是将热颗粒饲料冷却至略高于室温，同时降低饲料水分和增加颗粒硬度，以便安全储运。通常将其水分降至14%以下，温度降低至比气温高3－5 ℃以下。常见的颗粒饲料冷却器主要有立式双筒冷却器、立式逆流冷却器和带式冷却器等。

（4）破碎

在小颗粒饲料直径（≤2 mm）的生产过程中，为了节省电力，增加产量和提高质量，通常先将物料制成4.0－6.0 mm 的颗粒饲料，再根据畜禽饲用时的粒度用破碎机破碎成小颗粒（碎粒料）。

be destroyed and killed, and the liquid raw material components can be added. The quality of pellet feed products is directly determined by the quality of conditioning.

(2) Granulation

Granulation process is the process of making refined powdered compound feed into pellet feed, which can be extruded by a ring die or a flat die.

① Ring mold granulation. The uniform tempered material is continuously and evenly distributed between the pressing roller and the pressing die. The material enters the extrusion area from the feeding area and the pressing area, and is continuously squeezed by the pressing roller forceps to form columnar feed. As the pressing die turns, it is cut into granular feed by a cutting knife fixed outside the pressing die.

② Flat mold pelleting. After the mixed materials enter the pelletizing system, the rotary distributor at the upper part of the pelleting system evenly distributes the materials on the surface of the die, and then the rotary press roller presses the material into the die hole and presses it out from the bottom. The column feed coming out of the mold hole is cut into the required length by the cutter.

(3) Cooling

When the pellet feed just comes out of the granulator, the moisture content and temperature are as high as 16% - 18% and 75 - 90 ℃ respectively, which makes the pellet feed easy to be deformed and broken, and will also produce adhesion and mildew during storage. The function of the cooler is to cool the hot pellet to the temperature that is slightly higher than room temperature, reduce the moisture content of the feed and increase the hardness of the pellet for safe storage and transportation. The moisture content is usually reduced to below 14% and the temperature is lowered to 3 - 5 ℃ higher than the air temperature. The common pellet feed coolers are vertical double tube coolers, vertical counter current coolers and belt coolers.

(4) Crushing

In the production process of small granular materials (less than 2 mm), in order to save power, increase production and improve quality, the materials are usually made into 4.0 - 6.0 mm pellet feed, and then are crushed into small particles (crushed materials) by crusher according to the particle size of livestock and poultry.

（5）筛分

冷却后的颗粒饲料或碎粒饲料进行分级筛选,去除其中的部分粉末、凝块等不符合要求的物料。因此,破碎后的颗粒饲料需要筛分成颗粒整齐并且大小均匀一致的粒度合格的颗粒饲料,保证成品质量。

2. 膨化工艺

广义的膨化包括挤压膨化和气体热压膨化,后者很少应用。挤压膨化即将饲料从螺杆推进、增压和增温处理后膨化。膨化配合饲料是将饲料输入膨化机的调质器中,调至水分为 25%–30%,然后进入挤压机进行挤压膨化处理生产的饲料。挤压膨化技术是集混合、加热、冷却和成型等多种作业为一体的加工过程,物料在膨化机内经过剧烈的挤压、搅拌、剪切作用,被细化、均化后由粉状变成糊状,发生如淀粉糊化、蛋白质变性、酶和微生物的失活等一系列的物理、化学变化,改善了饲料的营养品质、提高了饲料的消化率。挤压膨化加工工艺分为干法膨化和湿法膨化两种。干法膨化的饲料进入膨化机前不需要调质,而是在物料中加入适量的水分和油脂,经膨化冷却处理为膨化饲料,如全脂大豆的膨化。湿法膨化多用于生产水产、断奶仔猪和宠物等膨化颗粒饲料,饲料送进膨化机前进行加水或蒸汽进行调质,再通过膨化制出多孔状的颗粒饲料。饲料原料经过膨化加工后,原料细胞壁的主要成分纤维和木质素被破坏。与普通颗粒饲料相比,膨化饲料具有适口性更好、消化率更高和外形多样等特点。

(5) Screening

The cooled pellet feed or granulated feed is graded and screened to remove unqualified materials such as powder and clot. Therefore, the crushed pellet feed needs to be sieved into qualified pellet feed with neat and uniform size to ensure the quality of finished products.

2. Expansion process

The general expansion includes extrusion and gas hot pressing expansion. The latter one is rarely used. Extrusion is that the feed is pushed from the screw, pressurized and heated before expansion. Expanded compound feed is a kind of feed that is fed into the conditioner of the extruder, adjusted to 25%–30% of moisture, and then extruded in the extruder. Extrusion technology is a processing that integrates mixing, heating, cooling, molding and other operations. The materials are extruded, stirred and sheared sharply in the extruder, and they change from powder to paste after being refined and homogenized. A series of physical and chemical changes occur, such as starch gelatinization, protein denaturation, inactivation of enzymes and microorganisms, which improve the nutritional quality and increase the digestibility of feed. Extrusion processing technology is divided into dry expansion and wet expansion. Dry expansion feed does not need to be tempered before entering the extruder. Instead, proper amount of water and oil are added into the materials, and then it is processed into expansion feed after expansion cooling treatment, such as the expansion of full fat soybean. Wet expansion is mostly used to produce expanded pellet feed for aquatic products, weaned piglets and pets. The feed is quenched and tempered with water or steam before being sent to the extruder, and then porous pellet feed is produced by expansion. The fiber and lignin of raw materials, which is the main components of cell wall, are destroyed after expansion processing. Compared with ordinary pellet feed, expanded feed has the characteristics of better palatability, higher digestibility and diverse appearance.

六、包装与贮存
Packaging and storage

配合饲料成品称量包装是其生产工艺流程中最后一个工段,包括饲料产品的称重、装袋、缝口、贴标签和运送。成品包装有人工打包计量包装及缝口和机械定量包装两种形式。现代化的饲料厂的成品包装常采用成套机械包装设备,包装机械由自动定量包装秤、夹袋机构、缠袋装置和输送装置等组成。配合饲料成品主要在成品库中暂时性保管,贮存在干燥、通风性能良好的仓库。同时,应注意防潮、防雨和防虫害、鼠害,也要防止被有害有毒物质污染,确保使用安全。在良好条件下贮藏的时间会相对长些,一般可保存90天左右。

Weighing and packaging of compound feed products is the last section in the production process, including weighing, bagging, sewing, labeling and transportation of feed products. There are two forms of finished product packaging: manual packaging and metering packaging and seam and, mechanical quantitative packaging. In modern feed factories, complete sets of mechanical packaging equipment are often used for finished product packaging. The packaging machine is composed of automatic quantitative packaging scale, bag clamping mechanism, bag wrapping device and conveying device. The finished products of formula feed are mainly kept temporarily in the finished product warehouse and stored in a dry and good ventilated warehouse. At the same time, it is necessary to pay attention to moisture proof, rain proof, pest and rodent prevention, as well as to prevent pollution by harmful and toxic substances to ensure safety in use. Under good conditions, the storage time will be relatively long, generally about 90 days.